BIO-INSPIRED MATERIALS FOR BIOMEDICAL ENGINEERING

WILEY–SOCIETY FOR BIOMATERIALS SERIES

Bio-inspired Materials for Biomedical Engineering • Anthony B. Brennan and Chelsea M. Kirschner, Editors

BIO-INSPIRED MATERIALS FOR BIOMEDICAL ENGINEERING

Edited by

ANTHONY B. BRENNAN

CHELSEA M. KIRSCHNER

Society For Biomaterials
Giving life to a world of materials

WILEY

Published by John Wiley & Sons, Inc., Hoboken, New Jersey.
Published simultaneously in Canada.

Library of Congress Cataloging-in-Publication Data:

Bio-inspired materials for biomedical engineering / edited by Anthony B. Brennan, Chelsea M. Kirschner.
p. ; cm.
"Society for Biomaterials."
Includes bibliographical references and index.
ISBN 978-1-118-36936-4 (cloth) – ISBN 978-1-118-84362-8 (ePDF) – ISBN 978-1-118-84343-7 (ePub) – ISBN 978-1-118-84342-0 (eMobi) – ISBN 978-1-118-84349-9 (oBK)
I. Brennan, Anthony B., editor of compilation. II. Kirschner, Chelsea M., editor of compilation. III. Society for Biomaterials.
[DNLM: 1. Biocompatible Materials. 2. Biomedical Engineering. QT 37]
R857.M3
610.28–dc23

2013049566

Printed in the United States of America

10 9 8 7 6 5 4 3 2 1

CONTENTS

CONTRIBUTORS

Marta Alves da Silva, 3B's Research Group – Biomaterials, Biodegradables, and Biomimetics, University of Minho, Headquarters of the European Institute of Excellence on Tissue Engineering and Regenerative Medicine, Guimarães, Portugal; ICVS/3B's – PT Government Associate Laboratory, Braga/Guimarães, Portugal

Michael Ansorge, Universität Leipzig, Institute of Biochemistry, Leipzig, Germany

Ezgi Antmen, BIOMATEN, METU Center of Excellence in Biomaterials and Tissue Engineering, Department of Biotechnology, Ankara, Turkey

Anthony Atala, Wake Forest Institute for Regenerative Medicine, Wake Forest School of Medicine, Winston-Salem, NC, USA

Stephen F. Badylak, McGowan Institute for Regenerative Medicine, Department of Surgery, University of Pittsburgh, Pittsburgh, PA, USA

Erkan Türker Baran, BIOMATEN, METU Center of Excellence in Biomaterials and Tissue Engineering, Ankara, Turkey

Liming Bian, Division of Biomedical Engineering, Department of Mechanical and Automation Engineering, The Chinese University of Hong Kong, Shatin, New Territories, Hong Kong

Anthony B. Brennan, Department of Materials Science and Engineering, J. Crayton Pruitt Family Department of Biomedical Engineering, University of Florida, Gainesville, FL, USA; Sharklet Technologies, Inc., Aurora, CO, USA

Ana Costa-Pinto, 3B's Research Group – Biomaterials, Biodegradables, and Biomimetics, University of Minho, Headquarters of the European Institute of Excellence on Tissue Engineering and Regenerative Medicine, Guimarães, Portugal; ICVS/3B's – PT Government Associate Laboratory, Braga/Guimarães, Portugal

Christopher L. Dearth, McGowan Institute for Regenerative Medicine, Department of Surgery, University of Pittsburgh, Pittsburgh, PA, USA

Tuğba Dursun, BIOMATEN, METU Center of Excellence in Biomaterials and Tissue Engineering, Department of Biotechnology, Ankara, Turkey

Adam J. Engler, Material Science Program, Department of Bioengineering, Sanford Consortium for Regenerative Medicine, University of California–San Diego, La Jolla, CA, USA

Menekse Ermis, BIOMATEN, METU Center of Excellence in Biomaterials and Tissue Engineering, Department of Biomedical Engineering, Ankara, Turkey

Adam W. Feinberg, Department of Biomedical Engineering, Department of Materials Science and Engineering, Carnegie Mellon University, Pittsburgh, PA, USA

Anthony W. Frei, Diabetes Research Institute, Leonard M. Miller School of Medicine, University of Miami, Miami, FL, USA; Department of Biomedical Engineering, College of Engineering, University of Miami, Coral Gables, FL, USA

A. Jon Goldberg, Reconstructive Sciences, University of Connecticut Health Center, Farmington, CT, USA

Vasif Hasirci, BIOMATEN, METU Center of Excellence in Biomaterials and Tissue Engineering, Departments of Biological Sciences, Biomedical Engineering, and Biotechnology, Ankara, Turkey

Jay D. Humphrey, Department of Biomedical Engineering, Yale University, New Haven, CT, USA

Emily Jacobs, Reconstructive Sciences, University of Connecticut Health Center, Farmington, CT, USA

Quentin Jallerat, Department of Biomedical Engineering, Carnegie Mellon University, Pittsburgh, PA, USA

Benjamin G. Keselowsky, J. Crayton Pruitt Family Department of Biomedical Engineering, University of Florida, Gainesville, FL, USA

Hye Young Kim, Departments of Chemical and Biological Engineering and Molecular Biology, Princeton University, Princeton, NJ, USA

Chelsea M. Kirschner, Sharklet Technologies, Inc., Aurora, CO, USA

Elizabeth W. Kollar, McGowan Institute for Regenerative Medicine, University of Pittsburgh, Pittsburgh, PA, USA

Liisa T. Kuhn, Reconstructive Sciences, University of Connecticut Health Center, Farmington, CT, USA

Elaine L. Lee, Department of Biomedical Engineering, Boston University, Boston, MA, USA

Sang Jin Lee, Wake Forest Institute for Regenerative Medicine, Wake Forest School of Medicine, Winston-Salem, NC, USA

Christopher A. Lemmon, Department of Biomedical Engineering, Virginia Commonwealth University, Richmond, VA, USA

Jamal S. Lewis, J. Crayton Pruitt Family Department of Biomedical Engineering, University of Florida, Gainesville, FL, USA

Albino Martins, 3B's Research Group – Biomaterials, Biodegradables, and Biomimetics, University of Minho, Headquarters of the European Institute of

Excellence on Tissue Engineering and Regenerative Medicine, Guimarães, Portugal; ICVS/3B's – PT Government Associate Laboratory, Braga/Guimarães, Portugal

Eike Müller, Leibniz Institute of Polymer Research Dresden, Max Bergmann Center of Biomaterials, Dresden, Germany

Celeste M. Nelson, Departments of Chemical and Biological Engineering and Molecular Biology, Princeton University, Princeton, NJ, USA

Nuno M. Neves, 3B's Research Group – Biomaterials, Biodegradables, and Biomimetics, University of Minho, Headquarters of the European Institute of Excellence on Tissue Engineering and Regenerative Medicine, Guimarães, Portugal; ICVS/3B's – PT Government Associate Laboratory, Braga/Guimarães, Portugal

Matthew G. Ondeck, Material Science Program, University of California–San Diego, La Jolla, CA, USA

Dennis P. Orgill, Division of Plastic Surgery, Brigham and Women's Hospital, Harvard Medical School, Boston, MA, USA

Shawn C. Owen, The Donnelly Centre for Cellular and Biomolecular Research, Department of Chemical Engineering and Applied Chemistry, Institute of Biomaterials and Biomedical Engineering, University of Toronto, Toronto, Ontario, Canada

Tilo Pompe, Leibniz Institute of Polymer Research Dresden, Max Bergmann Center of Biomaterials, Dresden; Universität Leipzig, Institute of Biochemistry, Leipzig, Germany

Rui L. Reis, 3B's Research Group – Biomaterials, Biodegradables, and Biomimetics, University of Minho, Headquarters of the European Institute of Excellence on Tissue Engineering and Regenerative Medicine, Guimarães, Portugal; ICVS/3B's – PT Government Associate Laboratory, Braga/Guimarães, Portugal

James F. Schumacher, J. Crayton Pruitt Family Department of Biomedical Engineering, University of Florida, Gainesville, FL, USA

Molly S. Shoichet, The Donnelly Centre for Cellular and Biomolecular Research, Department of Chemical Engineering and Applied Chemistry, Institute of Biomaterials and Biomedical Engineering, Department of Chemistry, University of Toronto, Toronto, Ontario, Canada

David D. Simon, Department of Biomedical Engineering, Yale University, New Haven, CT, USA

Cherie L. Stabler, Diabetes Research Institute, Department of Surgery, Leonard M. Miller School of Medicine, University of Miami, Miami, FL, USA; Department of Biomedical Engineering, College of Engineering, University of Miami, Coral Gables, FL, USA

John M. Szymanski, Department of Biomedical Engineering, Carnegie Mellon University, Pittsburgh, PA, USA

Roger Y. Tam, The Donnelly Centre for Cellular and Biomolecular Research, Department of Chemical Engineering and Applied Chemistry, Institute of Biomaterials and Biomedical Engineering, University of Toronto, Toronto, Ontario, Canada

Hongjun Wang, Department of Chemistry, Chemical Biology and Biomedical Engineering, Stevens Institute of Technology, Hoboken, NJ, USA

Carsten Werner, Leibniz Institute of Polymer Research Dresden, Max Bergmann Center of Biomaterials, Technische Universität Dresden, Center for Regenerative Therapies Dresden, Dresden, Germany

Joyce Y. Wong, Departments of Biomedical Engineering and Materials Science and Engineering, Boston University, Boston, MA, USA

Meng Xu, Department of Chemistry, Chemical Biology and Biomedical Engineering, Stevens Institute of Technology, Hoboken, NJ, USA

PREFACE

Natural materials exhibit highly sophisticated properties selected through evolution to achieve specific functions efficiently. As engineers, biological scientists, and physicians strive to recapitulate natural biological processes, such as wound healing or tissue regeneration, they incorporate bio-inspired approaches. These strategies have been implemented in the rational design of biomedical devices and biomaterials both to treat patients in the clinic and to probe the fundamental mechanisms of cellular interactions with biomaterials. In this effort, our community endeavors not only to copy the complex hierarchical structures present in nature, but to harness the power of natural processes to create dynamic, bioactive materials.

This book, the first in the Wiley–Society For Biomaterials book series, aims to introduce the reader to bio-inspired strategies that provide elegant solutions for contemporary biomedical engineering challenges. The intended audience is multidisciplinary and includes students and practitioners of materials science, engineering, biology, chemistry, physics, medicine, dentistry, and veterinary medicine. This level of diversity is pervasive throughout the biomaterials community and essential for innovation in that it allows researchers to approach modern biomedical engineering challenges from a variety of perspectives. The chapters that comprise this book originate from authors all over the world with expertise from a variety of disciplines and specialties. The text is divided into two parts—Engineering Bio-inspired Material Microenvironments and Bio-inspired Tissue Engineering—in an effort to introduce the reader to fundamental concepts in biomaterials science and engineering, as well as provide a perspective on the clinical application of these technologies.

These goals could only be achieved with a tremendous amount of support and involvement from the biomaterials community. First, we must introduce and acknowledge the Society For Biomaterials as the sponsor and inspiration for this book. The theme for this text originated as the topic for a general session organized by the editors at the annual Society For Biomaterials meeting in Orlando, Florida. The Society For Biomaterials provides the opportunity for engineers and scientists from industry and academia to collaborate with physicians and business professionals to promote advancement in all aspects of biomaterial science, education, and professional standards to enhance human health and quality of life. The editors also acknowledge and extend their sincere gratitude to the authors who generously devoted their valuable time to contribute their expert knowledge and experience to this book. The active and professional support of the editorial and production staff at Wiley is also appreciated.

We hope that this book will empower the reader to think beyond current paradigms when relating bio-inspired engineering concepts to the translation of biomaterials science into medicine, and in turn, contribute to the evolution and expansion of bio-inspired materials in medicine.

Anthony B. Brennan
Chelsea M. Kirschner

INTRODUCTION

SANG JIN LEE and ANTHONY ATALA
Wake Forest Institute for Regenerative Medicine, Wake Forest School of Medicine, Winston-Salem, NC, USA

Restoration and maintenance of normal function of injured or damaged tissues and organs with biological substitutes is the primary objective of tissue engineering, a major component of regenerative medicine that adheres to the standards established in the areas of cell transplantation, materials science, and engineering. Engineered tissue constructs that are composed of biomaterial scaffolds preseeded with tissue-specific cells are among the most promising approaches to generate biologically and/or mechanically functional tissue replacements. Clinical applications of tissue engineering and regenerative medicine technologies, however, have been relatively restricted due to the limitation in clinically approved biomaterials. The varieties of biomaterials, which have been developed in recent years, have encountered delays in translation to clinical practice. Many investigators have resorted to using biodegradable synthetic polymers that were first approved for use in humans more than 30 years ago. During normal development, tissue morphogenesis is greatly influenced by the interactions between cells and the extracellular matrix (ECM) proteins; however, these simple polymers, which have been used historically to provide architectural support for neotissue development, poorly mimic the complex interactions between tissue-specific cells and the tissue-specific ECMs that promote functional tissue regeneration. Consequently, tissue engineering and regenerative medicine strategies will advance as biomaterials that actively participate in functional tissue regeneration are developed.

Nature provides numerous systems that possess exceptional properties and performance that might be replicated for many biomedical applications. Thus, scientists have observed phenomena of nature, learned the principles, and incorporated various characteristics to mimic biological systems into materials. For example, the fabric hook and loop fastener was inspired by the seeds of the burdock plant. Adhesive biomaterials have been fabricated by using microfabrication techniques inspired by the feet of the gecko, which has extraordinary climbing ability. Multifunctional biomaterials that mimic the chemical composition, physical structure, and biologically functional moieties of natural living systems could contribute to the development of new biomaterials for tissue engineering applications. Accordingly, recent progress in tissue engineering strategies has led to a paradigm shift in biomaterials

---

Bio-inspired Materials for Biomedical Engineering, First Edition. Edited by Anthony B. Brennan and Chelsea M. Kirschner.

research, whereby the concepts of bio-inspiration and biomimetics play a more active role in small-scale structures and their time-dependent biological interactions with the host.

Throughout this book, the bio-inspired materials are defined as types of biomaterials that contribute to tissue engineering and regenerative medicine strategies to achieve multifunctional and integrated tissues or organs, which incorporate biological, chemical, mechanical, topographical, and electrical cues derived from nature, as well as offer space as a scaffold. By reproducing principles or structures of biological systems, bio-inspired materials offer several aspects of higher level integration in biological systems: sophistication, miniaturization, hierarchical organizations, hybridization, resistance, and adaptability, which could actively participate in functional tissue regeneration in tissue engineering applications. This collection of contributions is divided into two overarching themes: Part 1, Engineering Bio-inspired Material Microenvironments and Part 2, Bio-inspired Tissue Engineering.

In Part 1, three major approaches to bio-inspiration are introduced: (1) ECM-mimetic bioactive materials, (2) physicochemical signals for controlling cell fates, and (3) scaffolds derived from natural materials. In the tissue regeneration process, the interactions between cells and tissue-specific ECMs are critical since cell attachment to the ECMs regulates various cellular functions such as proliferation, migration, and differentiation. A variety of synthetic biodegradable polymers have been used as tissue-engineered scaffolds because they possess adjustable mechanical properties, biodegradability, and functionality. Synthetic polymers, however, often lack biological recognition. Naturally derived hydrogels, such as fibrin and collagen, have biological functions that enhance cell adhesion, proliferation, and differentiation, which are lacking in synthetic polymers. Therefore, synthetic polymers have been modified to mimic ECM functions using short peptide sequences derived from the bioactive domains of ECM components, including adhesive peptides, enzyme-sensitive peptides, and growth factors. Bioactive molecules (or signaling molecules), including proteins and small molecules, involved in tissue regeneration also play an important role in controlling the microenvironment *in vivo*. Chemotactic signals from bioactive molecules are responsible for inducing host cell mobilization; moreover, an anatomic destination is identified according to certain concentration gradients of chemicals produced at injured sites within the microenvironment. Bioactive molecules are able to regulate host cell migration, proliferation, and differentiation, and allow cells to interact via specific receptors for chemical recognition with their surrounding microenvironment. Thus, incorporation of a suitable bioactive molecule through the design of a tissue-engineered scaffold can promote tissue regeneration by stimulating the transplanted cells or adjacent host cells. The first chapter describes how ECM-inspired chemical cues can be incorporated into scaffolds for tissue engineering while the second details how dynamic materials can be used to recapitulate signals from the ECM.

Cells are exposed to tissue-specific microenvironments *in vivo*, and they respond to numerous chemical and physical stimuli. As a result, cellular functions like cell adhesion and proliferation, protein synthesis, and cytoskeletal architecture are critically influenced by the microenvironment of the cells. To maintain the phenotype expression and differentiation of tissue-specific cells, numerous cues have been applied to biomaterial scaffolds to mimic natural ECM microenvironments. Especially, the growth and specific differentiation of stem cells (embryonic or adult) are significantly affected by their microenvironments, including chemical, topographical, mechanical, and electrical cues. The interactions of cells with biomaterials are critically important for the successful outcome of tissue engineering applications. Thus, the behavior of cells grown on a biomaterial surface, including adhesion to the material, development of appropriate cellular structures, and maintenance of proper cell phenotype and function, must be investigated in order to obtain insight into the characteristics of the biomaterial. Substrate modulus and external mechanical stimuli are important for maintaining phenotype, controlling stem cell differentiation, and establishing functional tissues. Tissues can experience compressive forces or tensile forces, such as mechanical loading, or stretch and fluid-applied forces, such as shear flow. These phenomena are explored in two chapters in Part 1.

Cells also encounter topographical cues in the form of the physical features of their surrounding microenvironment. The topography of the surface of a biomaterial can directly influence cell adhesion and proliferation, which further affects cellular functions. For this reason, one chapter in the first section focuses on engineering bio-inspired topographic cues into materials for biomedical engineering and the resulting cellular responses. Finally, this section is completed with two distinct chapters that describe how naturally derived materials can be exploited as tissue engineering scaffolds, and one chapter is devoted to the role of one specific ECM protein, fibronectin.

In Part 2 of Bio-inspired Materials for Biomedical Engineering, these approaches to engineering bio-inspired cellular microenvironments are applied to tissue engineering. The chapters in this section highlight the implementation of bio-inspired design to create epithelial tissue, cardiac tissue, musculoskeletal tissue, and connective tissue and to elicit specific immune responses. Bio-inspired materials are desirable as tissue-engineered scaffolds because they mimic the native microenvironment of the ECM. Development of tissue engineering strategies continues to be a critical component of the research pursuits in the field of regenerative medicine. Future advances in tissue engineering strategies rely on the development of bio-inspired materials that actively participate in functional tissue regeneration. Bio-inspired materials could provide biological, chemical, mechanical, and structural functions that are inspired from nature. Desirable bio-inspired materials could be designed as scaffolds that mimic the natural biological system and integrate the necessary structural and biological properties. Solid understanding of materials science combined with extensive knowledge of the clinical challenges and cell biology is vital for the development of clinically applicable biomaterials to be used in tissue engineering. Therefore, interdisciplinary collaboration between material scientists, engineers, cell biologists, physiologists, and clinicians should be encouraged to develop novel bio-inspired materials for tissue-engineering applications that might enhance or improve current regenerative medicine therapies.

Acknowledgment

The authors thank Dr. Heather Hatcher for editorial assistance.

PART I

ENGINEERING BIO-INSPIRED MATERIAL MICROENVIRONMENTS

CHAPTER 1

ECM-INSPIRED CHEMICAL CUES: BIOMIMETIC MOLECULES AND TECHNIQUES OF IMMOBILIZATION

ROGER Y. TAM and SHAWN C. OWEN

The Donnelly Centre for Cellular and Biomolecular Research, Department of Chemical Engineering and Applied Chemistry, Institute of Biomaterials and Biomedical Engineering, University of Toronto, Toronto, Ontario, Canada

MOLLY S. SHOICHET

The Donnelly Centre for Cellular and Biomolecular Research, Department of Chemical Engineering and Applied Chemistry, Institute of Biomaterials and Biomedical Engineering, Department of Chemistry, University of Toronto, Toronto, Ontario, Canada

1.1 INTRODUCTION

The extracellular matrix (ECM) is a complex environment that provides chemical and physical support to cells, Figure 1.1 [1,2]. The composition of native ECM differs based on its location within the body [3–6], but it is generally comprised of proteins (fibronectin, laminin, and collagen), polysaccharides (hyaluronan and chondroitin sulfate proteoglycans [CSPGs]) and various growth factors [6]. Components of the ECM play important roles in controlling cell function. Molecules such as collagen [7] and elastin [8] function as the structural scaffold to support cell growth, whereas fibronectin, laminin, glycosaminoglycans (GAGs), and growth factors act as ligands to promote cell adhesion, proliferation, differentiation, and migration [9].

Cells can remodel the ECM in a dynamic fashion [9,10]. For example, cells can secrete proteases that can degrade the ECM to promote cell migration, which is important in tissue repair, such as neuroblast migration following traumatic brain injury (TBI) [11], as well as in disease states, such as cancer metastasis [12]. Cells

Bio-inspired Materials for Biomedical Engineering, First Edition. Edited by Anthony B. Brennan and Chelsea M. Kirschner.

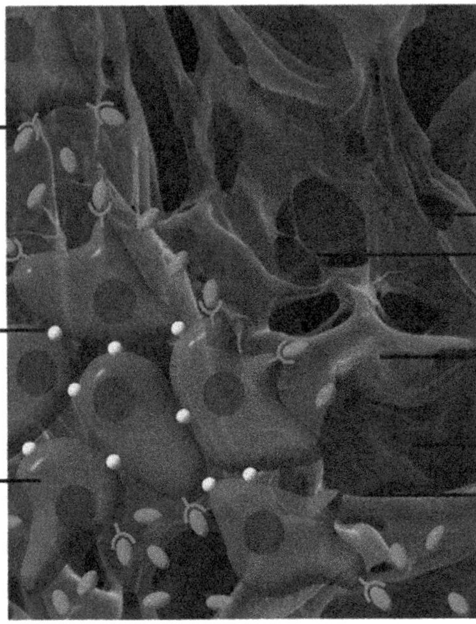

FIGURE 1.1 The complex 3-D cellular environment provides mechanical and biochemical signals that guide cell function. The components of the ECM dictate the stiffness of matrix and the types of cell–matrix interactions. The matrix composition determines the ease with which nutrients diffuse through tissues and the ability with which cells migrate through the matrix. Nonstructural factors such as cell density, cell–cell interactions, and bound or secreted signaling proteins are important in guiding cell differentiation and function. (Reproduced with permission from Owen, S.C., Shoichet, M.S. *Journal of Biomedical Materials Research A* **2011**, *94A*(4). Copyright 2013 Wiley Periodicals Inc.)

can also secrete their own ECM molecules on top of the existing ECM to provide new cues affecting both self and neighboring cells [10].

The increased understanding of the role of native ECM on cellular function and interactions has resulted in extensive research into biomimetic materials for applications in tissue engineering [13,14]. Hydrogels represent a class of biomaterials that have been used for this purpose. These highly hydrated polymers provide structural scaffolds and permit diffusion of molecules throughout. Matrigel® (BD Biosciences, San Jose, CA), a decellularized ECM derived from the Engelbreth–Holm–Swarm (EHS) mouse sarcoma, is a common hydrogel used to mimic the three-dimensional (3-D) properties of the ECM [15]. This material has been shown to promote various bioactivities such as cell adhesion, differentiation, viability, and invasion in a variety of cell types; however, for studies that require a more defined 3-D environment (such as those for mechanistic elucidation studies), the use of Matrigel® is nonideal as it is ill-defined in composition and the results are often difficult to reproduce. As such, a bottom-up approach is desirable where researchers begin with a blank palette in terms of cellular interactions and then *paint* in desirable features, such as cell adhesion, proliferation, and migration through both chemical and physical designs. Efforts to synthesize biomimetic ECMs with defined components were significantly advanced by the discovery that short peptide sequences (e.g., RGD, YIGSR, IKVAV, etc.), derived from native ECM proteins (fibronectin and laminin, respectively), promote cell adhesion and outgrowth. It was shown that RGD interacted with extracellular integrin receptors with affinity similar to that of native fibronectin [16]. Since this discovery, a large number of studies have been conducted to immobilize this and other biomimetic sequences to various biomaterials with the intention of promoting cell adhesion on nonadherent surfaces and biomaterials, thereby increasing the posttransplantation cell viability in tissue regeneration applications, and studying cell behavior in model biomimetic systems.

This chapter is focused on recent advances in the techniques used to incorporate ECM-inspired chemical signaling molecules into different hydrogels, and their effects on cellular interactions in 3-D. While similar approaches are used in multiple areas of biology, we highlight many examples applicable to neurobiology.

1.2 DEVELOPMENT AND IMMOBILIZATION OF BIOMIMETIC CUES IN 3-D BIOMATERIALS

The discovery that short peptide sequences showed comparable activity to their respective native ECM proteins from which they were derived has resulted in significant efforts to design peptide-modified biomaterials with which to study cellular interactions [17,18]. Biomaterial modification with these peptide sequences (typically 3–10 amino acids) results in inherently better-defined systems than the corresponding protein-modified systems due to the shorter sequence length and resulting 3-D structure. To take full advantage of ligand-containing biomaterials to study complex cellular interactions, it is imperative that the ligand is chemically conjugated to the biomaterial in a reproducible and specific manner in order to optimize cellular interaction. For example, conjugation at the active site of the peptide may diminish receptor binding and therefore limit bioactivity. An important consideration in designing biomimetic molecules is to include a specific functional group that has a selective chemical reactivity toward the material to which it will be conjugated.

The emergence of *click* chemistry as a method to conjugate molecules to biomaterials has proved to be a powerful technique to allow efficient conjugation with both defined chemical reactivities and orientation [19–22]. These orthogonal reactions are specific and occur with high yield and efficiency. While detailed discussion about this topic is beyond the scope of this chapter, Figure 1.2 shows a brief summary of different conjugation reactions that have been used to immobilize various peptides and proteins to biomaterials. The following section will describe the conjugation of various peptides to different biomaterials using these techniques. While most chemical conjugations have focused on the irreversible conjugation of molecules, recent work has enabled a versatile approach to forming reversible conjugations, which has the potential to synthesize dynamic biomimetic systems [23].

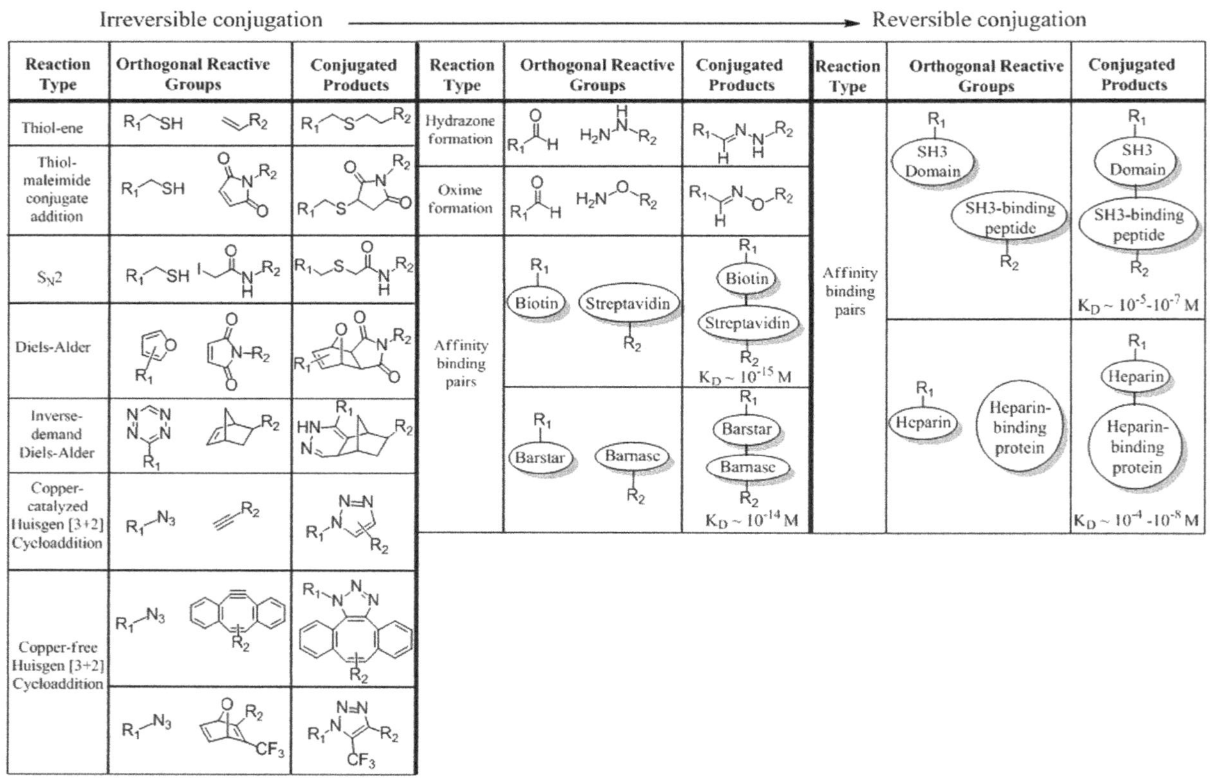

FIGURE 1.2 Typical orthogonal chemical reactions enable biomaterials to be synthesized with defined chemical and physical properties. Shown are a series of orthogonal *click* reactions which form covalent, irreversible (left panel) [27, 65, 79, 101–106] to near-covalent (middle panel) [63, 65, 107] to reversible (right panel) [52, 69, 70, 108, 109] bonds. For peptide or protein immobilization, this approach results in biomaterials that promote specific cellular responses, such as adhesion, proliferation, migration, and/or differentiation, depending on the biomolecule immobilized.

1.2.1 Synthetic Peptides Derived from Fibronectin, Laminin, and Collagen

The fibronectin-derived RGD peptide sequence is among the most studied peptide sequences for cell adhesion, and has been reviewed extensively [17,24,25]. Fibronectin is a ubiquitous protein that binds to different integrin receptors and promotes cell adhesion and cell survival. Immobilizing this sequence to biomaterials using bond-forming chemistries such as 1-ethyl-3-dimethylaminopropylcarbodiimide and N-hydroxysuccinimide (EDC/NHS) is problematic because the carboxylate-containing aspartic acid (D) participates in a competing reaction with the C-terminal carboxylate, thereby complicating the orientation of the sequence immobilized [26]. Bio-orthogonal conjugation chemistries have been used to overcome this problem and have resulted in effective adhesion for a variety of cell types to different types of modified biomaterials (Table 1.1) [17,27,28].

Derivatives of the linear RGD sequence have been synthesized in efforts to increase its binding to integrin receptors. Studies show that the RGD sequence in native fibronectin resides at the tip of a loop, which provides it with structural rigidity and a favorable conformation for integrin binding [29,30]. These structural characteristics have inspired the synthesis of cyclic RGD sequences [31,32]. Synthetic cyclic RGD peptides provide comparable conformational characteristics to facilitate integrin binding, and their conjugation to biomaterials have shown greater bioactivity compared with linear RGD sequences [33,34]. For example, cyclic RGD was recently conjugated to poly(4-methylpent-1-ene) (TPX) membranes for use as artificial lung supports. Endothelial cells cultured on this material showed significant cell adhesion, which is important for hemostasis use in these devices [35].

Another consideration for improved integrin interaction with immobilized RGD peptides is the distance between the peptide and the polymer backbone. A peptide that is bound too close to the polymer backbone may be hindered by steric interactions to efficiently bind with the receptors. Wilson *et al.* recently reported RGD peptides with a PEG linker containing greater than 27 ethylene oxide, repeat units showed significant adhesion of telomerase-immortalized human corneal epithelial cells (hTCEpi) [36].

Another ECM protein that has been extensively studied is laminin. Similar to fibronectin, laminin plays important roles in the ECM such as facilitating cell adhesion, differentiation, and migration. The most widely studied synthetic peptides for laminin are YIGSR [37] and IKVAV [38]. YIGSR has been shown to promote cell adhesion to the laminin-binding receptors, while IKVAV has been shown to promote primarily adhesion and neurite outgrowth of dorsal root ganglia (DRGs) [39], as well as differentiation of neural progenitor cells (NPCs) [40].

Collagens are another important class of proteins found in the ECM. Collagens provide structural support and also interact with receptors to mediate cell adhesion, migration, and proliferation [41,42]. The general structure of collagen consists of a triple helix, formed by three polypeptide strands, which can further assemble to supramolecular structures such as planar sheet-like networks, fibrils, and fibers [42]. There are 28 isoforms of collagen, with types I and IV being the most predominant in the ECM. Early collagen-mimetic synthetic sequences included the repeating

TABLE 1.1 Biomimetic Peptides of Common ECM Proteins and Methods of Immobilization

ECM molecule	Synthetic peptide mimic	Polymer	Method of peptide conjugation (polymer–peptide)	Bioactivity	Reference
Fibronectin	RGD	Gellan gum	Furan–maleimide	Increased neural stem/progenitor cell viability and adhesion	[27]
		Methyl cellulose	Thiol–maleimide	Increased oligodendrocyte differentiation from NSPCs when also treated with PDGF	[65]
		Agarose	Thiol–maleimide	Increased cell adhesion of NSPCs	[28]
		PEG	Thiol–acrylate	Increased endothelial cell adhesion and migration	[110]
			Cu (I) catalyzed azide–alkyne	Increased fibroblast viability and adhesion	[104]
			Oxyamine–ketone	Increased MSC viability and adhesion	[111]
			Tetrazine–norbornene	Increased MSC viability and adhesion	[103]
		Hyaluronan	Thiol–acrylate	Increased fibroblast viability and adhesion	[112]
		Elastin-mimetic polypeptide	tetrakis(hydroxymethyl) phosphonium chloride (THPC)	Increased dorsal root ganglia neurite outgrowth	[113]
	Cyclic RGD	PEG diacrylate	NHS–amine	Increased endothelial cell adhesion	[33]
		Elastin-mimetic polypeptide	Activated carboxylic acid (PyBOP)—amine	Osteoblast adhesion	[114]
		poly(4-methylpent-1-ene) (TPX)	Inverse electron demand Diels–Alder	Increased endothelial cell adhesion	[35]

Laminin	YIGSR	PEG monoacrylate	NHS–amine	Increased β-cell viability	[115]
		Dextran	Acrylate–thiol	Increased DRG neurite outgrowth	[39]
	IKVAV	PEG monoacrylate	NHS–amine	Increased β-cell viability	[115]
		Dextran	Acrylate–thiol	Increased DRG neurite outgrowth	[39]
Collagen	GFOGER	PEG	Acrylate–thiol	Increase chondrogenic differentiation	[44]
Antithrombin III	K(βA)FAKLAARLYRKA	Fibrin	transglutaminase factor XIIIa	Binding to NGF to promoted neurite outgrowth	[108]
Vitronectin	CGKKQRFRHRNRKG	Polyacrylamide	Maleimide–thiol	Maintained pluripotency of human ES cells	[55]
Human natural killer cells (HNKCs)	FLHTRLFV	Collagen	EDC–amine	Motor neuron neurite outgrowth	[60]
Polysialic acid	SSVTAWTTG	Collagen	EDC–amine	Motor neuron neurite outgrowth and Schwann cell neurite extension	[60]

tripeptide unit (Gly-X-Y), where X and Y were pre-dominantly conformationally rigid prolines to facilitate the formation of a triple helix. Subsequent work by Farndale and coworkers showed that the synthetic sequence (GFOGER, where O is hydroxyproline) derived from collagen I has high affinity for the $\alpha_2\beta_1$ integrin. Garcia *et al.* have also synthesized a peptide with the GFOGER hexapeptide flanked with the triple helical sequence $(GPP)_5$ to promote the formation of the triple helix [43]. On conjugating to various surfaces, HT1080 cells showed dose-dependent cell adhesion, and MC3T3-E1 cells showed vinculin staining, which suggests focal adhesion through integrin binding. Conjugation of a similar peptide to PEG resulted in increased chondrogenic differentiation of human mesenchymal stem cells compared with cells cultured in controls of PEG alone [44].

1.2.2 Carbohydrate-Binding Peptides

Carbohydrates play a significant role in cell recognition and binding. The chemical structures of carbohydrate complexes (glycans) are diverse and complex. They consist of numerous monosaccharide units (up to 200 total units) covalently bonded to each other linearly or as branched structures, with each structure providing a unique binding affinity to other molecules [45]. GAGs are a class of linear anionic polysaccharides that can be posttranslationally conjugated to proteins in the Golgi complex to form glycoproteins. Glycoproteins that are transported to the cell membrane function as transmembrane proteins, whereby the glycan is exposed to the extracellular space and participates in cellular recognition and protein binding [46].

Heparin is a GAG that is commonly found either in the ECM or conjugated to a transmembrane protein (proteoglycan). Heparin binds with high affinity to a variety of proteins such as antithrombin III (AT III) [47], bFGF [48], VEGF [49], and BMP-2 [50], and presents the proteins for enhanced bioactivity [51]. Thus, the conjugation of heparin to biomaterials is useful for applications that require interactions with heparin-binding proteins (HBPs). Sakiyama-Elbert and Hubbell reported that covalent conjugation of the AT III-derived sequence K(βA)FAKLAARLYRKA to fibrin matrices strongly bound to heparin [52]. A short peptide sequence (NQEQVSP) was also incorporated into the N-terminus to enable enzymatic peptide ligation to the fibrin hydrogel by transglutaminase factor XIIIa [53].

Another important role of transmembrane GAGs is to recognize chemical signals from the surrounding environment. Keissling *et al.* have discovered that the vitronectin-derived peptide sequence CGKKQRFRHRNRKG binds to GAGs expressed on the cell surface of human embryonic stem cells (hESCs), and can maintain their expression of pluripotent markers after 3 months [54]. Moreover, hESCs cultured on polyacrylamide hydrogels conjugated with this sequence both proliferated and maintained greater pluripotency than cells cultured on gels containing the integrin-binding sequence CRGDS [55].

1.2.3 Glycomimetic Peptides

As described earlier, carbohydrates play a significant role in cell recognition and binding. Efforts to study the interaction between glycans and cells using chemical analogs have been limited by the inability to readily and efficiently chemically synthesize complex polysaccharides, which are challenging synthetic targets due to the multiple glycosylation steps, and the need to preserve the numerous carbohydrate stereocenters. While antibodies can be used to bind to carbohydrate receptors, their size and stability have limited large-scale use.

Interestingly, synthetic peptides have been discovered that mimic the chemical structures of several complex polysaccharides. These peptides occupy a similar chemical space as the parent polysaccharides, and therefore can bind to similar polysaccharide receptors. For example, a peptide sequence (FLHTRLFV) that mimics glycans found on the cell surface of human natural killer cells (HNKCs) was discovered using phage display and antibody-binding assays [56]. Motor neurons cultured in the presence of the HNKC glycomimetic peptide showed significantly longer neurite outgrowth compared with those cultured in the absence of HNKC-peptides [56]. Masand *et al.* recently conjugated this peptide to collagen hydrogels using EDC chemistry, and as demonstrated, these hydrogels also increased neurite outgrowth and length of motor neurons compared with cells cultured on collagen alone.

Another important glycan group is polysialic acid (PSA), which is naturally found conjugated to a variety of different transmembrane proteins including neural cell adhesion molecules (NCAMs). The PSA is hypothesized to be involved in cell migration of neural cells and cancer cells. Novel PSA-mimetic peptides have been discovered, and delivery of these PSA-mimetic peptides into the brain and spinal cord showed improved functional recovery and tissue regeneration in various injury models [57–59]. Masand *et al.* have also conjugated this PSA-mimetic peptide to collagen hydrogels, and demonstrated an increase in neurite length of cultured dorsal root ganglion and motor neurons, and increased Schwann cell proliferation compared with cells cultured on collagen alone [60]. However, a mixture of both PSA and HNKC peptides to collagen hydrogels yielded neither an additive effect for neurite outgrowth nor proliferation, emphasizing the importance of understanding the underlying mechanism for synergistic effects.

1.2.4 Growth Factors

Recently, larger molecules such as proteins and growth factors have been conjugated to biomaterials in a site-specific manner. Previous methods used nonspecific conjugation of large proteins to hydrogel scaffolds through amide linkage chemistry, such as EDC coupling. This approach is problematic due to the presence of multiple amines and carboxylates found in many proteins; random amide bond formation may decrease or even block protein activity. The limitation of nonspecific amidation has been overcome by exploiting site-specific modification, including protein modification to include *click* moieties discussed earlier [61], or by noncovalently incorporating proteins through high-affinity binding with complementary peptides/proteins immobilized to the hydrogel [62,63].

Various genetic modifications have enabled the site-specific incorporation of sequences and functional groups that can interact with bio-orthogonal partners. Protein biotinylation is a widely studied posttranslational modification and can be selectively incorporated into a protein that has been modified with the biotin-ligase recognition sequence (GLNDIFEAQ*K*IEWHE) [64]. Biotin ligase selectively and covalently binds a biotin moiety to the primary amine of the lysine (K) residue in this sequence. The biotinylated protein is subsequently immobilized to streptavidin-containing biomaterials through high-affinity binding ($K_D \sim 10^{-15}$ M). Tam *et al.* recently reported a thiolated derivative of methylcellulose conjugated to maleimide–streptavidin, followed by immobilization of biotin-containing platelet-derived growth factor (PDGF) [65]. This material was shown to increase the differentiation of rat neural stem/progenitor cells into oligodendrocytes *in vitro*, and also promote functional and tissue repair in rat models of spinal cord injury [66].

Another method to immobilize proteins to biomaterials is to incorporate growth factor-binding domains derived from larger proteins such as fibronectin. The fibronectin domain FN III 12-14 was shown to have high affinity for several growth factors such as VEGF, PDGF-BB and BMP-2 [67]. Martino *et al.* demonstrated that incorporation of this domain into a fibrin hydrogel with these growth factors significantly increased cell proliferation and migration of endothelial cells (ECs), smooth muscle cells (SMCs) and mesenchymal stem cells (MSCs), respectively *in vitro*, as well as improved wound and bone tissue healing *in vivo* [68].

Growth factors have been incorporated into hydrogels through other modified polypeptide-based binding pairs. Stoller *et al.* reported that SH3-binding domains have variable binding affinities for short hydrophobic peptides [69]. Building on this work, Vulic *et al.* expressed SH3 and bFGF as a fusion protein and immobilized the complementary SH3 peptide-binding domain to methyl cellulose [70]. Using this approach, bFGF was noncovalently bound to hydrogels and the release rate of bFGF from the scaffold was tuned based on the SH3 protein–SH3 peptide dissociation constant. In a separate approach, Ehrbar *et al.* developed two fusion peptides: one containing the glutamine acceptor substrate (NQEQVSPL) (Gln) and a synthetic analog of Protein A (ZZ), the second containing interleukin-4 (IL-4) and the fragment crystallizable (Fc) region of immunoglobulin G (IgG) antibodies [71]. The Gln-ZZ construct was conjugated to PEG hydrogels via enzymatic ligation to lysine donors on the hydrogel backbone. High-affinity binding (4.8×10^{-8} M^{-1}) between Protein A and Fc led to the incorporation of IL4 into the PEG hydrogels. The activity of immobilized IL4 was preserved as evidenced by a cell-based fluorescent reporter.

Several additional bio-orthogonal partners have been investigated as a means to noncovalently control the extent and duration of growth factor presentation in hydrogels (Table 1.2). Potentially, any growth factor that can be stably expressed as a fusion protein can be adapted for one of these approaches.

1.3 SPATIAL ORIENTATION AND DYNAMIC DISPLAY

During tissue development, tissue repair, and many disease states, the composition of the ECM is dynamic [72]. Specifically, growth factors and adhesive molecules are often presented transiently and localized in specific locations or as gradients within the ECM. Therefore, the next major challenge for incorporating ECM-inspired molecules into hydrogels is to allow user-defined temporal and spatial control over the presentation of biomimetic cues discussed above [73].

TABLE 1.2 Growth Factors Immobilized to Hydrogels Through Bio-Orthogonal Partner Binding

Growth factor	Polymer	Binding partners	Bioactivity	Reference
bFGF	Methyl cellulose	SH3 protein—SH3binding peptide	NSPC survival	[70]
	PEG	bFGF—bFGF-binding peptide	Promoted differentiation	[116]
	Agarose	Albumin—albumin-binding domain	Neurosphere assay	[80]
BMP-2	Fibrin	FN III 12-14 binding domain	Increased MSC migration	[68]
CNTF	Agarose	Biotin-streptavidin	Promoted differentiation of retinal cells	[63]
IFN	methacrylate	Biotin-streptavidin	Promoted differentiation to neurons	[117]
IL4	PEG	Fc—synthetic protein A	Activation of reporter	[71]
PDGF	Methyl cellulose	Biotin-streptavidin	Increased NSPC differentiation into oligodendrocytes,increased functional recovery in rats with spinal cord injury	[65, 66]
	Agarose	Thiol–maleimide	Increased NSPC differentiation into oligodendrocytes	[28]
	Fibrin	FN III 12-14 binding domain	Increased smooth muscle cell (SMC) and MSC proliferation and migration	[68]
SHH	Agarose	Barstar–barnase	Increased NSPC migration	[63]
VEGF	Fibrin	FN III 12-14 binding domain	Increased endothelial cell proliferation and migration	[68]
	Agarose	Thiol–maleimide	Increased EC and RSPC migration	[78]

1.3.1 Spatially Controlled Display

Incorporating biomimetic cues into engineered scaffolds is essential. As discussed above, significant progress has been made in immobilizing ECM-inspired molecules into hydrogels. The majority of examples incorporate biomolecules into hydrogels uniformly. Recently, photochemical patterning has emerged as a powerful approach to control the spatial immobilization of biomolecules within 3-D hydrogels [74–76].

The majority of *photopatterning* or *photolithography* has been focused on the immobilization of cell adhesive molecules, such as RGD and IKVAV, by directly conjugating these peptides to the hydrogel backbone using light-activated ligation

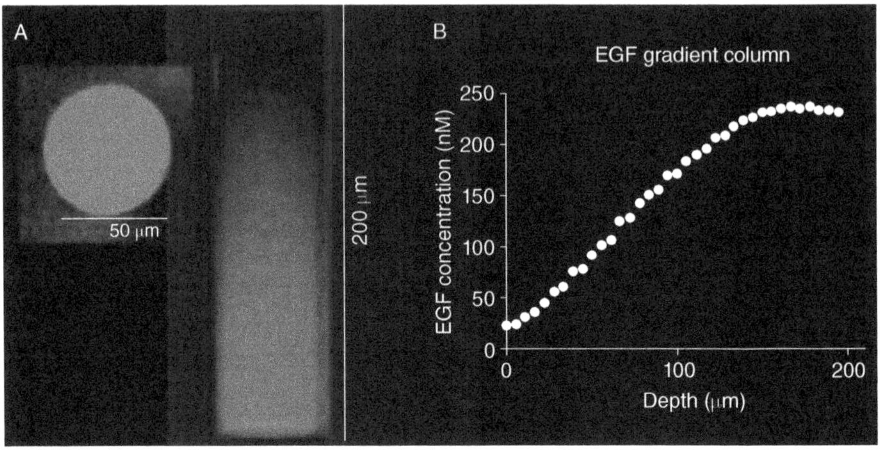

FIGURE 1.3 3-D photopatterning of EGF within a hyaluronic acid–PEG hydrogel. (A) Creation of a linear immobilized gradient of EGF. From the top of the hydrogel, the number of scans by the multiphoton laser are increased as it penetrates into the sample, corresponding to an increase in fluorescence intensity, and hence, an increase in protein immobilization. (B) The concentration of immobilized protein in the gradient was quantified by fluorescence intensity, showing a change in concentration from 25 nM at the top of the hydrogel to 250 nM at a depth of 150 μm in the hydrogel. EGF, epidermal growth factor. (Reproduced with permission from Owen, S.C., Fisher, S.A., Tam, R.Y., Nimmo, C.M., Shoichet, M.S. *Langmuir* **2013**. Copyright 2013 American Chemical Society.) *(See insert for color representation of the figure.)*

[78]. In this manner, Aizawa *et al.* patterned gradients of VEGF into agarose hydrogels and demonstrated that the immobilized proteins remained bioactive and effectively guided endothelial cell migration and tubulogenesis. Successful photopatterning has also been demonstrated in natural–synthetic hybrid hydrogels. For example, Owen *et al.* adapted coumarin-based photochemistry [63] into hyaluronic acid hydrogels to immobilize epidermal growth factor (EGF) gradients, Figure 1.3 [80]. Wylie *et al.* advanced photopatterning technology to simultaneously immobilize multiple proteins into hydrogels through bio-orthogonal paired *click* reactions [63,81]. The authors first conjugated one-half of the bio-orthogonal pair (barnase or streptavidin) directly to the hydrogel backbone. The complementary binding partner was coexpressed or ligated to a functional protein, in this case CNTF–biotin and sonic hedgehog (SHH)–barstar, and then mixed in the hydrogels where they bound specifically with their immobilized partners. Using this approach, both of the functional proteins were incorporated in spatially defined gradients within the hydrogels. It was further demonstrated that cultured neural progenitor cells responded to both CNTF and SHH gradients.

Photochemistry can also be exploited to remove biomimetic cues. Anseth and coworkers incorporated photodegradable cross-linking technology into PEG hydrogels in order to selectively degrade the scaffold [82] or remove adhesive biomolecules [83]. Significantly, the bio-*clip* removal of biomimetic peptides was recently combined with photopatterning technology to allow reversible presentation [84]. Recently, Kasko and coworkers have significantly expanded the number of different photocleavable linkers to facilitate conjugation of bioactive compounds [85].

Additional techniques are being investigated to control the immobilization of biomolecules in defined regions of interest or as gradients. For example, Turturro *et al.* employed perfusion-based frontal photopolymerization to create biofunctional gradients of RGD in matrix metalloproteinase (MMP)-sensitive hydrogels [86]; Wan *et al.* used electrical currents in conducting hydrogels to induce gradients of protein deposition and subsequent cell density [87,88].

1.3.2 Stimuli-Sensitive Dynamic Display

A second approach to control the display of biomimetic cues is to utilize hydrogels, which are sensitive to external stimuli such as temperature, pH, or enzymes. A number of stimuli-sensitive hydrogels have been developed and are reviewed extensively elsewhere [88–91]. For tissue engineering applications, the ability to control the presentation and the removal of specific cell adhesion proteins or growth factors may permit the controlled proliferation and differentiation of stem and progenitor cells.

Using enzymatic cleavage to release appended growth factors [62], Zisch *et al.* conjugated VEGF to PEG hydrogels by an MMP-cleavable synthetic peptide linker [92]. Similarly, Lutolf *et al.* utilized similar PEG hydrogels modified with pendant recombinant human bone morphogenetic protein-2 (rhBMP-2) also cleaved by MMP [93]. In both of these approaches, the growth factors are presented on the hydrogel surface but only released upon local cellular demand.

Many recent endeavors are more specifically focused on utilizing external stimuli to control the exposure of biomolecules. Okano and coworkers pioneered the use of temperature-sensitive poly(N-isopropylacrylamide) (poly(NIPAAm)) hydrogels to engineer *cell sheets* from several cell types [94–96]. Recently, Okano's group developed a modified version of poly(NIPAAm) by covalently tethering heparin onto poly(N-isopropylacrylamide-co-2-carboxyisopropylacrylamide) hydrogel surfaces. As discussed, heparin possesses an affinity for a number of growth factors, including VEGF and bFGF, and is important in stabilizing these proteins. As such, the heparin-functionalized hydrogels readily bind these growth factors on gels at 37°C and release the growth factors upon swelling of poly(NIPAAm-co-CIPAAm) chains at 20°C. Okano's group first incubated bFGF in heparin-modified gels and then cultured NIH/3T3 cells on the surface, demonstrating that heparin-bound bFGF enhanced growth factor-specific cell attachment. In a separate study, Tekin *et al.* used poly(NIPAAm) hydrogels to promote the formation and retrieval of stable cell aggregates for various applications [97].

Additional stimuli-sensitive hydrogels are in the early stages of development for tissue engineering applications. For example, Maynard and coworkers synthesized a novel temperature sensitive hydrogel based on poly(triethylene glycol methacrylate) (poly(TEGMA)) [98]. Significantly, the spatial deposition of these hydrogels can be controlled using e-beam radiation lithography. Further adaption of these hydrogels with biomimetic factors could provide user control over multiple cell-instructive cues.

1.4 FUTURE PERSPECTIVES

The ability to systematically re-create the 3-D microenvironment that defines cellular function and organization has profound implications for applications such as tissue regeneration strategies and understanding disease progression, including elucidating complex cellular mechanisms and functions. Current technologies such as immunochemistry and phage display have provided high-throughput methods to discover molecules that are involved in cellular interactions and techniques to develop synthetic peptide analogs. These strategies can now be combined with high-content information achieved with biomimetic strategies of the cell niche.

The combination of bio-*click* and bio-*clip* photopatterning are particularly exciting as they enable well-defined chemical cues that both guide and respond to cell fate. Although not discussed in this chapter, the mechanical properties of the ECM also influence cell fate, in concert with the chemical and physical properties. There is evidence that both the underlying elasticity of the ECM [99] and the tethering of cells to the ECM [100] play significant roles in stem cell growth and differentiation. As such, modular hydrogels, with independent user control over physical and biochemical properties, will facilitate our ability to understand and guide cell behavior. The cell itself has significant impact over its environment, and thus the culture of multiple cell types at different cell densities are additional considerations in the design of a bioengineered 3-D hydrogel matrix.

Abbreviations

AT III	Antithrombin III
bFGF	Basic fibroblast growth factor
BMP-2	Bone morphogenetic protein 2
CNTF	Ciliary neurotrophic factor
DRG	Dorsal root ganglion
ECM	Extracellular matrix
ECs	Endothelial cells
EGF	Epidermal growth factor
ES cells	Embryonic stem cells
HBP	Heparin-binding protein
HNKCs	Human natural killer cells
HT1080	Fibrosarcoma cell line
IFN	Interferon
IgG	immunoglobulin G
IKVAV	Synthetic peptide derived from the α-1 chain of laminin
IL-4	Interleukin-4
MC3T3-E1	Osteoblast precursor cell line
MMP	Matrix metalloproteinase
MSC	Mesenchymal stem cells

NCAM	Neural cell adhesion molecule
NGF	Nerve growth factor
NSPCs	Neural stem/progenitor cells
PDGF	Platelet-derived growth factor
PEG	Polyethylene glycol
PSA	Polysialic
RGD	(Arginine–glycine–aspartic acid): synthetic peptide derived from fibronectin
rhBMP-2	Recombinant human bone morphogenetic protein-2
RSPCs	Retinal stem/progenitor cells
SHH	Sonic hedgehog
VEGF	Vascular endothelial growth factor
YIGSR	(Tyr-ile-gly-ser-arg): synthetic peptide derived from β-1 chain of laminin

REFERENCES

[1] Baudino, T.A., Bowers, S.L.K., Banerjee, I. *J. Mol. Cell. Cardiol.* **2010**, *48*, 474–482.
[2] Klein, G. *Experientia* **1995**, *51*, 914–926.
[3] Divoux, A., Clement, K. *Obes. Rev.* **2011**, *12*, e494–e503.
[4] Ruoslahti, E. *Glycobiology* **1996**, *6*, 489–492.
[5] Martinez-Hernandez, A., Amenta, P.S. *FASEB J.* **1995**, *9*, 1401–1410.
[6] Zimmermann, D.R., Dours-Zimmermann, M.T. *Histochem. Cell Biol.* **2008**, *130*, 635–653.
[7] Hulmes, D.J. *J. Struct. Biol.* **2002**, *137*, 2–10.
[8] Kielty, C.M., Sherratt, M.J., Shuttleworth, C.A. *J. Cell Sci.* **2002**, *115*, 2817–2828.
[9] Bosman, F.T., Stamenkovic, I. *J. Pathol.* **2003**, *200*, 423–428.
[10] Lu, P.F., Weaver, V.M., Werb, Z. *J. Cell Biol.* **2012**, *196*, 395–406.
[11] Lee, S.R., Kim, H.Y., Rogowska, J., Zhao, B.Q., Bhide, P., Parent, J.M., Lo, E.H. *J. Neurosci.* **2006**, *26*, 3491–3495.
[12] Hiratsuka, S., Nakamura, K., Iwai, S., Murakami, M., Itoh, T., Kijima, H., Shipley, J.M., Senior, R.M., Shibuya, M. *Cancer Cell* **2002**, *2*, 289–300.
[13] Lanza, R.P., Langer, R.S., Vacanti, J. *Principles of Tissue Engineering*; Elsevier Academic Press, Boston: 2007.
[14] Owen, S.C., Shoichet, M.S. *J. Biomed. Mater. Res. A* **2010**, *94A*, 1321–1331.
[15] Kleinman, H.K., Martin, G.R. *Semin. Cancer Biol.* **2005**, *15*, 378–386.
[16] Pierschbacher, M.D., Ruoslahti, E. *Nature* **1984**, *309*, 30–33.
[17] Hersel, U., Dahmen, C., Kessler, H. *Biomaterials* **2003**, *24*, 4385–4415.
[18] Zhu, J.M. *Biomaterials* **2010**, *31*, 4639–4656.
[19] Azagarsamy, M.A., Anseth, K.S. *ACS Macro Lett.* **2013**, *2*, 5–9.
[20] Nimmo, C.M., Shoichet, M.S. *Bioconjug. Chem.* **2011**, *22*, 2199–2209.
[21] Becer, C.R., Hoogenboom, R., Schubert, U.S. *Angew. Chem. Int. Ed.* **2009**, *48*, 4900–4908.
[22] Crescenzi, V., Cornelio, L., Di Meo, C., Nardecchia, S., Lamanna, R. *Biomacromolecules* **2007**, *8*, 1844–1850.
[23] Kalia, J., Raines, R.T. *Angew. Chem. Int. Ed.* **2008**, *47*, 7523–7526.
[24] Schaffner, P., Dard, M.M. *Cell Mol. Life Sci.* **2003**, *60*, 119–132.
[25] Perlin, L., MacNeil, S., Rimmer, S. *Soft Matter* **2008**, *4*, 2331–2349.

[26] Lin, H.B., Sun, W., Mosher, D.F., Garciaecheverria, C., Schaufelberger, K., Lelkes, P.I., Cooper, S.L. *J. Biomed. Mater. Res.* **1994**, *28*, 329–342.

[27] Silva, N.S., Cooke, M.J., Tam, R.Y., Sousa, N., Salgado, A.J., Reis, R.L., Shoichet, M.S. *Biomaterials* **2012**, *33*, 6345–6354.

[28] Aizawa, Y., Leipzig, N., Zahir, T., Shoichet, M.S. *Biomaterials* **2008**, *29*, 4676–4683.

[29] Main, L.A., Harvey, T.S., Baron, M., Boyd, J., Campbell, I.D. *Cell* **1992**, *71*, 671–678.

[30] Baron, M., Main, L.A., Driscoll, P.C., Mardon, H.J., Boyd, J., Campbell, I.D. *Biochemistry* **1992**, *31*, 2068–2073.

[31] Haubner, R., Schmitt, W., Holzemann, G., Goodman, S.L., Jonczyk, A., Kessler, H. *J. Am. Chem. Soc.* **1996**, *118*, 7881–7891.

[32] Haubner, R., Gratias, R., Diefenbach, B., Goodman, S.L., Jonczyk, A., Kessler, H. *J. Am. Chem. Soc.* **1996**, *118*, 7461–7472.

[33] Zhu, J., Tang, C., Kottke-Marchant, K., Marchant, R.E. *Bioconjug. Chem.* **2009**, *20*, 333–339.

[34] Porte-Durrieu, M.C., Guillemot, F., Pallu, S., Labrugere, C., Brouillard, B., Bareille, R., Amedee, J., Barthe, N., Dard, M., Baquey, C. *Biomaterials* **2004**, *25*, 4837–4846.

[35] Moller, L., Hess, C., Palecek, J., Su, Y., Haverich, A., Kirschning, A., Drager, G. *Beilstein J. Org. Chem.* **2013**, *9*, 270–277.

[36] Wilson, M.J., Liliensiek, S.J., Murphy, C.J., Murphy, W.L., Nealey, P.F. *Soft Matter* **2012**, *8*, 390–398.

[37] Graf, J., Yamada, Y., Robey, F.A., Sasaki, M., Ogle, R.C., Iwamoto, Y., Kleinman, H.K., Martin, G.R. *Biochemistry* **1987**, *26*, 6896–6900.

[38] Tashiro, K.-I., Sephel, G.C., Weeks, B., Sasaki, M., Martin, G.R., Kleinman, H.K., Yamada, Y. *J. Biol. Chem.* **1989**, *264*, 16174–16182.

[39] Levesque, S.G., Shoichet, M.S. *Biomaterials* **2006**, *27*, 5277–5285.

[40] Silva, G.A., Czeisler, C., Niece, K.L., Beniash, E., Harrington, D.A., Kessler, J.A., Stupp, S.I. *Science* **2004**, *303*, 1352–1355.

[41] Farndale, R.W., Lisman, T., Bihan, D., Hamaia, S., Smerling, C.S., Pugh, N., Konitsiotis, A., Leitinger, B., De Groot, P.G., Jarvis, G.E., Raynal, N. *Biochem. Soc. Trans.* **2008**, *36*, 241–250.

[42] Fallas, J.A., O'Leary, L.E.R., Hartgerink, D. *Chem. Soc. Rev.* **2010**, *39*, 3510–3527.

[43] Reyes, C.D., Garcia, A.J. *J. Biomed Mater. Res. A* **2003**, *65A*, 511–523.

[44] Liu, S.Q., Tian, Q., Hedrick, J.L., Hui, J.H.P., Ee, P.L.R., Yang, Y.Y. *Biomaterials* **2010**, *31*, 7298–7307.

[45] Dumitriu, S. *Polysaccharides: Structural Diversity and Functional Versatility*; Marcel Dekker, New York: 2005.

[46] Iozzo, R.V. *Annu. Rev. Biochem.* **1998**, *67*, 609–652.

[47] Smith, J.W., Knauer, D.J. *J. Biol. Chem.* **1987**, *262*, 11964–11972.

[48] Schlessinger, J., Plotnikov, A.N., Ibrahimi, O.A., Eliseenkova, A.V., Yeh, B.K., Yayon, A., Linhardt, R.J., Mohammadi, M. *Mol. Cell.* **2000**, *6*, 743–750.

[49] Zhao, W.J., McCallum, S.A., Xiao, Z.P., Zhang, F.M., Linhardt, R.J. *Biosci. Rep.* **2012**, *32*, 71–81.

[50] Ruppert, R., Hoffmann, E., Sebald, W. *Eur. J. Biochem.* **1996**, *237*, 295–302.

[51] Conrad, H.E. *Heparin-Binding Proteins*; Academic Press, San Diego, CA: 1998.

[52] Sakiyama-Elbert, S.E., Hubbell, J.A. *J. Control. Release* **2000**, *65*, 389–402.

[53] Schense, J.C., Hubbell, J.A. *Bioconjug. Chem.* **1999**, *10*, 75–81.

[54] Klim, J.R., Li, L., Wrighton, P.J., Piekarczyk, M.S., Kiessling, L.L. *Nat. Methods* **2010**, *7*, 989–994.

[55] Musah, S., Morin, S.A., Wrighton, P.J., Zwick, D.B., Jin, S., Kiessling, L.L. *ACS Nano.* **2012**, *6*, 10168–10177.

[56] Simon-Haldi, M., Mantei, N., Franke, J., Voshol, H., Schachner, M. *J. Neurochem.* **2002**, *83*, 1380–1388.

[57] Torregrossa, P., Buhl, L., Bancila, M., Durbec, P., Schafer, C., Schachner, M., Rougon, G. *J. Biol. Chem.* **2004**, *279*, 30707–30714.

[58] Mehanna, A., Mishra, B., Kurschat, N., Schulze, C., Bian, S., Loers, G., Irintchev, A., Schachner, M. *Brain* **2009**, *132*, 1449–1462.

[59] Marino, P., Norreel, J.-C., Schachner, M., Rougon, G., Amoureaux, M.-C. *Exp. Neurol.* **2009**, *219*, 163–174.

[60] Masand, S.N., Perron, I.J., Schachner, M., Shreiber, D.I. *Biomaterials* **2012**, *33*, 790–797.

[61] Foley, T.L., Burkart, M.D. *Curr. Opin. Chem. Biol.* **2007**, *11*, 12–19.

[62] Lee, K., Silva, E.A., Mooney, D.J. *J. R. Soc. Interface* **2011**, *8*, 153–170.

[63] Wylie, R.G., Ahsan, S., Aizawa, Y., Maxwell, K.L., Morshead, C.M., Shoichet, M.S. *Nat. Mater* **2011**, *10*, 799–806.

[64] Beckett, D., Kovaleva, E., Schatz, P.J. *Protein Sci.* **1999**, *8*, 921–929.

[65] Tam, R.Y., Cooke, M.J., Shoichet, M.S. *J. Mater. Chem.* **2012**, *22*, 19402–19411.

[66] Mothe, A.J., Tam, R.Y., Zahir, T., Tator, C.H., Shoichet, M.S. *Biomaterials* **2013**, *34*, 3775–3783.

[67] Martino, M.M., Hubbell, J.A. *FASEB J.* **2010**, *24*, 4711–4721.

[68] Martino, M.M., Tortelli, F., Mochizuki, M., Traub, S., Ben-David, D., Kuhn, G.A., Müller, R., Livne, E., Eming, S.A., Hubbell, J.A. *Sci. Transl. Med.* **2011**, *3*, 100ra189.

[69] Stollar, E.J., Garcia, B., Chong, P.A., Rath, A., Lin, H., Forman-Kay, J.D., Davidson, A.R. *J. Biol. Chem.* **2009**, *284*, 26918–26927.

[70] Vulic, K., Shoichet, M.S. *J. Am. Chem. Soc.* **2012**, *134*, 882–885.

[71] Lienemann, P.S., Karlsson, M., Sala, A., Wischhusen, H.M., Weber, F.E., Zimmermann, R., Weber, W., Lutolf, M.P., Ehrbar, M. *Adv. Healthc. Mater.* **2013**, *2*, 292–296.

[72] Laurencin, C.T., Nair, L.S. *Nanotechnology and Tissue Engineering: the Scaffold*; CRC Press, Boca Raton, FL: 2008.

[73] Tibbitt, M.W., Anseth, K .S. *Sci. Transl. Med.* **2012**, *4*, 160ps124.

[74] Wosnick, J.H., Shoichet, M.S. *Chem. Mater.* **2008**, *20*, 55–60.

[75] DeForest, C.A., Polizzotti, B.D., Anseth, K.S. *Nat. Mater.* **2009**, *8*, 659–664.

[76] Lee, S.H., Moon, J.J., West, J.L. *Biomaterials* **2008**, *29*, 2962–2968.

[77] Luo, Y., Shoichet, M.S. *Nat. Mater.* **2004**, *3*, 249–253.

[78] Aizawa, Y., Shoichet, M.S. *Biomaterials* **2012**, *33*, 5198–5205.

[79] Owen, S.C., Fisher, S.A., Tam, R.Y., Nimmo, C.M., Shoichet, M.S. *Langmuir* **2013**, *29*, 7393–7400.

[80] Wylie, R.G., Shoichet, M.S. *Biomacromolecules* **2011**, *12*, 3789–3796.

[81] Tibbitt, M.W., Kloxin, A.M., Dyamenahalli, K.U., Anseth, K.S. *Soft Matter* **2010**, *6*, 5100–5108.

[82] Kloxin, A.M., Kasko, A.M., Salinas, C.N., Anseth, K.S. *Science* **2009**, *324*, 59–63.

[83] DeForest, C.A., Anseth, K.S. *Angew. Chem. Int. Ed.* **2012**, *51*, 1816–1819.

[84] Griffin, D.R., Schlosser, J.L., Lam, S.F., Nguyen, T.H., Maynard, H.D., Kasko, A.M. *Biomacromolecules* **2013**, *14*, 1199–1207.

[85] Turturro, M.V., Christenson, M.C., Larson, J.C., Young, D.A., Brey, E.M., Papavasiliou, G. *PLoS ONE* **2013**, *8*, e58897.

[86] Wan, A.M.D., Schur, R.M., Ober, C.K., Fischbach, C., Gourdon, D., Malliaras, G.G. *Adv. Mater.* **2012**, *24*, 2501–2505.

[87] Wan, A.M.D., Brooks, D.J., Gumus, A., Fischbach, C., Malliaras, G.G. *Chem. Commun.* **2009**, 5278–5280.

[88] Ulijn, R.V., Bibi, N., Jayawarna, V., Thornton, P.D., Todd, S.J., Mart, R.J., Smith, A.M., Gough, J.E. *Mater. Today* **2007**, *10*, 40–48.
[89] Lendlein, A., Shastri, V.P. *Adv. Mater.* **2010**, *22*, 3344–3347.
[90] Stuart, M.A.C., Huck, W.T.S., Genzer, J., Muller, M., Ober, C., Stamm, M., Sukhorukov, G.B., Szleifer, I., Tsukruk, V.V., Urban, M., Winnik, F., Zauscher, S., Luzinov, I., Minko, S. *Nat. Mater.* **2010**, *9*, 101–113.
[91] Hoffman, A.S. *Adv. Drug Deliv. Rev.* **2013**, *65*, 10–16.
[92] Zisch, A.H., Lutolf, M.P., Ehrbar, M., Raeber, G.P., Rizzi, S.C., Davies, N., Schmokel, H., Bezuidenhout, D., Djonov, V.D., Zilla, P., Hubbell, J.A. *FASEB J.* **2003**, *17*, 2260–2262.
[93] Lutolf, M.R., Weber, F.E., Schmoekel, H.G., Schense, J.C., Kohler, T., Muller, R., Hubbell, J.A. *Nat. Biotechnol.* **2003**, *21*, 513–518.
[94] Shimizu, H., Ohashi, K., Utoh, R., Ise, K., Gotoh, M., Yamato, M., Okano, T. *Biomaterials* **2009**, *30*, 5943–5949.
[95] Matsuda, N., Shimizu, T., Yamato, M., Okano, T. *Adv. Mater.* **2007**, *19*, 3089–3099.
[96] Yang, J., Yamato, M., Shimizu, T., Sekine, H., Ohashi, K., Kanzaki, M., Ohki, T., Nishida, K., Okano, T. *Biomaterials* **2007**, *28*, 5033–5043.
[97] Tekin, H., Anaya, M., Brigham, M.D., Nauman, C., Langer, R., Khademhosseini, A. *Lab Chip* **2010**, *10*, 2411–2418.
[98] Kolodziej, C.M., Maynard, H.D. *J. Am. Chem. Soc.* **2012**, *134*, 12386–12389.
[99] Trappmann, B., Gautrot, J.E., Connelly, J.T., Strange, D.G.T., Li, Y., Oyen, M.L., Stuart, M.A.C., Boehm, H., Li, B.J., Vogel, V., Spatz, J.P., Watt, F.M., Huck, W.T.S. *Nat. Mater.* **2012**, *11*, 642–649.
[100] Engler, A.J., Sen, S., Sweeney, H.L., Discher, D.E. *Cell* **2006**, *126*, 677–689.
[101] Lowe, A.B. *Polym. Chem. UK* **2010**, *1*, 17–36.
[102] Nimmo, C.M., Owen, S.C., Shoichet, M.S. *Biomacromolecules* **2011**, *12*, 824–830.
[103] Alge, D.L., Azagarsamy, M.A., Donohue, D.F., Anseth, K.S. *Biomacromolecules* **2013**, *14*, 949–953.
[104] Liu, S.Q., Ee, P.L.R., Ke, C.Y., Hedrick, J.L., Yang, Y.Y. *Biomaterials* **2009**, *30*, 1453–1461.
[105] Agard, N.J., Prescher, J.A., Bertozzi, C.R. *J. Am. Chem. Soc.* **2004**, *126*, 15046–15047.
[106] Krause, A., Kirschning, A., Drager, G. *Org. Biomol. Chem.* **2012**, *10*, 5547–5553.
[107] Dirksen, A., Dawson, P.E. *Bioconjug. Chem.* **2008**, *19*, 2543–2548.
[108] Sakiyama-Elbert, S.E., Hubbell, J.A. *J. Control. Release* **2000**, *69*, 149–158.
[109] Taylor, S.J., McDonald, J.W., 3rd, Sakiyama-Elbert, S.E. *J. Control. Release* **2004**, *98*, 281–294.
[110] Fittkau, M.H., Zilla, P., Bezuidenhout, D., Lutolf, M.P., Human, P., Hubbell, J.A., Davies, N. *Biomaterials* **2005**, *26*, 167–174.
[111] Grover, G.N., Lam, J., Nguyen, T.H., Segura, T., Maynard, H.D. *Biomacromolecules* **2012**, *13*, 3013–3017.
[112] Shu, X.Z., Ghosh, K., Liu, Y.C., Palumbo, F.S., Luo, Y., Clark, R.A., Prestwich, G.D. *J. Biomed. Mater. Res. A* **2004**, *68A*, 365–375.
[113] Lampe, K.J., Antaris, A.L., Heilshorn, S.C. *Acta Biomater.* **2012**, *9*, 5590–5599.
[114] Kaufmann, D., Fiedler, A., Junger, A., Auernheimer, J., Kessler, H., Weberskirch, R. *Macromol. Biosci.* **2008**, *8*, 577–588.
[115] Weber, L.M., Hayda, K.N., Haskins, K., Anseth, K.S. *Biomaterials* **2007**, *28*, 3004–3011.
[116] Lin, C.-C., Anseth, K.S. *Adv. Funct. Mater.* **2009**, *19*, 2325–2331.
[117] Leipzig, N.D., Xu, C., Shoichet, M.S. *J. Biomed. Mater. Res. A* **2010**, *92A*, 625–633.

CHAPTER 2

DYNAMIC MATERIALS MIMIC DEVELOPMENTAL AND DISEASE CHANGES IN TISSUES

MATTHEW G. ONDECK
Material Science Program, University of California–San Diego, La Jolla, CA, USA

ADAM J. ENGLER
Material Science Program, Department of Bioengineering, Sanford Consortium for Regenerative Medicine, University of California–San Diego, La Jolla, CA, USA

2.1 INTRODUCTION

A growing body of evidence suggests that cells are influenced not just by soluble growth factors, but also by the extracellular matrix (ECM), a three-dimensional (3-D) fibrillar material to which cells adhere [1,2], and its intrinsic properties [3], for example, structural [4,5], biochemical [6], and mechanical [1,7]. Not only are cells influenced by intrinsic aspects of the ECM, they can also modify ECM to change its properties [8, p. 226] and thereby direct the behavior of adjacent cells [3]. For example, the ECM is composed predominantly of three major proteins, that is, collagen, fibronectin, and laminin, and each has a very specific distribution in the body [8, pp. 349–359] that contributes to developmental process, for example, fibronectin is made by mesodermal cells for the endoderm to use as a substrate on which to migrate during gastrulation [9]. Cells also establish an ECM to influence morphology [10], stiffness [11], lineage [1], and durotaxis (directed motility as a result of a change in stiffness in the microenvironment) [12,13] of adjacent cells. By better understanding the dynamics of these changes, we will be able to engineer materials that will better mimic the dynamic nature of cells and tissues *in vivo*. While many aspects of the matrix change with time as cells remodel it, this chapter will mainly focus on the cell and dynamic tissue stiffness and questions of when, why,

Bio-inspired Materials for Biomedical Engineering, First Edition. Edited by Anthony B. Brennan and Chelsea M. Kirschner.

and how they change. We will do this in the context of several examples where we will detail how the environment changes with time and how that impacts cell behavior.

Prior to that, it is important to first understand the molecular mechanisms of how and why cells respond to physical stimuli, including passive stimuli such as stiffness. Cells perceive and *feel* the passive properties of their surroundings, for example, matrix stiffness, through physical signals it receives from transmembrane receptors called integrins, which are converted into biochemical signals that the nucleus or other organelles can interpret in a process called mechanotransduction. Mechanisms proposed to act within or surrounding perinuclear regions of the cell can cause nuclei to get stiffer as stem cells differentiate [14] and such changes can be dependent on cytoskeletal coupling to the nucleus via SUN- and LINK-domain-containing proteins [15]. Many other mechanisms have been proposed within focal adhesions, which are dynamic cell–matrix linking structures composed of clustered integrins located at the end of actin stress fibers [16,17]. They are integral in coordinating cell migration among many other processes [18]. Focal adhesion assembly and disassembly allows for cell movement along the matrix and is regulated by the tyrosine kinase called focal adhesion kinase (FAK) which is located at cell adhesion sites [16,18,19]. The expression of FAK has shown to be upregulated in metastatic tumors [16,20,21] and is essential in fibroblast spreading [22] and migration [23]. Rho GTPase-dependent cell signaling changes [13,24] directly impact actomyosin function and thus regulate contractile force [25,26], which plays a prominent role in cell spreading and adhesion [27]. Changes in stretch-activated channels to regulate calcium influx have been proposed and could also directly alter actomyosin function [15]. However, what directly transmits actomyosin contractions and relays signals back to the cell are cell–matrix adhesions, which have been an area of great interest for mechanotransduction [28]. Within these complexes, integrins bind to matrix outside the cell membrane, and on their cytoplasmic side, bind to actin via talin, which itself has many regions that unfold with force to induce signaling events [29]. Talin unfolds on application of 12 pN force, unveiling vinculin binding sites, which have mechanical/chemical signaling implications [15,29,30]. Proteins within this complex that change confirmation under force and expose cryptic sites have been termed *molecular strain gauges* [15]. Transduction within these complexes can be bidirectional where the cell generates a chemical signal in reaction to the mechanics of the ECM, which ultimately reinforces the attachment strength of the adhesions on ECM proteins [31]. This process is highly dynamic as the forces subjected to an adhesion fluctuate with both space and time, even within a single adhesion [32], and thus the intensity of signaling itself can scale with the stability of focal adhesions [33]. Through these different mechanisms, cells can *feel* the mechanics of their microenvironment and can respond by changes that are evident in cell morphology, motility, and lineage.

With these sensing mechanisms in mind, we now review the scaffold types and materials employed for cell culture and tissue engineering, both synthetic and natural. Unlike other reviews herein, we will pay particular attention to the dynamic nature of the scaffolds and the influence that that has on cell form and function.

2.2 CELL SCAFFOLDS, THEIR INTRINSIC PROPERTIES, AND THEIR EFFECTS ON CELLS

A biomimetic environment that promotes cell adhesion, proliferation, and directed lineage is very important to developing the appropriate cell behavior *in vitro*. However, cells are often cultured *in vitro* on substrates that are less biomimetic, for example, tissue culture plastic. Rigid materials do not simulate the *in vivo* microenvironment of a cell [34,35], and hydrogels are commonly substituted because of their capacity to mimic many intrinsic properties of the ECM by their high water content, mechanical properties, structure, adhesivity (when functionalized), and so on [36]. For example, mechanical properties, specifically Young's modulus (E_Y; measured in pascals, Pa) or stiffness (as it is referred to in biological contexts), of ECM is known to have a strong influence on the lineage of stem cells; polyacrylamide (PA) hydrogels with a stiffness that mimics muscle encourages mesenchymal stem cells (MSCs) to begin to commit toward a myogenic lineage in the presence of serum-containing media [1]. Hydrogels are made from a number of different sources, both naturally derived, for example, glycosaminoglycans (GAGs), and synthetically produced polymers. In this section, we will review the basic types of polymer, the unique biological, chemical, and mechanical properties each offers, and the resulting effects on cells.

2.2.1 Natural Polymers and Their Properties

A number of polymers that are used for cell scaffolds are naturally derived from biological sources and can be found in the body. These biopolymers are naturally biocompatible, can form many different structures depending on their assembly, and have many different functions that augment their role as a cell scaffold. Collagen is a biopolymer that comes from over 100 different genes, but assembles into 29 distinct types, each with its own distribution and signaling properties *in vivo* [8, pp. 5–22,37]. Gelatin, an irreversibly hydrolyzed, denatured form of collagen [38], is often derived from mammalian skin and used in hydrogels when cross-linked by chain entanglement at higher temperatures. Other matrix proteins, for example, fibrin [39], require cells to aide in their assembly, but together, these polymers are very flexible with low persistence lengths and small globular structures Unmodified natural polymers are typically mechanically inferior to synthetic polymers, in terms of stiffness, achieving an E_Y of up to only a few kilopascals (kPa) [40], which limits the types of tissue it can emulate. Conversely, polysaccharides, including anionic alginate [41], GAGs, and hyaluronan (HA) [42–44], have large amino sugars, making them less flexible. Via chemical functionalization, these polymers can be covalently linked and can provide properties for cells that more appropriately mimic native niches. For example, HA polysaccharide structure has several functional groups that can be chemically modified, through mechanisms such as thiolation [42] and methacrylation [43,44], allowing one to tune the mechanical properties toward a particular application. HA also acts as a ligand for CD44, allowing for cells to adhere without the use of adhesive peptides [45,46]. Given the diversity of intrinsic properties from their many ligands to varied stiffnesses, natural polymers

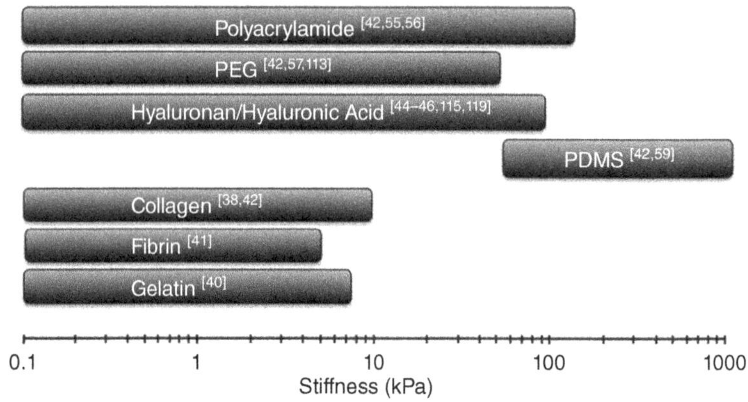

FIGURE 2.1 Range of stiffness from different synthetic and natural polymers. Note that citations in parentheses refer to authors in the reference list at the end of this chapter.

can provide many different microenvironments, often in 3-D, but their ability to predictably and reproducibly create these niches is not the same as synthetic polymers.

2.2.2 Synthetic Polymers and Their Properties

Synthetic polymers are advantageous for cell scaffolding applications because of the various ways in which they can be polymerized, as well as the larger range of intrinsic properties that can be achieved, for example, different structures (fibers [47], foams [48], hydrogels [1], etc.) and stiffnesses among others [40]. Synthetic polymers are typically biologically inert and require the use of adhesive peptides to allow for cell attachment [40]. For example, polyethylene glycol (PEG) [49,50], polydimethylsiloxane (PDMS) [51,52], and PA [35,53] hydrogels do not have cell-binding receptors, and therefore require a protein layer to encourage cell adhesion. Additionally, PEG [54], PA [35,55], and PDMS [56] are capable of reaching a variety of biological stiffnesses (Figure 2.1). However, most synthetic polymers are composed of very small monomers, generating a polymer with smaller pore size, which prevents any cell migration through the polymer, which limits their usage in cell encapsulation studies [40].

PA and PEG, when chemically modified with acrylate groups to form polyethylene glycol diacrylate (PEGDA), for example, can be polymerized by different free radical addition reaction: photopolymerization via an ultraviolet (UV) photoinitiator such as Irgacure 2959 (BASF—The Chemical Company, Minden, Germany) [43,44,57,58] or chemical methods using ammonium persulfate (APS) in combination with N,N,N′,N′-tetramethyl-ethylenediamine (TEMED) [12]. PDMS polymerizes by a curing process at 60°C in the presence of a cross-linker, and can reach stiffnesses in the megapascal (MPa) range [56].

2.2.3 The Effects of Scaffolds on Cells

It is important to recognize that the intrinsic properties mentioned in previous parts of this section are presented initially to cells. As the cells interact with the material, as described below, the cells are not only receiving signals from the scaffold, but they also remodel the scaffold. This dynamic state is often referred to as *dynamic reciprocity* [59] and can be summarized by the idea that while intrinsic ECM properties influence cells, the cell in turn remodels the matrix, thus shaping the signals that it receives. For progenitor cells, this process regulates many behaviors as well as cell fate. While this local remodeling presents active changes to the cell that is assembling or changing the matrix, the other cells around it respond to the passive changes that it induces. These sections to follow highlight how cells respond to passive matrix properties, but ultimately, we are concerned with how their collective actions together result in developmental changes that can be mimicked by dynamic matrices.

2.2.3.1 Stem Cells As stated previously, one of the ECM's intrinsic properties, stiffness, plays a significant role in determining cell lineage [1,60,61]. It has been shown that stem cells are highly perceptive to the mechanics of their microenvironment [1,7]. MSCs cultured on PA hydrogels of different stiffnesses expressed different lineage-specific markers. β3 Tubulin, a neurogenic marker, MyoD, a myogenic marker, and CBFα1, an osteogenic marker, were expressed on 1 kPa (soft), 11 kPa (intermediate), and 34 kPa (stiff) hydrogels, respectively (Figure 2.2) [1]. Stiffness has shown to have an important influence on neural progenitor cell fate as well. Neural stem cells (NSCs) cultured on hydrogels with stiffness between 0.1 and 10 kPa exhibited different neural lineages; neuron differentiation decreases with increasing hydrogel stiffness, while glial differentiation increases [62]. Likewise, myogenic differentiation is affected by scaffold stiffness. MSCs undergo myogenic differentiation in the presence of transforming growth factor beta (TGF-β), a promoter of myogenic differentiation, and also on hydrogels with stiffness more similar to firm muscle, for example, 11 kPa. Cells also exhibit more cell spreading, proliferation, and stress fibers [63].

Cells also respond to different spatially defined patterns on the ECM [64–66]. Using microcontact printing to pattern the scaffold with ECM proteins permits cells to adhere to specific patterned regions of the ECM [66,67]. This allows for control of cell morphology, which in turn controls cell traction forces, and ultimately determines cell lineage. Using different shapes and aspect ratios of micropatterns allows one to finely tune stem cell lineage, by mimicking the shape of the desired cell type. For example, using a star shape without additive media pushes cells toward an adipogenic fate [64]. This idea has shown to be a viable method to creating cell constructs, such as myotubes [68–71]. Hydrogels with a myogenic elasticity, coated with collagen strips, have shown to develop fully formed myotubes [70], while others have used mechanically patterned acrylamide hydrogels with distinct soft and stiff regions that promote durotaxis, cell alignment, and fusion, resulting in myotube formation of adipose-derived stem cells (ASCs) (Figure 2.3) [68]. It is hypothesized that these cellular responses are likely due to the remodeling of the cell cytoskeleton

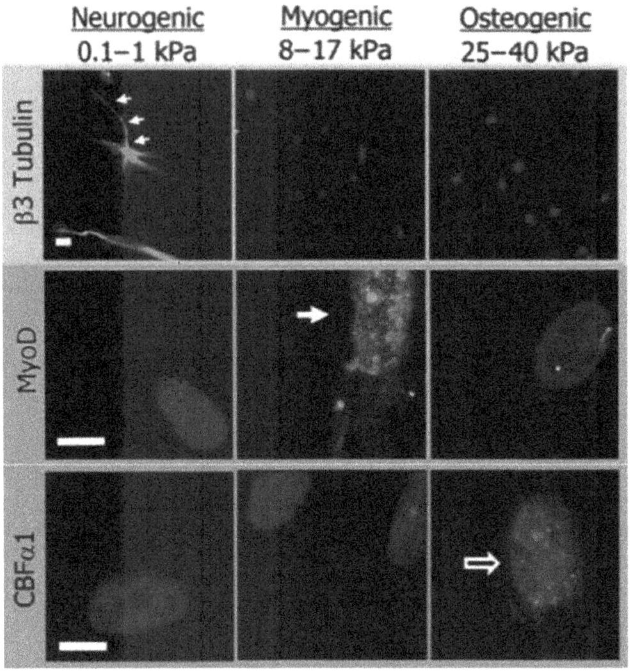

FIGURE 2.2 β3 Tubulin, MyoD, CBFα1 (neurogenic, myogenic, and osteogenic, respectively) differentiation markers are visible on respective PA hydrogels. (Reproduced with permission from Reference 1.)

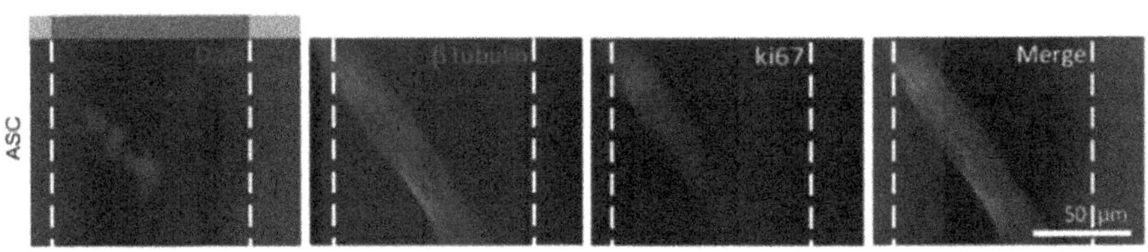

FIGURE 2.3 ASCs undergo myotube fusion when cultured on mechanically patterned hydrogels with soft and stiff regions. (Reproduced with permission from Reference 68.)

and focal adhesions in reaction to the mechanics and features of the microenvironment [61].

Nanopatterned topographical features on hydrogels, as well as nanopores, have been shown to influence cell lineage [72–74]. Electrospinning polymer cell scaffolds have shown to be an effective option for creating a cell-responsive nanoporous surface. MSCs cultured on chitosan electrospun scaffold align along the fibers and adopt a myogenic lineage [73], while MSCs cultured on electrospun methacrylated hyaluronic acid express increased chondrogenic markers with increasing scaffold adhesivity [74]. However, electrospun samples must have appropriate pore size;

when too small, cells cannot infiltrate the nanosized pores [75,76]. Other methods such as using nanopatterned scaffold have emerged as an effective way of differentiating stem cells. MSCs plated on PDMS scaffolds, patterned with 350 nm wide grids, have shown to align along the grids and undergo neurogenesis [72], while PDMS patterned with a 600-nm wide grating similarly observed embryonic stem cell (ESC) alignment [77], indicating the importance of mechanics and topographical features on cellular activity.

2.2.3.2 Cancer Cells Likewise, the ECM can significantly influence the onset of cancer and tumor metastasis. Malignant transformation begins in normal tissue at a healthy stiffness, but as part of the transformation to a tumor, overproduction of ECM, that is, fibrosis, often results. Thus the onset of cancer usually occurs with tissue stiffening, leading to enhanced integrin signaling and aberrant cell growth [78–80]. Though this is a hallmark of breast cancer in that you can manually palpate mammary tissue and *feel* a stiff lump, fibrosis occurs in many situations including in cirrhosis of the liver and fibrosarcoma. ECM stiffening directly promotes metastasis of the tumor and indirectly assists in tumor angiogenesis and inflammation [79]. A stiffer ECM can also disrupt the vasculature from normal tissue and leads to tumor growth [81].

While the influence of stiffness on tumor progression is still a relatively new topic and is likely very dynamic, the effect of step changes in static ECM stiffness on mammary epithelial cells has been extensively studied. In healthy mammary tissue, mammary epithelial cells develop into 3-D hollow spheres called *acini*. There are two stages to the tumor development; premalignancy and malignancy. When the tumor enters the premalignant stage, the ECM stiffens threefold from healthy tissue, from approximately 100 to 350 Pa [82]. Once the tumor becomes malignant, the ECM undergoes a 10-fold stiffness increase to 1500 kPa [35,82]. During this stage, the acinar structure of the mammary epithelial cells fail to develop and instead, the polarity is compromised. The cells can become highly invasive cells that mirror breast cancer cell metastasis (Figure 2.4) [35]. The composition of the ECM can also play a role in acini formation in causing a redistribution of the actin filaments, effecting morphology and acini formation [83].

2.2.3.3 Embryonic and Progenitor Cells The surrounding microenvironment plays a critical role in regulating ESC fate during embryogenesis [84]; cells are destined to either undergo self-renewal, that is, the process where two identical daughter cells are generated from a single parental cell, or develop into mature tissues via differentiation. For the latter case, tissues result from tightly controlled spatial and temporal presentation of growth factors to comprise the cell niche, and from this extracellular set of signals, stem cells must integrate their response [85,86]. Examples of such developmental programs can be found for virtually every tissue, and such regulation has even been shown to reprogram the fate of cells to transform from one lineage to another [87]. Similar phenomena are seen in different species including, mice, chicks, and zebrafish, where notch signaling is vital in deciding cell fate [88]. Notch signaling describes the time-sensitive developmental pathway that is responsible for guiding stem cell fate at different stages in development [89,90].

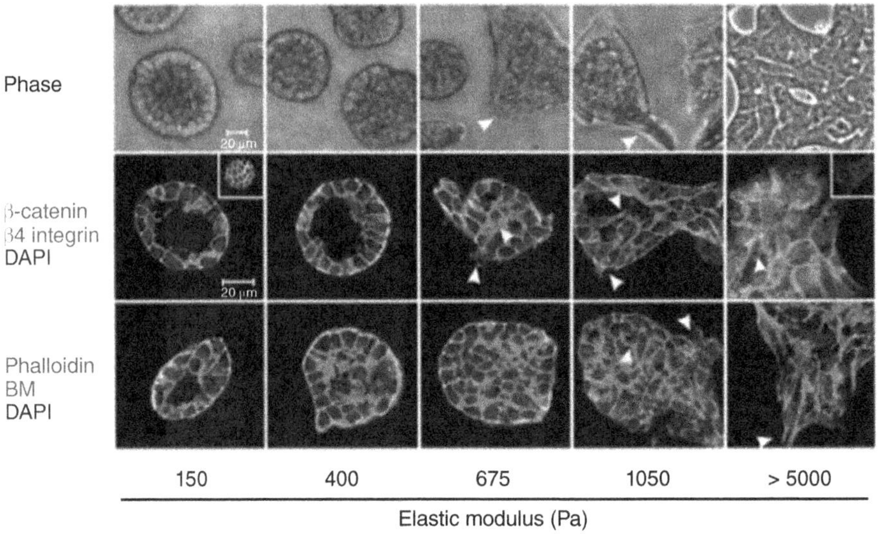

FIGURE 2.4 MCF10A acini polarization responds to extracellular matrix stiffness and is disturbed on being cultured on increased hydrogel stiffness. (Reproduced with permission from Reference 35.) *(See insert for color representation of the figure.)*

Notch has been demonstrated in the *Drosophila* and vertebrate models [91,92] and has shown to promote neural differentiation in ESCs [92]. It has also been implicated in central nervous system development by guiding neural progenitor fate to either glial or neural cell development [93].

2.3 ECM IS A DYNAMIC TISSUE

ECM is dynamic by nature; matrix is constantly being remodeled and maintains a balance of structural synthesis and destruction [8, pp. 179–251]. ECM is degraded by specific matrix metalloproteinases (MMPs), which cleave structural proteins such as collagen, resulting in degradation, and if not immediately filled by the cell, new matrix is likely synthesized to replace the degraded material. This process starts in the cytoplasm where new ECM components are synthesized and then secreted from the cell to rebuild the matrix [94]. When this balance of construction and destruction is disrupted, cellular diseases result when ECM is overproduced, resulting in a stiffer matrix and fibrosis. Excessive or aberrant production of MMPs is known to be a key contributor to tumor progression as well as heart disease [95,96]. Specific MMPs are known to supply critical growth factors to promote tumor growth and to degrade the ECM proteins, which results in tumor metastasis [97]. Likewise, there is an increased level of MMPs (collagenase and gelatinase) during the early stages of heart failure and in atherosclerotic plaque formation [98,99].

While dynamic stiffening characterizes many critical disease processes, for example, cancer metastasis [35,78,80,82] and myocardial infarction/heart failure [100], it is also critical in the developmental stages of embryogenesis [101–103]. In this case, however, ECM degradation is an integral part of the remodeling process, as it allows for cells to incorporate and regenerate the necessary ECM proteins to construct their own ECM [82]. In the heart for example, this requires the down-regulation of laminins and fibronectin in favor of type 1 collagen, which has been shown to assist in stiffening chick myocardium as it develops from E3 to E14 *in vivo* [42]. If one considers that all three germ layers, for example, ecto-, endo-, and mesoderm, are mechanically different but remain soft relative to mature tissues [101], all tissues must therefore undergo stiffening, albeit likely at different rates, to result in soft adipose, firm muscle, and rigid bone [1]. In addition to temporal changes in stiffness, stiffness gradients are physiologically important and are naturally found throughout the body, including the interface between ligament and bone [104], and in heart tissue, where there is a collagen orientation change between the endocardium and epicardium [105,106]. In the heart tissue, these gradients are essential to cell migration, motility, and signaling [105,106]. They are also found at the interfaces of healthy and diseased tissues, as seen in tumors and heart disease, where the disease tissue is overall stiffer than the surrounding tissue [35,100]. Thus, when synthesizing an experimental cell scaffold, it is important not just to consider specific, intrinsic matrix properties but also how those properties change with space and time.

2.4 DYNAMIC SCAFFOLDS

As was stated previously, synthetic polymers (i.e., PA, PEGDA, etc.) and natural polymers (i.e., collagen, hyaluronan, etc.) are typically used to simulate the cells' microenvironment. Despite this, these scaffolds are largely uniform and static and are not able to mimic the dynamic niche that is the ECM. Additionally, static gels are unable to emulate cellular processes that result in ECM fibrosis or degradation, which are of significant interest to cell biology and engineering research.

2.4.1 Dynamically Stiffening Scaffolds

Several different methods have very recently been employed to simulate dynamic stiffening *in vitro*. One approach is to use a slow Michael-type addition reaction (acrylate–thiol cross-linking chemistry) to generate time-dependent stiffening. Thiolated HA has been cross-linked with PEGDA to mimic the development of myocardial tissue *in vivo* (1–10 kPa) over 100 hours; larger molecular weight cross-linkers decreased cross-linking time due to improved diffusion of PEGDA [42]. Embryonic cells isolated from cardiac tissue cultured on the thiolated HA showed greater expression of troponin-T and decreased NKX-2.5 in comparison with a static PA gel, indicating mature cardiac development. The dynamic HA microenvironment was able to generate cardiomyocytes that more closely resemble those grown on

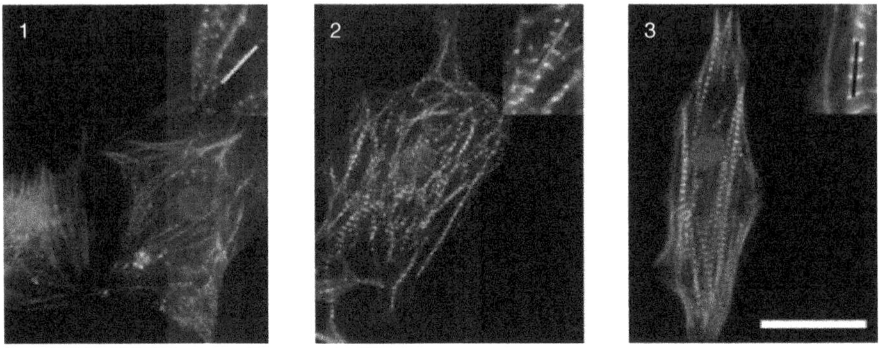

FIGURE 2.5 Immunofluorescent images of embryonic cardiomyocytes cultured on dynamic thiolated HA hydrogels at different developmental stages: premyofibril stage (1), maturing myofibrils (2), and mature cardiomyocytes (3). (Reproduced with permission from Reference 42.) *(See insert for color representation of the figure.)*

developing heart tissue (Figure 2.5) [42]. However, the use of free thiols *in vivo* has been limited to less vascularized regions due to oxidation of the thiols by macrophages, which are more prevalent in vascularized tissues [107].

Ideally, a dynamic material would be able to match the stiffness of the tissue in a time-dependent fashion. However, there are limited ways in which one can achieve this. UV photopolymerizable hydrogels have become viable options to achieve pseudotime-dependent stiffening, in which the hydrogel stiffness increases by a step function [43,44,108]. Guvendiren *et al.* utilized a two-step acrylate cross-linking process to achieve a stiffness increase from a base stiffness hydrogel using methacrylated hyaluronic acid (MeHA). Stem cells plated on soft 3-kPa gels displayed a round morphology. On stiffening with a UV-mediated radical addition reaction (30 kPa), the cells spread and increase in surface area, resembling the morphology of cells plated on a stiff scaffold and exhibit osteogenic markers [43]. Though these studies have been performed in 2-D, use of 3-D static and dynamic materials have more recently been proposed. Stiffness in ionically cross-linked hydrogels allow for cells to significantly alter the matrix [109], whereas for covalently cross-linked matrices, one needs to degrade the matrix first to allow for cell spreading [110]. Khetan *et al.* developed a 3-D dynamic matrix composed of hyaluronic acid functionalized with maleimides and acrylates, while cross-linked with an MMP degradable cross-linker. The multiple functional groups allow for the cross-linked gel, encapsulated with MSCs, to be either stiffened or degraded and allows one to observe cellular responses to the dynamic environmental mechanics in a 3-D environment [110]. Regardless of dimensionality, these systems employ the use of UV light and a photoinitiator as a cross-linking agent, which may limit their biocompatibility. Although studies have shown that small doses of UV at longer wavelengths (350 nm) do not kill stem cells or fibroblasts and do not result in DNA mutation [43,111], translation *in vivo* will be limited as UV light will not penetrate tissue as deeply as may be required, and free radicals generated by photoinitators are potentially cytotoxic [112]. Despite this drawback, UV-activated hydrogels have been made using photoactive peptides and DNA [113–115] in sequential polymerization steps. Thus, these hydrogels may still have utility as a cell biology tool in a variety of applications.

Rather than using UV light or even a chemical reaction to stiffen a hydrogel, there are other *in vitro* mechanical approaches that allow one to investigate cell behavior in response to increasing ECM stiffness. Atomic force microscope (AFM) cantilevers have been used to stretch a biological network to the point where it undergoes strain stiffening [116,117]. This method creates a feedback loop between the force the cell feels from its microenvironment and the force that is generated from the AFM cantilever, allowing for one to observe the recovery of a cell's cytoskeletal mechanics when a significant step change in stiffness occurs. Fibroblasts on the stiffening exhibit elevated traction forces and contraction velocities in a matter of seconds, and these behaviors are reversible when the matrix is relaxed [116]. Rapid recovery emphasizes the importance of myosin and the cytoskeleton when exposed to dynamic strain [117], unlike previous methods where stiffness was irreversibly changed from dynamically cross-linking the matrix. Yet it is important to note that each of these methods have drawbacks, such that they cannot currently be translated into an animal model, setting up the need for biocompatible, dynamic materials.

2.4.2 Degradable Scaffolds

Degradable scaffolds mimic the natural softening of a variety of processes but also permit the release of growth factors and drugs. Most degradable hydrogels have degradable cross-links of some form or fashion; and here we will highlight degradable matrices, their mechanism of degradation, and their impact on cells. One of the most common degradation mechanisms is hydrolysis. Hydrolysis involves the cleavage of bonds/functional groups by water when the nucleophilic water attacks carbonyl groups. This type of degradation occurs in esters, amides, anhydrides, carbonates, and so on (Figure 2.6) [118]. Polymers that undergo hydrolysis include polylactic acid [119], polyglycolic acid [120], and poly(ester amides) [121].

By adjusting the percent functionalization, the timescale of ester bond hydrolysis can be tuned to achieve a suitable degradation profile. PEG-based hydrogels, with acrylate and α-hydroxy acid functionality, allow for spontaneous degradability to take place between 1 day and 4 months [122]. Natural polymers, chitosan, or HA, are widely used for degradable hydrogels because they generally degrade on a shorter timescale than synthetic polymers. The degradation of chitosan can be tuned by adjusting the degree of deacetylation of the monomer [123,124], while hyaluronan is degraded by the incorporation of functional groups that result in a degradable ester bond [42,125]. Osteoblasts, as well as MSCs, have been encapsulated in hydrolytically degrading hydrogels and resulted in increased collagen expression [126] and increased cell chondrogenic differentiation [125], respectively. Though this method can be advantageous due to its spontaneity, it takes place over several days and does not allow one to exert as much control on the polymer system, as do other more novel methods.

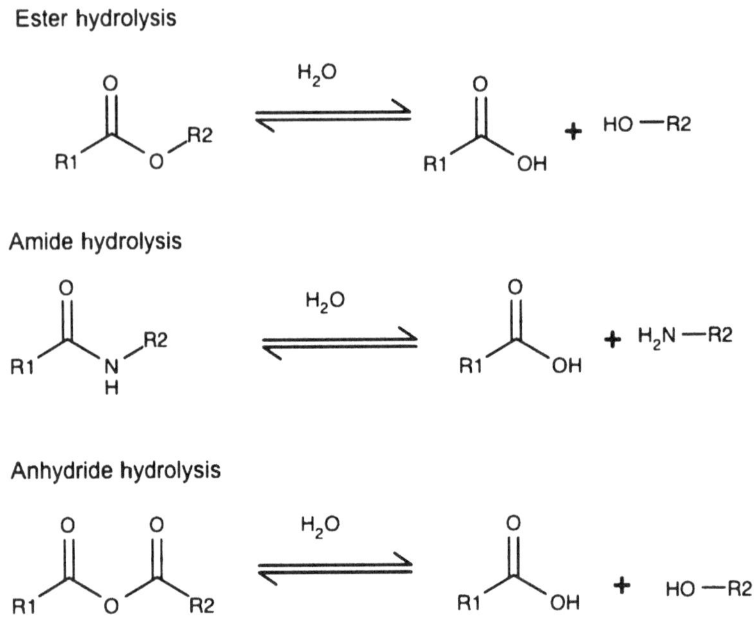

FIGURE 2.6 Hydrolysis examples of esters, amides, and anhydrides.

pH-sensitive and temperature-sensitive polymers offer a unique approach to degradation. pH-sensitive hydrogels with acidic comonomer functionalization have been developed that deprotonate on increased pH, resulting in electrostatic repulsion of the functional groups and hydrogel swelling [127]. In combination with enzymatically degradable crosslinks, these hydrogels can be used for *in vivo* drug delivery applications by adjusting pH accordingly [127]. However, these systems may be more applicable to drug delivery rather than promoting cell viability, due to the harsh conditions of extreme pH.

Typical polymers become more soluble at elevated temperatures, for example, 37°C. However polymers with a lower critical solution temperature (LCST) decrease their solubility, making them an interesting option for controlled drug delivery applications. Polymers with LCST are typically hydrophobic and tend to shrink at raised temperatures where hydrophobic interactions between the chains dominate [128]. Previously, LCST polymers have been made degradable using a block tri and copolymer-system composed PEG-poly(lactic acid-co-glycolic acid) (PLGA) [129,130] and copolymer system using polyphosphazenes [131]. Others have chosen to use temperature-sensitive methylcellulose cross-linkers [132] or different polymer chain conformation to resemble the protein motif coiled coil, which undergoes temperature-induced collapse [133].

Just as UV light can be an effective experimental *in vitro* means to stiffening a polymer system, it is also useful for degrading polymers, due to the amount of control the user can exert on the system. An example of this are PEG-acetate-based

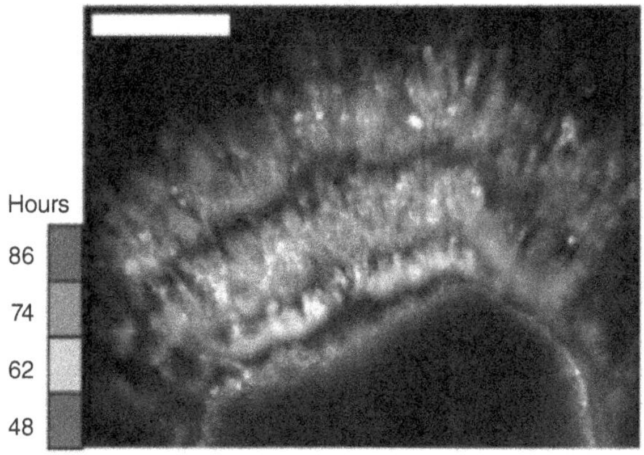

FIGURE 2.7 Encapsulated endothelial cells exhibit sprouting after progressive collagenase soaking. (Reproduced with permission from Reference 57.)

hydrogels, which are cross-linked using UV light at a wavelength of 350 nm and can be degraded at a wavelength of 254 nm to allow for fine tuning of hydrogel mechanics [134].

Several polymers incorporate photocleavable nitrobenzyl cross-linkers that degrade the hydrogel network once exposed to UV light [135,136]. Ramanan *et al.* constructed a hydroxyethacrylate hydrogel using a photocleavable cross-linker, nitrobenzyl acrylate, which results in a stiffness reduction of nearly 80%, with 15 minutes of UV exposure [136]. This method has advanced to cell encapsulation in a 3-D environment, where hMSC migration is shown to increase in PEGDA hydrogels once partially degraded by UV light [135]. Spatial and temporal degradative stiffness gradients can also be generated with this system using a light blocking photomask [135].

Using MMPs as a polymer degradation catalyst is an alternative method that typically involves the cross-linking of a specific protein into the polymer that is to be cleaved. One approach is to add an exogenous enzyme, such as lipase [137], or collagenase [57,138], to degrade the scaffold. Exogenous collagenase has been used to degrade a PEG-(collagen)-diacrylate hydrogel, to promote angiogenesis in embryonic endothelial cells, by cleaving the peptide bonds in collagen inside the polymer. Soaking in a collagenase solution degrades a significant amount of the hydrogel's net weight and results in cell angiogenic sprouting (Figure 2.7) [57]. Likewise, collagenase has also been used to degrade surface collagen on a PEG-dimethacrylate-based hydrogel with embedded chondrocytes, allowing for cells to maintain their morphology and integrate themselves into damaged tissue [138]. Similarly, semi-interpenetrating polymers have been constructed with MMP proteolytically cleavable peptides, which promotes osteoblast infiltration [139].

A secondary approach is to allow for cell-secreted MMPs to degrade the hydrogel networks, without the use of exogenously added enzymes, allowing for the degradation of the system to be completely spontaneous [49,50,140–142]. MSCs have been encapsulated in cell degradable PEG-thiol-peptide based hydrogels [140,141] and were shown to be more proliferative and contain a greater expression level of differentiation markers in a degradable PEG-based hydrogel as compared with a static PEG-based gel [140]. Similar hydrogels have also been used to investigate tumor invasiveness and migration of HT1080 fibrosarcoma cells [142].

2.5 CONCLUSION

In summary, the ECM is a dynamic tissue that is constantly being degraded then rebuilt [8, pp. 179–251]. ECM stiffening, degradation, and spatial stiffness gradients are very important in development and disease. While dynamic scaffolds are currently being developed to replicate the *in vivo* response *in vitro* using hyaluronan [42–44,110,114], PEG [108,135,137], and PA-based scaffolds [143], the majority of them lack the biocompatibility to extend their utility beyond an *in vitro* cell biology tool because these scaffolds require UV light, photoinitiators, or free functional groups that are toxic *in vivo*. New techniques need to be developed to allow for *in vivo* translation and must involve materials that intrinsically degrade/stiffen inside the body. There are experimental and healthcare applications for *in vivo* dynamic hydrogels in drug delivery as well as in diseases. *In vivo* degradable materials must incorporate hydrolyzable functional groups or MMP degradable hydrogels, similar to spontaneously degrading acrylate, maleimide functionalized hyaluronic acid developed by Khetan *et al.* [110]. Others use MMP degradable peptides, which presents another feasible degradation option [139–142]. Stiffening materials need to be created that will cross-link over time without creating an immune response in an *in vivo* setting or require light sources not possible *in vivo*. Despite these restraints, both degrading and stiffening hydrogels need to be able to exhibit some extent of user control in terms of stiffness/degradation profile to be truly useful in the biology and engineering fields.

Acknowledgments

The authors would like to thank the National Institutes of Health (DP02OD006460 to A.J.E.) and the National Science Foundation Graduate Research Fellowship (to M.G.O.) for financial support.

REFERENCES

[1] Engler, A.J., Sen, S., Sweeney, H.L., Discher, D.E. *Cell* **2006**, *126*, 677–689.

[2] Kleinman, H.K., Philp, D., Hoffman, M.P. *Current Opinion in Biotechnology* **2003**, *14*, 526–532.

[3] Reilly, G.C., Engler, A.J. *Journal of Biomechanics* **2010**, *43*, 55–62.

[4] Dalby, M.J., Gadegaard, N., Tare, R., Andar, A., Riehle, M.O., Herzyk, P., Wilkinson, C.D.W., Oreffo, R.O.C. *Nature Materials* **2007**, *6*, 997–1003.

[5] McBeath, R., Pirone, D.M., Nelson, C.M., Bhadriraju, K., Chen, C.S. *Developmental Cell* **2004**, *6*, 483–495.

[6] Ott, H.C., Matthiesen, T.S., Goh, S.-K., Black, L.D., Kren, S.M., Netoff, T.I., Taylor, D.A. *Nature Medicine* **2008**, *14*, 213–222.
[7] Engler, A.J., Wong, J., Picart, C. *Surface Science* **2004**, *570*, 142–154.
[8] Hay, E.D. *Cell biology of extracellular matrix*, 2nd edn; Plenum, New York: 1991.
[9] Darribère, T., Schwarzbauer, J.E. *Mechanisms of Development* **2000**, *92*, 239–250.
[10] Yeung, T., Georges, P.C., Flanagan, L.A., Marg, B., Ortiz, M., Funaki, M., Zahir, N., Ming, W., Weaver, V., Janmey, P. *Cell Motility and the Cytoskeleton* **2005**, *60*, 24–34.
[11] Tee, S.-Y., Fu, J., Chen, C.S., Janmey, P.A. *Biophysical Journal* **2011**, *100*, L25–L27.
[12] Vincent, L.G., Choi, Y.S., Alonso-Latorre, B., del Álamo, J.C., Engler, A.J. *Biotechnology Journal* **2013**, *8*, 472–484.
[13] Lo, C.M., Wang, H.B., Dembo, M., Wang, Y.L. *Biophysical Journal* **2000**, *79*, 144–152.
[14] Pajerowski, J.D., Dahl, K.N., Zhong, F.L., Sammak, P.J., Discher, D.E. *PNAS* **2007**, *104*, 15619–15624.
[15] Holle, A.W., Engler, A.J. *Current Opinion in Biotechnology* **2011**, *22*, 648–654.
[16] Chan, K.T., Cortesio, C.L., Huttenlocher, A. *The Journal of Cell Biology* **2009**, *185*, 357–370.
[17] Schoenwaelder, S.M., Burridge, K. *Current Opinion in Cell Biology* **1999**, *11*, 274–286.
[18] Mitra, S.K., Hanson, D.A., Schlaepfer, D.D. *Nature Reviews. Molecular Cell Biology* **2005**, *6*, 56–68.
[19] Parsons, J.T., Horwitz, A.R., Schwartz, M.A. *Nature Reviews. Molecular Cell Biology* **2010**, *11*, 633–643.
[20] Cance, W.G., Harris, J.E., Iacocca, M.V., Roche, E., Yang, X., Chang, J., Simkins, S., Xu, L. *Clinical Cancer Research* **2000**, *6*, 2417–2423.
[21] Lark, A.L., Livasy, C.A., Dressler, L., Moore, D.T., Millikan, R.C., Geradts, J., Iacocca, M., Cowan, D., Little, D., Craven, R.J., Cance, W. *Modern Pathology* **2005**, *18*, 1289–1294.
[22] Zimerman, B., Volberg, T., Geiger, B. *Cell Motility and the Cytoskeleton* **2004**, *159*, 143–159.
[23] Tilghman, R.W., Slack-Davis, J.K., Sergina, N., Martin, K.H., Iwanicki, M., Hershey, E.D., Beggs, H.E., Reichardt, L.F., Parsons, J.T. *Journal of Cell Science* **2005**, *118*, 2613–2623.
[24] Peyton, S.R., Putnam, A.J. *Journal of Cellular Physiology* **2005**, *204*, 198–209.
[25] Ridley, A.J., Hall, A. *Cell* **1992**, *70*, 389–399.
[26] Hall, A. *Science* **1998**, *279*, 509–514.
[27] Huang, S., Chen, C.S., Ingber, D.E. *Molecular Biology of the Cell* **1998**, *9*, 3179–3193.
[28] Chen, C.S., Tan, J., Tien, J. *Annual Review of Biomedical Engineering* **2004**, *6*, 275–302.
[29] del Rio, A., Perez-Jimenez, R., Liu, R., Roca-Cusachs, P., Fernandez, J.M., Sheetz, M.P. *Science* **2009**, *323*, 638–641.
[30] Golji, J., Lam, J., Mofrad, M.R.K. *Biophysical Journal* **2011**, *100*, 332–340.
[31] Alberts, B., Johnson, A., Lewis, J., Raff, M., Roberts, K., Walter, P. *Molecular biology of the cell*, 5th edn; Garland Science, New York: 2002.
[32] Plotnikov, S.V., Pasapera, A.M., Sabass, B., Waterman, C.M. *Cell* **2012**, *151*, 1513–1527.
[33] Choquet, D., Felsenfeld, D.P., Sheetz, M.P. *Cell* **1997**, *88*, 39–48.

[34] Roskelley, C.D., Desprez, P.Y., Bissell, M.J. *PNAS* **1994**, *91*, 12378–12382.
[35] Paszek, M.J., Zahir, N., Johnson, K.R., Lakins, J.N., Rozenberg, G.I., Gefen, A., Reinhart-King, C.A., Margulies, S.S., Dembo, M., Boettiger, D., Hammer, D.A., Weaver, V.M. *Cancer Cell* **2005**, *8*, 241–254.
[36] Tibbitt, M.W., Anseth, K.S. *Biotechnology and Bioengineering* **2009**, *103*, 655–663.
[37] Raub, C.B., Putnam, A.J., Tromberg, B.J., George, S.C. *Acta Biomaterialia* **2010**, *6*, 4657–4665.
[38] Wang, L.-S., Du, C., Chung, J.E., Kurisawa, M. *Acta Biomaterialia* **2012**, *8*, 1826–1837.
[39] Duong, H., Wu, B., Tawil, B. *Tissue Engineering. Part A* **2009**, *15*, 1865–1876.
[40] Gribova, V., Crouzier, T., Picart, C. *Journal of Materials Chemistry* **2011**, *21*, 14354.
[41] Banerjee, A., Arha, M., Choudhary, S., Ashton, R.S., Bhatia, S.R., Schaffer, D.V., Kane, R.S. *Biomaterials* **2009**, *30*, 4695–4699.
[42] Young, J.L., Engler, A.J. *Biomaterials* **2011**, *32*, 1002–1009.
[43] Guvendiren, M., Burdick, J.A. *Nature Communications* **2012**, *3*, 792.
[44] Marklein, R.A., Soranno, D.E., Burdick, J.A. *Soft Matter* **2012**, *8*, 8113–8120.
[45] Goodison, S., Urquidi, V., Tarin, D. *Molecular Pathology* **1999**, *52*, 189–196.
[46] Kim, Y., Lee, Y., Choe, J., Lee, H., Kim, Y., Jeoung, D. *The Journal of Biological Chemistry* **2008**, *283*, 22513–22528.
[47] Li, W.J., Laurencin, C.T., Caterson, E.J., Tuan, R.S., Ko, F.K. *Journal of Biomedical Materials Research* **2002**, *60*, 613–621.
[48] Viswanathan, P., Chirasatitsin, S., Ngamkham, K., Engler, A.J., Battaglia, G. *Journal of the American Chemical Society* **2012**, *134*, 20103–20109.
[49] Lutolf, M.P., Lauer-Fields, J.L., Schmoekel, H.G., Metters, A.T., Weber, F.E., Fields, G.B., Hubbell, J.A. *PNAS* **2003**, *100*, 5413–5418.
[50] Lutolf, M.P., Raeber, G.P., Zisch, A.H., Tirelli, N., Hubbell, J.A. *Advanced Materials* **2003**, *15*, 888–892.
[51] Li, B., Chen, J., Wang, J.H. *Journal of Biomedical Materials Research Part A* **2006**, *79*, 989–998.
[52] Wipff, P.J., Majd, H., Acharya, C., Buscemi, L., Meister, J.J., Hinz, B. *Biomaterials* **2009**, *30*, 1781–1789.
[53] Tse, J.R., Engler, A.J. *Current Protocols in Cell Biology* **2010**, Chapter 10, Unit 10.16.
[54] Nemir, S., Hayenga, H.N., West, J.L. *Biotechnology and Bioengineering* **2010**, *105*, 636–644.
[55] Tse, J.R., Engler, A.J. *PLoS One* **2011**, *6*, e15978.
[56] Brown, X.Q., Ookawa, K., Wong, J.Y. *Biomaterials* **2005**, *26*, 3123–3129.
[57] Miller, J.S., Shen, C.J., Legant, W.R., Baranski, J.D., Blakely, B.L., Chen, C.S. *Biomaterials* **2010**, *31*, 3736–3743.
[58] Nguyen, K.T., West, J.L. *Biomaterials* **2002**, *23*, 4307–4314.
[59] Bissell, M.J., Hall, H.G., Parry, G. *Journal of Theoretical Biology* **1982**, *99*, 31–68.
[60] Nava, M.M., Raimondi, M.T., Pietrabissa, R. *Journal of Biomedicine & Biotechnology* **2012**, *2012*, 797410.
[61] Guilak, F., Cohen, D.M., Estes, B.T., Gimble, J.M., Chen, C.S. *Cell Stem Cell* **2010**, *5*, 17–26.
[62] Saha, K., Keung, A.J., Irwin, E.F., Li, Y., Little, L., Schaffer, D.V., Healy, K.E. *Biophysical Journal* **2008**, *95*, 4426–4438.
[63] Park, J.S., Chu, J.S., Tsou, A.D., Diop, R., Wang, A., Li, S. *Biomaterials* **2012**, *32*, 3921–3930.
[64] Kilian, K.A., Bugarija, B., Lahn, B.T., Mrksich, M. *PNAS* **2010**, *107*, 4872–4877.

[65] Bhatia, S.N., Yarmush, M.L., Toner, M. *Journal of Biomedical Materials Research* **1997**, *34*, 189–199.
[66] Raghavan, S., Chen, C.S. *Advanced Materials* **2004**, *16*, 1303–1313.
[67] Théry, M., Piel, M. *Cold Spring Harbor Protocols* **2009**, *2009*, prot5255.
[68] Choi, Y.S., Vincent, L.G., Lee, A.R., Kretchmer, K.C., Chirasatitsin, S., Dobke, M.K., Engler, A.J. *Biomaterials* **2012**, *33*, 6943–6951.
[69] Huang, N.F., Patel, S., Thakar, R.G., Wu, J., Hsiao, B.S., Chu, B., Lee, R.J., Li, S. *Nano Letters* **2006**, *6*, 537–542.
[70] Engler, A.J., Griffin, M.A., Sen, S., Bönnemann, C.G., Sweeney, H.L., Discher, D.E. *The Journal of Cell Biology* **2004**, *166*, 877–887.
[71] Molnar, P., Wang, W., Natarajan, A., Rumsey, J.W., Hickman, J.J. *Biotechnology Progress* **2007**, *23*, 265–268.
[72] Yim, E., Pang, S., Leong, K. *Experimental Cell Research* **2008**, *313*, 1820–1829.
[73] Dang, J., Leong, K. *Advanced Materials* **2008**, *19*, 2775–2779.
[74] Kim, I.L., Khetan, S., Baker, B.M., Chen, C.S., Burdick, J.A. *Biomaterials* **2013**, *34*, 5571–5580.
[75] Baker, B.M., Gee, A.O., Metter, R.B., Nathan, A.S., Marklein, L., Burdick, J.A., Mauck, R.L. *Biomaterials* **2009**, *29*, 2348–2358.
[76] Nam, J., Huang, Y., Agarwal, S., Lannutti, J. *Tissue Engineering* **2007**, *13*, 2249–2257.
[77] Gerecht, S., Bettinger, C.J., Zhang, Z., Borenstein, J.T., Vunjak-Novakovic, G., Langer, R. *Biomaterials* **2007**, *28*, 4068–4077.
[78] Paszek, M.J., Weaver, V.M. *Journal of Mammary Gland Biology and Neoplasia* **2004**, *9*, 325–342.
[79] Lu, P., Weaver, V.M., Werb, Z. *The Journal of Cell Biology* **2012**, *196*, 395–406.
[80] Levental, K.R., Yu, H., Kass, L., Lakins, J.N., Erler, J.T., Fong, S.F.T., Csiszar, K., Giaccia, A., Yamauchi, M., Gasser, D.L., Weaver, V.M. *Cell* **2009**, *139*, 891–906.
[81] Padera, T.P., Stoll, B.R., Tooredman, J.B., Capen, D., di Tomaso, E., Jain, R.K. *Nature* **2004**, *427*, 695.
[82] Cox, T.R., Erler, J.T. *Disease Models & Mechanisms* **2011**, *4*, 165–178.
[83] Lance, A., Yang, C., Swamydas, M., Dean, D., Deitch, S., Burg, K.J.L., Dréau, D. *Journal of Tissue Engineering and Regenerative Medicine* **2013**, *14*, doi: 10.1002/term.
[84] Brenner, C.A., Adler, R.R., Rappolee, D.A., Pedersen, R.A., Werb, Z. *Genes & Development* **1989**, *3*, 848–859.
[85] Gregor, T., Tank, D.W., Wieschaus, E.F., Bialek, W. *Cell* **2007**, *130*, 153–164.
[86] Lewis, J. *Science* **2008**, *322*, 399–403.
[87] Mariani, F.V., Martin, G.R. *Nature* **2003**, *423*, 319–325.
[88] Palmeirim, I., Henrique, D., Ish-Horowicz, D., Pourquié, O. *Cell* **1997**, *91*, 639–648.
[89] Bray, S.J. *Nature Reviews. Molecular Cell Biology* **2006**, *7*, 678–689.
[90] Ehebauer, M., Hayward, P., Martinez-Arias, A. *Science's STKE: Signal Transduction Knowledge Environment* **2006**, *2006*, cm7.
[91] Ahmad, I., Zagouras, P., Artavanis-Tsakonas, S. *Mechanisms of Development* **1995**, *53*, 73–85.
[92] Lowell, S., Benchoua, A., Heavey, B., Smith, A.G. *PLoS Biology* **2006**, *4*, e121.
[93] Louvi, A., Artavanis-Tsakonas, S. *Nature Reviews. Neuroscience* **2006**, *7*, 93–102.
[94] Woessner, J.F., Jr. *The FASEB Journal* **1991**, *5*, 2145–2154.
[95] Malemud, C.J. *Frontiers in Bioscience* **2006**, *11*, 1696–1701.
[96] Stetler-Stevenson, W.G. *American Journal of Pathology* **1996**, *148*, 1345–1350.
[97] Fingleton, B. *Expert Opinion on Therapeutic Targets* **2003**, *7*, 385–397.

[98] Danielsen, C.C., Wiggers, H., Andersen, H.R. *Journal of Molecular and Cellular Cardiology* **1998**, *30*, 1431–1442.

[99] Galis, Z.S., Sukhova, G.K., Lark, M.W., Libby, P. *The Journal of Clinical Investigation* **1994**, *94*, 2493–2503.

[100] Berry, M.F., Engler, A.J., Woo, Y.J., Pirolli, T.J., Bish, L.T., Jayasankar, V., Morine, K.J., Gardner, T.J., Discher, D.E., Sweeney, H.L. *American Journal of Physiology. Heart and Circulatory Physiology* **2006**, *290*, H2196–H2203.

[101] Krieg, M., Arboleda-Estudillo, Y., Puech, P.-H., Käfer, J., Graner, F., Müller, D.J., Heisenberg, C.-P. *Nature Cell Biology* **2008**, *10*, 429–436.

[102] Hamburger, V., Hamilton, H.L. A series of normal stages in the development of the chick embryo. *Journal of Morphology* **1951**, *88*, 49e62.

[103] Hamburger, V., Hamilton, H.L. *Journal of Morphology* **1951**, *88*, 231–272.

[104] Phillips, J.E., Burns, K.L., Le Doux, J.M., Guldberg, R.E., García, A.J. *PNAS* **2008**, *105*, 12170–12175.

[105] Pope, A.J., Sands, G.B., Smaill, B.H., LeGrice, I.J. *American Journal of Physiology. Heart and Circulatory Physiology* **2008**, *295*, H1243–H1252.

[106] Sant, S., Hancock, M.J., Donnelly, J.P., Iyer, D., Khademhosseini, A. *The Canadian Journal of Chemical Engineering* **2010**, *88*, 899–911.

[107] Young, J.L., Tuler, J., Braden, R., Schüp-Magoffin, P., Schaefer, J., Kretchmer, K., Christman, K.L., Engler, A.J. *Acta Biomaterialia* **2013**, *9*, 7151–7157.

[108] Hahn, M.S., Miller, J.S., West, J.L. *Advanced Materials* **2006**, *18*, 2679–2684.

[109] Huebsch, N., Arany, P.R., Mao, A.S., Shvartsman, D., Ali, O.A., Bencherif, S.A., Rivera-Feliciano, J., Mooney, D.J. *Nature Materials* **2010**, *9*, 518–526.

[110] Khetan, S., Guvendiren, M., Legant, W.R., Cohen, D.M., Chen, C.S., Burdick, J.A. *Nature Materials* **2013**, *12*, 1–8.

[111] Bryant, S.J., Anseth, K.S. *Journal of Biomedical Materials Research Part A* **2003**, *64*, 70–79.

[112] Sabnis, A., Rahimi, M., Chapman, C., Nguyen, K. *Journal of Biomedical Materials Research Part A* **2009**, *91*, 52–59.

[113] Nowatzki, P.J., Franck, C., Maskarinec, S.A., Ravichandran, G., Tirrell, D.A. *Macromolecules* **2008**, *41*, 1839–1845.

[114] Khetan, S., Katz, J.S., Burdick, J.A. *Soft Matter* **2009**, *5*, 1601–1606.

[115] Jiang, F.X., Yurke, B., Schloss, R.S., Firestein, B.L., Langrana, N.A. *Biomaterials* **2010**, *31*, 1199–1212.

[116] Webster, K.D., Crow, A., Fletcher, D.A. *PLoS One* **2011**, *6*, e17807.

[117] Crow, A., Webster, K.D., Hohlfeld, E., Ng, W.P., Geissler, P., Fletcher, D.A. *Biophysical Journal* **2012**, *102*, 443–451.

[118] Nair, L.S., Laurencin, C.T. *Progress in Polymer Science* **2007**, *32*, 762–798.

[119] Lee, S., Kim, S.O.O.H., Han, Y., Kim, Y.H.A. *Journal of Polymer Science Part A: Polymer Chemistry* **2000**, *39*, 973–985.

[120] Gunatillake, P.A., Adhikari, R. *European Cells & Materials* **2003**, *5*, 1–16.

[121] Paredes, A., Rodrigues-Galan, N., Puiggali, J. *Journal of Polymer Science Part A: Polymer Chemistry* **1997**, *36*, 1271–1282.

[122] Sawhney, J.A.H.A.S., Pathak, C.P. *Macromolecules* **1993**, *26*, 581–587.

[123] Li, J., Du, Y., Liang, H. *Polymer Degradation and Stability* **2007**, *92*, 515–524.

[124] Mao, J.S., Cui, Y.L., Wang, X.H., Sun, Y., Yin, Y.J., Zhao, H.M., De Yao, K. *Biomaterials* **2004**, *25*, 3973–3981.

[125] Sahoo, S., Chung, C., Khetan, S., Burdick, J.A. *Biomacromolecules* **2008**, *9*, 1088–1092.

[126] Benoit, D.S.W., Durney, A.R., Anseth, K.S. *Tissue Engineering* **2006**, *12*, 1663–1673.

[127] Ghandehari, H., Kopecková, P., Kopecek, J. *Biomaterials* **1997**, *18*, 861–872.
[128] Qiu, Y., Park, K. *Advanced Drug Delivery Reviews* **2001**, *53*, 321–339.
[129] Jeong, B., Bae, Y.H., Kim, S.W. *Journal of Controlled Release* **2000**, *63*, 155–163.
[130] Jeong, B., Kibbey, M.R., Birnbaum, J.C., Won, Y.-Y., Gutowska, A. *Macromolecules* **2000**, *33*, 8317–8322.
[131] Lee, B.H., Lee, Y.M., Sohn, Y.S., Song, S.-C. *Polymer International* **2002**, *51*, 658–660.
[132] Liang, H.-F., Hong, M.-H., Ho, R.-M., Chung, C.-K., Lin, Y.-H., Chen, C.-H., Sung, H.-W. *Biomacromolecules* **2004**, *5*, 1917–1925.
[133] Wang, C., Stewart, R.J., Kopecek, J. *Nature* **1999**, *397*, 417–420.
[134] Andreopoulos, F.M., Beckman, E.J., Russell, A.J. *Journal of Polymer Science Part A: Polymer Chemistry* **2000**, *38*, 1466–1476.
[135] Kloxin, A.M., Tibbitt, M.W., Kasko, A.M., Fairbairn, J.A., Anseth, K.S. *Advanced Materials* **2010**, *22*, 61–66.
[136] Ramanan, V.V., Katz, J.S., Guvendiren, M., Cohen, E.R., Marklein, R.A., Burdick, J.A. *Journal of Materials Chemistry* **2010**, *20*, 8920–8926.
[137] Rice, M.A., Sanchez-Adams, J., Anseth, K.S. *Biomacromolecules* **2006**, *7*, 1968–1975.
[138] Rice, M.A., Homier, P.M., Waters, K.R., Anseth, K.S. *Journal of Tissue Engineering and Regenerative Medicine* **2008**, *2*, 418–429.
[139] Kim, S., Chung, E.H., Gilbert, M., Healy, K.E., Al, K.I.M.E.T. *Journal of Biomedical Materials Research Part A* **2005**, *75*, 73–88.
[140] Anderson, S.B., Lin, C.-C., Kuntzler, D.V., Anseth, K.S. *Biomaterials* **2011**, *32*, 3564–3574.
[141] Fairbanks, B.D., Schwartz, M.P., Halevi, A.E., Nuttelman, C.R., Bowman, C.N., Anseth, K.S. *Advanced Materials* **2009**, *21*, 5005–5010.
[142] Schwartz, M.P., Fairbanks, B.D., Rogers, R.E., Rangarajan, R., Zaman, M.H., Anseth, K.S. *Integrative Biology* **2010**, *2*, 32–40.
[143] Frey, M.T., Wang, Y.L. *Soft Matter* **2009**, *5*, 1918–1924.

CHAPTER 3

THE ROLE OF MECHANICAL CUES IN REGULATING CELLULAR ACTIVITIES AND GUIDING TISSUE DEVELOPMENT

LIMING BIAN

Division of Biomedical Engineering, Department of Mechanical and Automation Engineering, The Chinese University of Hong Kong, Shatin, New Territories, Hong Kong

3.1 INTRODUCTION

Mechanical cues impact virtually every organ system in our bodies including musculoskeletal, cardiovascular, pulmonary, dental, and neural tissues. These mechanical cues can be in the form of compression, tension, torsion, and fluid shear. An increasing amount of research is being devoted to understanding the effect of mechanical signals on cell behavior and tissue morphogenesis. Significant progress has been made in elucidating the molecular mechanisms by which individual cells sense these mechanical signals and translate them into cascading cellular signaling events and ultimately, gene expression, a process that is known as mechanotransduction. In this chapter, we will discuss the major discoveries in mechanotransduction and the role of mechanical signals in guiding tissue development in various physiological systems.

3.2 MECHANOTRANSDUCTION

Cells possess a plethora of mechanosensitive entities on their surface such as ion channels, extracellular matrix (ECM) binding receptors, and specialized cell

Bio-inspired Materials for Biomedical Engineering, First Edition. Edited by Anthony B. Brennan and Chelsea M. Kirschner.

membrane structures [1–3]. The downstream signaling cascades triggered by mechanical sensing lead to changes in gene expression in the cell nuclei [4,5]. Because mechanical signals propagate through long distances via physical cellular structures in the cells, the speed of the mechanically induced signaling is predicted to be significantly faster compared with signaling mediated by diffusion of soluble factors [6,7]. Along the propagation path of mechanical signals from cell surface to intranuclear genetic machinery, there are two critical steps of mechanotransduction; firstly, signal transfer from extracellular space to intracellular cytoplasm and secondly, from cytoplasm to intranuclear genetic machinery.

3.2.1 Mechanotransduction from Extracellular Matrix to Cytoplasmic Structures

Cells possess a number of membrane-based sensory structures that probe and detect external forces (also reviewed by DuFort, Paszek, and Weaver) [8]. This force sensing, also termed mechanosensing, is generally mediated by conformational changes in force-sensitive cellular molecules or structures, such as cadherin complexes in cell–cell adhesions, mechanosensitive ion channels, G protein-coupled receptors, kinase receptors, and integrins [3,9–11]. The detected mechanical cues generate specific signals that are propagated through cytoplasmic molecular structures leading to the activation of mechanosensitive intracellular signaling pathways. For instance, integrins are capable of aggregating and binding to extracellular molecules such as fibronectin and vitronectin to form focal adhesion (FA) complexes [12]. External mechanical stains can be transmitted through these integrin-mediated adhesions to modulate intracellular molecules such as focal adhesion kinase (FAK) [13] (Figure 3.1). These molecular changes initiate a cascade of signaling events including the activation of Rho-family GTPases (enzymes that hydrolyze guanosine triphosphate (GTP)), such as RhoA, which promotes actin restructuring, induces protein phosphorylation, and regulates gene expression by influencing transcription factors [14]. Through received external mechanical signals, integrins also impact the mitogen-activated protein kinase and extracellular signal-regulated kinase (MAPK–ERK) pathway, which has been implicated in cell proliferation and differentiation [5,15].

3.2.2 Mechanotransduction by Cell–ECM Adhesions

One class of the most-studied mechanosensory complexes is the FAs developed at the binding sites of integrins to ECM molecules [16]. The FA complexes consist of not only integrins but also an array of cytoplasmic adapter and signaling proteins, including vinculin and talin, which connect the intracellular domain of integrins to actomyosin cytoskeleton [17] (Figure 3.1). These cross membrane molecular complexes serve as the link between the ECM structures and the intracellular cytoskeleton and allow transmission of mechanical signals from the outside into the cellular space in response to physical forces. The FA complexes are dynamic rather than static, and external forces can lead to conformational changes in proteins within the FAs, including the transmembrane integrin and adapter proteins such as talin, effectively converting mechanical signals into biochemical ones [18,19]. The effects of

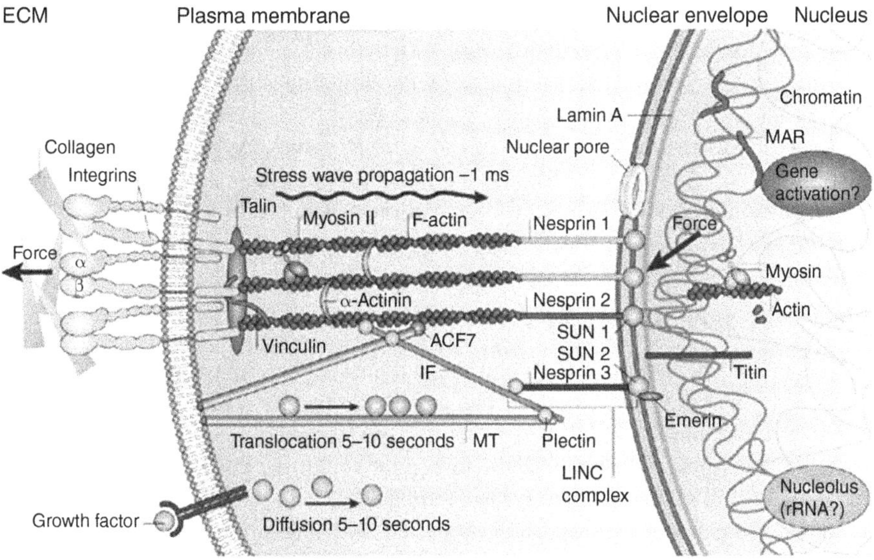

FIGURE 3.1 Propagation of mechanical signals from ECM to genetic machinery in the cell nucleus. (Reproduced with permission from Reference 44.)

these force-induced protein conformational changes on protein–protein interactions depend on the specific molecules involved. For example, when integrins bind to the ECM molecule fibronectin, the conformational change in integrins generally stabilizes the binding. In contrast, the force-induced conformational change in talin helps unravel the molecule, leading to the exposure of otherwise inaccessible vinculin-binding sites [17,20]. Talin–vinculin binding in response to mechanical signals promotes the clustering of integrins and the aggregation of adhesion plaque proteins, which enhances the activation of signal transduction molecules at the cytoplasmic side of the FA complexes [5].

3.2.3 Mechanotransduction by Cell–Cell Adhesions

Cell–cell adhesions, including adherens junctions (AJs), tight junctions, and gap junctions, also play an important role in mechanotransduction [21–23] (also reviewed by Chen, Tan, and Tien) [24]. Among these different cell–cell adhesions, the AJs have received the most research attention for their capability to directly transmit external mechanical signals into intracellular actomyosin activities. The AJs are mediated by the homotypic binding of extracellular domains of cadherins, a family of transmembrane Ca^{2+}-dependent adhesion molecules, expressed on the surfaces of the neighboring cells [25]. On formation of AJs, scaffolding proteins are recruited by the cytoplasmic domains of cadherins and help anchor the actin cytoskeleton. FAs and AJs share many similar attributes. Firstly, both are formed by transmembrane receptors that are capable of transmitting external mechanical signals to intracellular actin cytoskeleton. Furthermore, in both FAs and AJs, the transmembrane receptors form clusters and recruit, with their intracellular domains, signaling, and structural molecules that can interact with either the actin cytoskeleton or other relevant molecules. In addition, many of these recruited intracellular proteins, some of which are shared by these two types of adhesions such as vinculin, can undergo conformational changes in response to perceived mechanical cues and thereby convert mechanical signals into biochemical ones, which ultimately influence transcriptional activities in the nuclei [26].

3.2.4 Intracellular Molecules

On ligand binding, integrins form clusters and initiate recruitment of actin filaments to the integrin cytoplasmic domain. The actin recruitment is facilitated by the direct or indirect binding of a number of structural and signaling proteins including talin, vinculin, α-actinin, and filamin to the integrin cytoplasmic domain. Talin, vinculin, and FAK (or PTK2, a cytosolic tyrosine kinase) play a key role in the mechanical integrity of FA complexes and are correlated with the strength of the adhesions [27]. Talin is a large dimeric protein that attaches integrins to the actin cytoskeleton and is indispensable to the structural integrity of FAs [28]. Furthermore, ligand-binding activity of integrins and the recruitment of signaling proteins like FAK are also regulated by talin [29]. Vinculin does not bind integrins directly, but facilitates FA formation indirectly by connecting talin and α-actinin to the actin cytoskeleton and recruiting additional proteins such as paxillin and vinexin [30]. The amount of vinculin recruited to FAs correlates with the extent of local tension [15,31]. FAK is one of the molecules in FAs that respond to mechanical signals the quickest. Increasing cytoskeletal tension or application of extracellular forces can enhance FAK phosphorylation, leading to increased cell proliferation [32]. Force-induced FAK tyrosine phosphorylation also plays an important role in cell migration in response to mechanical cues such as fluid shear [33]. β-Catenin, which binds to the cytoplasmic domain of cadherins, is a structural element in AJs which links cadherins to the actin cytoskeleton with the help of other adapter proteins [34]. β-Catenin can dissociate from AJs on phosphorylation and translocates to the nuclei to interact with transcription factors leading to modulation of gene expressions [35,36].

3.2.5 Adhesion-Mediated Signaling Pathways

Cytoplasmic signaling from integrin-mediated adhesion is typically characterized by two phases. Early adhesions lead to activation of pathways, such as those involving Rac and Cdc42 that stimulate actin polymerization and membrane protrusion, whereas established adhesions activate pathways such as the RhoA signaling pathway that promote cytoskeletal contractility and transmission of tension to the FAs [37]. Cdc42, Rac, and RhoA all belong to the Rho family of GTPases. Cdc42 and Rac regulate the development of filopodia and lamellipodia during cell adhesion and spreading on ECM, respectively [38]. The tensile forces that cells exert against their adhesions have been shown to positively correlate with externally applied forces or cell spreading [39,40]. RhoA, through its effector Rho kinase (ROCK), plays a key role in the formation of stress fibers and generation of cytoskeletal tension in adherent cells [41]. ROCK promotes phosphorylation of myosin II regulatory light chain (MLC) to increase myosin II contractility and intracellular tension, leading to cell proliferation [42,43].

3.3 MECHANOTRANSDUCTION FROM CYTOPLASM TO NUCLEUS

The mechanotransduction in cytoplasm and nucleus has been comprehensively reviewed previously [44]. Recent studies identified a specialized structure, known as the linker of nucleoskeleton and cytoskeleton (LINC) complex, which helps anchor cytoskeletal structures such as actin filaments to the nuclear membrane. The LINC complex contains adapter molecules such as nesprins, SUN, and lamin proteins [45] (Figure 3.1). Nesprins isoforms (nesprin 1 and nesprin 2) are rod-like nuclear membrane proteins that connect actin filaments to SUN1 located on the inner nuclear membrane with their amino and carboxyl terminals, respectively [46]. SUN1 connects this anchoring structure to lamin A on internal nuclear scaffold, as well as nuclear pore complexes, and therefore may be involved in mechanical regulation of nuclear pores [47].

Lamins A, B, and C are structure proteins that form a molecular network on the nucleoplasmic surface of the inner nuclear membrane. Lamins are also components of internal nuclear scaffold and play a vital role in the control of nuclear organization and regulation of gene expressions [48]. Lamin A and lamin C (A-type) have been shown to be more closely involved in mechanotransduction compared with lamin B (B-type) [49]. It is postulated that lamins connect to the genetic machinery both directly and indirectly by partnering with other nuclear proteins such as emerin [48,50,51]. Emerin binds to the LINC complex through nesprins and lamins, and together with barrier-to-autointegration factor (BAF), connect lamins to polymerized nuclear actin, resulting in enhanced actin polymerization [52,53]. In addition, actin and myosin are also involved in modulating nuclear structure and nuclear functions, such as chromosome movements and transcription [54,55]. Therefore, it is likely that actomyosin interactions also facilitate force propagation and transduction in the nucleus.

The structure and organization of the genome itself is organized into loops and is influenced by attachment to nuclear matrix attachment regions (MARs). External factors and induced cell differentiation can change the genome organization and nuclear matrix [56]. Many nuclear proteins could have a dual role not only in the regulation of DNA transcription but also in the contribution to the mechanical properties of nucleoskeleton. For instance, Runx2, a critical transcription factor required during osteogenesis, also seems to function as a nuclear structural protein and links different regions of chromosomes, thereby potentially assisting in the regulation of combinatorial gene transcription [57,58]. Therefore, mechanical signals propagating through cytoskeleton and then the LINC complex can directly impact crucial DNA regulatory enzymes and binding factors via the interconnected nuclear scaffold.

3.4 ROLE OF MECHANICAL CUES IN DEVELOPMENTAL BIOLOGY

Since the advent of modern molecular biology, an increasing emphasis in developmental biology research has been placed on the understanding of cellular events based on molecular signaling cascades, with little consideration for the impact of physical cues from the larger mechanical context at tissue and organ level, in which the cells are residing. As discussed in the previous sections, this mechanical context is the physical continuum from tissue and organ level to cell level as a collective outcome of cell–cell adhesions, cell–ECM interactions, and linking of these two adhesions to intracellular cytoskeletal and nucleoskeletal structures in tandem. Cell behavior and therefore tissue morphogenesis are strongly influenced by the mechanical signals from the environment, including mechanical information in the form of biophysical stimuli such as compression, tension, fluid shear, and so on.

The effect of mechanical signals on tissue and organ development, especially on skeletogenesis, has clearly been demonstrated in established animal models. For example, immobilization of chick embryos using neuromuscular blocking agents abrogates muscle dynamic contractions and lowers muscle forces. This induced paralysis had a dramatic effect on animal bone development and resulted in abnormal curvature of the mandible, neck, and spine, and retarded growth of clavicle, femur, tibia, and humerus [59]. Hind limb muscular atrophy, which was induced by excising the neural tube, resulted in significantly short femur and tibiotarsus compared with controls. The bulk bone stiffness and bending strength of tibiotarsi of immobilized chick embryos were also significantly reduced due to the lack of muscle contraction [60]. The dramatic effects of muscle contraction-induced mechanical stimulation on mammalian bone development were also clearly demonstrated in genetically modified mice with absent or noncontractile muscles [61–63]. In mice with double knockouts of Myf5 and MyoD, the normal development of the anatomical structure and geometry in cervical vertebrae, spine, mandible, palate, clavicle, and sternum was substantially compromised [61]. Numerous studies have also shown that muscle contraction is critical to joint cavitation. Cavitation of the knee, ankle, hip, and toe joints was minimal or absent in animal treated with drugs that blocked muscle contraction. When muscle force was eliminated by the surgical excision, effects on joint cavitation similar to those in the drug-treated animals were observed. Furthermore, immobilization of embryos after the formation of joint cavity resulted in the loss of the cavity, and wider joint cavities in the hip, shoulder, knee, hands, and feet developed after muscle force was pharmacologically restored [31,64]. It was found that changes in ECM molecules were responsible for the morphological abnormalities due to immobilization [65]. Studies showed altered patterns of tenascin-C, extracellular-regulated kinase, fibroblast growth factor (FGF)-2, and collagen-XII in immobilized limbs and joints [66,67]. The removal of muscle contractions also affected the structures associated with the joint. In immobilized embryos, even though the presence of meniscus was evident at day 8, they started to degenerate by day 10 and disappeared by day 11 or 12, and the plantar tarsal sesamoid completely failed to form [68]. This indicated that mechanical loading is essential for sesamoid formation and late stages of meniscal development and maintenance [68].

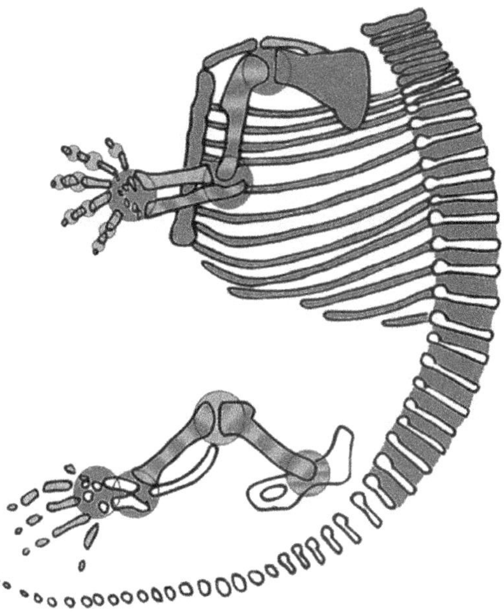

FIGURE 3.2 Effects of abnormal muscle force on skeletogenesis in mouse models. Red indicates effect on rudiment or joint due to abnormal muscle, green indicates no effect, striped red and green indicates findings of affected and unaffected aspects, and white indicates no data available. (Reproduced with permission from Reference 69.) *(See insert for color representation of the figure.)*

Dynamic patterns of biophysical stimuli in the joint colocalize with the development of patella and articular cartilages. Changes in joint shape may have been due to changes in the tissue properties of immobilized joints, since immobilization seems to differentially affect the mechanical properties, glycosaminoglycan (GAG) (proteoglycan), and total collagen content in different areas of joint epiphyses.

The key role of mechanical forces played in bone and joint formation, as demonstrated by developmental models with abnormal skeletal muscle, has been thoroughly summarized and reviewed by Nowlan *et al.* [69] (Figure 3.2). However, relatively little is known about how biophysical stimuli are translated into gene regulation via molecular mechanisms. Most of the previous studies of mechanotransduction signaling were conducted in the classic two-dimensional (2-D) monolayer culture, which may not fully reflect the complexity in the three-dimensional (3-D) multicellular tissue environment in which cells reside. Refinements on the animal models, such as more sophisticated genetic manipulation, generation of more realistic *in vitro* 3-D tissue models, and development of novel imaging technologies, will be instrumental in promoting our understanding of this important missing link. A greater understanding of these mechanisms could in turn facilitate advancements in regenerative medicine.

3.5 APPLICATIONS OF MECHANICAL STIMULATION IN REGENERATIVE MEDICINE

Regenerative medicine or stem cell tissue engineering has emerged as a promising approach to treat a variety of diseases. Recent clinical success in several cases has underscored the importance of the enhanced functional performance of the engineered tissue implants to support and maintain the impaired function of diseased tissues and organs. Mechanical conditioning regimens have been shown to improve the functional properties of an array of engineered tissues, such as bone, cartilage, ligament/tendon, cardiac muscles, blood vessels, and so on.

3.5.1 Articular Cartilage

Mechanical loading of various forms has been shown to regulate the normal maintenance of articular cartilage *in vivo* [70,71] and *in vitro* [72–79]. Investigators using dynamic or cyclic loading conditions have shown that the biosynthetic response is strongly dependent on the magnitude and frequency of the applied load [75,76,79–82], with slow frequency loading generally resulting in suppression of proteoglycan (PG) synthesis, while more rapid loading frequencies result in stimulated synthesis [83]. Furthermore, dynamic compressive loading also improved the functional properties, that is, mechanical stiffness of the tissue-engineered cartilage using either chondrocytes or stem cells as the cell source [84–86]. For example, chondrogenic differentiation and cartilage matrix production and distribution by mesenchymal stem cells (MSCs) encapsulated in biomaterial scaffolds were significantly enhanced due to dynamic compressive loading [87,88]. Mechanical signals promoted the expression of chondrogenic genes and synthesis of chondrogenic growth factors by bone marrow-derived MSCs [89,90]. Furthermore, loading also suppressed the hypertrophic differentiation of chondrogenically differentiated MSCs, resulting in reduced abnormal neocartilage calcification [91]. Mechanical signals have been shown to regulate the expression of Indian hedgehog (IHH), parathyroid hormone-related protein (PTHrP), transforming growth factor beta (TGF-β), and its receptor [92–96]. The IHH–PTHrP negative feedback pathway regulates hypertrophic differentiation of both articular and growth plate chondrocytes to maintain the zonal structure of cartilage [97–100]. Our unpublished data and previous work also showed that TGF-β is an inhibitor of MSC hypertrophy [101]. These mechanically responsive molecular pathways could have contributed to the observed suppression in the gene expression of hypertrophic markers and mineralization by MSCs. In addition, other forms of mechanical conditioning, such as tension and hydrostatic pressure, can also regulate chondrogenic differentiation and cartilage matrix synthesis by MSCs [102].

3.5.2 Tendon/Ligament

The transitional interface from ligaments or tendons, which are soft tissues that consist of highly aligned type I collagen and fibroblasts, to the hard bone tissue exhibits a gradation in cell phenotype, tissue organization, tissue composition, and tissue mechanical properties. The graded transition is vital to the effective transfer of load between two materials with distinct mechanical properties by reducing the potentially damaging stress concentrations [103,104]. It is postulated that the spatial variation in mechanical strain at the insertions may have contributed to the graded changes in cell differentiation, cell morphology, tissue composition, and subsequent tissue properties [105–107].

Tensile stress has been shown to upregulate the expression of scleraxis, an essential tenogenic maker, and promoted tendon-specific matrix elaboration and cell alignment along the direction of tension by MSCs encapsulated in collagen gels [108,109]. In contrast, fibrocartilaginous tissue with increased proteoglycan content was found in regions of tendon under compression, such as those that wrap around bony pulleys [106,110]. It was found that the compressive but not tensile forces can induce expression of the chondrogenic transcription factor Sox9, which may have led to the transdifferentiation by tendon fibroblasts to fibrochondrocytes [107,111]. Another study showed that cyclic pure tension promoted alignment of spindle-shaped cells along the direction of tension and upregulated scleraxis and type I collagen expression. Tension with a single component of compressive stress resulted in random orientation of cells with rounded morphology and increase cartilaginous matrix synthesis in the presence of growth factor [112].

3.5.3 Bone

Compressive strain and fluid shear are the two primary forms of biomechanical stimuli that regulate the bone homeostasis, the dynamic balance between catabolic and anabolic activities in the bone. As first indicated by Wolff's law, bone trabecular remodeling is dictated by the mechanical forces applied. It was further shown that dynamic, rather than static, strains are directly correlated with bone adaptation in animal models [113,114]. Fluid flow in bone is generated by interstitial fluid movement within bone microstructure as a result of mechanical loading and is essential to the diffusion of oxygen and nutrients to bone cells. *In vitro* experiments employing pulsatile or oscillatory flow have demonstrated upregulation of osteogenic genes including bone sialoprotein (BSP) and osteocalcin (OC) [115]. It was found that sustained application of mechanical signals such as fluid shear desensitize the cell, leading to a muted response. Addition of a low shear period in between high shear flows reduced the effect of desensitization and promoted the expression levels of osteopontin (OPN) and prostaglandin E_2 (PGE_2) [116–118]. A number of previous studies also showed that fluid shear promoted osteogenesis and mineral matrix production by MSCs [119–121]. Tensile strain has been shown to upregulate bone morphogenetic protein (BMP-2) expression in MSCs embedded 3-D collagen matrices, suggesting that strain alone can induce osteogenic differentiation [122].

3.5.4 Blood Vessels

Pulsatile radial stress at biomimetic frequency applied by customized bioreactors leads to faster tissue remodeling, enhanced histological organization, and superior contractility and mechanical properties in tissue-engineered blood vessels [123–

125]. Furthermore, collagen and elastin synthesis and orientation in engineered blood vessels can also be improved by cyclic circumferential strain, resulting in better match to the properties of native blood vessels [126–128]. Recently, studies done in pulsatile flow bioreactors showed that mechanical stimulation promoted the formation of a confluent monolayer of endothelial cells, which is critical to the functional performance and maintenance of tissue engineered blood vessels [129].

3.6 SUMMARY

Various well-established animal models demonstrated that mechanical forces are indispensible guiding cues in the development, maturation, and maintenance of a plethora of tissues and organs in the human body. Cells in these tissues and organs can sense mechanical forces via cell–cell adhesions (AJ, gap junction, tight junction, etc.), cell–ECM adhesions (FA), and other mechanosensitive cellular structure and molecules. Mechanotransduction is the process of mechanical forces getting converted to biochemical signaling cascades in the cell cytoplasm and nucleus, resulting in modulation of cellular activities including proliferation, migration, differentiation, and so on. Many of these findings have already been applied to developing strategies to promote the functional performance of engineered tissues including cartilage, bone, blood vessels, and so on. However, there are still many unanswered questions in the role of mechanical cues in cellular regulation and organ morphogenesis. Future research will focus on further elucidating the molecular mechanism of mechanotransduction in subcellular level, such as the emerging field of mechanically activated microRNAs and other epigenetic regulations by mechanical cues. At tissue or organ level, the active remodeling of ECM by cells subjected to mechanical stresses and the role of mechanotransduction on organ morphogenesis need to be thoroughly investigated. The collective findings from the basic research will not only lead to the discovery of new drugs, but also guide the design and development of novel biomedical materials and devices, and thereby promote the translational research for regenerative medicine.

REFERENCES

[1] van der Flier, A., Sonnenberg, A. *Cell Tissue Res.* **2001**, *305*, 285.

[2] Lansman, J.B., Hallam, T.J., Rink, T.J. *Nature* **1987**, *325*, 811.

[3] Tzima, E., Irani-Tehrani, M., Kiosses, W.B., Dejana, E., Schultz, D.A., Engelhardt, B., Cao, G., DeLisser, H., Schwartz, M.A. *Nature* **2005**, *437*, 426.

[4] Chen, K.D., Li, Y.S., Kim, M., Li, S., Yuan, S., Chien, S., Shyy, J.Y. *J. Biol. Chem.* **1999**, *274*, 18393.

[5] Vogel, V., Sheetz, M.P. *Curr. Opin. Cell Biol.* **2009**, *21*, 38.

[6] Poh, Y.C., Na, S., Chowdhury, F., Ouyang, M., Wang, Y., Wang, N. *PLoS ONE* **2009**, *4*, e7886.

[7] Na, S., Collin, O., Chowdhury, F., Tay, B., Ouyang, M., Wang, Y., Wang, N. *Proc. Natl. Acad. Sci. U.S.A.* **2008**, *105*, 6626.

[8] DuFort, C.C., Paszek, M.J., Weaver, V.M. *Nat. Rev. Mol. Cell Biol.* **2011**, *12*, 308.
[9] Hayakawa, K., Tatsumi, H., Sokabe, M. *J. Cell Sci.* **2008**, *121*, 496.
[10] Felsenfeld, D.P., Schwartzberg, P.L., Venegas, A., Tse, R., Sheetz, M.P. *Nat. Cell Biol.* **1999**, *1*, 200.
[11] Meyer, C.J., Alenghat, F.J., Rim, P., Fong, J.H., Fabry, B., Ingber, D.E. *Nat. Cell Biol.* **2000**, *2*, 666.
[12] Tamkun, J.W., DeSimone, D.W., Fonda, D., Patel, R.S., Buck, C., Horwitz, A.F., Hynes, R.O. *Cell* **1986**, *46*, 271.
[13] Leopoldt, D., Yee, H.F., Jr., Rozengurt, E. *J. Cell. Physiol.* **2001**, *188*, 106.
[14] Matthews, B.D., Overby, D.R., Mannix, R., Ingber, D.E. *J. Cell Sci.* **2006**, *119*, 508.
[15] Jaalouk, D.E., Lammerding, J. *Nat. Rev. Mol. Cell Biol.* **2009**, *10*, 63.
[16] Geiger, B., Spatz, J.P., Bershadsky, A.D. *Nat. Rev. Mol. Cell Biol.* **2009**, *10*, 21.
[17] del Rio, A., Perez-Jimenez, R., Liu, R., Roca-Cusachs, P., Fernandez, J.M., Sheetz, M.P. *Science* **2009**, *323*, 638.
[18] Galbraith, C.G., Yamada, K.M., Sheetz, M.P. *J. Cell Biol.* **2002**, *159*, 695.
[19] Friedland, J.C., Lee, M.H., Boettiger, D. *Science* **2009**, *323*, 642.
[20] Lee, S.E., Kamm, R.D., Mofrad, M.R. *J. Biomech.* **2007**, *40*, 2096.
[21] Vasioukhin, V., Fuchs, E. *Curr. Opin. Cell Biol.* **2001**, *13*, 76.
[22] Balda, M.S., Matter, K. *J. Cell Sci.* **1998**, *111*(Pt 5), 541.
[23] Kumar, N.M., Gilula, N.B. *Cell* **1996**, *84*, 381.
[24] Chen, C.S., Tan, J., Tien, J. *Annu. Rev. Biomed. Eng.* **2004**, *6*, 275.
[25] McNeill, H., Ryan, T.A., Smith, S.J., Nelson, W.J. *J. Cell Biol.* **1993**, *120*, 1217.
[26] Nagafuchi, A. *Curr. Opin. Cell Biol.* **2001**, *13*, 600.
[27] Gallant, N.D., Michael, K.E., Garcia, A.J. *Mol. Biol. Cell* **2005**, *16*, 4329.
[28] Zhang, X., Jiang, G., Cai, Y., Monkley, S.J., Critchley, D.R., Sheetz, M.P. *Nat. Cell Biol.* **2008**, *10*, 1062.
[29] Critchley, D.R. *Annu. Rev. Biophys.* **2009**, *38*, 235.
[30] Ziegler, W.H., Liddington, R.C., Critchley, D.R. *Trends Cell Biol.* **2006**, *16*, 453.
[31] Mitrovic, D. *Acta Anat. (Basel)* **1982**, *113*, 313.
[32] Numaguchi, K., Eguchi, S., Yamakawa, T., Motley, E.D., Inagami, T. *Circ. Res.* **1999**, *85*, 5.
[33] Wang, H.-B., Dembo, M., Hanks, S.K., Wang, Y.-l. *Proc. Natl. Acad. Sci. U.S.A.* **2001**, *98*, 11295.
[34] Gumbiner, B.M. *Curr. Opin. Cell Biol.* **1995**, *7*, 634.
[35] Piedra, J., Martinez, D., Castano, J., Miravet, S., Dunach, M., de Herreros, A.G. *J. Biol. Chem.* **2001**, *276*, 20436.
[36] Tetsu, O., McCormick, F. *Nature* **1999**, *398*, 422.
[37] DeMali, K.A., Wennerberg, K., Burridge, K. *Curr. Opin. Cell Biol.* **2003**, *15*, 572.
[38] Price, L.S., Leng, J., Schwartz, M.A., Bokoch, G.M. *Mol. Biol. Cell* **1998**, *9*, 1863.
[39] Wang, N., Butler, J.P., Ingber, D.E. *Science* **1993**, *260*, 1124.
[40] Tan, J.L., Tien, J., Pirone, D.M., Gray, D.S., Bhadriraju, K., Chen, C.S. *Proc. Natl. Acad. Sci. U.S.A.* **2003**, *100*, 1484.
[41] Ridley, A.J., Hall, A. *Cell* **1992**, *70*, 389.
[42] Kimura, K., Ito, M., Amano, M., Chihara, K., Fukata, Y., Nakafuku, M., Yamamori, B., Feng, J., Nakano, T., Okawa, K., Iwamatsu, A., Kaibuchi, K. *Science* **1996**, *273*, 245.
[43] Geiger, B., Bershadsky, A. *Curr. Opin. Cell Biol.* **2001**, *13*, 584.
[44] Wang, N., Tytell, J.D., Ingber, D.E. *Nat. Rev. Mol. Cell Biol.* **2009**, *10*, 75.
[45] Crisp, M., Liu, Q., Roux, K., Rattner, J.B., Shanahan, C., Burke, B., Stahl, P.D., Hodzic, D. *J. Cell Biol.* **2006**, *172*, 41.

[46] Haque, F., Lloyd, D.J., Smallwood, D.T., Dent, C.L., Shanahan, C.M., Fry, A.M., Trembath, R.C., Shackleton, S. *Mol. Cell. Biol.* **2006**, *26*, 3738.
[47] Worman, H.J., Gundersen, G.G. *Trends Cell Biol.* **2006**, *16*, 67.
[48] Dechat, T., Pfleghaar, K., Sengupta, K., Shimi, T., Shumaker, D.K., Solimando, L., Goldman, R.D. *Genes Dev.* **2008**, *22*, 832.
[49] Lammerding, J., Fong, L.G., Ji, J.Y., Reue, K., Stewart, C.L., Young, S.G., Lee, R.T. *J. Biol. Chem.* **2006**, *281*, 25768.
[50] Sakaki, M., Koike, H., Takahashi, N., Sasagawa, N., Tomioka, S., Arahata, K., Ishiura, S. *J. Biochem.* **2001**, *129*, 321.
[51] Lee, K.K., Haraguchi, T., Lee, R.S., Koujin, T., Hiraoka, Y., Wilson, K.L. *J. Cell Sci.* **2001**, *114*, 4567.
[52] Holaska, J.M., Kowalski, A.K., Wilson, K.L. *PLoS Biol.* **2004**, *2*, E231.
[53] Lattanzi, G., Cenni, V., Marmiroli, S., Capanni, C., Mattioli, E., Merlini, L., Squarzoni, S., Maraldi, N.M. *Biochem. Biophys. Res. Commun.* **2003**, *303*, 764.
[54] Pederson, T. *J. Cell Biol.* **2008**, *180*, 1061.
[55] Ye, J., Zhao, J., Hoffmann-Rohrer, U., Grummt, I. *Genes Dev.* **2008**, *22*, 322.
[56] Nickerson, J.A., Blencowe, B.J., Penman, S. *Int. Rev. Cytol.* **1995**, *162A*, 67.
[57] Stein, G.S., Lian, J.B., van Wijnen, A.J., Stein, J.L., Javed, A., Montecino, M., Choi, J.Y., Vradii, D., Zaidi, S.K., Pratap, J., Young, D. *Adv. Enzyme Regul.* **2007**, *47*, 242.
[58] Durst, K.L., Hiebert, S.W. *Oncogene* **2004**, *23*, 4220.
[59] Hall, B.K., Herring, S.W. *J. Morphol.* **1990**, *206*, 45.
[60] Wong, M., Germiller, J., Bonadio, J., Goldstein, S.A. *Prog. Clin. Biol. Res.* **1993**, *383B*, 587.
[61] Rot-Nikcevic, I., Reddy, T., Downing, K.J., Belliveau, A.C., Hallgrimsson, B., Hall, B.K., Kablar, B. *Dev. Genes Evol.* **2006**, *216*, 1.
[62] Nowlan, N.C., Bourdon, C., Dumas, G., Tajbakhsh, S., Prendergast, P.J., Murphy, P. *Bone* **2009**, *46*, 1275.
[63] Rot-Nikcevic, I., Downing, K.J., Hall, B.K., Kablar, B. *Histol. Histopathol.* **2007**, *22*, 51.
[64] Ruano-Gil, D., Nardi-Vilardaga, J., Teixidor-Johe, A. *Acta Anat. (Basel)* **1985**, *123*, 90.
[65] Mikic, B., Wong, M., Chiquet, M., Hunziker, E.B. *J. Orthop. Res.* **2000**, *18*, 406.
[66] Bastow, E.R., Lamb, K.J., Lewthwaite, J.C., Osborne, A.C., Kavanagh, E., Wheeler-Jones, C.P., Pitsillides, A.A. *J. Biol. Chem.* **2005**, *280*, 11749.
[67] Kavanagh, E., Church, V.L., Osborne, A.C., Lamb, K.J., Archer, C.W., Francis-West, P.H., Pitsillides, A.A. *Dev. Dyn.* **2006**, *235*, 826.
[68] Mikic, B., Johnson, T.L., Chhabra, A.B., Schalet, B.J., Wong, M., Hunziker, E.B. *J. Rehabil. Res. Dev.* **2000**, *37*, 127.
[69] Nowlan, N.C., Sharpe, J., Roddy, K.A., Prendergast, P.J., Murphy, P. *Birth Defects Res. C Embryo Today* **2010**, *90*, 203.
[70] Palmoski, M., Perricone, E., Brandt, K.D. *Arthritis Rheum.* **1979**, *22*, 508.
[71] Setton, L.A., Mow, V.C., Howell, D.S. *J. Orthop. Res.* **1995**, *13*, 473.
[72] Gray, M.L., Pizzanelli, A.M., Grodzinsky, A.J., Lee, R.C. *J. Orthop. Res.* **1988**, *6*, 777.
[73] Guilak, F., Meyer, B.C., Ratcliffe, A., Mow, V.C. *Osteoarthritis Cartilage* **1994**, *2*, 91.
[74] Jones, I.L., Klamfeldt, A., Sandstrom, T. *Clin. Orthop.* **1982**, *165*, 283.
[75] Sah, R.L., Kim, Y.J., Doong, J.Y., Grodzinsky, A.J., Plaas, A.H., Sandy, J.D. *J. Orthop. Res.* **1989**, *7*, 619.
[76] Sah, R.L., Doong, J.Y., Grodzinsky, A.J., Plaas, A.H., Sandy, J.D. *Arch. Biochem. Biophys.* **1991**, *286*, 20.
[77] Valhmu, W.B., Stazzone, E.J., Bachrach, N.M., Saed-Nejad, F., Fischer, S.G., Mow, V.C., Ratcliffe, A. *Arch. Biochem. Biophys.* **1998**, *353*, 29.

[78] Wong, M., Siegrist, M., Cao, X. *Matrix Biol.* **1999**, *18*, 391.

[79] Buschmann, M.D., Kim, Y.J., Wong, M., Frank, E., Hunziker, E.B., Grodzinsky, A.J. *Arch. Biochem. Biophys.* **1999**, *366*, 1.

[80] Palmoski, M.J., Brandt, K.D. *Arthritis Rheum.* **1984**, *27*, 675.

[81] Parkkinen, J.J., Lammi, M.J., Helminen, H.J., Tammi, M. *J. Orthop. Res.* **1992**, *10*, 610.

[82] Kim, Y.J., Sah, R.L., Grodzinsky, A.J., Plaas, A.H., Sandy, J.D. *Arch. Biochem. Biophys.* **1994**, *311*, 1.

[83] Guilak, F., Sah, R., Setton, L. in *Basic Orthopaedic Biomechanics* (Eds: V.C. Mow, W.C. Hayes), Lippincott-Raven, Philadelphia: 1997, p. 179.

[84] Bian, L., Fong, J.V., Lima, E.G., Stoker, A.M., Ateshian, G.A., Cook, J.L., Hung, C.T. *Tissue Eng. A* **2009**, *16*, 1781.

[85] Hung, C.T., Mauck, R.L., Wang, C.C., Lima, E.G., Ateshian, G.A. *Ann. Biomed. Eng.* **2004**, *32*, 35.

[86] Mauck, R.L., Byers, B.A., Yuan, X., Tuan, R.S. *Biomech. Model. Mechanobiol.* **2007**, *6*, 113.

[87] Huang, A.H., Farrell, M.J., Kim, M., Mauck, R.L. *Eur. Cell. Mater.* **2010**, *19*, 72.

[88] Huang, C.Y., Hagar, K.L., Frost, L.E., Sun, Y., Cheung, H.S. *Stem Cells* **2004**, *22*, 313.

[89] Campbell, J.J., Lee, D.A., Bader, D.L. *Biorheology* **2006**, *43*, 455.

[90] Pelaez, D., Huang, C.Y., Cheung, H.S. *Stem Cells Dev.* **2009**, *18*, 93.

[91] Bian, L., Zhai, D.Y., Zhang, E.C., Mauck, R.L., Burdick, J.A. *Tissue Eng. A* **2012**, *18*, 715.

[92] Wu, Q., Zhang, Y., Chen, Q. *J. Biol. Chem.* **2001**, *276*, 35290.

[93] Tang, G.H., Rabie, A.B., Hagg, U. *J. Dent. Res.* **2004**, *83*, 434.

[94] Tanaka, N., Ohno, S., Honda, K., Tanimoto, K., Doi, T., Ohno-Nakahara, M., Tafolla, E., Kapila, S., Tanne, K. *J. Dent. Res.* **2005**, *84*, 64.

[95] Chen, X., Macica, C.M., Nasiri, A., Broadus, A.E. *Arthritis Rheum.* **2008**, *58*, 3788.

[96] Huang, C.Y., Reuben, P.M., Cheung, H.S. *Stem Cells* **2005**, *23*, 1113.

[97] Zhang, M., Xie, R., Hou, W., Wang, B., Shen, R., Wang, X., Wang, Q., Zhu, T., Jonason, J.H., Chen, D. *J. Cell Sci.* **2009**, *122*, 1382.

[98] Guo, J., Chung, U.I., Yang, D., Karsenty, G., Bringhurst, F.R., Kronenberg, H.M. *Dev. Biol.* **2006**, *292*, 116.

[99] St-Jacques, B., Hammerschmidt, M., McMahon, A.P. *Genes Dev.* **1999**, *13*, 2072.

[100] Vortkamp, A., Lee, K., Lanske, B., Segre, G.V., Kronenberg, H.M., Tabin, C.J. *Science* **1996**, *273*, 613.

[101] Mello, M.A., Tuan, R.S. *J. Orthop. Res.* **2006**, *24*, 2095.

[102] McMahon, L.A., Campbell, V.A., Prendergast, P.J. *J. Biomech.* **2008**, *41*, 2055.

[103] Thomopoulos, S., Williams, G.R., Gimbel, J.A., Favata, M., Soslowsky, L.J. *J. Orthop. Res.* **2003**, *21*, 413.

[104] Liu, Y., Birman, V., Chen, C., Thomopoulos, S., Genin, G.M. *J. Eng. Mater. Technol.* **2010**, *133*(1), 011006.

[105] Benjamin, M., Kumai, T., Milz, S., Boszczyk, B.M., Boszczyk, A.A., Ralphs, J.R. *Comp. Biochem. Physiol. A. Mol Integr. Physiol.* **2002**, *133*, 931.

[106] Benjamin, M., Ralphs, J.R. *J. Anat.* **1998**, *193*(Pt 4), 481.

[107] Matyas, J.R., Anton, M.G., Shrive, N.G., Frank, C.B. *J. Biomech.* **1995**, *28*, 147.

[108] Altman, G.H., Horan, R.L., Martin, I., Farhadi, J., Stark, P.R.H., Volloch, V., Richmond, J.C., Vunjak-Novakovic, G., Kaplan, D.L. *FASEB J.* **2002**, *16*(2), 270.

[109] Juncosa-Melvin, N., Matlin, K.S., Holdcraft, R.W., Nirmalanandhan, V.S., Butler, D.L. *Tissue Eng.* **2007**, *13*, 1219.

[110] Evanko, S.P., Vogel, K.G. *Arch. Biochem. Biophys.* **1993**, *307*, 153.

[111] Li, K.W., Lindsey, D.P., Wagner, D.R., Giori, N.J., Schurman, D.J., Goodman, S.B., Smith, R.L., Carter, D.R., Beaupre, G.S. *Tissue Eng.* **2006**, *12*, 2611.
[112] Thomopoulos, S., Das, R., Birman, V., Smith, L., Ku, K., Elson, E.L., Pryse, K.M., Marquez, J.P., Genin, G.M. *Tissue Eng. A* **2010**, *17*, 1039.
[113] Liskova, M., Hert, J. *Folia Morphol. (Praha)* **1971**, *19*, 301.
[114] Rubin, C.T., McLeod, K.J. *Clin. Orthop. Relat. Res.* **1994**, (298), 165.
[115] Sharp, L.A., Lee, Y.W., Goldstein, A.S. *Ann. Biomed. Eng.* **2009**, *37*, 445.
[116] Partap, S., Plunkett, N.A., Kelly, D.J., O'Brien, F.J. *J. Mater. Sci. Mater. Med.* **2010**, *21*, 2325.
[117] Kreke, M.R., Huckle, W.R., Goldstein, A.S. *Bone* **2005**, *36*, 1047.
[118] Vance, J., Galley, S., Liu, D.F., Donahue, S.W. *Tissue Eng.* **2005**, *11*, 1832.
[119] Riddle, R.C., Taylor, A.F., Genetos, D.C., Donahue, H.J. *Am. J. Physiol. Cell Physiol.* **2006**, *290*, C776.
[120] Knippenberg, M., Helder, M.N., Doulabi, B.Z., Semeins, C.M., Wuisman, P.I., Klein-Nulend, J. *Tissue Eng.* **2005**, *11*, 1780.
[121] Kapur, S., Baylink, D.J., Lau, K.H. *Bone* **2003**, *32*, 241.
[122] Sumanasinghe, R.D., Bernacki, S.H., Loboa, E.G. *Tissue Eng.* **2006**, *12*, 3459.
[123] Niklason, L.E., Gao, J., Abbott, W.M., Hirschi, K.K., Houser, S., Marini, R., Langer, R. *Science* **1999**, *284*, 489.
[124] Seliktar, D., Black, R.A., Vito, R.P., Nerem, R.M. *Ann. Biomed. Eng.* **2000**, *28*, 351.
[125] Seliktar, D., Nerem, R.M., Galis, Z.S. *Tissue Eng.* **2003**, *9*, 657.
[126] Bulick, A.S., Munoz-Pinto, D.J., Qu, X., Mani, M., Cristancho, D., Urban, M., Hahn, M.S. *Tissue Eng.* **2009**, *15*, 815.
[127] Thomopoulos, S., Fomovsky, G.M., Holmes, J.W. *J. Biomech. Eng.* **2005**, *127*, 742.
[128] Schutte, S.C., Chen, Z., Brockbank, K.G., Nerem, R.M. *Tissue Eng.* **2010**, *16*, 3149.
[129] Villalona, G.A., Udelsman, B., Duncan, D.R., McGillicuddy, E., Sawh-Martinez, R.F., Hibino, N., Painter, C., Mirensky, T., Erickson, B., Shinoka, T., Breuer, C.K. *Tissue Eng. B Rev.* **2010**, *16*, 341.

CHAPTER 4

CONTRIBUTION OF PHYSICAL FORCES ON THE DESIGN OF BIOMIMETIC TISSUE SUBSTITUTES

MENEKSE ERMIS

BIOMATEN, METU Center of Excellence in Biomaterials and Tissue Engineering, Department of Biomedical Engineering, Ankara, Turkey

ERKAN TÜRKER BARAN

BIOMATEN, METU Center of Excellence in Biomaterials and Tissue Engineering, Ankara, Turkey

TUĞBA DURSUN

BIOMATEN, METU Center of Excellence in Biomaterials and Tissue Engineering, Department of Biotechnology, Ankara, Turkey

EZGI ANTMEN

BIOMATEN, METU Center of Excellence in Biomaterials and Tissue Engineering, Department of Biotechnology, Ankara, Turkey

VASIF HASIRCI

BIOMATEN, METU Center of Excellence in Biomaterials and Tissue Engineering, Departments of Biological Sciences, Biomedical Engineering, and Biotechnology, Ankara, Turkey.

4.1 INTRODUCTION

Cells have to adhere onto biomaterial surfaces in order for the biomaterial to serve its function. Otherwise, the tissue and the implant stay as separate phases with insufficient interaction, which might eventually lead to extrusion of the biomedical devices. In order to improve the interaction, the mechanisms of cell adhesion to their

Bio-inspired Materials for Biomedical Engineering, First Edition. Edited by Anthony B. Brennan and Chelsea M. Kirschner.

environment have to be known, as the biomaterial itself cannot contribute to this interaction actively, except by presenting its chemical and physical surface to the biological entities.

4.1.1 Molecular Mechanisms of Cell Adhesion

Cells are always in contact with their surroundings and this involves communication with each other and interaction with the microenvironment. These interactions allow cells to organize in the form of tissues and function as a member of a group of cells acting in harmony. This is what differentiates the cells in an *in vitro* medium from those in an *in vivo* setting. The types of cells, the extracellular matrix (ECM) that surrounds them, the presence of nerves and blood vessels, and above all, the microarchitecture of the organ in which these cells reside define these interactions. In order to be able to communicate with their environment, the cells use a number of molecules.

The main family of transmembrane molecules involved in these contacts is called the integrins. They are positioned in the cell membrane and serve as a relay center between the cell interior and exterior. In humans, about 20 integrins are involved in the interactions with different kinds of molecules. Internally, integrins are linked to the actin cytoskeleton. Externally, integrins interact with the ECM, other cells, and the substrates such as implants. On the ECM, the main counterpart for the integrins is fibronectin, a protein with cell adhesive properties. On the cytoplasm side of the cell, the integrin molecule binds to a number of adapter proteins that connect to the actin filaments, thus, integrins do an inside–outside bridging function.

Binding between cells is achieved mainly through gap junctions, tight junctions, adherens junctions, and desmosomes. Gap junctions are a group of channels that connect the aqueous compartments of neighboring cells that allow the exchange of small molecules and ions. Tight junctions are regions of the plasma membranes fused together to form a tight seal between the cells, which does not allow transportation of ions and other solutes. Adherens junctions are more extensive connections; they connect the actin filaments of neighboring cells. Desmosomes are similar to adherens junctions in that the intermediate filaments of neighboring cells are connected. Through these interactions, the cells perform their metabolic activities and maintain homeostasis.

4.1.2 Cell Adhesion to Substrates

The main molecules involved in triggering cell adhesion to substrates belong to the integrin family. Their α and β subunits are expressed on the surface of the membranes. Each consists of a large extracellular domain, a transmembrane section, and a short tail in the cytoplasm linked to the intermediate filaments instead of actin filaments. Binding of paxillin to cytoplasmic tails has been demonstrated, and this binding is reported to regulate cell spreading, migration, and stress fiber formation (association between α_4 integrin cytoplasmic tail and nonmuscle myosin IIA regulates cell migration) [1]. Integrins cluster to form *matrix adhesions*, which are extremely dynamic and complex structures with different sizes, compositions, and orientations. The largest, most mature structures are referred to as focal adhesions (FAs), and are involved in the attachment of cells onto substrates by way of the cytoskeleton. The β-subunit of integrin is linked through talin and vinculin to α-actinin and actin filaments, which are linked to the nuclear membrane [2]. The focal adhesion consists of an assembly of proteins including integrins, talin, tensin, vinculin and paxillin, α-actinin, Src, and the enzyme focal adhesion kinase (FAK). Vinculin and paxillin are responsible for driving cell migration, and vinculin, tensin, and talin for linkage with the actin-based cytoskeleton [3]. These adhesions and the molecules involved are in contact with the external cellular membrane through the integrins and attach to the surface of a biomaterial or ECM through fibronectin or the specific amino acid sequences on collagen such as arginine–glycine–aspartic acid (RGD) tripeptide sequences. When cells are placed on implants, they are confronted with a substrate, the architecture and the chemistry of which is completely different, and the absence of the protective ECMs of the natural tissue is a major disadvantage. The cell has to adapt to this new poor environment. If the cell in question is a stem cell, it has to differentiate into the right cell type for the targeted organ or tissue in addition to proliferation.

4.2 PHYSICAL FORCES

4.2.1 Mechanical Forces

4.2.1.1 The Mechanism of Mechanical Stimulation on Tissue Regeneration

Mechanobiology investigates the mechanisms of signal transduction from the cell membrane to the cytoskeleton as well as mechanotransduction to the nucleus by way of sequential stimulation of adhesion complexes and secondary messenger proteins on actin filaments (Figure 4.1). Integrins, cadherins, and Ca^{2+} channels are known to be involved in the transfer of mechanical signals into the cell, which are subsequently propagated via cellular signal transduction [4]. Although the signal transduction mechanism is not understood entirely, there are studies in which integrins have been implicated in a remarkable range of mechanotransduction phenomena. Cellular responses to stretching, elevated hydrostatic pressure, fluid shear stress, and osmotic forces increase the tension on the integrins and lead to recruitment of vinculin, zyxin, and probably other focal adhesion components [5]. Multiple studies have also shown that application of strain to adherent cells triggers activation of focal adhesion kinase and c-terminal Src kinase (c-src) [6]. Mimicking the principles of mechanotransduction may become a potential strategy for tissue regeneration in guiding cells to desired phenotypes.

4.2.2 Thermal Forces (NIPAM)

4.2.2.1 Thermal Forces and Thermo Responsiveness in Tissue Engineering

A typical tissue-engineered product consists of a scaffold and cells and some additives such as growth factors, hydroxyapatite (HA), elastin-like peptides, and so

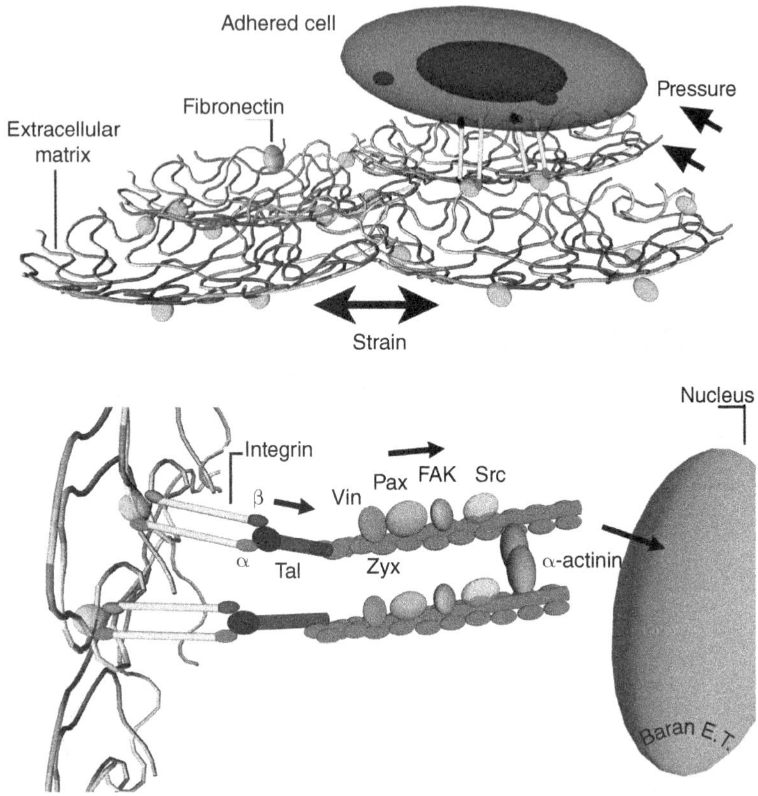

FIGURE 4.1 Molecular pathways mediating mechanotransduction signaling in a cell. In this pathway, mechanical forces such as, stretching, hydrostatic pressure, and shear stress stimulate the integrins on the cell membrane via extracellular matrix. In turn, the stimuli is transduced into the nucleus by engagement of anchorage proteins talin (tal), vinculin (vin), paxillin (pax), and α-actinin and signaling proteins FAK, Src, and zyxin (zyx). *(See insert for color representation of the figure.)*

on. A novel approach to tissue engineering is called *cell sheet engineering*, in which transplantable monolayers of cells are produced. To produce a cell sheet, the most commonly used strategy is to use a thermoresponsive polymer to culture the cells on. This polymer substrate changes its properties (expands or contracts) with temperature and the cell sheets that lift off are harvested [7]. Most of these thermoresponsive polymers are hydrogels. The term sol–gel transition refers to the transition between a solution and a gel form. Some hydrogels exhibit a separation from solution and solidification above a certain temperature called the lower critical solution temperature (LCST); below this temperature, the polymer is soluble. Above LCST, polymers are very hydrophobic and insoluble. For a polymer exhibiting an LCST, increasing the temperature results in negative free energy change which makes the water–polymer interaction unfavorable and the polymer–polymer interactions favorable. This is also called the hydrophobic effect [8]. Thermoresponsive polymers can be divided into two subcategories: natural polymers and derivatives (i.e., cellulose derivatives, chitosan, dextran, xyloglucan, gelatin), and synthetic polymers (N-isopropylacrylamide (NIPAM), poly(ethylene oxide) (PEO)/poly (p-phenylene oxide) (PPO), polyethylene glycol (PEG)/polyester copolymers, poly(organophosphazenes)) [9].

Many techniques are available for synthesis of thermoresponsive polymers. For example, Duarte *et al.* polymerized poly(N-isopropylacrylamide) (PNIPAM) by supercritical fluid foaming [10]. They also produced poly(D,L-lactic acid) (P(D,L-LA)) and P(D,L-LA)/PNIPAM foams with this technique. Biocompatibility and viability tests with L929 cells showed that these foams were highly biocompatible. This approach to production of thermoresponsive three-dimensional (3-D) scaffolds appears to be a cost-effective method of scaffold production. Kobayashi *et al.* designed PNIPAM surface coatings to produce cell sheets in a cell culture dish [11]. They showed that, epidermal keratinocyte sheets for use in autologous transplantation to treat skin defects, and limbal stem cell sheets for use in autologous transplantation in the treatment of corneal defects, could be produced using this technique. In another study, aminated alginate (AAlg) was copolymerized with PNIPAAm in a comb-like fashion to produce injectable hydrogels [12] and human bone marrow mesenchymal stem cells (hBMSCs) were encapsulated in them. These thermoresponsive hydrogels were shown to be not cytotoxic and the cells entrapped in them maintained their viability.

4.2.3 Electromagnetic Forces (Continuous, Pulsatile)

4.2.3.1 Electrical Forces in Tissue Engineering All cells are known to have a voltage difference across their membranes called the *membrane potential*. This potential is a result of the ion concentration differences between the two sides of the membrane. For every ion, the potential difference generated can be calculated using the *Nernst equation*:

$$EMF = \frac{RT}{zF} ln \frac{[A]_{in}}{[A]_{out}} \tag{4.1}$$

where EMF is the electromotive force (mV), R the ideal gas constant, T temperature in Kelvin, F is the Faraday constant, z charge of the ion, and $[A]_{in}$ and $[A]_{out}$ are concentrations inside and outside the cell membrane.

Equation 4.1 is applicable only when the membrane is in thermodynamic equilibrium. Due to the presence of active ion pumps on the cell membrane, there is never an equilibrium between the inside and the outside of the membrane, and therefore, the membrane potential cannot be calculated from the *Nernst equation*. In such a case, the resting potential of a cell can be calculated using the *Goldman–Hodgkin–Katz equation*:

$$EMF_M = \frac{RT}{F} ln \frac{\sum_i^N P_{M_i^+}[M_i^+]_{out} + \sum_j^M P_{A_j^-}[A_j^-]_{in}}{\sum_i^N P_{M_i^+}[M_i^+]_{in} + \sum_j^M P_{A_j^-}[A_j^-]_{out}} \tag{4.2}$$

where P_{ion} is a permeability factor for each ion and the subscripts i and j are for the positive and negative charged species.

These equations take into account that cellular electromotive forces are generated by the ionic gradients across the membrane. Although all cells have a potential across their membranes, some cells are specialized in generating and conducting these electrical forces [13]. Most important examples of such cells are neurons, cells of the conducting system of the heart, and the cardiac and skeletal muscles. These cells, unlike other cells of the organism, use ion channels and cellular junctions actively not just to create a membrane potential but also to manipulate them to elicit action potentials.

Tissue engineering constructs are designed to mimic different physical, chemical, and biological aspects of a given tissue to generate a product that can meet the functional and structural demands of the original tissue to be assisted or substituted. From this point of view, electrical forces are used in tissue engineering applications in two ways: (a) to mimic the original (i.e., nervous or cardiac tissue), by generating a scaffold that can be used to transmit an externally applied or internally generated potential across its surface to align and orient cells and (b) to make a design so that even if the engineered tissue itself does not have the intrinsic electrical force generating capacity, the potential applied may lead to the changes that would guide the cells in the desired direction (i.e., differentiation).

Nanowires (NWs) are commonly used for these applications. For example, Long *et al.* prepared poly(3,4-ethylenedioxythiophene) (PEDOT) NWs over templates of track-etched poly(carbonate) membranes [14]. NWs with 190 nm, 95–100 nm, 35–40 nm, and 20–25 nm diameters were tested for their conductivity and were found to have different characteristics (critical, metallic, and insulating), in relation to their diameters. Carbon nanotubes are commonly used in conductive/semiconductive composites. Single-walled carbon nanotubes (swCNTs) behave as semiconductors or metals depending on their diameter, while multiwalled carbon nanotubes (mwCNTs) are always conductive. In a study by Sulong *et al.*, mwCNTs were functionalized by carboxylation and octadecylation [15]. Results show that unfunctionalized carbon nanotubes are more suitable for electric applications because chemical functionalization of CNTs significantly decrease electrical conductivity. Among the polymers, polypyrrole, polyaniline, and poly(3,4-ethylenedioxythiophene) are the most commonly used conductive polymers [16].

Nerve injury, whether central or peripheral, is a catastrophic event. Most of the approaches to nerve injury today include surgical repair of the spinal cord or the peripheral nerves. Surgical repair options include end-to-end suturing and nerve grafts, with quite promising results, but these techniques also have their shortcomings [17]. Tissue-engineered nerve conduits are being increasingly used in their repair; the choice of the right material is very important and their biochemical functionalization or topographic modifications are frequently used in material improvement [18].

From the materials point of view, several polymers and nanocomposites have been studied for their electrical conductivity or suitability for conversion into conductive materials when they are normally nonconducting. Using electrical forces to direct, align, or orient nerve cells to achieve neuronal function was the subject of many recent studies. Bechara *et al.* extruded poly(caprolactone) (PCL) through nanoporous membranes to obtain NWs, which were coated with the conductive polymer poly(pyrrole) (PPy) [19]. C17.2 murine neural stem cells (NSCs) were cultured on these NWs, and physiological levels of electrical stimulation was applied. Results showed that PPy coating decreased the resistivity of PCL NW. Quantitative microscopic analysis of NSC using calcein-AM staining demonstrated an increase of more than twofold in proliferation using PPy–NW surfaces instead of only NW. Schmidt *et al.* synthesized oxidized PPy films and laminated them with poly(lactic-co-gylcolic acid) (PLGA) to increase their mechanical stability [20]. Rat PC-12 cells and sciatic nerve explants from 16-day-old chick embryos were cultured on these substrates. As judged by the neurite lengths of cells stimulated with 100 mV for 2 hours, the application of electrical stimulus enhanced PC-12 cell differentiation. In another study with PPy films, surfaces were bathed with a fibronectin solution to increase cell adhesion, and potential was applied [21]. It was observed that electrical force enhanced neurite outgrowth on the fibronection-coated surfaces.

According to the World Health Organization 2008 statistics, the number one leading cause of death is ischemic heart disease [22]. Ischemic heart disease is characterized by the reduced blood flow (and thus oxygen) to the cardiac muscle due to atherosclerosis, which causes the thickening and narrowing of the coronary arteries. When the blood supply to the heart muscle ceases, myocardial cells die and a necrotic area is formed. This necrotic area heals through fibrosis and scar tissue formation; the scar tissue lacks contractility [23]. Conventional treatment options are either prevention or early intervention by revascularization. If ischemia is prolonged and tissue viability is lost, only palliative measures can be taken. Cardiac tissue engineering is the current approach which aims to replace necrotic or scar tissue with a substitute carrying functioning cardiomyocytes. In a study by Barash *et al.*, alginate scaffolds were seeded with ventricular cardiomyocytes isolated from 1- to 4-day-old neonatal Sprague-Dawley rats [24]. Scaffolds were either stimulated with 5 V, bipolar, 2 ms pulses at 1 Hz, or were cultured in a perfusion bioreactor where electrical stimulation was applied. They concluded that electrical stimulation, together with perfusion in a reactor, was able to produce thick, functional cardiac patches. Pedrotty *et al.* used skeletal myoblasts in order to demonstrate that under similar conditions, like electrical stimulus or soluble factors derived from cardiac myocytes, proliferation, or differentiation of skeletal myoblasts, could be controlled and might be suitable for cellular cardiomyoplasty [25]. They seeded these cells on polyglycolic acid meshes and cultured it in a culture chamber equipped with platinum electrodes to apply a potential of 100 mV with an upstroke of 1 ms. The skeletal myoblasts cultured together with electrical stimulation showed an increased number of cells. Tandon *et al.* cultured neonatal rat ventricle myocytes, isolated from 2-day-old neonatal Sprague-Dawley rats, on Ultrafoam collagen sponges [26]. The culture conditions included electrical stimulation; effects of electrode material, amplitude (1–6 V/cm), duration (0.25–10 ms), and frequency (1, 3, and 5 Hz) were studied. They concluded that the optimum amplitude for stimulations was 2–3 V/cm and constructs stimulated at 3–4 V/cm had improved functionality. It was shown that stimulation amplitude had more effect than frequency. Functional performance, cell elongation, and tissue compactness were improved in the stimulated group. The researchers noted that to obtain a synchronously contracting thick and uniform tissue, perfusion reactors were required.

Another use of electrical stimulation is with tissues, which do not have intrinsic electrical force generating capacity. An example for this is the study by Supronowicz *et al.* where a poly(lactic acid) and carbon nanotube composite scaffold was prepared to study the effect of electrical stimulation on proliferation capability of osteoblasts [27]. Under stimulation with alternating current (AC), osteoblasts showed a 46% increase in proliferation and 307% increase in calcium deposition on the scaffolds.

4.2.3.2 Magnetic Forces in Tissue Engineering

The force exerted on a charged particle defines a magnetic field. For a given charge q in a given electric field E, the force is calculated according to Equation 4.3:

$$F = qE \tag{4.3}$$

Unfortunately, this equation cannot be extended to charged particles in motion. *Lorentz force law* (4.4) accounts for the force correctly:

$$F = q(E + v\text{x}B) \tag{4.4}$$

where B is the magnetic field vector and v is the velocity of the particle.

For tissue engineering applications, magnetic particles are used under a magnetic field. The most commonly used materials are ferro fluids, which by use of a surfactant can be made into particles. Unfortunately, ferro fluids may not retain magnetization when the external field is discontinued. Magnetization of a particle requires generation of a net magnetic dipole moment and depending on the AC or direct current (DC) used, the response of the ferro fluids differs. When a DC is used to create the field, phases of magnetic field and magnetization of the particles cancel each other and no net moment is experienced. When an AC is used, a phase lag is created between the magnetic field and magnetization of the particles [28].

For bone tissue engineering, magnetic forces are used to apply a preload to the cells seeded on scaffolds to mimic the loading process in the original tissue. Bock *et al.* studied three ferro fluids prepared by aqueous dispersion of magnetic nanoparticles (200 nm diameter): (a) FF-DXS (magnetite nanoparticles coated with dextransulfate and functionalized with sodium sulfate functional groups), (b) FF-PAA (magnetite nanoparticles coated with poly-$_{DL}$-aspartic acid and functionalized with sodium carboxylate), and (c) FF-DP (magnetite nanoparticles coated with starch and functionalized with phosphate groups) with different surface modifications [29]. Collagen scaffolds carrying HA were dipped into these solutions. Viability of the cells within the scaffolds evaluated using MTT assay showed a 25% increase on day 5 for all types of scaffolds and on day 10, 30% more than on day 5. In another study, Tampieri *et al.* synthesized ferrous HA powder from a suspension containing calcium hydroxide, $Ca(OH)_2$, and Fe ions (Fe^{2+}/Fe^{3+}) [30]. Adult rabbit osteoblasts were cultured with Fe–HA and HA–magnetite mixtures. Exposure to a magnetic field resulted in a much higher hyperthermia in the Fe–HA phase than with the HA–magnetite mixtures, on which the researchers concluded that the biomaterial was suitable for anticancer therapy due to the intense local temperature change that could be achieved. Fe–HA powder was also found to be biocompatible for the osteoblasts. With magnetic and biocompatibility properties, this new ferromagnetic particle was concluded to be suitable for bone tissue engineering applications.

4.2.4 Hydrodynamic Forces (Shear; Pulsatile, Compression; Continuous)

4.2.4.1 Fluid Flow, Shear and Hydrostatic Stress-Based Systems Shear stress is a force of friction, and in the case of cells in the body, it is the force exerted on cells by a fluid flowing through channels or narrow spaces in tissues. *In vivo* bone constantly remodels in response to mechanical stresses. It is hypothesized that *in vivo*, these stresses are transmitted to bone cells mainly via fluid shear stresses as the interstitial fluid flows through the lacunar–canalicular pores surrounding the osteocytes, and the shear stress is sensed by differentiated osteoblasts or osteocytes [31]. Chondrocytes can also sense the changes in their hydrodynamic environments resulting from the difference in substrate architectures (e.g., porosity), via $\alpha_5\beta_1$ integrins [32]. This indicates the importance of characterizing spatial and temporal shear stresses applied to osteoblasts and chondrocytes when designing substrates for bone and cartilage tissue engineering. The shear stress resulting from blood flow has been found to affect gene expression and cellular function of endothelial cells (ECs) such as proliferation, apoptosis, migration, permeability, cell alignment, and mechanical properties [33]. Shear also has a dramatic effect on mesenchymal stem cell (MSC) differentiation, and therefore, perfusion bioreactors are designed to regulate shear in 3-D constructs [34]. This was shown in a study to be especially important in the development and differentiation of stem cells; for example, significant changes were detected in the expression of various genes in the mitogen-activated protein kinase (MAPK) signaling pathway [35]. Mechanical stimulation through fluid shear stresses was shown to be influential on bone differentiation and mineralization [36].

Cells present in the blood vessels are exposed to both shear stress and radial stress. The response of MSCs to shear forces in the vasculature is important for vascular tissue engineering. In one study, a pulsatile pressure between 40 and 120 mmHg was applied to MSCs seeded on flexible silicone substrates causing a shear stress of 1 Pa [37]. These MSCs exhibited a response similar to that of ECs, that is, the MSCs oriented in the direction of flow and adapted a morphology similar to that of the ECs.

Perfusion bioreactors are designed to use shear stress as a source for mechanical stress because the medium is forced through the pores or channels of scaffold materials (Figure 4.2). For vascular tissue engineering, bioreactors were used to create pulsatile flow conditions by using a modified four-well Labtek® Chamber-Slide (Sigma-Aldrich, St. Louis, MO) culture system. Such culture systems mimic *in vitro* and *in vivo* environment closely and facilitate the differentiation of stem cells into the cell types needed for tissue-engineered vascular grafts [38]. When human progenitor-derived endothelial cells (PDECs) were exposed to laminar pulsatile physiological shear stress, they spread on the substrates and the number of genes

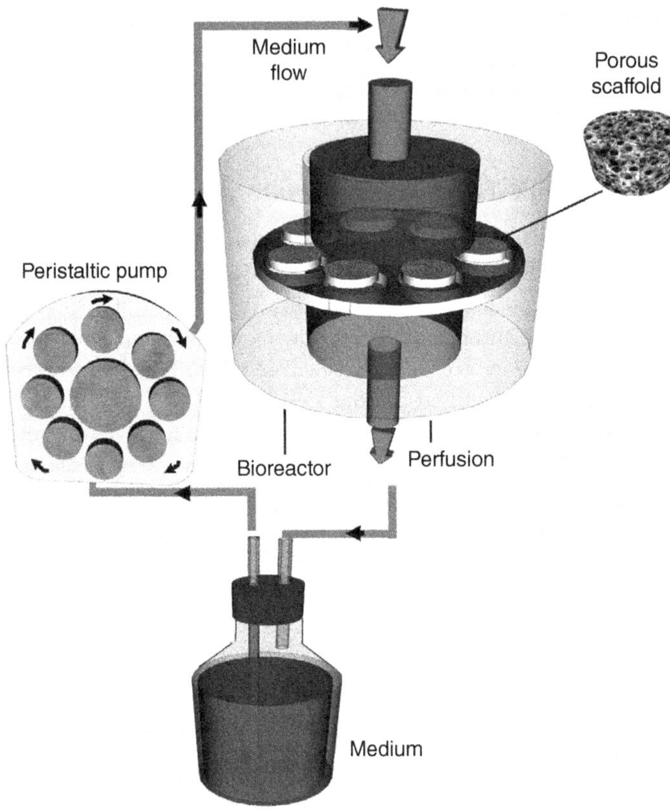

FIGURE 4.2 Perfusion scheme designed for culturing cell in interconnected porous scaffolds. A peristaltic pump circulates and purges culture medium into a perfusion chamber where porous scaffolds are placed in a sealed holder. The culture medium is forced through the pores of the scaffold and the perfused medium is collected at the lower side of the chamber to be recycled or disposed.

engaged in antithrombogenic activity (thrombomodulin and tissue factor) were overexpressed [39]. The comparison of the effect of orbital and laminar shear stress on EC morphology, proliferation, and apoptosis revealed that orbital shear stress increases EC proliferation by 29% and (^{3}H)-thymidine intake twice, compared with 16% and 38% decreases, respectively, observed with laminar shear stresses [40]. Cells in the periphery of the culture well were aligned in the direction of shear stress similar to the shape change seen with laminar shear stress, whereas the ECs in the center of the well appeared unaligned, similar to ECs not exposed to shear stress. Shear stress preconditioning and scaffold surface modification for the construction of functional vascular networks with channel patency, were also used to promote biomaterial endothelialization [41]. It was reported that a transient increase in the shear stress at the appropriate time is the key to enhancing endothelialization of the channels of the scaffold.

Shear stress-generating bioreactors have also been adopted for cartilage tissue engineering. For instance, 3-D maturation of autologous respiratory epithelial cells and differentiation of MSCs such as BMSCs into chondrocytes was induced by hydrodynamic stimuli and adequate mass transport provided by the double-chamber rotating bioreactor [42]. These cells were seeded onto a decellularized human donor tracheal matrix and cultured within the perfusion bioreactor. The bioreactor permitted efficient repopulation of the matrix and the transplantation of the tissue-engineered trachea into a patient was successful. Similarly, epithelial cell and chondrocyte suspension (obtained from biopsies) were perfused directly through the pores of a 3-D scaffold, and the tissues grown for 2 weeks were viable and homogeneously cartilaginous with biomechanical properties approaching those of native cartilage [43].

4.2.4.2 Compression Based Systems Mechanical stimuli play a crucial physiologic role in bone formation and osteochondral repair and regeneration. It is well known that pathological atrophy occurs in circumstances of space flight and long immobilization, because the bone and skeletal tissue experience no gravitational and/or not enough physical forces. Learning from this, it was deduced that the application of mechanical forces, especially dynamic compression, could be used to induce and control osteo/chondral cell differentiation. The hypothesis that selective enhancement of chondrogenic and/or osteogenic differentiation in human periosteal cells under dynamic compression has recently been suggested and verified by the elevated expression of bone and cartilage specific markers [44]. In this light, fully differentiated osteoblastic cells were seeded onto open cell foam scaffolds of polyurethane and exposed to a cyclic compressive loading to study the mechanical modulation of bone matrix formation [45]. The expression of messenger ribonucleic acid (mRNA) for type I collagen, osteopontin, and osteocalcin increased after a single round of loading.

Mechanical compression in a bioreactor can mimic *in vivo* physiologic dynamic forces, which are exerted on the muscoloskeletal tissues by cyclic compression of elastic and microporous scaffolds (Figure 4.3A). Compressive cyclic loading on articular chondrocytes seeded onto ceramic substrates was carried out [46] and a single application of cyclic loading on chondrocytes increased matrix accumulation and enhanced the mechanical properties of the *in vitro* formed tissue. Cyclic forces applied after 24 hours of cell culture led to increased collagen and proteoglycan synthesis (48% and 49%, respectively) and increased mechanical properties (twofold increase in equilibrium stress and modulus).

Recently, the clinical potential of mechanical stimulation in the transplantation of matrix associated autologous chondrocyte to enhance its mechanical and biological properties was studied using long-term, continuous, compressive loading. The samples were collagen type I hydrogels seeded with human chondrocytes harvested from knee joints of patients [47]. Histological and histomorphometric evaluation revealed that this treatment significantly increased homogenous collagen type II and proteoglycan. The role of intermittent dynamic compression in chondrogenic and osteogenic differentiation of human bone marrow stromal cells (hMSC) encapsulated in PEG hydrogels carrying RGD moieties was studied [48]. The loading

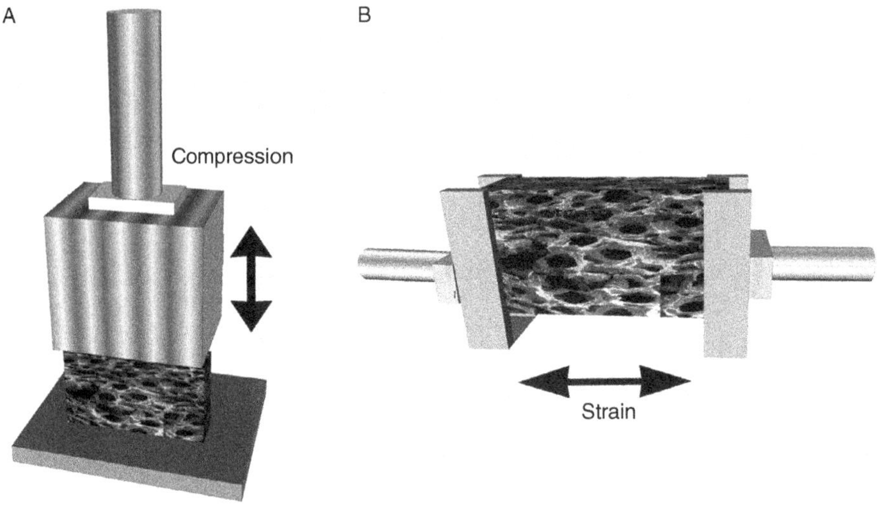

FIGURE 4.3 (A) Compression and (B) strain mechanical setups that mimic biomechanics of the scaffolds in bioreactors. In the compression bioreactors, the load cell applies a periodic load on the scaffolds with seeded cells. Similarly, the scaffold cells can be strained periodically in the bioreactor by extending the structure with a gripping load cell.

regime applied for 14 days, however, appeared to have an inhibitory effect on chondrogenesis and osteogenesis probably due to excessive loading of the differentiating hMSCs before the production of sufficient pericellular matrix and/or due to excessively large strains, particularly for osteogenically differentiating hMSC. The influence of loading level on bone formation was also studied by a mechanobiological model, which was used to determine the influence of vascular network development and tissue growth inside a scaffold [49]. Low levels of mechanical loading (0.01–0.1 MPa) stimulated bone formation while high levels (0.1–2 MPa) inhibited it along with capillary growth.

Complex limb movements cannot be exactly mimicked in the common piston-based cyclic loading probes used in conventional biomechanical bioreactors. In order to better mimic the *in vivo* environment, sliding-type biomechanical stimuli, which move more like natural joints have been recommended, especially in the regeneration and maintenance of functional articular surfaces. In one such study, bovine chondrocytes were seeded onto polyurethane scaffolds and subjected to dynamic compression via a ceramic ball to generate both compressive and slide loading [50]. This resulted in increased synthesis of collagen type II and aggrecan, and appeared to increase cell proliferation. In a similar study, sliding indentation was used to induce tensile strain on periosteum explants embedded between two agarose layers [51]. Application of sliding indentation enhanced the production of collagen type I, and led to the formation of fibrous tissue without any evidence of cartilage formation. However, when stimulated by sliding indentation and transforming growth factor beta-1 (TGF-β1), collagen production and expression of the chondrogenic cell phenotype were enhanced.

During mandibular movement, the condyle is subjected to repetitive compression and the mandibular condylar chondrocytes (MCCs) can detect and respond to this biomechanical action. In order to mimic mandibular conditions, rabbit MCCs were compressed under a continuous hydraulic pressure of 90 kPa for 1 and 6 hours [52]. The cellular processes were observed to be elongated and voluminous, and expression of aggrecan mRNA was increased after 1-hour treatment. After 6 hours, however, the aggrecan mRNA was decreased and some cells showed signs of apoptosis. A similar treatment was applied to fibrochondrocytes seeded in alginate under unconfined compression conditions [53]. The loaded samples presented 2- to 3.2-fold increases in the ECM content and 1.8- to 2.5-fold increases in the compressive modulus compared with those treated under static conditions.

4.2.4.3 Strained (Tensile) Systems The stretching of scaffolds is an effective way of mimicking dynamic conditions of vascular, cardiac, and musculoskeletal tissues (Figure 4.3B). In the blood vessels, cells are subjected to hemodynamic forces in the form of cyclic stretch and shear due to the pulsatile blood flow. In comparison with ECs, smooth muscle cells are more sensitive to the cyclic stretch resulting from pulsatile pressure. Elastomeric biomaterials provide the required biological conditioning and the ability to transmit the biomechanical stimuli to vascular smooth muscle cells (VSMCs). In a study to mimic these conditions, uniaxial cyclic mechanical strain was applied to VSMCs on an elastomeric polyurethane scaffold for 4 weeks and increased proliferation (DNA mass), increased cell area and deeper penetration of the cells into the scaffold was reported [54]. In addition, the samples demonstrated improved tensile mechanical properties, suggesting the presence of more ECMs within the constructs. Gould *et al.* demonstrated the effect of controlled anisotropic strain on valvular interstitial fibroblast cells in 3-D engineered tissues [55]. Increasing the anisotropy of the biaxial strain resulted in increased cellular orientation and collagen fiber alignment along the direction of strain and cell orientation was found to precede fiber reorganization. Collectively, these results suggest that strain anisotropy is an independent regulator of fibroblast cell phenotype, turnover, and matrix reorganization. In a similar study, a tubular scaffold was implanted into the peritoneal cavity of sheep model and was cyclically stretched to generate autologous arterial grafts with increased mechanical strength [56]. Collagen organization in the circumferential direction increased the expression of vimentin and F-actin. Additionally, increased mechanical failure strength and strain were observed when the tubular scaffolds were implanted and pulsated *in vivo* for 10 days.

The stimulation of differentiation of MSCs into cardiac cells requires mimicking of myocardium structure and biomechanics. In a related study, cell alignment was attained by statically stretching tissue constructs 25% during culture, along with induction of the MSCs to differentiate into a cardiac lineage [57]. Increasing the strain to 75% increased the degree of 3-D cell alignment and expression of cardiac markers (GATA4, Nkx2.5, and MEF2C). It was also reported that the differentiated cells developed calcium channels, which are required for electrophysiological properties of the cardiac cells.

In hard tissue engineering applications, cyclic stretching was used to induce bone or cartilage formation. In one such study, cell-seeded collagen type I constructs were mechanically stretched by a daily cyclic uniaxial strain application [58]. Stretching the matrices to a level observed in healing bone increased cell proliferation and slightly elevated the expression of nearly all osteogenic genes. Cyclic compression of porous poly(L-lactic acid) scaffolds seeded with rat bone cells and cultured for up to 3 weeks under continuous perfusion led to detectable mineralized nodules [59]. The long-term application of cyclic hydrostatic pressure could be used to improve the functional properties of engineered cartilaginous tissues using bone marrow-derived MSCs. For instance, the long-term application of 10 MPa of cyclic hydrostatic pressure for 1 h/day for 5 days a week enhanced collagen and glycosaminoglycan (GAG) accumulation in agarose hydrogels loaded with MSCs [60]. Electrospun poly(L-lactide-co-D,L-lactide) scaffolds possessing a wave pattern similar to collagen organization were seeded with bovine fibroblasts and mechanically stimulated under dynamic uniaxial tension with the goal of constructing a ligament [61]. A 10% strain increased only the collagen synthesis, while a 20% strain increased both the collagen and sulfated proteoglycan synthesis. Additionally, these fibroblasts formed bundles that resemble fascicles, a characteristic hierarchical feature of the native ligament.

4.2.4.4 Hybrid Mechanical Systems In the tissues, cells are rarely exposed to a single type of mechanical stimulus, and therefore, a bioreactor must ideally simulate simultaneous exposure to a variety of biomechanical forces including hydrostatic, compressive, stretching, and shear. In order to mimic this complex array of forces and apply a variety of mechanical loads to the cells residing in scaffolds, modular and multifunctional bioreactor systems have been proposed. For instance, a pressurized chamber system for sustained and dynamic application of hydrostatic pressure was designed for shear- and strain-sensitive vessel and cardiac tissues. This pressurized chamber had dynamic airflow and extension of an elastomeric membrane [62]. Cyclic stretching of elastic tubular constructs subjected to pressure by culture medium was used to develop a tissue-engineered heart valve (TEHV), which possessed improved tensile and compositional properties [63]. The resultant TEHV possessed the tensile stiffness and stiffness anisotropy of leaflets of sheep pulmonary valves and could withstand cyclic pulmonary pressures with expansion similar to that of a sheep pulmonary artery. In a similar approach, pulsatile circulation of the medium for 48 hours stimulated EC proliferation on decellularized heart valve scaffolds [64]. As a result of rotation and shear stress, a monolayer of cells covered the inner surface of the valve, which expressed von Willebrand factor (vWF), indicating their endothelial origin.

A multiple force generating bioreactor was used in treating BMSCs seeded on flexible silicone substrates. They were cultured under two mechanical stimuli physiologically relevant to heart valves: cyclic flexure and fluid shear stress [65]. Cyclic flexure and laminar flow synergistically accelerated BMSC-mediated tissue formation with expression of endothelial-associated markers, increased collagen content by 75%, and achieved an effective stiffness (E) of 948 ± 233 kPa.

Computer-controlled bioreactors capable of applying axial compressive and shear deformations, individually or simultaneously, at various regimes of strain and frequency were tested in cartilage tissue engineering [66]. The viability, proliferation, and fibro-cartilaginous differentiation of the hBMSC cultured on the polyurethane-based meniscal scaffolds were investigated during the perfusion and mechanical stimulation process [67]. It was indicated that the perfusion and on–off cyclic compressions maintained viability, promoted proliferation of hBMSC, increased the equilibrium modulus of the tissue-engineered construct, and increased type I procollagen production 1.85- and 3.02-fold, respectively.

Human dermal fibroblasts and human urothelial cells without exogenous scaffolding were subjected to dynamic flow and hydrostatic pressure for up to 2 weeks with the aim of *in vitro* terminal urothelium differentiation for a human genitourinary substitute [68]. Dynamic conditions showed well-established stratified urothelium and basement membrane formation, whereas no stratification was observed in the static culture. Mechanical stimuli induced expression of the major uroplakin transcripts, but expression was low or undetectable in static culture. In addition, permeation studies showed that mechanical stimuli significantly improved the barrier function compared with static conditions and were comparable with native urothelium.

4.3 CONCLUSION

As can be seen from the examples presented earlier, engineering tissues is a complex task and researchers have come up with a variety of approaches for the different kinds of tissues to regenerate. The absence of a perfect tissue-engineered product is an indicator that there is need for more extensive research, but there is hope as the improvements in the techniques, as well as our knowledge, is improved with every new research result.

Acknowledgments

E.T.B. gratefully acknowledges the The Scientific and Technological Research Council of Turkey (TÜBİTAK) for the Post Doctoral 2232 BIDEP Fellowship. T.D. and E.A. also gratefully acknowledge the TÜBİTAK Fellowship 2205 BIDEP. M.E. gratefully acknowledges the State Planning Organization (SPO) support from the project DPT 2011 K 120350. We also acknowledge the grant from SPO for the establishment of BIOMATEN, the Center of Excellence in Biomaterials and Tissue Engineering.

REFERENCES

[1] Rosado, L.A.R., Horn, T.A., McGrath, S.C., Cotter, R.J., Yang, J.T. *J. Cell Sci.* **2011**, *124*(3), 483–492.

[2] Maniotis, A.J., Chen, C.S., Ingber, D.E. *Proc. Natl. Acad. Sci. U.S.A.* **1997**, *94*(3), 849–854.

[3] Deakin, N.O., Turner, C.E. *J. Cell Sci.* **2008**, *121*, 2435–2444.

[4] Liedert, A., Kaspar, D., Blakytny, R., Claes, L., Ignatius, A. *Biochem. Biophys. Res. Commun.* **2006**, *349*, 1.
[5] Schwartz, M.A., DeSimone, D.W. *Curr. Opin. Cell Biol.* **2008**, *20*, 551.
[6] Katsumi, A., Orr, A.W., Tzima, E., Schwartz, M.A. *J. Biol. Chem.* **2004**, *279*, 12001.
[7] Yamato, M., Okano, T. *Mater. Today* **2004**, *7*(5), 42–47.
[8] Southall, N., Dill, K.A., Haymet, A.D.J. *J. Phys. Chem. B* **2002**, *106*, 521–533.
[9] Klouda, L., Mikos, A.G. *Eur. J. Pharm. Biopharm.* **2008**, *68*(1), 34–45.
[10] Duarte, A.R.C., Mano, J.F., Reis, R.L. *Acta Biomater.* **2011**, *7*, 526–529.
[11] Kobayashi, J., Okano, T. *Sci. Technol. Adv. Mater.* **2010**, *11*, 1–12.
[12] Tana, R., Sheb, Z., Wangb, N., Fanga, Z., Liua, Y., Fenga, Q. *Carbohydr. Polym.* **2012**, *87*, 1515–1521.
[13] Guyton, A.C., Hall, J.E. in *Textbook of Medical Physiology* (Eds: A.C. Guyton, J.E. Hall), Elsevier Saunders, Philadelphia: 2006.
[14] Long, Y.Z., Duvail, J.L., Li, M.M., Gu, C., Liu, Z., Ringer, S.P. *Nanoscale Res. Lett.* **2010**, *5*, 237–242.
[15] Sulong, A.B., Muhamad, N., Sahari, J., Ramli, R., Deros, B.M., Park, J. *Eur. J. Sci. Res.* **2009**, *29*(1), 13–21.
[16] Ghasemi-Mobarakehl, L., Prabhakaran, P.M., Morsed, M., Nasr-Esfahani, M.H., Baharvand, H., Kiani, S., Al-Deyab, S.S., Ramakrishna, S. *J. Tissue Eng. Regen. Med.* **2011**, *5*, 17–35.
[17] Flores, A.J., Lavernia, C.J., Owens, P.W. *Am. J. Orthop.* **2000**, *3*, 167–173.
[18] Strauch, B. *Hand Clin.* **2000**, *16*(1), 123–130.
[19] Bechara, S., Wadman, L., Popat, K.C. *Acta Biomater.* **2011**, *7*, 2892–2901.
[20] Schmidt, C.E., Shastri, V.R., Vacanti, J.P., Langer, R. *Proc. Natl. Acad. Sci. U.S.A.* **1997**, *94*, 8948–8953.
[21] Kotwal, A., Schmidt, C.E. *Biomaterials* **2001**, *22*, 1055–1064.
[22] WHO Factsheet No:310, (June 2011), http://www.who.int/mediacentre/factsheets/fs310/en/index.html (accessed November 19, 2012).
[23] Burns, K.D., Kumar, V. in *Robbin's Basic Pathology* (Eds: V. Kumar, R.S. Cotran, S.L. Robbins), Saunders, Philadelphia: 2003.
[24] Barash, Y., Dvir, T., Tandeitnik, P., Ruvinov, E., Guterman, H., Cohen, S. *Tissue Eng. C.* **2012**, *16*(6), 1417–1426.
[25] Pedrotty, D.M., Koh, J., Davis, H.B., Taylor, D.A., Wolf, P., Niklason, L.E. *Am. J. Physiol. Heart Circ. Physiol.* **2005**, *288*, 1620–1626.
[26] Tandon, N., Marsanol, A., Maidhof, R., Wan, L., Park, H., Vujnak-Novakovic, G. *J. Tissue Eng. Regen. Med.* **2011**, *5*, 115–125.
[27] Supronowicz, P.R., Ajayan, P.M., Ulmann, K.R., Arulanandam, B.P., Metzger, W., Bizios, R. *J. Biomed. Mater. Res. A.* **2002**, *59*(3), 499–506.
[28] Franklin, T.A. Ferrofluid flow phenomena. MSc. Dissertation, Massachusetts Institute of Technology, Massachusetts, USA, 2003.
[29] Bock, N., Riminucci, A., Dionigi, C., Russo, A., Tampieri, A., Landi, E., Goranov, V.A., Marcacci, M., Dediu, V. *Acta Biomater.* **2010**, *6*, 786–796.
[30] Tampieri, A., D'Alessandro, T., Sandri, M., Sprio, S., Landi, E., Bertinetti, L., Panseri, S., Pepponi, G., Goettlicher, J., Banobre-Lopez, M., Rivas, J. *Acta Biomater.* **2012**, *8*, 843–851.
[31] Rubin, J., Rubin, C., Jacobs, C.R. *Gene* **2006**, *367*, 1.
[32] Spiteri, C.G., Young, E.W.K., Simmons, C.A., Kandel, R.A., Pilliar, R.M. *Biomaterials* **2008**, *29*, 2477.
[33] Cao, L., Wu, A., Truskey, G.A. *J. Biomech.* **2011**, *44*, 2150.
[34] Yeatts, A.B., Choquette, D.T., Fisher, J.P. *Biochim. Biophys. Acta* **2012**, *1830*, 2470.

[35] Glossop, J.R., Cartmell, S.H. *Gene Expr. Patterns* **2009**, *9*, 381.
[36] Yeatts, A.B., Fisher, J.P. *Bone* **2011**, *48*, 171.
[37] O'Cearbhaill, E.D., Punchard, M.A., Murphy, M., Barry, F.P., McHugh, P.E., Barron, V. *Biomaterials* **2008**, *29*, 1610.
[38] Abilez, O., Benharash, P., Mehrotra, M., Miyamoto, E., Gale, A., Picquet, J., Xu, C., Zarins, C. *J. Surg. Res.* **2006**, *132*, 170.
[39] Thébaud, N.B., Bareille, R., Daculsi, R., Bourget, Ch., Rémy, M., Kerdjoudj, H., Menu, P., Bordenave, L. *Acta Biomater.* **2010**, *6*, 1437.
[40] Dardik, A., Chen, L., Frattini, J., Asada, H., Aziz, F., Kudo, F.A., Sumpio, B.E. *J. Vasc. Surg.* **2005**, *41*, 869.
[41] Kang, T.Y., Hong, J.M., Kim, B.J., Cha, H.J., Cho, D.W. *Acta Biomater.* **2013**, *9*, 4716.
[42] Asnaghi, M.A., Jungebluth, P., Raimondi, M.T., Dickinson, S.C., Rees, L.E.N., Go, T., Cogan, T.A., Dodson, A., Parnigotto, P.P., Hollander, A.P., Birchall, M.A., Conconi, M.T., Macchiarini, P., Mantero, S. *Biomaterials* **2009**, *30*, 5260.
[43] Santoro, R., Olivares, A.L., Brans, G., Wirz, D., Longinotti, C., Lacroix, D., Martin, I., Wendt, D. *Biomaterials* **2010**, *31*, 8946.
[44] Bonzani, I.C., Campbell, J.J., Knight, M.M., Williams, A., Lee, D.A., Bader, D.L., Stevens, M.M. *J. Mech. Behav. Biomed. Mater.* **2012**, *11*, 72.
[45] Sittichockechaiwut, A., Scutt, A.M., Ryan, A.J., Bonewald, L.F., Reilly, G.C. *Bone* **2009**, *44*, 822.
[46] Waldman, S.D., Couto, D.C., Grynpas, M.D., Pilliar, R.M., Kandel, R.A. *OsteoArthr. Cartilage* **2006**, *14*, 323.
[47] Nebelung, S., Gavenis, K., Lüring, C., Zhou, B., Mueller-Rath, R., Stoffel, M., Tingart, M., Rath, B. *Ann. Anat.* **2012**, *194*, 351–358.
[48] Steinmetz, N.J., Bryant, S.J. *Acta Biomater.* **2011**, *7*, 3829.
[49] Checa, S., Prendergast, P.J. *J. Biomech.* **2010**, *43*, 961.
[50] Grad, S., Loparic, M., Peter, R., Stolz, M., Aebi, U., Alini, M. *OsteoArthr. Cartilage* **2012**, *20*, 288.
[51] Kock, L.M., Ravetto, A., van Donkelaar, C.C., Foolen, J., Emans, P.J., Ito, K. *OsteoArthr. Cartilage* **2010**, *18*, 1528.
[52] Chen, Y.J., Zhang, M., Wang, J.J. *Arch. Oral Biol.* **2007**, *52*, 173.
[53] Ballyns, J.J., Bonassar, L.J. *J. Biomech.* **2011**, *44*, 509.
[54] Sharifpoor, S., Simmons, C.A., Labow, R.S., Paul Santerre, J. *Biomaterials* **2011**, *32*, 4816.
[55] Gould, R.A., Chin, K., Santisakultarm, T.P., Dropkin, A., Richards, J.M., Schaffer, C.B., Butcher, J.T. *Acta Biomater.* **2012**, *8*, 1710.
[56] Stickler, P., De Visscher, G., Mesure, L., Famaey, N., Martin, D., Campbell, J.H., Van Oosterwyck, H., Meuris, B., Flameng, W. *Acta Biomater.* **2010**, *6*, 2448.
[57] Guan, J., Wang, F., Li, Z., Chen, J., Guo, X., Liao, J., Moldovan, N.I. *Biomaterials* **2011**, *32*, 5568.
[58] Ignatius, A., Blessing, H., Liedert, A., Schmidt, C., Neidlinger-Wilke, C., Kaspar, D., Friemert, B., Claes, L. *Biomaterials* **2005**, *26*, 311.
[59] Baas, E., Kuiper, J.H., Yang, Y., Wood, M.A., El Haj, A.J. *J. Biomech.* **2010**, *43*, 733.
[60] Meyer, E.G., Buckley, C.T., Steward, A.J., Kelly, D.J. *J. Mech.. Behav. Biomed. Mater.* **2011**, *4*, 1257.
[61] Surrao, D.C., Fan, J.C.Y., Waldman, S.D., Amsden, B.G. *Acta Biomater.* **2012**, *8*, 3704.
[62] Hasel, C., Düurr, S., Bruderlein, S., Melzner, I., Möller, P. *J. Biomech.* **2002**, *35*, 579.
[63] Syedain, Z.H., Tranquillo, R.T. *Biomaterials* **2009**, *30*, 4078.
[64] Lichtenberg, A., Tudorache, I., Cebotari, S., Ringes-Lichtenberg, S., Sturz, G., Hoeffler, K., Hurscheler, C., Brandes, G., Hilfiker, A., Haverich, A. *Biomaterials* **2006**, *27*, 4221.

[65] Engelmayr, G.C., Sales, V.L., Mayer, J.E., Sacks, M.S. *Biomaterials* **2006**, *27*, 6083.
[66] Yusoff, N., Abu Osman, N.A., Pingguan-Murphy, B. *Med. Eng. Phys.* **2011**, *33*, 782.
[67] Liu, C., Abedian, R., Meister, R., Haasper, C., Hurschler, C., Krettek, C., von Lewinski, G., Jagodzinski, M. *Biomaterials* **2012**, *33*, 1052.
[68] Cattan, V., Bernard, G., Rousseau, A., Bouhout, S., Chabaud, S., Auger, F.A., Bolduc, S. *Eur. Urol.* **2011**, *60*, 1291.

CHAPTER 5

CELLULAR RESPONSES TO BIO-INSPIRED ENGINEERED TOPOGRAPHY

CHELSEA M. KIRSCHNER

Sharklet Technologies, Inc., Aurora, CO, USA

JAMES F. SCHUMACHER

J. Crayton Pruitt Family Department of Biomedical Engineering, University of Florida, Gainesville, FL, USA

ANTHONY B. BRENNAN

Department of Materials Science and Engineering, J. Crayton Pruitt Family Department of Biomedical Engineering, University of Florida, Gainesville, FL, USA; Sharklet Technologies, Inc., Aurora, CO

5.1 INTRODUCTION

Biological tissues are composite materials that exhibit hierarchical structure on size scales ranging from nanometers to centimeters. A closer look at these structures reveals cells, which are on the order of tens to hundreds of micrometers, organized in distinct structures within an extracellular matrix (ECM). The ECM is a dynamic microenvironment consisting of large molecules synthesized by cells that provides biochemical cues through both presentation of adhesive ligands and sequestration, storage, and presentation of soluble regulatory molecules [1]. It also imparts biophysical cues via cellular interactions with protein structures. Cell-secreted proteins such as collagens and elastins provide physical support for cell anchorage, while proteins such as fibronectin and laminin present biochemical adhesive signals. While both biochemical and biophysical cues guide tissue formation, the biophysical cues sensed by cells *in vivo*, which range from microscale to nanoscale, are critical for guiding cell shape and orientation to form specialized structures that are adapted to the particular function of the tissue [2]. This chapter begins with a brief history of

Bio-inspired Materials for Biomedical Engineering, First Edition. Edited by Anthony B. Brennan and Chelsea M. Kirschner.

the foundational and pioneering work in investigating cellular responses to biophysical cues and then presents an overview of natural surface structures and their bio-inspired counterparts, including the fabrication and application of these surfaces. Advances in the design of biomaterials for cell culture have led to new ways for scientists and bioengineers to probe and answer fundamental questions about how cells receive information from, and respond to, their natural microenvironments.

5.1.1 Historical Introduction to Cellular Responses to Physical Cues

The first evidence that physiochemical cues such as substratum topography influence cell morphology was reported over a century ago. Harrison, in 1914, first recognized this phenomenon when he observed that fibroblasts from the embryonic nervous tissue of frogs elongated while cultured on spider silk [3]. Weiss coined the term *contact guidance* in 1945 to describe how cells orient, elongate, and migrate in response to structures in their microenvironment [4]. Through hundreds of experiments, Weiss and colleagues showed that neuronal cells adhered, elongated, and migrated along scratches in the surface of mica and glass fiber structures ranging from 10 to 30 μm in diameter [4,5]. Curtis and Varde subsequently investigated contact guidance, contact inhibition, and cell spreading of chick heart fibroblasts on silica substrates with varying surface topographies, and concluded that cell behavior and morphology were controlled by both cell density and substratum topography [6]. The pioneering investigations of cellular responses to physical cues primarily concentrated on grooved/ridged topographies or fibers due to both the marked cellular responses observed and the ease of fabrication. Microfabrication techniques developed in the electronics industry, such as photolithography and electron beam lithography, began to gain popularity in biomaterials research to produce cell culture substrates with micro- and nanoscaled topographies of various shapes and spatial arrangements approximately 20 years ago [7]. The application of microfabrication techniques in biomedical engineering has resulted in tremendous progress toward elucidating cellular responses to both chemical and topographic patterning, as reviewed previously [8,9]. This chapter will specifically examine the bio-inspired, engineered surface topographies that have emerged as a result.

5.1.2 Physical Cues in Nature

In vivo cells receive biochemical and biophysical cues through interactions with the ECM, soluble factors, and neighboring cells (Figure 5.1A). The ECM consists of a variety of components with specific properties that combine synergistically to form tissues. Collagen provides structural support, elastin and proteoglycans impart elasticity to the matrix, and structural glycoproteins adhere these components together. These components collectively provide mechanical support for cell anchorage, determine the orientation and shape of the cell, control cell growth, promote cell differentiation and sequester, store, and present soluble signaling molecules [10]. The ECM also facilities long-range force transfer between cells to coordinate cellular and tissue responses such as wound healing over large areas [10].

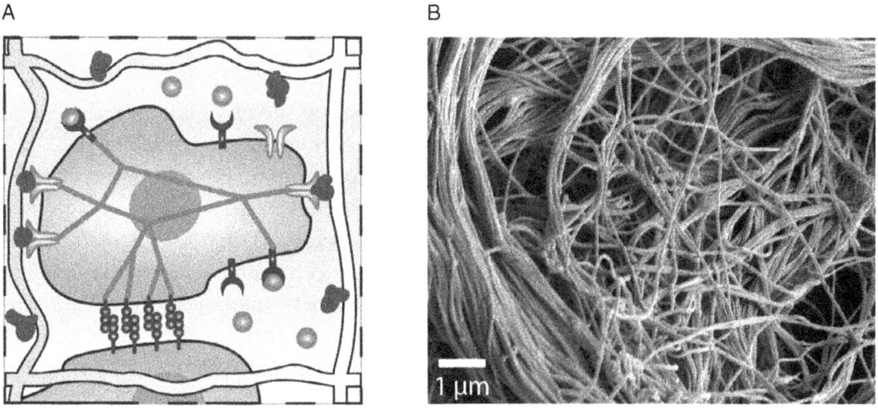

FIGURE 5.1 (A) *In vivo* cells receive biochemical and biophysical cues through interactions with the ECM, soluble factors, and neighboring cells. Integrins (pink) in the cell membrane bind to ECM proteins (purple), soluble growth factors (green) bind to surface receptors (blue), and cell–cell junctions are formed with adjacent cells (purple). These signals influence cell function through signaling pathways that involve the cytoskeletal (orange fibers) arrangement and focal adhesion placement. (B) A scanning electron micrograph reveals the complex architecture of the basement membrane of a porcine urinary bladder. The fibrous ECM guides cell shape and function within the tissue. (Courtesy of Christopher Carruthers and Denver Faulk of the Badylak Research Group.) *(See insert for color representation of the figure.)*

Collagen is the most abundant protein found in the human ECM. It is secreted by cells, and after modification by proteolytic enzymes outside the cell, collagen molecules can self-assemble into fibrils with diameters ranging from 10 to 300 nm [11]. These fibrils are subsequently stabilized by disulfide and other covalent cross-links that form between collagen molecules. Fibrils can then aggregate into collagen fibers that reach diameters of several microns [11]. These fibers are an integral component of the basement membrane, which is best described as the basal laminae, thin flexible mats of specialized ECM (40- to 120-nm thick) that underlie or surround many cell types, and the layer of collagen fibrils connecting it to the underlying connective tissue [11]. The basement membranes of different tissues perform specialized functions so each has its own complex nano- to microscale architectures ranging from 5 to 200 nm [9], which include structures such as pits, pores, ridges, and fibers [8]. Both the composition and the structure of ECM are highly correlated with cell phenotype and tissue or organ function. Badylak *et al.* illustrated the relationship between structure and function by characterizing the ultrastructure and molecular composition of porcine urinary bladder, small intestine, and liver [12]. The results demonstrate that each surface of every ECM scaffold exhibits unique structural and compositional characteristics. Here, Figure 5.1B depicts the complex fibrous architecture of the basement membrane of a porcine urinary bladder. Bladders were collected from animals ($\sim$120 kg) at a local abattoir (Thoma's Meat Market, Saxonburg, PA), frozen ($>$16 hours at -80°C), and thawed completely before use. The basement membrane and underlying lamina propria were isolated and harvested from the bladders as previously described [12,13]. The tissue was then placed in 0.02% trypsin/0.05% ethylene glycol tetraacetic acid (EGTA) solution for 2 hours at 37°C with physical agitation to detach cells from the ECM.

The introduction of micro- and nanofabrication techniques into the field of biomedical engineering has allowed researchers to create highly ordered, precisely engineered topographic cues in biomaterials. Nanotopographies are structures on the same size scale as large proteins or small assemblies of molecules, while microstructures represent large aggregates of smaller structures such as collagen fibers. Throughout this chapter, the prefix micro will refer to structures >1 μm while nano will describe signals <100 nm. The ability to replicate natural topographic cues has led to pioneering work that has clearly demonstrated that, by altering substratum topography, it is possible to change cell morphology. The presentation of topographic cues has been exploited in recent years to regulate cell function. Topographies have been shown to influence cell adhesion [14], migration [15], proliferation [16,17], protein expression [18], gene regulation [16], and differentiation [19,20]. Cell shape, cell spreading, and cytoskeletal tension have been demonstrated to have a strong influence on the lineage commitment of human mesenchymal stem cells (hMSCs) [21]. Even though these patterning techniques and resulting cellular responses are now well established, very little is known about the mechanisms underlying cell sensing within the niches these cells naturally inhabit. The goal to advance our fundamental knowledge of how cells receive and process signals from their surroundings motivates researchers to recapitulate natural cellular microenvironments. The first step in this process is the design of the bio-inspired engineered topography.

5.2 DEFINITION OF ENGINEERED TOPOGRAPHY

The terms used to describe a topographically modified surface are numerous. These terms include: roughness, topography, pattern, architecture, structure, and texture. They are often used interchangeably and are typically combined with dimensional prefixes such as *nano* and *micro* or the word *surface*. Some of these combinations are surface topography, nanoroughness, microarchitecture, and microscale surface structure. Although the prefixes define and narrow the size scale of the surface, it does not provide a distinction between the root words.

In 2006, a new nomenclature was introduced called *engineered topography* [22]. This term was developed to be distinguishable in both definition and design from the topographies, patterns, and structures presented in previous research and patent art relating to the effect of surface characteristics on biological responses. This is not to be confused with the manufacturing term *engineered surface topography*, describing machined surfaces typically created by sandblasting or grinding [23].

Engineered topography, as implied by the prefix *engineered*, is a designed or predefined topography. The desired shape, arrangement, and dimensions of the features that comprise the topographical surface are prescribed before a fabrication technique is even considered. An engineered topography is not the consequence of a surface modification or fabrication technique. Surface features are selected based on the consideration of a specific biological response, such as the inhibition of bacterial settlement, the increase in attachment of proteins, or the alignment of cells. When a particular cell, organism, or biological system is selected, this topography becomes a *predefined* engineered topography, of which the geometry, size, and arrangement of the features are tailored to elicit a specific response from the living entity targeted.

In the same year, 2006, a related term was introduced by Charest *et al.* called *mechanical topography* and is defined as "a pattern of mechanical structures with regular and specifically designed size, shape, and periodicity" [24]. This definition was introduced by the authors to distinguish these distinct surfaces from *mechanical roughness* that was defined as "a group of mechanical features that exhibits randomness and polydispersity in terms of size, shape, and periodicity" [24]. Thus, for a succinct definition, engineered topography can be defined as a mechanical topography tailored for a specific biological response.

An engineered topography that results from the examination of a surface in nature that influences the design is termed a bio-inspired engineered topography. It is not meant to be an exact structural and dimensional copy of the natural surface, but a design that can be tailored and adjusted on the nano-, micro-, or macroscale in order to elicit the desired biological response.

5.3 SURFACE FABRICATION TECHNIQUES

5.3.1 Fabrication of Engineered Topography

Engineered topographies are created on the surface of material substrates, typically soft plastics, by replication of microfabricated silicon molds. Silicon molds are created using lithographic and dry etching techniques. First, a two-dimensional (2-D) pattern is designed, digitized, and printed to a photomask using electron beam (e-beam) lithography (Figure 5.2). Photolithographic techniques are then used to transfer the pattern present on the photomask to a photoresist-coated silicon wafer (Figure 5.3). Next, the patterned silicon wafer is etched to a desired depth using deep reactive ion etching (Figure 5.4). As a final step, the remaining photoresist is removed from the wafer by an oxygen plasma etch. The processed silicon wafer contains three-dimensional (3-D) features etched within the wafer surface in the designed pattern. This wafer serves as a mold for topographical replication. The replication of topographical features to substrate surfaces is accomplished via an iterative casting method. This section describes the typical materials, equipment, and methods to complete this fabrication process.

5.3.1.1 Pattern Design and Photomask Generation A pattern consists of 2-D geometric shapes arrayed over a specified area. For standard photolithography, patterns are not to contain any dimensions smaller than 1 μm. This dimensional limitation is defined based on the commonly known resolution limit of ultraviolet (UV) photolithography (~0.5 μm under ideal conditions). The pattern must be able

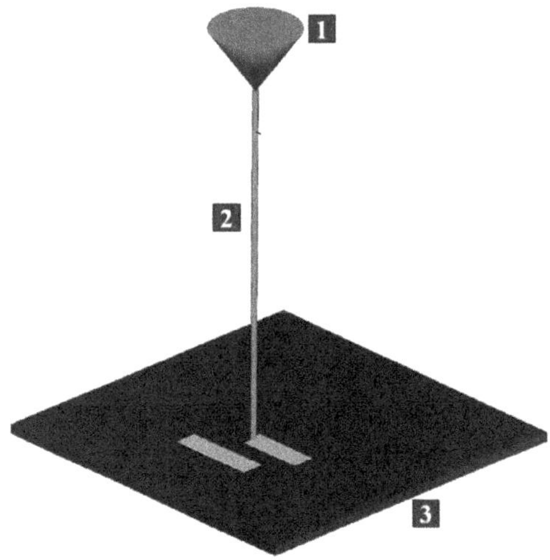

FIGURE 5.2 Schematic of the process of e-beam lithography used to print a photomask from a CAD drawing of a designed pattern. (1) Electron beam source. (2) Electron beam used to print pattern to a photomask. (3) Photoresist-coated photomask.

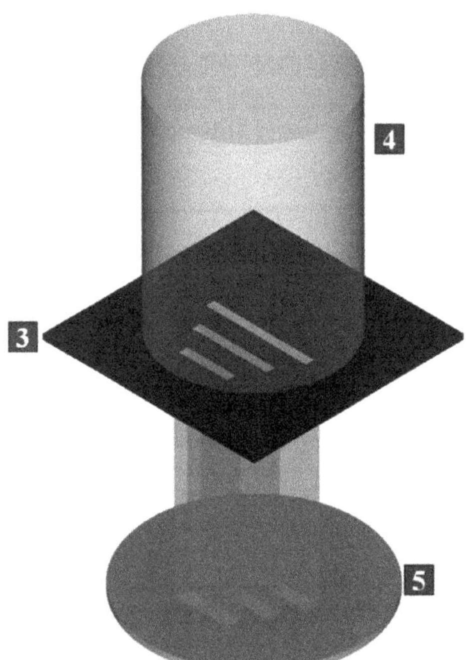

FIGURE 5.3 Schematic of the photolithography process used to transfer the pattern on a photomask to a photoresist-coated silicon wafer. (3) Photomask containing a pattern of clear features on a black background. (4) Ultraviolet light source. (5) Photoresist-coated silicon wafer.

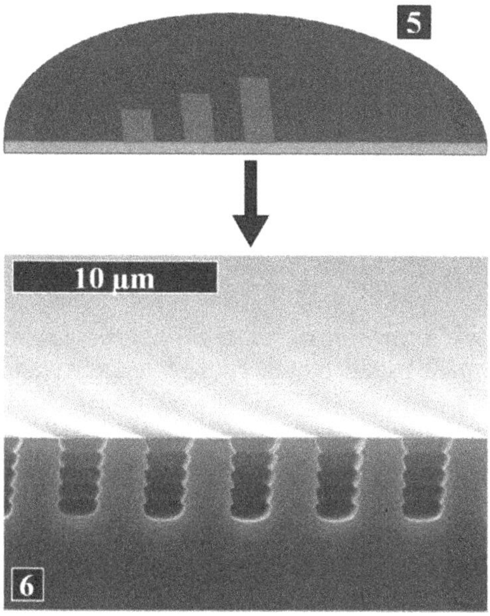

FIGURE 5.4 Etch processing of pattern silicon wafers using deep reactive ion etching. (5) Schematic of a cross-section of a patterned silicon wafer with bare silicon features surrounded by photoresist. Silicon covered in photoresist is not etched during the process. (6) Scanning electron micrograph of the cross-section of an etched silicon wafer showing features within the silicon surface.

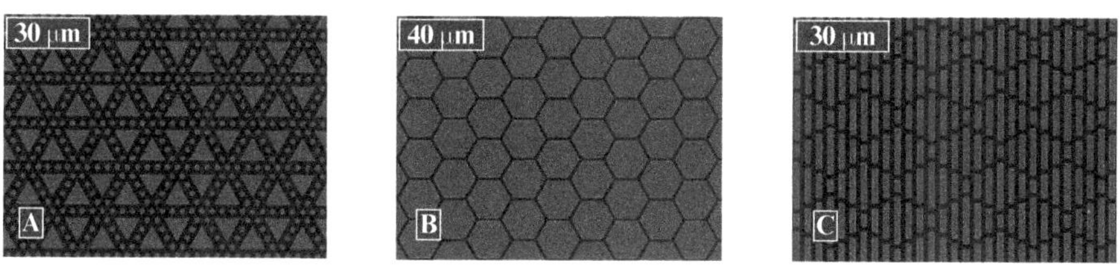

FIGURE 5.5 Designed patterns drawn in AutoCAD®. (A) 10-μm equilateral triangles surrounded by 2-μm (diameter) circles spaced by 2 μm. (B) 20-μm height hexagons spaced by 2 μm. (C) Sharklet AF™ design composed of 2-μm wide ribs of varying lengths including 4, 8, 12, and 16 μm. Ribs are spaced by 2 μm in all directions.

to be drawn and stored in a computer-aided design (CAD) program. This is necessary so that the pattern can be uploaded to computer-controlled lithography equipment and printed on a photomask. Also, patterns must not be too complex or robust such that digitized computer files exceed 1 GB as storage and uploading becomes a problem. Most geometric patterns that follow these guidelines can be successfully fabricated into 3-D topographical features. Some examples of designed patterns are included in Figure 5.5.

Most CAD software packages can be used to generate and digitize designed patterns. AutoCAD® (Autodesk, Inc., San Rafael, CA) is useful for defining the layout of multiple patterns on a single photomask. Once complete, the electronic drawing is transferred to a computer-controlled e-beam lithography system to be directly printed to a photomask. Fabricated photomasks are typically made of quartz (~0.1-in. thick) with a thin layer of chrome (<100 nm) on one side of the quartz surface. The designed pattern is etched within this chrome layer, such that the features of the pattern are clear (no chrome), surrounded by a dark background of chrome (Figure 5.3B).

5.3.1.2 Process of Photolithography The process of photolithography involves the transfer of the pattern present on the fabricated photomask to the surface of a silicon wafer. Prime silicon wafers are placed in an oven at ~150°C for at least 15 minutes to completely dry the wafers. Once cooled, hexamethyldisilazane (HMDS) is vapor deposited on each wafer for 5 minutes. HMDS promotes the adhesion of photoresist to the silicon surface.

Photoresist is coated on silicon wafers by a spin coating process, typically at 4000 RPM for at least 30 seconds, but varies by photoresist type and manufacturer. Photoresist-coated silicon wafers are then placed in an oven at ~90°C for 30 minutes to drive out all the solvent in the photoresist. This produces a completely dry photoresist layer in preparation for UV exposure. This preexposure bake is sometimes referred to as a prebake or softbake. The photoresist manufacturer will specify the exact time and temperature of the preexposure bake.

Patterns contained on the photomask are transferred to softbaked, photoresist-coated silicon wafer using a mask aligner. The photomask was attached to the mask holder, placed above the loading stage, and aligned over the center of the wafer chuck. The silicon wafer was placed on the wafer chuck and loaded into the mask aligner beneath the mask holder containing the photomask. The wafer chuck is raised so that the silicon wafer makes direct contact with the photomask. The photoresist is then exposed through the photomask using UV light (e.g., 405 nm wavelength) for time periods less than 30 seconds as dictated by resist thickness and type. The photoresist was exposed only in the areas of the pattern present on the photomask.

The exposed silicon wafer is developed by immersion in developer solution to rinse away exposure areas of the photoresist. Exposed wafers were completely immersed and agitated in developer solution for 40 seconds, removed, and immediately immersed and agitated in deionized for 60 seconds. The wafer can then be dried with compressed nitrogen.

As a final step, the developed and dried silicon wafer is placed in an oven at ~120°C for 30 minutes. This process is typically referred to as a hard bake. The hard bake enhances the adhesion and increases the etch resistance of the remaining photoresist to the silicon wafer.

5.3.1.3 Etching of Patterned Silicon Wafers Patterned silicon wafers are etched to a desired feature depth using processes such as deep reactive ion etching. Deep reactive ion etching utilizes the *Bosch* process that allows for the creation of

TABLE 5.1 Process Conditions Used to Etch Patterned Silicon Wafers

Process parameters	Passivation	Etching
C_4F_8	85 sccm	0 sccm
SF_6	0 sccm	130 sccm
RF power at stage	0 W	12 W
RF power from coil	600 W	600 W
Cycle time	5.0 seconds	7.0 seconds
Delay time	0.5 seconds	0.5 seconds

RF, radio frequency; sccm, standard cubic centimeter per minute.

TABLE 5.2 Process Conditions Used for Oxygen Plasma Etch

Process parameters	Plasma etch
O_2	45 sccm
RF power at stage	12 W
RF power from coil	800 W
Total time	5 minutes

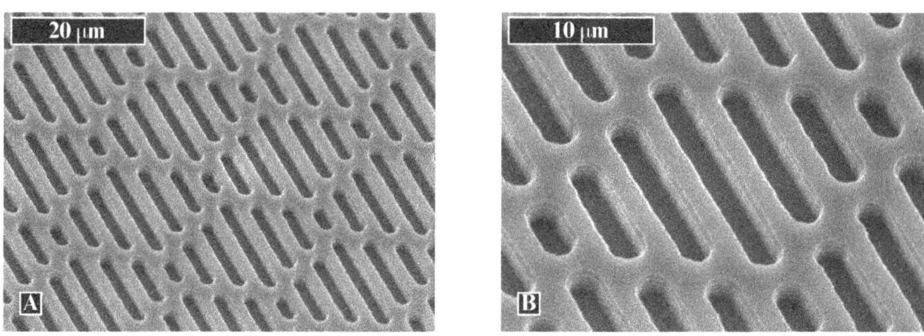

FIGURE 5.6 SEM image of a silicon wafer etched for a total time of 55 seconds using process parameters in Table 5.1.

high-aspect ratio features. A Surface Technology Systems (STS) multiplex reactive ion etcher is a capable system for this type of etching. An example process condition for an isotropic etch is listed in Table 5.1. The etch depth is controlled by the total time of the process. Approximated etch depths using 2-μm resolution patterns for different process times are listed in Table 5.1. Once etched, photoresist is cleaned from the wafer using oxygen plasma etch on the STS system using process conditions outlined in Table 5.2. The completed wafer contains features etched within the surface of the silicon wafer and served as a negative mold for topographical replication in a substrate (Figure 5.6).

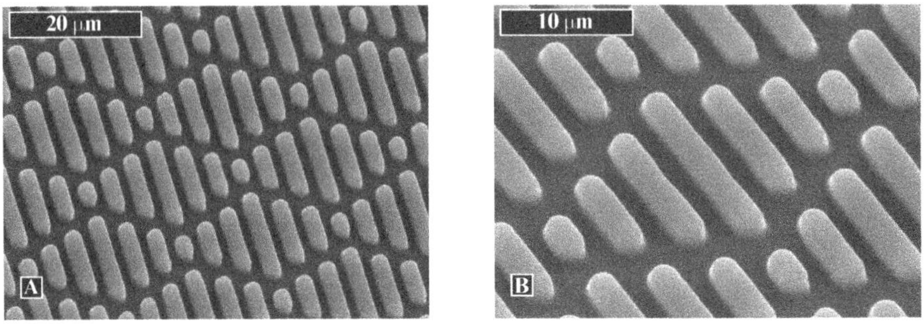

FIGURE 5.7 SEM image of the surface of PDMSe replicated from the etched silicon wafer pictured in Figure 5.6.

5.3.1.4 Replication of Patterned Silicon Wafers HMDS is vapor deposited on the processed silicon wafers to methylate the surfaces in order to prevent adhesion to most soft plastics. For an example, a platinum-catalyzed polydimethylsiloxane elastomer (PDMSe), SILASTIC® T-2 (Dow Corning Corporation, Midland, MI), can be used as the base material for topographical transfer. The elastomer was prepared, as specified by the manufacturer, by mixing 10 parts of the resin and 1 part of the curing agent by weight for 5 minutes. The mixture was degassed under vacuum (95–102 kPa) for 30 minutes, removed, and poured over the processed silicon wafer. After curing for 24 hours, the PDMSe film is carefully removed from the silicon wafer. The resultant topography on the PDMSe surface contain features projecting from the surface at heights respective of the etch depth (Figure 5.7). This casting fabrication process can be repeated to produce multiple copies of engineered topography on the surface of PDMSe; however, pattern fidelity must be monitored closely during this process.

5.3.1.5 Pattern Fidelity Evaluation of Engineered Topography The evaluation of pattern fidelity is critical for any engineered topographical surface replicated from the silicon master. Scanning electron microscopy (SEM) can be used to evaluate short-range pattern fidelity (~40 × 40 μm field of view) and to inspect individual features to define approximate dimensions such as feature width, length, and spacing. Feature height can be visualized and measured with SEM images of cross-sectional samples. Long-range pattern fidelity (~350 × 350 μm field of view) can be assessed using light microscopy.

SEM was useful when new patterns were first being fabricated or process modifications were investigated. Gross defects such as flopped features (Figure 5.8), overexposed features (Figure 5.9), and missing features (Figure 5.10) can be identified quickly and modifications to the process can be made. When a random sampling of SEM images from a particular engineered topography on the substrate surface using controlled process parameters show no evidence of gross defects (Figure 5.11), iterative casting can be used to create an inventory of replicated surfaces. Pattern fidelity (%) is defined as (1 − defective features/total features) × 100 for a given field of view.

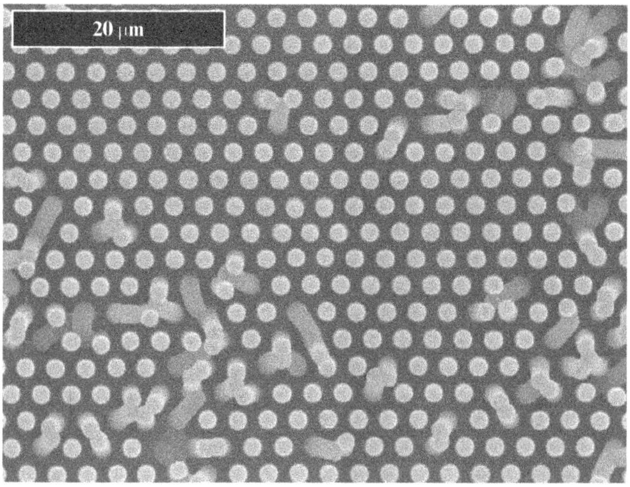

FIGURE 5.8 SEM image of an engineered topography on the surface of PDMSe containing flopped features. Flopped features are a result of too high of an aspect ratio (feature height/feature width).

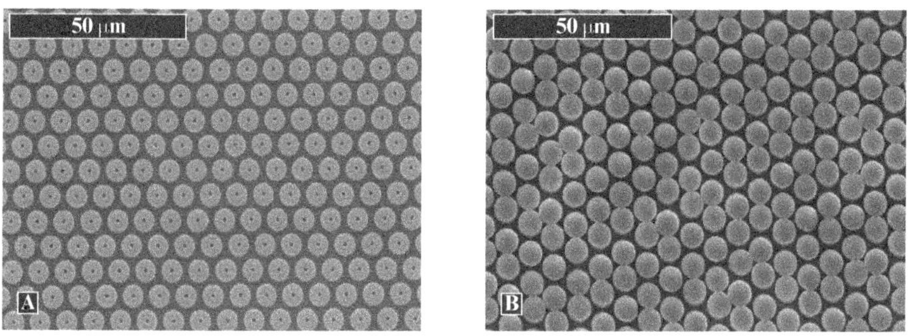

FIGURE 5.9 SEM images of a set of engineered topographies on the surface of PDMSe produced from the same pattern showing the effect of overexposed features. (A) 4-μm diameter pillars with 2-μm diameter hole in the center. (B) Same pattern in A, but overexposed during photolithography.

Light microscopy is used as a final pattern fidelity assessment on engineered topography samples prepared for biological assays. It is suggested that five random areas on each sample be imaged at 400× magnification. Samples are rejected and not tested if any of the five random areas have pattern fidelity value lower than 97% (Figure 5.12). An example of a high-fidelity (>97%) sample of an engineered topography is shown in Figure 5.13.

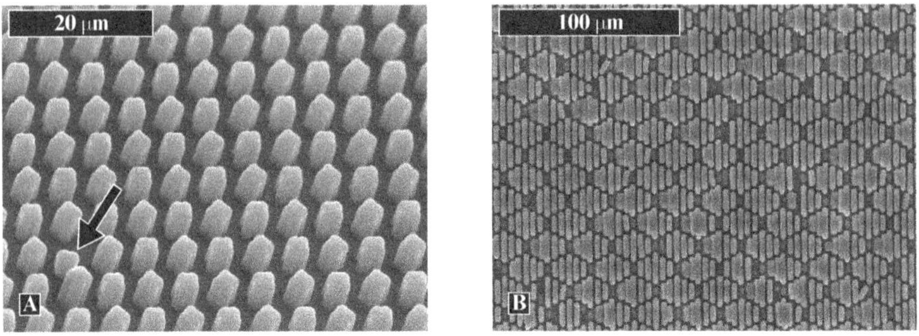

FIGURE 5.10 SEM images of engineered topographies on the surface of PDMSe with missing features. (A) Arrow indicates single missed feature on the surface. The surface was tilted at 35° when imaged. (B) Multiple missing features on the surface indicated by the dark, irregular spots across the viewable area. This is an example of pattern fidelity of <70%.

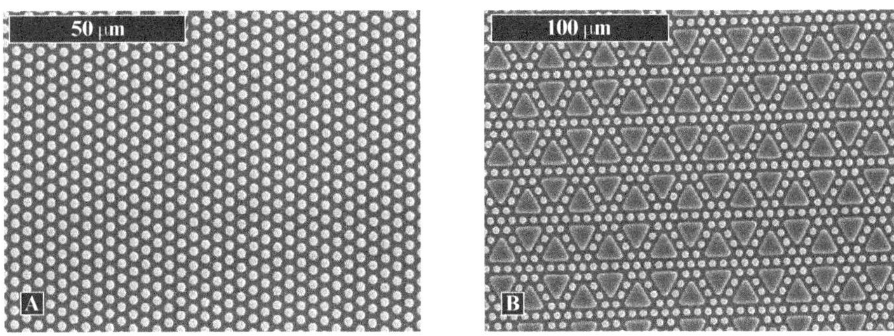

FIGURE 5.11 SEM images of engineered topographies on the surface of PDMSe with no defects across the viewable area. These are representative images of an engineered topography with pattern fidelity of >97% for the relative field of view.

5.4 CELLULAR RESPONSES TO 2-D ENGINEERED TOPOGRAPHIES

Since microfabrication techniques such as photolithography described in the previous sections began to gain popularity in biomaterials research nearly two decades ago, researchers have made strides toward illuminating how cells respond to specific parameters defined by engineered topographies. Broadly, it has been shown that isotropic topographies such as pillars tend to control more collective cell functions such as cell adhesion [25] and proliferation [16,26], while anisotropic topographies, like channels, more noticeably alter cell morphology and cytoskeletal organization [8]. Cell morphology, in turn, is known to influence cytoskeletal organization [9], gene expression [17], protein expression [27], and differentiation [28]. A combinatorial screening approach was recently used to evaluate the effect of 169 distinct pillar-based microtopographies on fibroblast proliferation and focal adhesion

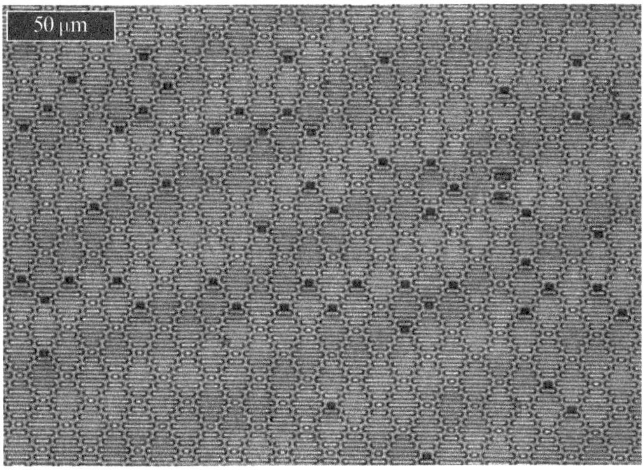

FIGURE 5.12 Light micrograph image of a rejected sample of an engineered topography on the surface of PDMSe taken during the final pattern fidelity assessment before biological testing. The defects appear as dark spots in the image, indicating that those features had flopped over during fabrication and sample preparation. This is an example of an engineered topography of pattern fidelity of <97%.

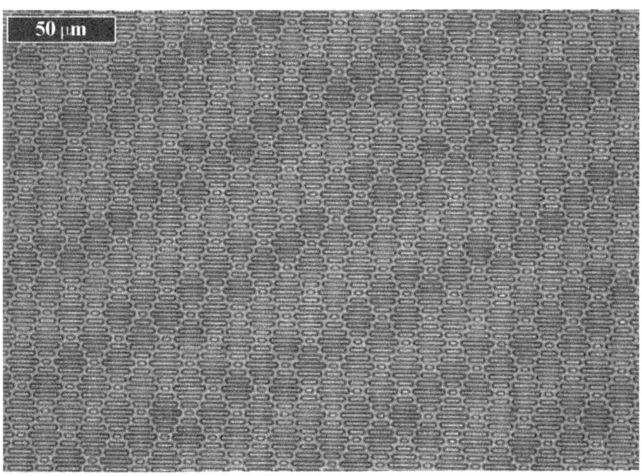

FIGURE 5.13 Light micrograph image of a high-fidelity sample (>99%) of an engineered topography on the surface of PDMSe taken during the final pattern fidelity assessment before biological testing. Unlike in Figure 5.12, no defects are visible within the field of view.

morphology. This study revealed that larger interpillar spacing reduced fibroblast proliferation and increased cellular elongation, while smaller interpillar gaps sizes induced proliferation comparable with cells on smooth surfaces [29].

The majority of studies using bio-inspired, engineered microtopographies have focused on cellular responses to anisotropic, ridged patterns to mimic collagen fibrils (10–300 nm in diameter), which constitute the main structural component of the

ECM (see reviews by Flemming *et al.* [9], Ross *et al.* [30], and Lim *et al.* [8]). For example, Yim *et al.* have used a combination of mathematical modeling and experimentation to determine which physical parameters influence cell elongation. hMSCs were cultured on engineered topographical gratings replicated in PDMSe with varying widths, heights, and rigidities, and the cellular aspect ratio was quantified as a measure of cellular elongation. Over the range of grating aspect ratio (0.035–2) and substrate stiffness (0.18–1.43 MPa) investigated, their work showed that both cell elongation and alignment were enhanced on smaller, stiffer gratings [31]. These results corroborate previous work with porcine vascular endothelial cells (PVECs) that showed there were statistical differences in the PVEC elongation on PDMSe substrates between the elastic modulus extremes (E = 0.3 MPa and E = 2.3 MPa), but not for other elasticity and topography combinations. This observation suggests that an increase in the height of microtopographies over nanotopographies overrides sensitivity to elastic modulus variations and that topography was a greater contributor to cell alignment than substrate mechanics for the range of elastic moduli and microtopography heights (1.5 and 5 μm) evaluated [32].

Cells interact with substrates through focal adhesions, which are mechanosensitive, signaling complexes that have been shown to grow and adapt in response to topographically modified substrates [33], resulting in intracellular tension and cellular anisotropy [19,21]. Although it seems intuitive that a channel pattern would induce the highest levels of cellular anisotropy, this is not the case. Discontinuous microtopographic features allow for focal adhesions to be precisely guided for a higher level of control over the morphology of a cell population.

Engineered topographies with high fidelity have been deliberately designed to modulate surface wettability and, thus, biological adhesion. The Brennan Research Group, in particular, has extensively studied and modeled the optimization of parameters that influence surface energy and in turn regulate biological adhesion to bio-inspired, engineered microtopographies. Systematic studies have related topographic feature geometry and surface chemistry to biological adhesion [34,35] and cellular morphology [35,36] in specific, predictable ways. For example, Brennan and colleagues showed that Sharklet microtopographies, inspired by the skin of fast-moving sharks, created with features protruding from the surface in a diamond-shaped pattern with a height of 1 μm and width and spacing of 2 μm, influence the morphology of two distinct cell types: PVECs [35] and human coronary artery smooth muscle cells (SMCs) [36]. The engineered topographies depicted in Figure 5.14A–E, based on the Sharklet pattern, have been designed to vary the number of unique features repeated in the pattern (n). Therefore, using the current nomenclature, the n-series surfaces are referred to as +1SK2x2_n4, where n is the number of unique features, ranging from 1 to 5. Results indicated that Sharklet patterns influenced cell shape by increasing the cellular aspect ratio—the ratio of the long axis of the cell to the short axis—compared with standard tissue culture polystyrene (TCPS). SMCs aligned and elongated with the engineered microtopographies after 24 hours in culture. Topographies with the greatest number of distinct features (n = 4 and n = 5) in the n-series resulted in the largest elongations and highest orientations (Figure 5.14F). In another study, discontinuous microchannels were shown to increase expression of proteins representative of the contractile phenotype in vascular SMCs

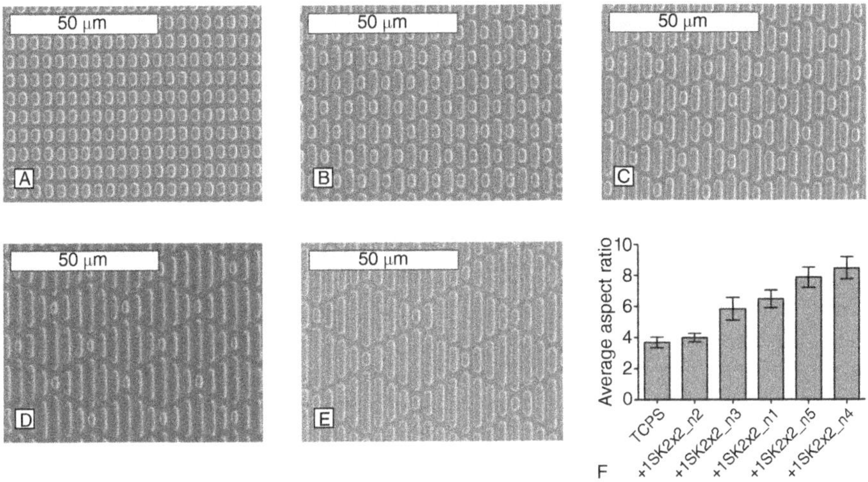

FIGURE 5.14 Sharklet engineered microtopographies varying in the distinct number of features (n) replicated in PDMSe. (A) +1SK2x2_n1, (B) +1SK2x2_n2, (C) +1SK2x2_n3, (D) +1SK2x 2_n4, (E) +1SK2x2_n5. (F) Average aspect ratio for SMCs cultured on PDMSe Sharklet microtopographies after 24 hours. Significant differences in cell morphology were induced by altering feature geometry and arrangement. Error bars, 95% CI.

[37]. Promoting the contractile SMC phenotype is one strategy to minimize intimal hyperplasia, or the excessive proliferation of SMCs that limits the clinical application of small-diameter vascular grafts [38]. Engineered microtopographies may be one solution. These patterns may trigger alignment and elongation, as well as some intracellular signal that triggers the expression of the contractile phenotype [38,39].

Cells have also been shown to sense and react to nanosized signals as small as approximately 10 nm [8,40]. Nanotopographies with dimensions less than 1 μm have been extensively studied using advanced photolithography techniques to create silicon dioxide master molds. In one example, silicon dioxide masters were used as templates for PDMSe stamps that were in turn employed to pattern optically clear polyurethane surfaces through soft lithography. Ridged patterns were replicated to mimic the size scale of the collagen fibrils in the basement membrane of the corneal stroma, that is, 200–2000 nm. Human corneal epithelial cells (HCECs) were cultured on these engineered substrates, and proliferation rates were quantified after 5 days in culture. Topographies with feature sizes below 1 μm significantly decreased proliferation [41]. In a similar study, the same ridge nanotopographies were shown to induce anisotropic HCEC migration, promoting cell movement along the long axes of the ridge topographies and inhibiting migration orthogonal to these features. Feature size impacted migration rates with the highest individual cell migration rates on 1600-nm ridges [42]. Nanogrooved (450-nm wide, 100- or 350-nm deep) substrates also made through soft lithography were replicated in both polystyrene and polyurethane to elucidate how substrate nanostructure and elasticity influence elongation and contraction of cardiomyocytes [43]. The orientation and elongation of cardiomyocytes was regulated by surface topography while contraction of the cells depended on both topography and rigidity of the substrate. Soft substrates with deeper (350 nm) grooves best maintained contractile function [43].

Inspired by the nanosized resorption pits left behind by osteoclasts, Dalby *et al.* have studied the influence of preprinted nanopits replicated in polycaprolactone (PCL) through hot embossing on both primary human osteoblasts [44] and hMSCs [19,28,45]. Through the rational design of engineered nanotopographies, results have shown that nanopits retained hMSC multipotency for 8 weeks in culture when presented in an ordered arrangement and promoted osteogenic differentiation when the same features were presented in a disordered array [19]. Nanopits with larger diameters, 30 and 40 μm, demonstrated higher osteogenic capacity than smaller diameter pits [44].

Collectively, this seminal body of work has used engineered microtopographies to gain a nascent understanding of how the biophysical properties of the cellular microenvironment regulate cell function; however, the static, 2-D cues employed in the majority of these experiments fail to reproduce the dynamic, hierarchical nature of the ECM.

5.5 CELLULAR RESPONSES TO DYNAMIC, ENGINEERED 2-D TOPOGRAPHIES

Recently, in order to better recapitulate the natural cellular microenvironment, researchers have designed dynamically responsive biomaterials [46] to elicit changes in cell morphology in real time. User-controlled mechanisms that trigger dynamic changes in 2-D surface topographies afford researchers unprecedented temporal control over this key experimental variable. This new level of control has been exploited to probe how cells respond to signals received from their extracellular microenvironment. Physical external stimuli including temperature, strain, and electromagnetic fields, as well as photoexcitation, have been used to trigger reversible changes in surface microstructure in the presence of cells.

Dynamic control of cell morphology has been achieved through exploiting the unique properties of shape–memory polymers, which change shape on exposure to an external stimulus such as temperature. In an early example of controlling cell morphology *in situ*, Davis *et al.* embossed reversible microgrooves into a polyurethane-based, thiol–ene cross-linked polymer system that were programmed to switch when exposed to temperatures experienced under standard cell culture conditions (30–37°C) [47]. Transitions from the microgrooved surface to the original smooth surface controlled fibroblast alignment and cytoskeletal organization [47]. Subsequently, Le and colleagues developed a dynamic cell culture substrate by first cross-linking a PCL-based shape–memory polymer in a mold to produce an initial shape and then mechanically deforming the primary shape to form a secondary shape at a temperature above the transition temperature of the system [48]. These treatments result in a substrate, which demonstrates the ability to transition between two predefined shapes, instead of simply changing from a smooth substrate to a topographically modified one to control hMSC morphology [48]. One disadvantage to switchable topographies produced using shape–memory polymer systems is that these substrates are limited to two transitions, one from the original shape to the secondary shape and another back to the original. In this way, it is only possible to study cellular responses to two specific topographies at a time.

To provide a complementary approach with the potential to overcome this limitation, Burdick *et al.* created strain-responsive lamellar patterns in PDMSe by uniaxially stretching PDMSe sheets and exposing them to ultraviolet/ozone (UVO) to stiffen the top surface of the material, resulting in buckling perpendicular to the strain [49]. The size of the lamellar patterns was controlled by selectively exposing sections of PDMSe to UVO through a photomask. When the patterned sheets were stretched back to the initial strain, the lamellar structure was released and the surface returned to smooth. Sequential stretching and UVO exposure led to the presentation of a series of topographies. These experiments demonstrated spatial and temporal control over hMSC morphology. Likewise, Engler and colleagues developed a soft polyacrylamide hydrogel composite containing magnetic nickel microwires in the surface of the substrate, that when exposed to a magnetic field, oriented resulting in changes in surface roughness [50]. Surface roughness was varied from 0.05 to 0.7 μm. Results showed that vascular SMC shape and spreading changed in response to acute changes in substrate roughness up to 0.7 μm, but that dynamic oscillation did not produce significant differences. Another dynamic, composite surface presented by Zhou and colleagues was composed of PCL and ferrous nanoparticles [51]. In this way, dynamic shape changes could be triggered by exposure to temperature changes or an alternating magnetic field. Micropillar topographies were fabricated through thermal embossing microimprint lithography and dynamic changes in surface topography were related to changes in adhesion, spreading, and alignment of rate bone marrow mesenchymal stem cells [51].

Recent efforts in the Anseth group have focused on the development of a cell culture substrate that affords the user both spatial and temporal control over the presentation of a series of topographical cues via photodegradation. Anseth *et al.* exploited the photolabile linkages in a polyethylene glycol (PEG) hydrogel platform to create topographies using precise spatial erosion by way of irradiation through a photomask [52]. To dynamically present a series of topographic cues to a cell population *in situ*, photodegradation was carried out under cytocompatible conditions. hMSCs were initially seeded onto smooth surfaces, which were in turn patterned sequentially using photolithography (365 nm, 10 mW/cm^2, 250 seconds) into channels and square patterns (Figure 5.15). Live imaging of fluorescently labeled cells 24 hours after patterning revealed that cell morphology and alignment had responded and changed based on the new underlying pattern. The introduction of channels, an anisotropic pattern, led to a statistically significant increase in the average cellular aspect ratio (analysis of variance (ANOVA), Tukey's test, $\alpha = 0.05$) and an increase (not significant, Marascuilo procedure $\alpha = 0.05$) in cellular alignment along the direction of the topographic features. The next patterning step returned the substrate to an isotropic surface (e.g., squares) and reversed these changes in cell morphology. Even though these dynamically tunable cell culture systems may serve as a powerful tool to improve our understanding of how cells respond to real time changes in topographic cues in their microenvironment, the native cellular microenvironment is 3-D, so to best recapitulate these niches, cell culture strategies must be translated to 3-D.

5.6 CONCLUSIONS AND FUTURE DIRECTIONS

Since the first description of contact guidance nearly 100 years ago, significant progress has been made in the methodical design of topographic cues for studying the fundamental interactions between cells and patterned surfaces. The introduction of the concept of engineered topography allows researchers to systematically alter key topographic parameters, such as feature geometry, size, and depth, to elucidate how cells receive and respond to topographic cues. When these designs are inspired by nature, it becomes possible to rationally design materials that more closely mimic the physiological extracellular microenvironment to further these research efforts. In order to deliberately design biomaterials for healthcare applications, it is important to better understand the natural cellular microenvironment, which is dynamic, 3-D, and influenced by the cells that inhabit it.

One important step toward this goal is the translation of precisely engineered biophysical cues into 3-D. Initial efforts in this area have succeeded in controlling cellular morphology. In one example, Khademhossieni *et al.* encapsulated endothelial cells in micropatterned gelatin methacrylate (GelMA) hydrogels with high aspect ratio rectangular features 50–150 μm in height [53,54]. Results showed that the endothelial cells aligned, elongated [54], and assembled cord structures [53] within these constructs. Toward more systematically controlling the presentation of biophysical cues, Burdick and coworkers introduced degradable and nondegradable network structures to encapsulated hMSCs by sequentially cross-linking a protease degradable, multiacrylate hyaluronic acid-based hydrogel. The secondary cross-linking step was carried out via a UV-initiated polymerization of unreacted acrylates from the first step, which resulted in increased density of nondegradable cross-links in specific volumes [55]. This technique is a powerful tool to control cell morphology both spatially and temporally in 3-D. It has also been shown that using this technique to restrict cell-mediated scaffold degradation reduced cell traction within the material and caused hMSCs to differentiate toward the adipogenic pathway. These results demonstrate that in 3-D, cell traction independent of cell morphology or matrix mechanics directs lineage commitment of hMSCs [56]. Recently, a very elegant approach to translate engineered biophysical cues into 3-D was reported by Vunjak-Novakovic *et al.* [57]. In this work, cells were encapsulated into 3-D GelMA-based hydrogel shapes with defined dimensions, which were in turn encapsulated through iterative sedimentation into a larger hydrogel construct. The size and shape of the microcarrier, as well as the cell density within each shape, were precisely controlled. Orienting the shapes within a secondary hydrogel structure created 3-D engineered physical cues that were used to spatially template outgrowth and migration of hMSCs [57]. Translating the concept of bio-inspired, engineered topographies into 3-D will advance our understanding of how cells interact with physiological biophysical cues and, hence, improve our ability to rationally design biomedical materials for clinical applications.

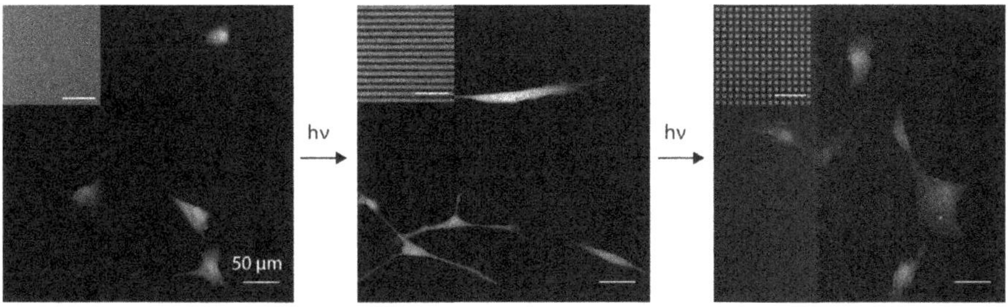

FIGURE 5.15 Fluorescent images of hMSCs labeled with CellTracker™ Green CMFDA (5-chloromethylfluorescein diacetate) (Life Technologies, Grand Island, NY), invitrogen, to track cellular morphology on sequentially presented dynamic microtopographies. hMSCs were first seeded onto smooth surfaces, which were patterned sequentially using photolithography (365 nm, 10 mW/cm^2, 250 seconds) into an anisotropic, channels pattern and then an isotropic, squares pattern *in situ*. Cells exhibited a rounded morphology on the smooth pattern, then on patterning, elongated along the features on the channels topography and returned to a more rounded morphology after presentation of the squares pattern. (Reprinted with permission from Reference 52.)

Acknowledgments

The authors acknowledge NIH (F32AR061923) from the National Institute of Arthritis and Musculoskeletal and Skin Diseases for fellowship funding to CMK. The content is solely the responsibility of the authors and does not necessarily represent the official views of the National Institute of Arthritis and Musculoskeletal and Skin Diseases or the National Institutes of Health.

REFERENCES

[1] Taipale, J., Keski-Oja, J. *The FASEB Journal* **1997**, *11*, 51.

[2] Schoen F.J., Mitchell R.N., Tissues, the extracellular matrix, and cell-biomaterial interactions, in *Biomaterials Science: An Introduction to Materials in Medicine*; Elsevier Academic Press, San Diego: 2004; pp. 260–281.

[3] Harrison, R. *The Journal of Experimental Zoology* **1914**, *17*, 521.

[4] Weiss, P. *Journal of Experimental Zoology* **1945**, *100*, 353.

[5] Weiss, P. *Yale Journal of Biology and Medicine* **1947**, *19*, 235.

[6] Curtis, A.S.G., Varde, M. *Journal of the National Cancer Institute* **1964**, *33*, 15.

[7] Voldman, J., Gray, M.L., Schmidt, M.A. *Annual Review of Biomedical Engineering* **1999**, *1*, 401.

[8] Lim, J.Y., Donahue, H.J. *Tissue Engineering* **2007**, *13*, 1879.

[9] Flemming, R.G., Murphy, C.J., Abrams, G.A., Goodman, S.L., Nealey, P.F. *Biomaterials* **1999**, *20*, 573.

[10] Ingber, D.E., Mow, V.C., Butler, D., Niklason, L., Huard, J., Mao, J., Yannas, I., Kaplan, D., Vunjak-Novakovic, G. *Tissue Engineering* **2006**, *12*, 3265.

[11] The extracellular matrix of animals, in *Molecular Biology of the Cell* (Eds: B. Alberts, A. Johnson, J. Lewis, M. Raff, K. Roberts, P. Walter), Garland Science, New York: 2002. Available at http://www.ncbi.nlm.nih.gov/books/NBK26810/.

[12] Brown, B.N., Barnes, C.A., Kasick, R.T., Michel, R., Gilbert, T.W., Beer-Stolz, D., Castner, D.G., Ratner, B.D., Badylak, S.F. *Biomaterials* **2010**, *31*, 428.

[13] Badylak, S.F., Freytes, D.O., Gilbert, T.W. *Acta Biomaterialia* **2009**, *5*, 1; Freytes, D.O., Stoner, R.M., Badylak S.F. *Journal of Biomedical Materials Research Part B: Applied Biomaterials* **2008**, *84B*, 408.

[14] Schulte, V.A., Diez, M., Moller, M., Lensen, M.C. *Biomacromolecules* **2009**, *10*, 2795; Hamilton, D., Wong, K., Brunette, D. *Calcified Tissue International* **2006**, *78*, 314.

[15] Ngalim, S.H., Magenau, A., Le Saux, G., Gooding, J.J., Gaus, K. *Journal of Oncology* **2010**, *2010*, 1–7; Tse, J.R., Engler, A.J. *PLoS ONE* **2011**, *6*, e15978.

[16] Kim, J.E., Boehm, C.A., Mata, A., Fleischman, A.J., Muschler, G.F., Roy, S. *Acta Biomaterialia* **2010**, *6*, 160.

[17] Ayala, P., Lopez, J.I., Desai, T.A. *Tissue Engineering: Part A* **2010**, *16*, 2519.

[18] Yang, Y., Kusano, K., Frei, H., Rossi, F., Brunette, D., Putnins, E. *Journal of Biomedical Materials Research. Part A* **2010**, *95A*, 294; Kloxin, A.M., Benton, J.A., Anseth, K.S. *Biomaterials* **2010**, *31*, 1.

[19] McMurray, R.J., Gadegaard, N., Tsimbouri, P.M., Burgess, K.V., McNamara, L.E., Tare, R., Murawski, K., Kingham, E., Oreffo, R.O.C., Dalby, M.J. *Nature Materials* **2011**, *10*, 637.

[20] Choi, Y.S., Vincent, L.G., Lee, A.R., Dobke, M.K., Engler, A.J. *Biomaterials* **2012**, *33*, 2482; Engler, A.J., Sen, S., Sweeney, H.L., Discher, D.E. *Cell* **2006**, *126*, 677.

[21] McBeath, R., Pirone, D.M., Nelson, C.M., Bhadriraju, K., Chen, C.S. *Developmental Cell* **2004**, *6*, 483; Kilian, K.A., Bugarija, B., Lahn, B.T., Mrksich, M. *PNAS* **2010**, *107*, 4872.

[22] Carman, M., Estes, T., Feinberg, A., Schumacher, J., Wilkerson, W., Wilson, L., Callow, M., Callow, J., Brennan, A. *Biofouling* **2006**, *22*, 11.

[23] Lopez, J., Hansali, G., Zahouani, H., Le Bosse, J., Mathia, T. *International Journal of Machine Tools & Manufacture* **1995**, *35*, 211.

[24] Charest, J., Eliason, M., Garcia, A., King, W. *Biomaterials* **2006**, *27*, 2487.

[25] Curtis, A.S.G., Gadegaard, N., Dalby, M.J., Riehle, M.O., Wilkinson, C.D., Aitchison, G. *IEEE Transactions on Nanobioscience* **2004**, *3*, 61.

[26] Yang, J.L., Liu, A.M., Zhou, C.R. *Journal of Biomaterials Science. Polymer Edition* **2011**, *22*, 919.

[27] Thomas, C.H., Collier, J.H., Sfeir, C.S., Healy, K.E. *Proceedings of the National Academy of Sciences of the United States of America* **2001**, *99*, 1972.

[28] McNamara, L.E., McMurray, R.J., Biggs, M.J., Kantawong, F., Oreffo, R.O., Dalby, M.J. *Journal of Tissue Engineering* **2010**, *2010*, 120623.

[29] Kolind, K., Dolatshahi-Pirouz, A., Lovmand, J., Pedersen, F.S., Foss, M., Besenbacher, F. *Biomaterials* **2010**, *31*, 9182.

[30] Ross, A.M., Jiang, Z., Bastmeyer, M., Lahann, J. *Small* **2012**, *8*, 336.

[31] Wong, S., Teo, S.-K., Park, S., Chiam, K.-H., Yim, E.F. *Biomechanics and Modeling in Mechanobiology* **2013**, *1*.

[32] Feinberg, A.W., Wilkerson, W., Seegert, C.A., Gibson, A.L. *Journal of Biomedical Materials Research. Part A* **2008**, *86A*, 522.

[33] Seo, C.H., Furukawa, K., Montagne, K., Jeong, H., Ushida, T. *Biomaterials* **2011**, *32*, 9568.

[34] Magin, C.M., Finlay, J.A., Clay, G., Callow, M.E., Callow, J.A., Brennan, A.B. *Biomacromolecules* **2011**, *12*, 915; Magin, C.M., Long, C.J., Cooper, S.P., Ista, L.K., Lopez, G.P., Brennan, A.B. *Biofouling* **2010**, *26*, 719; Long, C.J., Schumacher, J.F., Robinson, P., II, Finaly, J.A., Callow, M.E., Callow, J.A., Brennan, A.B. *Biofouling* **2010**, *26*, 411;

Schumacher, J.F., Carman, M.L., Estes, T.G., Feinberg, A.W., Wilson, L.H., Callow, M.E., Callow, J.A., Finlay, J.A., Brennan, A.B. *Biofouling* **2007**, *23*, 55.

[35] Carman, M.L., Estes, T.G., Feinberg, A.W., Schumacher, J.F., Wilkerson, W., Wilson, L.H., Callow, M.E., Callow, J.A., Brennan, A.B. *Biofouling* **2006**, *22*, 11.

[36] Magin, C. *Engineered Microtopographies and Surface Chemistries Direct Cell Attachment and Function*; University of Florida, Gainesville, FL: 2010.

[37] Cao, Y., Poon, Y.F., Feng, J., Rayatpisheh, S., Chan, V., Chan-Park, M.B. *Biomaterials* **2010**, *31*, 6228.

[38] Beamish, J.A., He, P., Kottke-Marchant, K., Marchant, R.E. *Tissue Engineering: Part B* **2010**, *16*(5), 467–491.

[39] Thakar, R.G., Ho, F., Huang, N.F., Liepmann, D., Li, S. *Biochemical and Biophysical Research Communications* **2003**, *307*, 883; Sarkar, S., Lee, G.Y., Wong, J.Y., Desai, T.A. *Biomaterials* **2006**, *27*, 4775.

[40] Kim, H.N., Jiao, A., Hwang, N.S., Kim, M.S., Kang, D.H., Kim, D.-H., Suh, K.-Y. *Advanced Drug Delivery Reviews* **2013**, *65*, 536.

[41] Liliensiek, S.J., Campbell, S., Nealey, P.F., Murphy, C.J. *Journal of Biomedical Materials Research. Part A* **2006**, *79A*, 185.

[42] Diehl, K.A., Foley, J.D., Nealey, P.F., Murphy, C.J. *Journal of Biomedical Materials Research. Part A* **2005**, *75A*, 603.

[43] Wang, P.-Y., Yu, J., Lin, J.-H., Tsai, W.-B. *Acta Biomaterialia* **2011**, *7*, 3285.

[44] Wilkinson, A., Hewitt, R.N., McNamara, L.E., McCloy, D., Dominic Meek, R.M., Dalby, M.J. *Acta Biomaterialia* **2011**, *7*, 2919.

[45] Tsimbouri, P.M., Murawski, K., Hamilton, G., Herzyk, P., Oreffo, R.O.C., Gadegaard, N., Dalby, M.J. *Biomaterials* **2013**, *34*, 2177.

[46] Kirschner, C.M., Anseth, K.S. *Acta Materialia* **2013**, *61*, 931.

[47] Davis, K.A., Burke, K.A., Mather, P.T., Henderson, J.H. *Biomaterials* **2011**, *32*, 2285.

[48] Le, D.M., Kulangara, K., Adler, A.F., Leong, K.W., Ashby, V.S. *Advanced Materials* **2011**, *23*, 3278.

[49] Guvendiren, M., Burdick, J.A. *Advanced Healthcare Materials* **2012**, *2*(1), 155–164.

[50] Kiang, J.D., Wen, J.H., del Álamo, J.C., Engler, A.J. *Journal of Biomedical Materials Research. Part A* **2013**, *101A*(8), 2313–2321.

[51] Zhou, S., Li, W.B., Gong, T., Chen, H.M., Wang, L., Li, J.R. *RSC Advances* **2013**, *3*, 9865–9874.

[52] Kirschner, C.M., Anseth, K.S. *Small* **2013**, *9*, 578.

[53] Nikkhah, M., Eshak, N., Zorlutuna, P., Annabi, N., Castello, M., Kim, K., Dolatshahi-Pirouz, A., Edalat, F., Bae, H., Yang, Y., Khademhosseini, A. *Biomaterials* **2012**, *33*, 9009.

[54] Aubin, H., Nichol, J.W., Hutson, C.B., Bae, H., Sieminski, A.L., Cropek, D.M., Akhyari, P., Khademhosseini, A. *Biomaterials* **2010**, *31*, 6941.

[55] Khetan, S., Burdick, J.A. *Biomaterials* **2010**, *31*, 8228.

[56] Khetan, S., Guvendiren, M., Legant, W.R., Cohen, D.M., Chen, C.S., Burdick, J.A. *Nature Materials* **2013**, *12*, 458–465.

[57] Eng, G., Lee, B.W., Parsa, H., Chin, C.D., Schneider, J., Linkov, G., Sia, S.K., Vunjak-Novakovic, G. *Proceedings of the National Academy of Sciences* **2013**, *110*(12), 4551–4556.

CHAPTER 6

ENGINEERING THE MECHANICAL AND GROWTH FACTOR SIGNALING ROLES OF FIBRONECTIN FIBRILS

CHRISTOPHER A. LEMMON
Department of Biomedical Engineering, Virginia Commonwealth University, Richmond, VA, USA

6.1 INTRODUCTION

The *in vivo* presentation of extracellular matrix (ECM) plays several critical roles in cell migration, survival, differentiation, and organization. The ECM provides a complex presentation of cell attachment sites, mechanical signals, and tethered growth factor signals. However, the effects of this complex presentation is often ignored in tissue engineering and regenerative medicine, oftentimes replaced by surfaces coated with adsorbed ECM molecules, tethered ECM protein fragments, or tethered peptides that contain sequences that facilitate cell attachment. These approaches fail to replicate the complex environment that cells would be exposed to *in vivo*. In this chapter, we will discuss the mechanical and signaling roles of one particular ECM protein, fibronectin (FN). FN is a soluble protein that is found at high concentration in the blood plasma [1]; it is stretched by cells into an extended conformation that is purported to expose cryptic binding sites to facilitate assembly of insoluble fibrils [2]. FN assembly is required for collagen assembly [3–5], fibrillin assembly [6], and tenascin assembly [7–9], and as such, is crucial to successful tissue engineering. We will first discuss the mechanisms of FN fibrillogenesis as well as the roles of FN fibrils in mechanotransduction, cell attachment and migration, and growth factor signaling. Subsequently, we will discuss *in vitro* mechanisms to assemble FN fibrils and highlight examples of FN matrix signaling in several tissue engineering applications.

Bio-inspired Materials for Biomedical Engineering, First Edition. Edited by Anthony B. Brennan and Chelsea M. Kirschner.

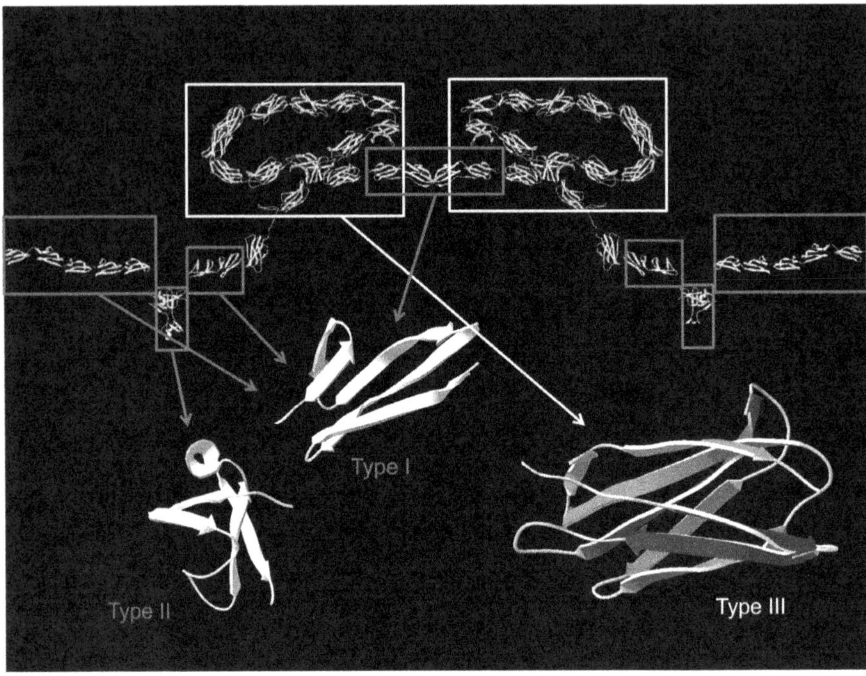

FIGURE 6.1 An assembled image of known FN structures. FN is a dimer of two 250-kDa monomers, each of which consists of 30–32 individually folded domains. These domains have one of three structures, referred to as type I, type II, or type III domains. Type I and type II domains have several internal disulfides and, are thus, unlikely to unfold when subjected to cell-derived forces. Type III domains have no internal disulfides, and previous studies have shown that these domains unfold under tension. The schematic shown in the figure represents the entire FN dimer, and is based on protein data base (PDB) files of structures for known domains of FN. The loop in the type III domains shows the interactions, which have been shown to exist in the compact conformation of FN. The RGD binding site, which is often substituted for the full molecule in tissue engineering applications, is contained on a single loop of the 10th type III domain. Compiled image was generated from PDB files: FNI^{1}: 1O9A [94]; $FNI^{2\text{-}3}$: 3CAL [95]; $FNI^{4\text{-}5}$: 2RL0 [95]; $FNI^{6}\text{-}II^{1}\text{-}II^{2}$: 1E88 [96]; $FNI^{8\text{-}9}$: 3GXE [97]; $FNIII^{1\text{-}2}$: 2HA1 [98]; $FNIII^{7\text{-}10}$: 1FNF [15]; $FNIII^{12\text{-}14}$: 1FNH [99]. Images for domains without published structures were generated in SPDBViewer (Swiss Institute of Bioinformatics, Basel, Switzerland) by matching the sequence to the structure of the same domain homology (i.e., $FNIII^{3\text{-}6}$ was generated by matching AA sequence to $FNIII^{7\text{-}10}$ structure).

6.2 STRUCTURE OF FIBRONECTIN

FN is a dimer of a 250-kDa protein that forms insoluble fibrils that are a critical component of ECM [10,11]. It is made up of a series of three domain types connected end to end [12,13] (Figure 6.1). The N-terminus of FN consists of nine FN-I and two FN-II domains, each of which contains an internal disulfide bond [12,14]. The C-terminus of FN consists of three FN-I domains and contains intermolecular disulfides that link the two FN monomers to form the dimer. The central portion of the FN molecule is made up of 15 tandem FN-III domains; these domains each consist of seven beta strands folded into a beta-sandwich structure and contain no internal disulfides [15,16]. Numerous studies have shown that these domains are capable of unfolding under both chemical and mechanical forces [17–19]. Interestingly, these domains have nearly identical structure, but have only roughly 30% sequence homology [20,21]. The significant variation in domain sequence most likely gives rise to the wide variance in chemical and mechanical stability that is observed between the domains [17,19,22,23]. FN in solution adopts a compact conformation in which the 2nd–4th FN-III domains (III-2–4) of one subunit interact with the 12th–14th FN-III domains (III-12–14) of the other, causing the FN dimer to fold onto itself [24].

6.3 ASSEMBLY OF FIBRONECTIN FIBRILS

FN is present in blood plasma at extremely high concentration in a soluble state. Cells bind to FN and stretch it by applying contractile forces. This stretching causes FN to assemble into insoluble, rope-like fibrils [25–28]. These fibrils are an extremely unique biopolymer, in that they are highly elastic and can be stretched to up to four times their resting length. The existing dogma for FN assembly is that cell-applied contractile forces stretch FN to expose a cryptic binding site; however, several studies over the course of 40+ years of research have yet to identify such a buried site. It is, however, well documented that FN assembly requires the application of cell-generated forces [2,29,30], and that FN is absolutely necessary for tissue formation; deletion of the FN gene is lethal at the embryonic stage in mammals [26]. Previous studies have demonstrated that the 70-kDa N-terminal fragment, which composes the nine FN-I and two FN-II domains, is necessary for FN fibrillogenesis; however, this 70-kDa N-terminus is not a cryptic FN–FN attachment site. The 70-kDa fragment is readily exposed in soluble FN [31–33], and as such, could not be a buried cryptic binding site. While a cryptic binding site has not been discovered, several studies have shed light onto potential mechanisms of FN assembly. Mosher and colleagues have shown that a domain from the streptococcus pyogenes protein adhesin F1 is capable of binding to the 70-kDa fragment of FN through beta-strand addition [34–36]. The 15 type III domains are made of seven beta-strand sandwiches, which have several features that block beta-strand addition, including twisting of the free-edge beta strands and orientation of side chains that sterically hinder addition of new beta strands [37]; however, steered molecular dynamics studies have indicated a stable intermediate of stretched type III domains in which the beta-strand-blocking features of the type III structure are removed [38,39]. Thus, it is feasible that cell-generated forces untwist type III domains and transition the domains to a stable intermediate that is capable of beta-strand addition with the 70-kDa N-terminus of new FN molecules (Figure 6.2). This could offer a possible explanation for the elusiveness of the cryptic FN–FN site; since each of the 15 type III domains shares a similar structure, it is possible that *all* 15 domains are capable of binding new FN molecules when exposed to cell-generated tension. This would also suggest an extremely elegant, mechanosensitive assembly process. Given the previously

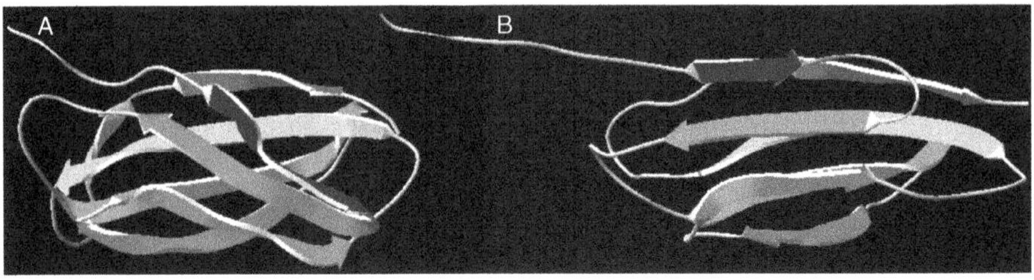

FIGURE 6.2 Untwisting of type III domains. Steered Molecular Dynamics (SMD) predicts a stable intermediate in which the beta strands of the type III domains have been straightened and aligned. In the unstretched state, nonspecific beta-strand addition is inhibited by the twisting of the beta strands, while in the stretched state, the beta strands along the domain edges are straightened, allowing for beta-strand addition. This image was made with VMD 1.8.7. software support. VMD (Visual Molecular Dynamics) is developed with NIH support by the Theoretical and Computational Biophysics group at the Beckman Institute, University of Illinois at Urbana–Champaign. NIH, National Institutes of Health.

discussed wide array of mechanical stabilities of the 15 type III domains, one could imagine a mechanism in which FN assembly is highly controlled by the magnitude of applied cellular forces; small forces would open only the weakest domains, generating only slight fibril assembly, while large forces would open many domains and drive significant fibril assembly.

6.4 MECHANICS OF FIBRONECTIN FIBRILS

While the exact mechanism of FN fibril formation is still unknown, we do know that insoluble FN fibrils are highly elastic structures and can be stretched to up to four times their resting length [40–43]. This elasticity requires cell contraction; disruption of cytoskeletal contraction relaxes stretched fibrils [41]. Two different mechanisms have been proposed to explain the elasticity of FN matrix fibrils [44]. One theory contends that the elasticity is a function of FN-III domains unfolding under tension [45,46]. These domains contain no internal disulfides and can be mechanically unfolded by atomic force microscopy [17,18]. It is unclear how relevant these atomic force microscopy (AFM) pulling forces are to the *in vivo* assembly of fibrils; the unfolding by AFM occurred at high forces on the order of 100 pN, due to high pulling rates. In cell-assembled fibrils, the pulling speed is nearly zero, so it is difficult to extrapolate the cell-applied force that would be needed to unfold these domains. In addition to mechanical unfolding of FN-III domains, transient opening of certain FN-III domains has been demonstrated in the absence of applied force during the assembly of *in vitro* assembled super-FN [47,48]. The second theory to explain fibril elasticity contends that FN molecules are in a compact conformation in relaxed fibrils, and are stretched to the extended conformation in elongated fibrils [49]. These two mechanisms are not exclusive; elasticity of fibrils may be a combination of the

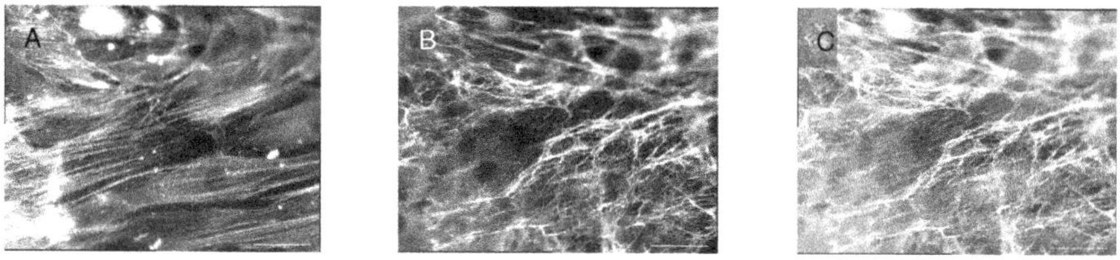

FIGURE 6.3 Assembly of FN matrix in hMSCs. Confluent layers of hMSCs assemble extensive FN matrices. (A) F-actin immunofluorescence; (B) FN labeled with anti-cellular FN antibody indicates that assembled FN was expressed in hMSCs; (C) Composite image of cell–matrix interactions. *(See insert for color representation of the figure.)*

two. Recent studies have suggested that a subset of the type III domains are stretched open in cell-assembled fibrils [19]. Unfolding of this subset cannot explain the fourfold extensibility; these domains would have to be completely extended in order to generate fibril extensions on this order, and forces needed to completely extend the domains are greater than the forces generated by attached cells.

Regardless of the mechanism, the fourfold extensibility of FN fibrils is extremely significant in cellular mechanosensing. There are a great number of recent studies that have shown that cells are extremely sensitive to the elastic modulus of the substrate to which they are attached: substrate rigidity regulates cell spreading [50], cell migration [51–55], cell survival [56–58], and cell differentiation [54,59–65]. The mechanism by which these mechanical signals is transduced to intracellular signaling pathways has yet to be elucidated; however, the majority of cell lines investigated in these studies are known to assemble FN fibrils *in vitro*. For example, human mesenchymal stem cells (hMSCs), which have been well characterized as exhibiting stiffness-dependent differentiation, assemble substantial FN fibrils (Figure 6.3). The presence of these fibrils not only has the potential to alter growth factor signaling (discussed in further detail below), but they also may serve to alter the ability of cells to generate contractile forces. Odde and colleagues have previously demonstrated that cells can generate larger contractile forces on soft surfaces, and they have explained this phenomenon through the use of an elegant computational model [66]. In this model, the myosin–actin adhesion-substrate system is modeled as a series of elastic spring elements that are capable of rupturing at the adhesion-substrate interface. On stiffer surfaces, the model predicts what the authors refer to as a *frictional slippage* regime, where the substrate stiffness results in frequent rupture events that prevent the establishment of large contractile forces. On softer surfaces, the deflection of the substrate reduces the frequency of adhesion-substrate rupture events, which allows for the generation of larger cell forces (referred to as a *load-and-fail* regime by the authors, as the model predicts the generation of large forces that are interspersed with infrequent but large rupture events). Assembly of FN fibrils may serve as an *intermediate spring* in this model; that is, the fibril would act as a spring between the substrate and adhesion. The fourfold extensibility of the fibril would thus create a situation in which cells are capable of generating large forces, even in the presence of a stiff surface. Thus, assembly of FN fibrils may facilitate larger cell forces, which have been shown to correlate with differentiation and migration.

6.5 ROLE OF FIBRONECTIN FIBRILS IN CELL ATTACHMENT

Assembled FN is not just a passive scaffold to which cells attach. It also triggers a wide array of signal cascades within the attached cells that regulate cell proliferation, migration, contractile force generation, and further ECM remodeling. FN binds to the transmembrane proteins, β_1 integrin and β_3 integrin, as well as the transmembrane heparan sulfate-containing protein syndecan on the surface of the cell, which facilitates the assembly of protein clusters known as focal adhesions. These adhesions serve as a signaling nexus that regulate cell migration and cell survival by triggering downstream pathways, and also feedback to regulate the contractile state of the cell. Attachment of cells via integrins is well documented and well studied; in fact, binding of integrins to FN is known to be mediated by a mere three amino acid sequence: an arginine–glycine–aspartic acid (RGD) loop found in the 10th type III domain of FN [67,68]. This amino acid sequence is capable of binding integrins even in the absence of the full FN molecule, and as such, is frequently used as a surface peptide to facilitate cell attachment. While the attachment of cells to FN via transmembrane integrins has been extensively described, attachment via transmembrane syndecans may play just as significant of a role. Syndecans bind to the 12th–14th type III domains via heparan sulfate chains on the syndecans. Several studies have demonstrated that syndecans regulated and modulate ECM assembly [9,69,70], while others have demonstrated that focal adhesion formation and cell contractile force generation are modified by syndecan pathways [71–74]. This suggests that recapitulation of *in vivo* signaling cannot be adequately accomplished by adding only the integrin-binding site of FN to engineered surfaces. FN also binds and/or regulates assembly of other ECM proteins that are critical for wound healing and tissue development, including fibrin [75–77], which is the other primary component of the provisional ECM in wound healing, collagen [78], and fibrillin, whose assembly is necessary for the assembly of elastin fibers [6]. As such, the use of FN in tissue engineering scaffolds is essential to drive native-like ECM assembly.

6.6 ROLE OF FIBRONECTIN FIBRILS IN GROWTH FACTOR SIGNALING

Several studies have shown that FN signaling is not limited to cell adhesion signaling cascades. FN is capable of binding over 40 different growth factors, including several members of the vascular endothelial growth factor (VEGF) family, transforming growth factor beta (TGF-β), and platelet-derived growth factor (PDGF), all with extremely high affinity [79]. These growth factors bind to a single region of FN located in the 12th–14th type III domains [79]. Binding of growth factors to this FN site does not inhibit interactions between the growth factors and their corresponding growth factor receptor; as such, it is hypothesized that FN acts as a growth factor delivery scaffold, localizing growth factors near the sites of cell attachment (this growth factor binding site is located in close proximity to III-10, the primary site of integrin attachment). Interestingly, the 12th–14th type III domains also bind to the 2nd–4th type III domains in the compact conformation of FN [80]. This would suggest a potential interaction between growth factor signaling and fibril formation: it is possible that elevated growth factor concentration may disrupt the compact conformation of FN and facilitate FN fibril assembly, which in turn acts as a scaffold that localizes growth factors to the cell surface and facilitates the activation of these pathways.

6.7 CELL-FREE MECHANISMS OF FIBRIL FORMATION

Assembly of FN fibrils *in vivo* requires the application of cell-generated contractile forces. However, several methods have been developed to apply shear to soluble FN in the absence of cells in order to facilitate cell-free fibril assembly. Soluble FN requires low shear in order to assemble; in fact, shear forces at an air–water interface are sufficient to create an accumulated layer at the surface. Fibrils can be generated by pulling a pipette tip through the air–water interface of an FN solution; this technique has been used to generate single fibrils in order to study the mechanical properties of assembled FN [81–83]. Fibrils can also be generated by plating the air–water interface-accumulated FN layer onto an array of micropillars; fibrils form in a cell-free manner between pillars on the addition of polyvinyl alcohol (PVA) [84]. It is also possible to generate FN matrices through the application of chemical denaturation [85]. Fibrils can also be generated through the addition of a fragment of FN known as anastellin [48,86]. Anastellin is a stable protein that is a fragment of the 1st type III domain of FN. Addition of anastellin to FN causes a fibrillar precipitate to form, which resembles assembled fibrils. Interestingly, anastellin also serves as an antiangiogenic signal; when added to tumors *in vivo*, anastellin inhibits tumor growth and block the assembly of new vasculature [87,88].

While all of these methods generate insoluble fibrils, it is unclear how similar these fibrils are to cell-derived fibrils; for example, pipette-pulled fibrils can be stretched up to 12 times their resting length, which is three times longer than the length that cell-derived fibrils can be stretched. While these fibrils may be distinct from cell-derived fibrils, they provide unique and novel ways to recreate a fibril environment in tissue engineering that better simulates the fibrillar composition of *in vivo* ECM.

6.8 CELL-DERIVED FIBRONECTIN MATRICES

Cell-derived FN matrices are easily generated by culturing mesenchymal or fibroblastic cells for extended periods of time (24 hours–14 days) on a culture surface. Cells can be removed from these matrices to generate a cell-free FN matrix [89,90]; however, these matrices are isotropic with no structural organization and tend to be

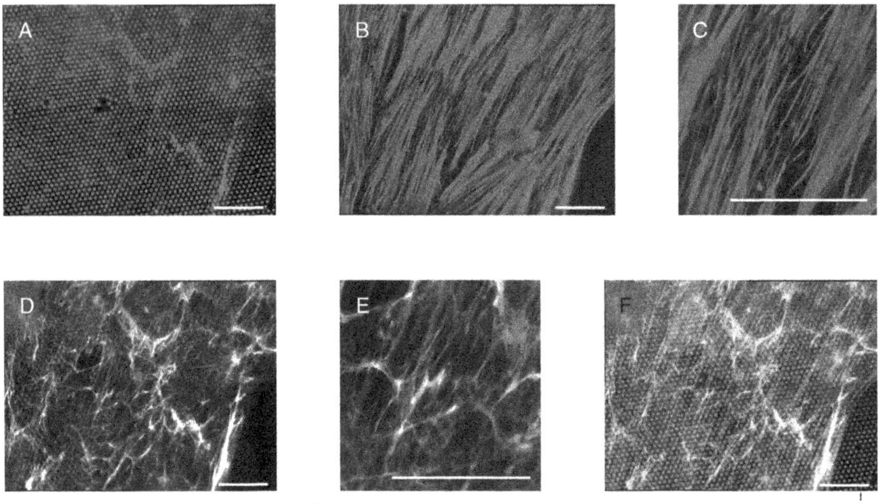

FIGURE 6.4 Assembly of FN matrix on a micropillar scaffold. Immunofluorescence images of a layer of human mesenchymal stem cells grown on a surface of micropillars for 10 days. (A) Fluorescently labeled pillars; (B) Actin cytoskeleton (red) (higher magnification shown in C); (D) Assembled fibronectin fibrils (higher magnification shown in E); (F) Composite image. Note that while there are visible spaces between cells in the actin image, they have formed a complete layer of ECM across the top surface of the pillars. Scale bar is 50 μm. *(See insert for color representation of the figure.)*

difficult to remove from a continuous surface. FN matrices can also be assembled on discontinuous surfaces, such as microfabricated pillar arrays (MPAs) [30] (Figure 6.4).

6.9 USE OF FIBRONECTIN IN TISSUE ENGINEERING APPLICATIONS

There are several examples of tissue engineering and regenerative medicine applications in which the unique mechanical properties and signaling pathways of FN fibrils have been exploited to facilitate tissue growth. Martino and colleagues exploited the growth factor binding capabilities of FN by ligating a fragment of FN containing the growth factor-binding domain into a fibrin scaffold [91]. Their results demonstrated that addition of this domain dramatically increased tissue regeneration in both a diabetic mouse model of chronic wounds and a rat model of bone regrowth. In the diabetic model, regeneration was increased through increased angiogenic activity, while in the bone model, regeneration was increased through the recruitment of mesenchymal stem cells. These two disparate mechanisms of tissue regrowth in distinct disease states and tissue type demonstrate the wide-ranging effects of FN signaling. Several recent studies have highlighted improved cell adhesion in vascular engineering applications, in which full length FN is covalently attached. Ravi *et al.* demonstrated the improved compatibility of human umbilical vein endothelial cells

(HUVECs) with elastin-like peptide (ELP) tissue scaffolds [92]. Others have shown that covalently linked FN dramatically improves the endothelialization of vascular grafts [93]. Taken together, these studies demonstrate that the use of full-length FN promotes signaling and mechanical cues that are not activated in FN fragment ligations.

6.10 CONCLUSIONS

In many tissue-engineering applications, cell adhesion is facilitated by coating surfaces with the RGD sequence of FN. While this does indeed facilitate integrin attachment, it fails to recapitulate the full mechanical and signaling role of assembled, insoluble, elastic FN fibrils. We have discussed here the numerous effects of FN fibrils on growth factor signaling, cell adhesion via nonintegrin mechanisms, elastic properties of fibrils, and the effects of full-length FN in tissue engineering applications. We have also reviewed mechanisms of generating cell-free and cell-derived FN matrices, which recreate the *in vivo* FN matrix to provide the full array of chemical and mechanical signals propagated by assembled FN.

REFERENCES

[1] Mosher, D.F. *Prog. Hemost. Thromb.* **1980**, *5*, 111.
[2] Zhong, C., Chrzanowska-Wodnicka, M., Brown, J., Shaub, A., Belkin, A.M., Burridge, K. *J. Cell Biol.* **1998**, *141*, 539.
[3] Shi, F., Harman, J., Fujiwara, K., Sottile, J. *Am. J. Physiol. Cell Physiol.* **2010**, *298*, C1265.
[4] Kadler, K.E., Hill, A., Canty-Laird, E.G. *Curr. Opin. Cell. Biol.* **2008**, *20*, 495.
[5] McDonald, J.A., Kelley, D.G., Broekelmann, T.J. *J. Cell. Biol.* **1982**, *92*, 485.
[6] Sabatier, L., Chen, D., Fagotto-Kaufmann, C., Hubmacher, D., McKee, M.D., Annis, D.S., Mosher, D.F., Reinhardt, D.P. *Mol. Biol. Cell* **2009**, *20*, 846.
[7] Van Obberghen-Schilling, E., Tucker, R.P., Saupe, F., Gasser, I., Cseh, B., Orend, G. *Int. J. Dev. Biol.* **2011**, *55*, 511.
[8] To, W.S., Midwood, K.S. *Matrix Biol.* **2010**, *29*, 573.
[9] Huang, W., Chiquet-Ehrismann, R., Moyano, J.V., Garcia-Pardo, A., Orend, G. *Cancer Res.* **2001**, *61*, 8586.
[10] Mosher, D.F., Sottile, J., Wu, C., McDonald, J.A. *Curr. Opin. Cell Biol.* **1992**, *4*, 810.
[11] Aguirre, K.M., McCormick, R.J., Schwarzbauer, J.E. *J. Biol. Chem.* **1994**, *269*, 27863.
[12] Potts, J.R., Campbell, I.D. *Matrix Biol.* **1996**, *15*, 313.
[13] Petersen, T.E., Thogersen, H.C., Skorstengaard, K., Vibe-Pedersen, K., Sahl, P., Sottrup-Jensen, L., Magnusson, S. *Proc. Natl. Acad. Sci. U.S.A.* **1983**, *80*, 137.
[14] Baron, M., Norman, D., Willis, A., Campbell, I.D. *Nature* **1990**, *345*, 642.
[15] Leahy, D.J., Aukhil, I., Erickson, H.P. *Cell* **1996**, *84*, 155.
[16] Huber, A.H., Wang, Y.M., Bieber, A.J., Bjorkman, P.J. *Neuron* **1994**, *12*, 717.
[17] Oberhauser, A.F., Badilla-Fernandez, C., Carrion-Vazquez, M., Fernandez, J.M. *J. Mol. Biol.* **2002**, *319*, 433.
[18] Abu-Lail, N.I., Ohashi, T., Clark, R.L., Erickson, H.P., Zauscher, S. *Matrix Biol.* **2006**, *25*, 175.

[19] Lemmon, C.A., Ohashi, T., Erickson, H.P. *J. Biol. Chem.* **2011**, *286*, 26375.
[20] Craig, D., Gao, M., Schulten, K., Vogel, V. *Structure* **2004**, *12*, 21.
[21] Kornblihtt, A.R., Umezawa, K., Vibe-Pedersen, K., Baralle, F.E. *EMBO J.* **1985**, *4*, 1755.
[22] Cota, E., Hamill, S.J., Fowler, S.B., Clarke, J. *J. Mol. Biol.* **2000**, *302*, 713.
[23] Cota, E., Clarke, J. *Protein Sci.* **2000**, *9*, 112.
[24] Johnson, K.J., Sage, H., Briscoe, G., Erickson, H.P. *J. Biol. Chem.* **1999**, *274*, 15473.
[25] Chen, L.B., Murray, A., Segal, R.A., Bushnell, A., Walsh, M.L. *Cell* **1978**, *14*, 377.
[26] George, E.L., Georges-Labouesse, E.N., Patel-King, R.S., Rayburn, H., Hynes, R.O. *Development* **1993**, *119*, 1079.
[27] Mosher, D.F., Fogerty, F.J., Chernousov, M.A., Barry, E.L. *Ann. N. Y. Acad. Sci.* **1991**, *614*, 167.
[28] Singer, I.I. *Cell* **1979**, *16*, 675.
[29] Halliday, N.L., Tomasek, J.J. *Exp. Cell. Res.* **1995**, *217*, 109.
[30] Lemmon, C.A., Chen, C.S., Romer, L.H. *Biophys. J.* **2009**, *96*, 729.
[31] Khan, M.Y., Medow, M.S., Newman, S.A. *Biochem. J.* **1990**, *270*, 33.
[32] Isaacs, B.S., Brew, S.A., Ingham, K.C. *Biochemistry* **1989**, *28*, 842.
[33] Ingham, K.C., Brew, S.A., Isaacs, B.S. *J. Biol. Chem.* **1988**, *263*, 4624.
[34] Maurer, L.M., Ma, W., Eickstaedt, N.L., Johnson, I.A., Tomasini-Johansson, B.R., Annis, D.S., Mosher, D.F. *J. Biol. Chem.* **2012**, *287*, 13303.
[35] Maurer, L.M., Tomasini-Johansson, B.R., Ma, W., Annis, D.S., Eickstaedt, N.L., Ensenberger, M.G., Satyshur, K.A., Mosher, D.F. *J. Biol. Chem.* **2010**, *285*, 41087.
[36] Tomasini-Johansson, B.R., Kaufman, N.R., Ensenberger, M.G., Ozeri, V., Hanski, E., Mosher, D.F. *J. Biol. Chem.* **2001**, *276*, 23430.
[37] Richardson, J.S., Richardson, D.C. *Proc. Natl. Acad. Sci. U.S.A.* **2002**, *99*, 2754.
[38] Gao, M., Craig, D., Lequin, O., Campbell, I.D., Vogel, V., Schulten, K. *Proc. Natl. Acad. Sci. U.S.A.* **2003**, *100*, 14784.
[39] Gao, M., Craig, D., Vogel, V., Schulten, K. *J. Mol. Biol.* **2002**, *323*, 939.
[40] Ohashi, T., Kiehart, D.P., Erickson, H.P. *Proc. Natl. Acad. Sci. U.S.A.* **1999**, *96*, 2153.
[41] Ohashi, T., Kiehart, D.P., Erickson, H.P. *J. Cell Sci.* **2002**, *115*, 1221.
[42] Sivakumar, P., Czirok, A., Rongish, B.J., Divakara, V.P., Wang, Y.P., Dallas, S.L. *J. Cell Sci.* **2006**, *119*, 1350.
[43] Baneyx, G., Baugh, L., Vogel, V. *Proc. Natl. Acad. Sci. U.S.A.* **2002**, *99*, 5139.
[44] Erickson, H.P. *J. Muscle Res. Cell Motil.* **2002**, *23*, 575.
[45] Erickson, H.P. *Proc. Natl. Acad. Sci. U.S.A.* **1994**, *91*, 10114.
[46] Smith, M.L., Gourdon, D., Little, W.C., Kubow, K.E., Eguiluz, R.A., Luna-Morris, S., Vogel, V. *PLoS Biol.* **2007**, *5*, e268.
[47] Ohashi, T., Augustus, A., Erickson, H. *Biochemistry* **2009**.
[48] Ohashi, T., Erickson, H.P. *J. Biol. Chem.* **2005**, *280*, 39143.
[49] Erickson, H.P., Carrell, N.A. *J. Biol. Chem.* **1983**, *258*, 14539.
[50] Han, S.J., Bielawski, K.S., Ting, L.H., Rodriguez, M.L., Sniadecki, N.J. *Biophys. J.* **2012**, *103*, 640.
[51] Shebanova, O., Hammer, D.A. *Biotechnol. J.* **2012**, *7*, 397.
[52] Sochol, R., Higa, A., Janairo, R., Li, S., Lin, L. *Micro Nano Lett.* **2011**, *6*, 323–326.
[53] Saez, A., Ghibaudo, M., Buguin, A., Silberzan, P., Ladoux, B. *Proc. Natl. Acad. Sci. U.S.A.* **2007**, *104*, 8281.
[54] Tse, J.R., Engler, A.J. *PLoS ONE* **2011**, *6*, e15978.
[55] Lo, C.M., Wang, H.B., Dembo, M., Wang, Y.L. *Biophys. J.* **2000**, *79*, 144.
[56] Choi, J.S., Harley, B.A.C. *Biomaterials* **2012**, *33*, 4460.
[57] Kniazeva, E., Putnam, A.J. *Am. J. Physiol. Cell Physiol.* **2009**, *297*, C179.
[58] Paszek, M.J., Weaver, V.M. *J. Mammary Gland Biol. Neoplasia* **2004**, *9*, 325.

[59] Markowski, M.C., Brown, A.C., Barker, T.H. *J. Biomed. Mater. Res. A* **2012**, *100A*, 2119.
[60] Leight, J.L., Wozniak, M.A., Chen, S., Lynch, M.L., Chen, C.S. *Mol. Biol. Cell* **2012**, *23*, 781.
[61] Olsen, A.L., Bloomer, S.A., Chan, E.P., Gaça, M.D.A., Georges, P.C., Sackey, B., Uemura, M., Janmey, P.A., Wells, R.G. *Am. J. Physiol. Gastroint. Liver Physiol.* **2011**, *301*, G110.
[62] Lam, W.A., Cao, L., Umesh, V., Keung, A.J., Sen, S., Kumar, S. *Mol. Cancer* **2010**, *9*, 35.
[63] Paszek, M.J., Zahir, N., Johnson, K.R., Lakins, J.N., Rozenberg, G.I., Gefen, A., Reinhart-King, C.A., Margulies, S.S., Dembo, M., Boettiger, D., Hammer, D.A., Weaver, V.M. *Cancer Cell* **2005**, *8*, 241.
[64] Fu, J., Wang, Y.-K., Yang, M.T., Desai, R.A., Yu, X., Liu, Z., Chen, C.S. *Nat. Methods* **2010**, *7*, 733.
[65] Engler, A.J., Sen, S., Sweeney, H.L., Discher, D.E. *Cell* **2006**, *126*, 677.
[66] Chan, C.E., Odde, D.J. *Science* **2008**, *322*, 1687.
[67] Yamada, K.M., Kennedy, D.W. *J. Cell Biol.* **1984**, *99*, 29.
[68] Pierschbacher, M., Hayman, E.G., Ruoslahti, E. *Proc. Natl. Acad. Sci. U.S.A.* **1983**, *80*, 1224.
[69] Stepp, M.A., Daley, W.P., Bernstein, A.M., Pal-Ghosh, S., Tadvalkar, G., Shashurin, A., Palsen, S., Jurjus, R.A., Larsen, M. *Exp. Cell Res.* **2010**, *316*, 2322.
[70] Yang, N., Mosher, R., Seo, S., Beebe, D., Friedl, A. *Am. J. Pathol.* **2011**, *178*, 325.
[71] Whiteford, J.R., Behrends, V., Kirby, H., Kusche-Gullberg, M., Muramatsu, T., Couchman, J.R. *Exp. Cell Res.* **2007**, *313*, 3902.
[72] Saoncella, S., Echtermeyer, F., Denhez, F., Nowlen, J.K., Mosher, D.F., Robinson, S.D., Hynes, R.O., Goetinck, P.F. *Proc. Natl. Acad. Sci. U.S.A.* **1999**, *96*, 2805.
[73] Woods, A., Longley, R.L., Tumova, S., Couchman, J.R. *Arch. Biochem. Biophys.* **2000**, *374*, 66.
[74] Wang, Z., Telci, D., Griffin, M. *Exp. Cell Res.* **2011**, *317*, 367.
[75] Sekiguchi, K., Hakomori, S. *Biochem. Biophys. Res. Commun.* **1980**, *97*, 709.
[76] Makogonenko, E., Tsurupa, G., Ingham, K., Medved, L. *Biochemistry* **2002**, *41*, 7907.
[77] Pereira, M., Rybarczyk, B.J., Odrljin, T.M., Hocking, D.C., Sottile, J., Simpson-Haidaris, P.J. *J. Cell Sci.* **2002**, *115*, 609.
[78] Sottile, J., Hocking, D.C. *Mol. Biol. Cell* **2002**, *13*, 3546.
[79] Martino, M.M., Hubbell, J.A. *FASEB J.* **2010**, *24*, 4711.
[80] Johnson, K.J., Sage, H., Briscoe, G., Erickson, H.P. *J. Biol. Chem.* **1999**, *274*, 15473.
[81] Chabria, M., Hertig, S., Smith, M.L., Vogel, V. *Nat. Commun.* **2010**, *1*, 135.
[82] Little, W.C., Smith, M.L., Ebneter, U., Vogel, V. *Matrix Biol.* **2008**, *27*, 451.
[83] Baneyx, G., Kubow, K., Vogel, V. *Faraday Discuss.* **2008**, *139*, 229.
[84] Ulmer, J., Geiger, B., Spatz, J.P. *Soft Matter* **2008**, *4*, 1998.
[85] Feinberg, A.W., Parker, K.K. *Nano Lett.* **2010**, *10*, 2184.
[86] Briknarova, K., Akerman, M.E., Hoyt, D.W., Ruoslahti, E., Ely, K.R. *J. Mol. Biol.* **2003**, *332*, 205.
[87] Yi, M., Sakai, T., Fassler, R., Ruoslahti, E. *Proc. Natl. Acad. Sci. U.S.A.* **2003**, *100*, 11435.
[88] Yi, M., Ruoslahti, E. *Proc. Natl. Acad. Sci. U.S.A.* **2001**, *98*, 620.
[89] Cukierman, E., Pankov, R., Stevens, D.R., Yamada, K.M. *Science* **2001**, *294*, 1708.
[90] Chernousov, M.A., Metsis, M.L., Koteliansky, V.E. *FEBS Lett.* **1985**, *183*, 365.
[91] Martino, M.M., Tortelli, F., Mochizuki, M., Traub, S., Ben-David, D., Kuhn, G.A., Müller, R., Livne, E., Eming, S.A., Hubbell, J.A. *Sci. Trans. Med.* **2011**, *3*, 100ra89.

[92] Ravi, S., Caves, J.M., Martinez, A.W., Haller, C.A., Chaikof, E.L. *J. Biomed. Mat. Res. A* **2013**, *101*, 1915.

[93] De Visscher, G., Mesure, L., Meuris, B., Ivanova, A., Flameng, W. *Acta Biomater.* **2012**, *8*, 1330.

[94] Schwarz-Linek, U., Werner, J.M., Pickford, A.R., Gurusiddappa, S., Kim, J.H., Pilka, E.S., Briggs, J.A.G., Gough, T.S., Höök, M., Campbell, I.D., Potts, J.R. *Nature* **2003**, *423*, 177.

[95] Bingham, R.J., Rudiño-Piñera, E., Meenan, N.A.G., Schwarz-Linek, U., Turkenburg, J.P., Höök, M., Garman, E.F., Potts, J.R. *Proc. Natl. Acad. Sci. U.S.A.* **2008**, *105*, 12254.

[96] Pickford, A.R., Smith, S.P., Staunton, D., Boyd, J., Campbell, I.D. *EMBO J.* **2001**, *20*, 1519.

[97] Erat, M.C., Sladek, B., Campbell, I.D., Vakonakis, I. *J. Biol. Chem.* **2013**, *288*, 17441.

[98] Vakonakis, I., Staunton, D., Rooney, L.M., Campbell, I.D. *EMBO J.* **2007**, *26*, 2575.

[99] Sharma, A., Askari, J.A., Humphries, M.J., Jones, E.Y., Stuart, D.I. *EMBO J.* **1999**, *18*, 1468.

CHAPTER 7

BIOLOGIC SCAFFOLDS COMPOSED OF EXTRACELLULAR MATRIX AS A NATURAL MATERIAL FOR WOUND HEALING

ELIZABETH W. KOLLAR
McGowan Institute for Regenerative Medicine, University of Pittsburgh, Pittsburgh, PA, USA

CHRISTOPHER L. DEARTH
McGowan Institute for Regenerative Medicine, Department of Surgery, University of Pittsburgh, Pittsburgh, PA, USA

STEPHEN F. BADYLAK
McGowan Institute for Regenerative Medicine, Department of Surgery, University of Pittsburgh, Pittsburgh, PA, USA

7.1 INTRODUCTION

The ideal substrate/scaffold for cells, tissues, and organs is that which exists in nature; that is, the extracellular matrix (ECM). The ECM represents the secreted product of resident cells and is a dynamic, *living* structure. Stated differently, the ECM is constantly changing in response to microenvironmental niche conditions such as pH, oxygen tension, nutrient availability, mechanical loading, and other external influences; a process aptly named *dynamic reciprocity* [1,2]. In addition, there are undoubtedly internal influences which affect the composition and ultrastructural organization of the matrix, such as the inherent genetic programming of resident cells, the influences of age, gender, and other endogenous factors. This ideal scaffold which we call the ECM is an extremely complex structure that is only partially understood. It is logical therefore, that a *bio-inspired material* is limited by

Bio-inspired Materials for Biomedical Engineering, First Edition. Edited by Anthony B. Brennan and Chelsea M. Kirschner.

the depth of our understanding of the composition, structure, and signaling mechanisms of the native matrix.

It is widely recognized that the ECM has a profound influence on cell differentiation, phenotype, and behavior [3–6]. Cell adhesion to matrix ligands triggers intracellular signaling pathways that regulate cell migration, cell cycle progression, and cell differentiation [7–9]. The transitional ECM in developing fetuses influences cell behavior through molecules such as tenascin-C, hyaluronic acid, and fibronectin [10]. Changes in ECM organization and mechanical properties affect cell phenotype [3]. Not only is our understanding of matrix structure–function relationships limited, but our attempts to mimic the ECM are wrought with such challenges. The technical limitations of reproducing all elements of the ECM, especially those which are dynamic and responsive to transitional external influences, are daunting.

Although a synthetic mimic of the ECM cannot be produced, individual components of the matrix such as collagen, fibronectin, laminin, and tenacin can be isolated and engineered to provide an instructive scaffold for tissue engineering and regenerative medicine applications. There currently exist many biologic scaffold materials composed of mammalian ECM (Table 7.1). Such naturally occurring *materials* are the subject of this chapter. An overview of these materials, which are typically regulated as surgical mesh medical devices by the Food and Drug Administration (FDA), is provided. In addition, the preparation of such materials will be briefly reviewed including the influence of various processing methods on clinical efficacy. Mechanisms by which the host responds to these materials is becoming increasingly clear and such mechanisms will be described herein.

7.2 PRODUCTS AND CLINICAL USE OF ECM

Scaffolds composed of biologic materials such as ECM, or components of ECM, are used in a variety of surgical applications. The advantages of biologic materials as compared with synthetic materials include a greater resistance to bacterial infection and a more constructive host healing response. However, biologic scaffold materials tend to be less strong than synthetic materials such as polypropylene, and are subject to biologic variability, which can affect mechanical and material properties to a small degree.

As can be seen in Table 7.1, the species from which biologic scaffold materials are harvested include: human, porcine, bovine, equine, and ovine. In addition, the tissues from which these materials are prepared vary widely and include small intestine, urinary bladder, dermis, pericardium, and fascia, among others. The wide variety of raw material sources for biologic scaffold materials suggests that there are favorable characteristics inherent within and conserved across all mammalian ECM. There are clear differences not only in the structure and composition of ECM derived from different species and different tissues, but more importantly, differences in the manufacturing processes can dramatically affect structure function relationships [11–13]. The manufacture of a biologic scaffold material requires decellularization of the source tissue while maintaining, as much as possible, the native structure and

TABLE 7.1 Clinically Available Products Composed of Extracellular Matrix (ECM)

Species	Organ source	Product (Company)	General uses
Human	Dermis	Allomax (Bard)	Soft tissue reconstruction: hernia and abdominal wall repairs
		Alloderm, Graftjacket (Life Cell)	Alloderm: hernia repairs, postmastectomy breast reconstruction, large abdominal wall defects. Graft jacket: superficial wounds, deep ulcers involving joint tissues
		Dermaspan (Biomet)	Diabetic foot ulcers
		FlexHD/FlexHD Diamond (Musculoskeletal Transplant Foundation)	Postmastectomy breast reconstruction, hernia repair
		Matrix HD (RTI Biologics)	Soft tissue reconstruction: hernia, breast, cartilage repairs; chronic skin wounds
		DermaMatrix (DePuy Synthes)	Soft tissue reconstruction: facial, intraoral, breast, abdominal wall repairs
	Amniotic membrane	Neox (Amniox Medical)	Surgical covering, wrap or barrier
		EpiFix (MiMedx)	Burns, plastic surgery, wound care
	Pericardium	IOPatch (IOP Ophthalmics Inc.)	Ophthalmologic surgical reconstruction

(Continued)

TABLE 7.1 *(Continued)*

Species	Organ source	Product (Company)	General uses
Porcine	Dermis	Collamend, XenMatrix (Bard)	Soft tissue reconstruction: hernia and abdominal wall repairs
		Permacol (Tissue Science Laboratories)	Soft tissue reconstruction
		Strattice (Life Cell)	Postmastectomy breast reconstruction
		EZ-Derm (AM Scientifics)	Partial thickness skin loss: skin donor harvest sites, skin ulcerations or abrasions
		XCM (Synthes)	Plastic and reconstructive surgeries; tissue reinforcement
		Conexa (Tornier)	Tendon repair: reinforcement of soft tissues repaired by sutures or suture anchors
	Small intestinal submucosa	Oasis (Cook Biotech Inc.)	Venous leg ulcers, diabetic foot ulcers; second degree burns
		Surgisis (Cook Biotech Inc.)	Hernia and abdominal wall repair
		CorMatrix ECM (CorMatrix Cardiovascular Inc.)	Pericardium repair and closure
	Urinary bladder matrix	MatriStem (Acell)	Plastic and reconstructive surgery, burns, wound care
	Pericardium	Vascutek (Terumo Group)	Cardiac and vascular reconstruction
	Heart valve	Hancock II, Mosaic, Freestyle (Medtronic Inc.)	Cardiac valve replacement
	Mesothelium	Meso BioMatrix (Kensey Nash)	Plastic and reconstructive surgery; dura substitute; suture line reinforcement
Bovine	Dermis	Primatrix (TEI Biosciences)	Wound care: pressure, diabetic, or venous ulcers; second degree burns
		SurgiMend (TEI Biosciences)	Plastic and reconstructive surgery; muscle flap reinforcement; hernia repair
		TissueMend (TEI Biosciences)	Orthopedic applications: broken bones, torn ligaments and tendons, bruised muscle, neck, or spinal injuries
	Pericardium	Veritas (Synovis Surgical Innovations)	Plastic and reconstructive surgery, hernia and pelvic floor repairs
Equine	Pericardium	OrthAdapt (Penta Biomedical)	Soft tissue reconstruction and reinforcement
Ovine	Forestomach	Endoform (Mesynthes)	Soft tissue reconstruction and reinforcement

composition of the ECM [14–17]. Decellularization methods vary widely and are the subject of several reviews [11,14]. Chemical, enzymatic, and physical methods of decellularization are utilized to various degrees. The host response following subsequent implantation of the resulting ECM is directly related to the efficacy of decellularization of the source tissue [18,19]. Standard criteria for defining effective decellularization have been proposed [14], and include several different quantitative measures: the absence of visible nuclear material on histologic examination with H&E or DAPI staining, DNA remnants less than 50 ng/mg of dry weight, and remnant DNA fragments must be 200 base pairs or less. Commercially available products have a wide range of residual cellular material [20,21], and this cellular debris likely plays a critical role in the host response. There have been descriptions of excellent functional and constructive remodeling outcomes following the use of biologic scaffold materials in several surgical applications, including esophageal [22,23], musculoskeletal [24–28], and the lower urinary tract [29–33], among others [34–36]. However, there have also been reports of unfavorable proinflammatory reactions following the use of biologic scaffold materials [37–39]. These unfavorable reactions include serous fluid accumulation, redness and swelling, and mechanical failure [21,40,41]. It is likely that the disparate outcomes are related to processing (i.e., decellularization) methods.

The use of chemical cross-linking agents to provide strength to biologic materials is commonplace [42–45]. Although chemical cross-linking does indeed provide strength, such treatment transforms the naturally occurring, degradable material, into a nondegradable (or very slowly degradable) material, which elicits a foreign body response [19,38,46]. In addition, the inherent growth factors and bioactive cryptic peptides, which are produced during the degradation process, are substantially inhibited [47–50]. In the opinion of the authors, the use of chemically cross-linked ECM transforms an inductive biologic material into one that is similar to most nondegradable synthetic materials.

In addition to efficacy of decellularization and the use of chemical cross-linking agents, methods of terminal sterilization such as ethylene oxide, electron beam, and gamma irradiation can affect mechanical, physical, and biologic properties of these naturally occurring materials [51–56]. Lyophilization, vacuum pressing, and storage conditions can also play a notable role in remodeling outcomes [57–60]. In summary, there are numerous potential sources of variability in the preparation of biologic scaffold materials. The appropriate and optimal use of such materials for various clinical applications requires an understanding not only of the methods of preparation for each product, but also the mechanisms by which the host responds to these materials (discussed in more detail below).

7.3 MECHANISMS OF ECM REMODELING

Appropriately prepared biologic scaffolds composed of ECM have been shown to facilitate constructive and site-appropriate remodeling when implanted into *in vivo* injury sites both in preclinical animal models and human clinical applications. The new host tissue that forms is grossly, histologically, and functionally similar to the native tissue it is intended to replace, but does not perfectly recapitulate the native tissue structure and function. Site-specific remodeling has been shown in a variety of tissues, including esophagus [22,23], blood vessels [61,62], skin and body wall [25,63], urinary tract [29,30], skeletal muscle [24–28,64–67], tendon and ligaments [27,68,69], myocardium [70,71], and bone [72,73], among others. In addition to the intact two- or three-dimensional form of ECM scaffolds, such materials can be enzymatically digested, solubilized, and polymerized to form a hydrogel, a network of polymer chains in which water is the dispersion medium [74–80]. ECM hydrogels retain many components found in the native tissue [81]. Injectable, *in situ* polymerizing hydrogels have been used for biomedical applications such as drug delivery and/or targeted stem cell delivery. Their ability to be delivered by minimally invasive techniques, conform to an irregular shape, and maintain robust biologic activity, make hydrogels optimal for these applications [82].

The use of ECM scaffolds at least partially transforms the natural fibrotic scarring host response injury toward one of constructive remodeling. Although the mechanism(s) by which ECM scaffolds promote constructive remodeling are not completely understood, rapid proteolytic scaffold degradation with the generation of bioactive cryptic molecules, referred to as *cryptic* because they are hidden until released by proteases, that have potent immunomodulatory, angiogenic, chemoattractant, mitogenic, and antimicrobial properties [47–50,83–91], recruitment of multipotent progenitor cells [27,49,92,93], site-appropriate mechanical loading [94–96], and modulation of the host innate immune response toward a regulatory and constructive phenotype [38,39,97], have all been shown to be important events in this process. Each of these mechanisms will be discussed in detail.

7.3.1 Biologic Scaffold Degradation and Recruitment of Stem/Progenitor Cells

Biologic scaffolds composed of ECM are rapidly infiltrated by host polymorphonuclear (PMN) and mononuclear cells *in vivo*, a process which initiates scaffold degradation. Macrophages in particular have been shown to be key facilitators of this degradation process, as their depletion prevents the degradation of ECM scaffolds [98]. Quantitative studies utilizing ECM scaffolds radiolabeled with ^{14}C labeled proline, an amino acid that is the building block for hydroxyproline which constitutes approximately 10% of each collagen molecule, were conducted to track their degradation profile *in vivo*. Since ^{14}C labeled proline is integrally bound to every collagen molecule, a decrease in the radioactivity per unit weight of tissue corresponds directly to a decrease in the mass of the scaffold. These studies showed that approximately 40% of the ECM scaffold was degraded after 28 days and virtually all of the ECM was replaced with host tissue by 90 days postimplantation [87–89]; a process partially dependent on the anatomic site of placement and the tissue source of the ECM.

Since the degradation of ECM scaffolds *in vivo* is critical to creating an optimal microenvironment that enhances constructive remodeling [48,49,91,99,100], a number of studies have evaluated the biologic activity of ECM degradation products. Several *in vitro* studies have shown that low molecular weight cryptic oligopeptides (5–16 kDa) are formed through ECM degradation [49,83,91]. These peptides show potent bioactivity not present in the parent ECM proteins. For example, these cryptic peptides have antibacterial activity against both gram-negative *Escherichia coli* and gram-positive *Staphylococcus aureus*. Several preclinical and clinical studies confirm resistance to deliberate bacterial contamination in a surgical site repaired with ECM scaffold materials [84,85].

Furthermore, ECM degradation products have shown chemotactic activity for a number of stem/progenitor cells *in vitro* including: multipotential progenitor cells [49], Sox2$^+$ cells [101], and multipotent perivascular progenitor cells [93,99,102,103]. Furthermore, endogenous recruitment of host multipotential progenitor cells (e.g., Sox2$^+$, Rex1$^+$, Sca1$^+$, CD133$^+$, CD34$^+$, and perivascular stem cells [CD146$^+$/NG2$^+$]) to the site of ECM scaffold degradation *in vivo* has been demonstrated [24,27,49,92,99,101–106]. Stated differently, modifying the microenvironmental niche of the wound site by placement of an ECM scaffold induces the endogenous recruitment of perivascular stem/progenitor cells. However, the contributions of ECM degradation products to tissue remodeling encompass more than just resistance to bacterial contamination and stem cell recruitment, as cellular differentiation can also be affected [27,48,49,91,93,99,107]. For example, a cryptic peptide isolated from collagen III has been shown to accelerate osteogenesis of perivascular stem/progenitor cells *in vitro* and can initiate new bone formation at the site of injury [100]. Another study showed that cells isolated from the injury site of animals treated with biologic scaffold degradation products had the ability to differentiate down both mesodermal and neuroectodermal lineages *in vitro*, whereas cells from control animals could only differentiate to a mesodermal lineage [99]. These studies suggest that specific products of biologic scaffold degradation may direct stem cells down a specific lineage pathway and/or accelerate differentiation. In addition to cryptic peptides, growth factors such as vascular endothelial growth factor (VEGF), basic fibroblast growth factor (bFGF), and transforming growth factor beta (TGF-β) can be liberated during the degradation of the ECM scaffold and subsequently exert their biologic effects (i.e., angiogenesis, mitogenesis, and/or cellular differentiation) to aid in the remodeling process [50,81,108].

In summary, unlike synthetic materials, the degradation of ECM scaffolds is an important and requisite process with bioactive consequences (i.e., release of cryptic peptides, growth factors, and cytokines) that work in concert to facilitate a microenvironment that enhances constructive tissue remodeling [48,49,91,99,100]. It is logical therefore that inhibition of scaffold degradation by chemical cross-linking of the ECM will eliminate the beneficial effects of ECM degradation products and ultimately result in a chronic inflammatory response and encapsulation.

7.3.2 The Effect of Site-Appropriate Mechanical Loading

It is widely accepted that passive range of motion and mechanical loading can have a positive and substantial effect on the endogenous healing response. Accordingly, several *in vivo* and *in vitro* studies have demonstrated that passive range of motion and tissue-specific biomechanical loading is essential for constructive remodeling of

ECM scaffold materials. For example, Hodde *et al.* conducted a study to investigate the effect of range of motion on constructive remodeling of ECM scaffolds when used as an Achilles tendon repair material in the rabbit. Following surgical manipulation, animals were subject to one of three groups for 4 weeks: (a) full range of motion, (b) partial range of motion (60–90° of flexion), or (c) no range of motion. Histological analysis of the groups with a partial or full range of motion showed dense collagenous connective tissue oriented along the longitudinal axis of the tendon. Spindle-shaped cells were distributed throughout the tendon and oriented along the longitudinal axis of the tendon. The ECM material in rabbits with no range of motion was associated with an increased presence of nonaligned fibroblasts, ECM, mononuclear cells, and blood vessels compared with native tendons. These findings demonstrate that applying physiologic range of motion to early postsurgically implanted ECM scaffold materials facilitates a constructive remodeling response. Additionally, a study by Boruch *et al.* used a Foley catheter to evaluate the effects of physiologic biomechanical loading on ECM scaffold remodeling following partial cystectomy in a canine model. The ECM scaffolds subjected to site-appropriate mechanical loading showed a constructive remodeling response (i.e., a differentiated urothelium and islands of smooth muscle cells), which starkly contrasted the fibrotic response seen in those patients who were deprived of mechanical loading. These findings show that early exposure of site-appropriate mechanical loading mediates a constructive remodeling response after ECM repair.

In vitro models have also been developed in an effort to increase our understanding of the role of mechanical loading in the constructive remodeling response observed with ECM scaffolds *in vivo*. Multiple *in vitro* studies have shown that mechanical loading of naturally derived ECM scaffolds changes the biologic properties of the scaffold and can alter cellular phenotypes and function. Cyclic uniaxial stretching of fibroblasts seeded on the small intestinal submucosa (SIS) ECM scaffold led to increased expression of collagen I, while the expression of collagen III decreased slightly [95]. A study by Borschel *et al.* showed that constructs composed of C2C12 myoblasts seeded onto an ECM scaffold harvested from the mouse extensor digitorum longus muscle produced specific forces that were approximately 5% of that observed for the native muscle after 3 weeks of culture. In a similar fashion, Christ and colleagues seeded primary human muscle precursor cells onto acellular bladder submucosa and subjected the constructs to cyclic strain for up to 4 weeks, followed by subcutaneous implantation of the cell-seeded construct onto the latissimus dorsi of the mice. This study showed that specific forces of 1% of that observed for native latissimus dorsi were produced when the constructs were preconditioned in an *in vitro* bioreactor.

In summary, site-appropriate mechanical loading has been shown to enhance the adhesion, proliferation, fusion, and functionality of progenitor cell populations on various ECM scaffolds *in vitro*, as well as promotes the constructive remodeling *in vivo*. Future studies may provide additional information on the effects of the mechanical environment of other cell types that have been shown to be important contributors to site-specific remodeling, such as macrophages [46] and multipotent progenitor cells [27,92,109].

7.3.3 Modulation of the Host Innate Immune Response Toward a Regulatory and Constructive Phenotype

The *inflammatory response* to surgical implant materials has been the topic of intense investigation for over two decades. Historically, the response of host immune cells to surgically implanted materials was considered to have deleterious consequences [38,39,46,110,111]. For example, in response to synthetic mesh materials (e.g., polypropylene), the host responds acutely with a robust inflammatory cell accumulation, which persists and ultimately results in the formation of foreign body giant cells, granulation tissue, and implant encapsulation. This long-accepted paradigm has resulted in the dogmatic view that the implantation of all nondegradable (or slowly degradable) surgical devices will inevitably lead to an adverse host immune response and a less than optimal tissue remodeling response.

Recently, however, it has been shown that the implantation of appropriately prepared ECM scaffold materials elicits a robust host immune response, yet results in a favorable constructive remodeling outcome (i.e., the formation of functional, site-appropriate tissue). A number of cell types, including PMN leukocytes (e.g., neutrophils) and monocytes/macrophages, are thought to be involved in these dynamic events [38,39,46,110,111]. Studies have shown that within the first few hours of implantation, an abundance of neutrophils invade the ECM scaffold, but persist for only a few days, at which point macrophages accumulate and remain throughout the subsequent 2–6 weeks and slowly decreases thereafter. Interestingly, the macrophage has received considerable attention recently due to its remarkable phenotypic and functional plasticity and ability to modulate tissue remodeling following tissue injury [38,39,46,110,111]. Briefly, the proinflammatory macrophage phenotype (i.e., M1) is characterized by cells that are associated with classic signs of inflammation, whereas the anti-inflammatory macrophage phenotype (i.e., M2) has been shown to promote tissue repair and constructive tissue remodeling [38,112–117]. The M2 population can be further subdivided into wound repair macrophages (M2a), Th2-focused macrophages (M2b), and regulatory macrophages (M2c) [118,119]; however, these polarization states are likely not mutually exclusive, rather represent a blending of phenotypes along a continuum [119].

ECM scaffold materials have been shown to promote a macrophage population enriched in M2 macrophages by 7–14 days postimplantation [38,39,98,110,111]. The constructive remodeling response associated with ECM scaffold materials has been correlated to the ability of the material to modulate macrophage phenotype by showing that the phenotypic profile of macrophages at early time points is a strong statistical predictor of the downstream remodeling events [38,39,98,110]. It has been hypothesized that an initial classically activated *proinflammatory* macrophage polarization state (M1) may be required for the removal of pathogens and necrotic cell debris, and to initiate progenitor cell proliferation; while a subsequent alternatively activated *anti-inflammatory* macrophage polarization state (M2) might then be required to coordinate the events necessary for constructive tissue remodeling (i.e., angiogenesis, progenitor cell differentiation) as the inflammatory response begins to resolve. The precise mechanisms which control the *in vivo* ECM scaffold-modulated macrophage polarization are currently unknown.

7.4 SUMMARY

In summary, this chapter provides an overview of the most important biological features of ECM scaffolds and their clinical applications. Historically, the production of ECM scaffolds has focused largely on maintaining the overall mechanical integrity of the material, with little attention paid to preserving the intrinsic biological properties of the material. It is increasingly being recognized that ECM scaffolds prepared by means that also preserve the innate biologic properties of the native material, along with the three-dimensional ultrastructure, will elicit a favorable host response *in vivo* and ultimately yield improved clinical success. Given the emerging interest in tissue engineering and regenerative medicine applications for ECM scaffolds, a more detailed understanding of the mechanisms that facilitate this ECM-mediated constructive remodeling will undoubtedly be paramount in guiding the development of the next generation of Mother Nature's idea substrate for a bio-inspired material.

Acknowledgment

The authors would like to acknowledge those colleagues whose work we could not cite due to space limitations.

REFERENCES

[1] Bissell, M.J., Hall, H.G., Parry, G. *J. Theor. Biol.* **1982**, *99*, 31.
[2] Bissell, M.J., Aggeler, J. *Prog. Clin. Biol. Res.* **1987**, *249*, 251.
[3] Barkan, D., Green, J.E., Chambers, A.F. *Eur. J. Cancer* **2010**, *46*, 1181.
[4] Vorotnikova, E., McIntosh, D., Dewilde, A., Zhang, J., Reing, J.E., Zhang, L., Cordero, K., Bedelbaeva, K., Gourevitch, D., Heber-Katz, E., Badylak, S.F., Braunhut, S.J. *Matrix Biol.* **2010**, *29*, 690.
[5] Nelson, C.M., Bissell, M.J. *Annu. Rev. Cell Dev. Biol.* **2006**, *22*, 287.
[6] Bornstein, P., Sage, E.H. *Curr. Opin. Cell. Biol.* **2002**, *14*, 608.
[7] Nelson, C.M., Bissell, M.J. *Semin. Cancer Biol.* **2005**, *15*, 342.
[8] Bissell, M.J., Radisky, D.C., Rizki, A., Weaver, V.M., Petersen, O.W. *Differentiation* **2002**, *70*, 537.
[9] Gassmann, P., Enns, A., Haier, J. *Onkologie* **2004**, *27*, 577.
[10] Calve, S., Odelberg, S.J., Simon, H.G. *Dev. Biol.* **2010**, *344*, 259.
[11] Gilbert, T.W., Sellaro, T.L., Badylak, S.F. *Biomaterials* **2006**, *27*, 3675.
[12] Brown, B., Lindberg, K., Reing, J., Stolz, D.B., Badylak, S.F. *Tissue Eng.* **2006**, *12*, 519.
[13] Sacks, M.S., Gloeckner, D.C. *J. Biomed. Mater. Res.* **1999**, *46*, 1.
[14] Crapo, P.M., Gilbert, T.W., Badylak, S.F. *Biomaterials* **2011**, *32*, 3233.
[15] Ott, H.C., Matthiesen, T.S., Goh, S.K., Black, L.D., Kren, S.M., Netoff, T.I., Taylor, D.A. *Nat. Med.* **2008**, *14*, 213.
[16] Uygun, B.E., Soto-Gutierrez, A., Yagi, H., Izamis, M.-L., Guzzardi, M.A., Shulman, C., Milwid, J., Kobayashi, N., Tilles, A., Berthiaume, F., Hertl, M., Nahmias, Y., Yarmush, M.L., Uygun, K. *Nat. Med.* **2010**, *16*, 814.

[17] Nakayama, K.H., Batchelder, C.A., Lee, C.I., Tarantal, A.F. *Tissue Eng. Part A* **2010**, *16*, 2207.
[18] Keane, T.J., Londono, R., Turner, N.J., Badylak, S.F. *Biomaterials* **2012**, *33*, 1771.
[19] Valentin, J.E., Badylak, J.S., McCabe, G.P., Badylak, S.F. *J. Bone Joint Surg. Am.* **2006**, *88*, 2673.
[20] Gilbert, T.W., Freund, J.M., Badylak, S.F. *J. Surg. Res.* **2009**, *152*, 135.
[21] Zheng, M.H., Chen, J., Kirilak, Y., Willers, C., Xu, J., Wood, D. *J. Biomed. Mater. Res. B Appl. Biomater.* **2005**, *73*, 61.
[22] Badylak, S.F., Hoppo, T., Nieponice, A., Gilbert, T.W., Davison, J.M., Jobe, B.A. *Tissue Eng. Part A* **2011**, *17*, 1643.
[23] Badylak, S.F., Vorp, D.A., Spievack, A.R., Simmons-Byrd, A., Hanke, J., Freytes, D.O., Thapa, A., Gilbert, T.W., Nieponice, A. *J. Surg. Res.* **2005**, *128*, 87.
[24] Turner, N.J., Yates, A.J., Jr., Weber, D.J., Qureshi, I.R., Stolz, D.B., Gilbert, T.W., Badylak, S.F. *Tissue Eng. Part A* **2010**, *16*, 3309.
[25] Valentin, J.E., Turner, N.J., Gilbert, T.W., Badylak, S.F. *Biomaterials* **2010**, *31*, 7475.
[26] Mase, V.J., Jr., Hsu, J.R., Wolf, S.E., Wenke, J.C., Baer, D.G., Owens, J., Badylak, S.F., Walters, T.J. *Orthopedics* **2010**, *33*, 511.
[27] Zantop, T., Gilbert, T.W., Yoder, M.C., Badylak, S.F. *J. Orthop. Res.* **2006**, *24*, 1299.
[28] Iannotti, J.P., Codsi, M.J., Kwon, Y.W., Derwin, K., Ciccone, J., Brems, J.J. *J. Bone Joint Surg. Am.* **2006**, *88*, 1238.
[29] Wood, J.D., Simmons-Byrd, A., Spievack, A.R., Badylak, S.F. *J. Am. Vet. Med. Assoc.* **2005**, *226*, 1095.
[30] Sutherland, R.S., Baskin, L.S., Hayward, S.W., Cunha, G.R. *J. Urol.* **1996**, *156*, 571.
[31] Jones, J.S., Vasavada, S.P., Abdelmalak, J.B., Liou, L., Ahmed, E.S., Zippe, C.D., Rackley, R.R. *Urology* **2005**, *65*, 1163.
[32] Jones, J.S., Rackley, R.R., Berglund, R., Abdelmalak, J.B., DeOrco, G., Vasavada, S.P. *BJU Int.* **2005**, *96*, 103.
[33] Mantovani, F., Trinchieri, A., Castelnuovo, C., Romano, A.L., Pisani, E. *Eur. Urol.* **2003**, *44*, 600.
[34] MacLeod, T.M., Sarathchandra, P., Williams, G., Sanders, R., Green, C.J. *Burns* **2004**, *30*, 431.
[35] Zhang, F., Zhu, C., Oswald, T., Lei, M.P., Lineaweaver, W.C. *Ann. Plast. Surg.* **2003**, *51*, 488.
[36] Helton, W.S., Fisichella, P.M., Berger, R., Horgan, S., Espat, N.J., Abcarian, H. *Arch. Surg.* **2005**, *140*, 549.
[37] Malcarney, H.L., Bonar, F., Murrell, G.A. *Am. J. Sports Med.* **2005**, *33*, 907.
[38] Badylak, S.F., Valentin, J.E., Ravindra, A.K., McCabe, G.P., Stewart-Akers, A.M. *Tissue Eng. Part A* **2008**, *14*, 1835.
[39] Brown, B.N., Valentin, J.E., Stewart-Akers, A.M., McCabe, G.P., Badylak, S.F. *Biomaterials* **2009**, *30*, 1482.
[40] Kasimir, M.T., Rieder, E., Seebacher, G., Nigisch, A., Dekan, B., Wolner, E., Weigel, G., Simon, P. *J. Heart Valve Dis.* **2006**, *15*, 278.
[41] Sandor, M., Xu, H., Connor, J., Lombardi, J., Harper, J.R., Silverman, R.P., McQuillan, D.J. *Tissue Eng. Part A* **2008**, *14*, 2021.
[42] Lee, M.S. *Orthopedics* **2004**, *27*, s151.
[43] Abraham, G.A., Murray, J., Billiar, K., Sullivan, S.J. *J. Biomed. Mater. Res.* **2000**, *51*, 442.
[44] Jarman-Smith, M.L., Bodamyali, T., Stevens, C., Howell, J.A., Horrocks, M., Chaudhuri, J.B. *J. Mater. Sci. Mater. Med.* **2004**, *15*, 925.

[45] van der Laan, J.S., Lopez, G.P., van Wachem, P.B., Nieuwenhuis, P., Ratner, B.D., Bleichrodt, R.P., Schakenraad, J.M. *Int. J. Artif. Organs.* **1991**, *14*, 661.

[46] Badylak, S.F., Gilbert, T.W. *Semin. Immunol.* **2008**, *20*, 109.

[47] Li, F., Li, W., Johnson, S., Ingram, D., Yoder, M., Badylak, S. *Endothelium* **2004**, *11*, 199.

[48] Beattie, A.J., Gilbert, T.W., Guyot, J.P., Yates, A.J., Badylak, S.F. *Tissue Eng. Part A* **2009**, *15*, 1119.

[49] Reing, J.E., Zhang, L., Myers-Irvin, J., Cordero, K.E., Freytes, D.O., Heber-Katz, E., Bedelbaeva, K., McIntosh, D., Dewilde, A., Braunhut, S.J., Badylak, S.F. *Tissue Eng. Part A* **2009**, *15*, 605.

[50] Hodde, J.P., Record, R.D., Liang, H.A., Badylak, S.F. *Endothelium* **2001**, *8*, 11.

[51] Hodde, J., Janis, A., Hiles, M. *J. Mater. Sci. Mater. Med.* **2007**, *18*, 545.

[52] Hodde, J., Janis, A., Ernst, D., Zopf, D., Sherman, D., Johnson, C. *J. Mater. Sci. Mater. Med.* **2007**, *18*, 537.

[53] Freytes, D.O., Stoner, R.M., Badylak, S.F. *J. Biomed. Mater. Res. B Appl. Biomater.* **2008**, *84*, 408.

[54] Sun, W.Q., Leung, P. *Acta Biomater.* **2008**, *4*, 817.

[55] Moreau, M.F., Gallois, Y., Basle, M.F., Chappard, D. *Biomaterials* **2000**, *21*, 369.

[56] Gouk, S.S., Lim, T.M., Teoh, S.H., Sun, W.Q. *J. Biomed. Mater. Res. B Appl. Biomater.* **2008**, *84*, 205.

[57] Freytes, D.O., Tullius, R.S., Valentin, J.E., Stewart-Akers, A.M., Badylak, S.F. *J. Biomed. Mater. Res. A* **2008**, *87*, 862.

[58] Hafeez, Y.M., Zuki, A.B., Yusof, N., Asnah, H., Loqman, M.Y., Noordin, M.M., Ainul-Yuzairi, M.Y. *Cell Tissue Bank.* **2005**, *6*, 85.

[59] Curtil, A., Pegg, D.E., Wilson, A. *Cryobiology* **1997**, *34*, 13.

[60] Gilbert, T.W., Gilbert, S., Madden, M., Reynolds, S.D., Badylak, S.F. *Ann. Thorac. Surg.* **2008**, *86*, 967.

[61] Dahan, N., Zarbiv, G., Sarig, U., Karram, T., Hoffman, A., Machluf, M. *Tissue Eng. Part A* **2012**, *18*, 411.

[62] Yang, D., Guo, T., Nie, C., Morris, S.F. *Ann. Plast. Surg.* **2009**, *62*, 297.

[63] Whitaker, I.S., Prowse, S., Potokar, T.S. *Ann. Plast. Surg.* **2008**, *60*, 333.

[64] Clarke, K.M., Lantz, G.C., Salisbury, S.K., Badylak, S.F., Hiles, M.C., Voytik, S.L. *J. Surg. Res.* **1996**, *60*, 107.

[65] Daly, K.A., Wolf, M., Johnson, S.A., Badylak, S.F. *Tissue Eng. Part C Methods* **2011**, *17*, 631.

[66] Agrawal, V., Brown, B.N., Beattie, A.J., Gilbert, T.W., Badylak, S.F. *J. Tissue Eng. Regen. Med.* **2009**, *3*, 590.

[67] Badylak, S., Kokini, K., Tullius, B., Simmons-Byrd, A., Morff, R. *J. Surg. Res.* **2002**, *103*, 190.

[68] Derwin, K.A., Badylak, S.F., Steinmann, S.P., Iannotti, J.P. *J. Shoulder Elbow Surg.* **2010**, *19*, 467.

[69] Youn, I., Jones, D.G., Andrews, P.J., Cook, M.P., Suh, J.K. *Clin. Orthop. Relat. Res.* **2004**, *223*.

[70] Robinson, K.A., Li, J., Mathison, M., Redkar, A., Cui, J., Chronos, N.A., Matheny, R.G., Badylak, S.F. *Circulation* **2005**, *112*, I135.

[71] Ota, T., Gilbert, T.W., Schwartzman, D., McTiernan, C.F., Kitajima, T., Ito, Y., Sawa, Y., Badylak, S.F., Zenati, M.A. *J. Thorac. Cardiovasc. Surg.* **2008**, *136*, 1309.

[72] Rapp, S.J., Jones, D.C., Gerety, P., Taylor, J.A. *Surgery* **2012**, *152*, 595.

[73] Turner, N.J., Badylak, J.S., Weber, D.J., Badylak, S.F. *J. Surg. Res.* **2011**, *176*, 490.

[74] Freytes, D.O., Martin, J., Velankar, S.S., Lee, A.S., Badylak, S.F. *Biomaterials* **2008**, *29*, 1630.
[75] DeQuach, J.A., Mezzano, V., Miglani, A., Lange, S., Keller, G.M., Sheikh, F., Christman, K.L. *PLoS ONE* **2010**, *5*, e13039.
[76] DeQuach, J.A., Yuan, S.H., Goldstein, L.S., Christman, K.L. *Tissue Eng. Part A* **2011**, *17*, 2583.
[77] Hong, Y., Huber, A., Takanari, K., Amoroso, N.J., Hashizume, R., Badylak, S.F., Wagner, W.R. *Biomaterials* **2011**, *32*, 3387.
[78] Seif-Naraghi, S.B., Salvatore, M.A., Schup-Magoffin, P.J., Hu, D.P., Christman, K.L. *Tissue Eng. Part A* **2010**, *16*, 2017.
[79] Singelyn, J.M., DeQuach, J.A., Seif-Naraghi, S.B., Littlefield, R.B., Schup-Magoffin, P.J., Christman, K.L. *Biomaterials* **2009**, *30*, 5409.
[80] Young, D.A., Ibrahim, D.O., Hu, D., Christman, K.L. *Acta Biomater.* **2011**, *7*, 1040.
[81] Voytik-Harbin, S.L., Brightman, A.O., Kraine, M.R., Waisner, B., Badylak, S.F. *J. Cell Biochem.* **1997**, *67*, 478.
[82] Van Vlierberghe, S., Dubruel, P., Schacht, E. *Biomacromolecules* **2011**, *12*, 1387.
[83] Davis, G.E., Bayless, K.J., Davis, M.J., Meininger, G.A. *Am. J. Pathol.* **2000**, *156*, 1489.
[84] Sarikaya, A., Record, R., Wu, C.C., Tullius, B., Badylak, S., Ladisch, M. *Tissue Eng.* **2002**, *8*, 63.
[85] Brennan, E.P., Reing, J., Chew, D., Myers-Irvin, J.M., Young, E.J., Badylak, S.F. *Tissue Eng.* **2006**, *12*, 2949.
[86] Voytik-Harbin, S.L., Brightman, A.O., Waisner, B., Lamar, C.H., Badylak, S.F. *In Vitro Cell Dev. Biol. Anim.* **1998**, *34*, 239.
[87] Record, R.D., Hillegonds, D., Simmons, C., Tullius, R., Rickey, F.A., Elmore, D., Badylak, S.F. *Biomaterials* **2001**, *22*, 2653.
[88] Gilbert, T.W., Stewart-Akers, A.M., Badylak, S.F. *Biomaterials* **2007**, *28*, 147.
[89] Gilbert, T.W., Stewart-Akers, A.M., Simmons-Byrd, A., Badylak, S.F. *J. Bone Joint Surg. Am.* **2007**, *89*, 621.
[90] Brennan, E.P., Tang, X.H., Stewart-Akers, A.M., Gudas, L.J., Badylak, S.F. *J. Tissue Eng. Regen. Med.* **2008**, *2*, 491.
[91] Agrawal, V., Tottey, S., Johnson, S.A., Freund, J.M., Siu, B.F., Badylak, S.F. *Tissue Eng. Part A* **2011**, *17*, 2435.
[92] Badylak, S.F., Park, K., Peppas, N., McCabe, G., Yoder, M. *Exp. Hematol.* **2001**, *29*, 1310.
[93] Crisan, M., Yap, S., Casteilla, L., Chen, C.W., Corselli, M., Park, T.S., Andriolo, G., Sun, B., Zheng, B., Zhang, L., Norotte, C., Teng, P.N., Traas, J., Schugar, R., Deasy, B.M., Badylak, S., Buhring, H.J., Giacobino, J.P., Lazzari, L., Huard, J., Peault, B. *Cell Stem. Cell* **2008**, *3*, 301.
[94] Hodde, J.P., Badylak, S.F., Shelbourne, K.D. *Adv. Exp. Med. Biol.* **1997**, *3*, 27.
[95] Gilbert, T.W., Stewart-Akers, A.M., Sydeski, J., Nguyen, T.D., Badylak, S.F., Woo, S.L. *Tissue Eng.* **2007**, *13*, 1313.
[96] Boruch, A.V., Nieponice, A., Qureshi, I.R., Gilbert, T.W., Badylak, S.F. *J. Surg. Res.* **2010**, *161*, 217.
[97] Allman, A.J., McPherson, T.B., Badylak, S.F., Merrill, L.C., Kallakury, B., Sheehan, C., Raeder, R.H., Metzger, D.W. *Transplantation* **2001**, *71*, 1631.
[98] Valentin, J.E., Stewart-Akers, A.M., Gilbert, T.W., Badylak, S.F. *Tissue Eng. Part A* **2009**, *15*, 1687.
[99] Agrawal, V., Johnson, S.A., Reing, J., Zhang, L., Tottey, S., Wang, G., Hirschi, K.K., Braunhut, S., Gudas, L.J., Badylak, S.F. *Proc. Natl. Acad. Sci. U.S.A.* **2010**, *107*, 3351.

[100] Agrawal, V., Kelly, J., Tottey, S., Daly, K.A., Johnson, S.A., Siu, B.F., Reing, J., Badylak, S.F. *Tissue Eng. Part A* **2011**, *17*, 3033.
[101] Agrawal, V., Siu, B.F., Chao, H., Hirschi, K.K., Raborn, E., Johnson, S.A., Tottey, S., Hurley, K.B., Medberry, C.J., Badylak, S.F. *Tissue Eng. Part A* **2012**, *18*, 1454.
[102] Tottey, S., Corselli, M., Jeffries, E.M., Londono, R., Peault, B., Badylak, S.F. *Tissue Eng. Part A* **2011**, *17*, 37.
[103] Tottey, S., Johnson, S.A., Crapo, P.M., Reing, J.E., Zhang, L., Jiang, H., Medberry, C.J., Reines, B., Badylak, S.F. *Biomaterials* **2011**, *32*, 128.
[104] Nieponice, A., Gilbert, T.W., Johnson, S.A., Turner, N.J., Badylak, S.F. *J. Surg. Res.* **2012**, *182*, e1.
[105] Turner, N.J., Johnson, S.A., Badylak, S.F. *Arch. Histol. Cytol.* **2010**, *73*, 103.
[106] Wolf, M.T., Daly, K.A., Reing, J.E., Badylak, S.F. *Biomaterials* **2012**, *33*, 2916.
[107] Siracusano, S., Ciciliato, S., Bernabei, M., Vatovani, V., Lampropoulou, N., Knez, R. *Eur. Urol.* **2008**, *7*, 147.
[108] Hammond, J.S., Gilbert, T.W., Howard, D., Zaitoun, A., Michalopoulos, G., Shakesheff, K.M., Beckingham, I.J., Badylak, S.F. *J. Hepatol.* **2011**, *54*, 279.
[109] Almarza, A.J., Yang, G., Woo, S.L., Nguyen, T., Abramowitch, S.D. *Tissue Eng. Part A* **2008**, *14*, 1489.
[110] Brown, B.N., Londono, R., Tottey, S., Zhang, L., Kukla, K.A., Wolf, M.T., Daly, K.A., Reing, J.E., Badylak, S.F. *Acta Biomater.* **2012**, *8*, 978.
[111] Brown, B.N., Ratner, B.D., Goodman, S.B., Amar, S., Badylak, S.F. *Biomaterials* **2012**, *33*, 3792.
[112] Mosser, D.M. *J. Leukoc. Biol.* **2003**, *73*, 209.
[113] Mantovani, A., Sica, A., Sozzani, S., Allavena, P., Vecchi, A., Locati, M. *Trends Immunol.* **2004**, *25*, 677.
[114] Stout, R.D., Jiang, C., Matta, B., Tietzel, I., Watkins, S.K., Suttles, J. *J. Immunol.* **2005**, *175*, 342.
[115] Anderson, C.F., Mosser, D.M. *J. Leukoc. Biol.* **2002**, *72*, 101.
[116] Gordon, S., Taylor, P.R. *Nat. Rev. Immunol.* **2005**, *5*, 953.
[117] Mantovani, A., Sica, A., Locati, M. *Immunity* **2005**, *23*, 344.
[118] Mantovani, A., Sozzani, S., Locati, M., Allavena, P., Sica, A. *Trends Immunol.* **2002**, *23*, 549.
[119] Lawrence, T., Natoli, G. *Nat. Rev. Immunol.* **2011**, *11*, 750.

CHAPTER 8

BIO-INSPIRED INTEGRATION OF NATURAL MATERIALS

ALBINO MARTINS, MARTA ALVES DA SILVA, ANA COSTA-PINTO, RUI L. REIS, and NUNO M. NEVES

3B's Research Group—Biomaterials, Biodegradables, and Biomimetics, University of Minho, Headquarters of the European Institute of Excellence on Tissue Engineering and Regenerative Medicine, Guimarães, Portugal; ICVS/3B's— PT Government Associate Laboratory, Braga/Guimarães, Portugal

8.1 INTRODUCTION

Natural extracellular matrices (ECMs) have been isolated and extracted from various tissues, such as small intestine submucosa, skin (from cadavers), pancreas, and breast [1]. Although these purified ECMs certainly have useful applications, their use is limited in scope, owing to the need for well-defined microenvironments in tissue regeneration and stem cell transplantation, in which animal by-products and contaminants must be limited. Moreover, it has been demonstrated that, besides the spatial framework, tissue-specific ECM cues are essential to regenerate an organ or tissue [2]. It is well established that ECM is dynamic and has an instructive role in building a tissue and in its regeneration after trauma or disease [2].

Tissue engineering (TE) has been recognized as a promising alternative to the use of autografts or allografts or even xenografts for tissue reconstruction and regeneration. Tissue engineering and regenerative medicine (TERM) aims at the development of biological substitutes that restore, maintain, or improve tissue function or a whole organ [3]. This approach utilizes cells, biomaterial scaffolds, and signaling molecules for the repair of diseased or damaged tissues. In the TERM strategies, the development of a man-made/synthetic ECM is a critical issue, since we need to learn how to engineer biomaterials that will help in recapitulating the early events of morphogenesis [4]. Currently, biomaterial scaffolds are designed to support cell and tissue growth, aiming at a macroscopic level to be compatible with the mechanical

Bio-inspired Materials for Biomedical Engineering, First Edition. Edited by Anthony B. Brennan and Chelsea M. Kirschner.

loading of the surrounding organs and tissues. However, to maintain the proper cell phenotype in an engineered biomaterial scaffold, it may be necessary to recreate the complexity and hierarchical organization observed in natural ECM, seeding/infiltration of cells into the biomaterial scaffold, and culturing the seeded scaffold with adequate nutrient supply [5].

Natural ECM materials provide physiologically relevant cellular environments, as they are a rich source of bioactive molecules. Therefore, those natural ECM-origin materials have been widely studied to engineer ECM analogs. However, to understand the molecular and biophysical mechanisms by which the ECM analogs elicit diverse effects on cellular differentiation and morphogenesis, it is crucial to use chemically and physically defined ECMs that can be reliably reproduced. In this respect, synthetic ECMs have been developed that feature defined and tunable compositions, organization, biomechanics, and ECM remodeling capabilities. Considering all these assumptions, with the present chapter, we intend to provide an updated overview on the applicability of natural materials, mainly the ones present in the ECM composition, to the processing of biomaterial scaffolds for TERM approaches.

8.1.1 Extracellular Matrix (ECM) Structure and Composition

The ECM of human tissues is a dynamic and hierarchically organized structure composed of water, proteins, and polysaccharides (such as the glycosaminoglycans [GAGs]: hyaluronic acid [HA], dermatan sulfate [DS], chondroitin sulfate [CS], heparin, heparan sulfate [HS], and keratan sulfate [KS]), proteins (such as collagen, elastin, fibronectin, and laminin), and proteoglycans (PGs) (including aggrecan, brevican, decorin, keratocan, lumican, neurocan, perlecan, syndecans, and versican) synthesized by the adjacent cells [6–8]. In this complex structure, the collagen fibers provide strength to the tissue and, more importantly, have many cell-adhesive peptide moieties intended to allow for cellular anchoring. This hydrated gel composed of PGs and other proteins fills the extracellular space, creating an appropriate microenvironment for ensuring the tissue maintenance and remodeling by cells in response to appropriate stimuli, while allowing for the diffusion of nutrients, metabolites, and signaling molecules. These components interact to form an interconnected nano- or microranged fibrous network bound to the membranes of cells. Indeed, tissue ECMs act as a scaffold to support and hold cells together, to control their structure, and to regulate cellular functions like adhesion, migration, proliferation, differentiation, and ultimately, tissue morphogenesis [8,9]. The ECM also serves as a storage depot and a controlled release system for growth factors and signaling molecules.

The ECM interacts with the adjacent cells both mechanically and chemically, remodeling the architecture of the tissues. The structure of different collagen types within the ECM determines its function as a structural element of the connective tissues [6]. Tendon ECM, for example, is composed of parallel and aligned collagen fibrils, while those found on the skin are mesh-like. In most connective tissues, the matrix macromolecules are secreted by fibroblastic cells into the extracellular space. In specialized types of connective tissues, such as cartilage and bone, cells of the fibroblast family (chondrocytes and osteoblasts, respectively) are responsible for

ECM deposition. The matrix either becomes calcified into the hard and tough structures of bone and teeth, or can form the transparent matrix of cornea. ECM can also adopt the cord-like organization that gives tendons their tensile strength and elasticity.

8.1.2 Fundamentals of Scaffolding Using Naturally Derived Materials

Efforts in the area of TERM have been directed to produce biomaterial scaffolds, which physically support cells and providing conditions for cell adhesion and growth, mimicking the native ECM of tissues [10]. Those scaffolds can be obtained from different materials, including biodegradable polymers, that is, synthetic or natural in origin, ceramics or composites containing both polymer and ceramic phases. Generally, those systems aim at being resorbable under physiological conditions. The degradation kinetics of an ideal scaffold should follow the tissue growth kinetics, in such a way that the material is completely degraded when the tissue is fully regenerated [3]. Moreover, appropriate biocompatibility, porosity, pore size, surface properties, and mechanical stability have been defined as being critical requirements of a biomaterial scaffold [11,12]. The biomaterial scaffold biocompatibility relies on the nontoxic effect of both the scaffold and of its degradation products, to ensure device safety [13]. Another requisite is the total scaffold porosity and the dimensions of the pores, in order to obtain enough surface area for cells attachment and consequent proliferation. The interconnectivity of those pores is a critical issue in ensuring opportunities for cell migration and further colonization of the scaffold surface. Furthermore, pore interconnectivity allows the scaffold to maintain cell viability by enabling diffusion into the scaffold of required nutrients and oxygen, and diffusion out of metabolic residues. The chemistry/bioactivity of the scaffold surface is important to provide the right cues for the cells to secrete ECM components, to allow the ECM to mature and to facilitate cellular remodeling of the construct. Another important aspect contributing to the efficacy of the biomaterial scaffold in mimicking the functionality of the natural ECM is the balance between the level of porosity and the mechanical properties [11]. Ideally, the mechanical properties should provide a good and stable integration of the populated construct with the surrounding host tissues to sustain the stresses applied on implantation. Ultimately, all these scaffolding requirements will drive the cells to build up fully functional adult tissues.

8.2 NATURALLY DERIVED MATERIALS

As previously mentioned, natural ECM is mainly composed of collagen, arranged in a hierarchical manner with laminin, fibronectin, and PGs in a complex topography at the nanometer range [14]. Three-dimensional (3-D) natural ECM biomaterials provide physiologically relevant cellular environments, as they are a rich source of bioactive molecules, so they have been widely studied to engineer tissues. ECM protein such as collagen or chitosan are very good candidates for these approaches and can be modified for specific cell types [14]. Natural-origin biomaterials also allow the diffusion of soluble molecules to the basal and apical surface [14]. These polymers are often combined with other natural or synthetic biomaterials to produce scaffolds for TE applications (Table 8.1). The following sections describe the origins and applications of many naturally derived materials.

8.2.1 Animal Origin

8.2.1.1 Collagen Collagen is the most abundant protein in animals, providing the main structural and mechanical support to the tissues such as cartilage, bone, and teeth [14]. This protein also forms molecular strands that strengthen the tendons and strong sheets that support the skin and internal organs [15]. Among the different collagen types, type I collagen is the most abundant component of the ECM and can be used as scaffolding material, promoting cell migration, wound healing, and tissue regeneration [16]. Collagen may be used for TE in the following applications: artificial skin, tendons or blood vessels, bone graft substitutes [17], dental [18], corneal, or stress urinary incontinence implants, and regeneration of nerve [19], cartilage [20], skin, or other organs [16]. Collagen-based biomaterials can be prepared from two fundamental techniques. One is to decellularize the collagen matrix, preserving the original tissue shape and ECM structure, while the other relies on extraction, purification, and polymerization of collagen and its components to produce a functional scaffold.

In a recent work, collagen type I scaffolds were produced by combining gel and electrospinning technologies to create a novel 3-D hybridized collagen implant with an aligned ultrastructure, aiming at tendon regeneration. The construct was implanted in rabbits, and results showed that the implanted construct was biodegradable, biocompatible, and possibly could be considered as a substitute for allografts in the near future [21].

Collagen-based scaffolds have been extensively used in clinical practice, such as Chondro-Gide® (Geistlich Pharma AG, Wolhusen, Switzerland), aimed at repairing cartilage defects [22]. Short-term results are very promising, however, long-term follow-up studies are needed to determine if the grafted area will maintain structural and functional integrity over time [22].

8.2.1.2 Fibronectin Fibronectin has several important functions in the ECM, such as providing structural support and signaling cues for cell survival, migration, differentiation, and growth [23], as well as inducing cell attachment and spreading through its cell binding sites and related synergy sites [24]. Fibronectin exists in a soluble form in the plasma and in an insoluble, fibrillar form in the ECM [25]. During tissue repair, fibronectin is converted into biologically active ECM fibrils through a cell-dependent process [25]. Fibronectin-based scaffolds and hydrogels have been used for different TE applications such as spinal cord injury [26] or to improve the functionality of other biomaterials [27].

Fibronectin matrix mimetic variants were designed and analyzed for their ability to support new ECM assembly. The ability of fibronectin matrix mimetics to direct cell–substrate interactions and regulate ECM assembly was demonstrated,

TABLE 8.1 Biomaterials Used in Tissue Engineering and Regenerative Medicine Approaches Derived from Different Natural Sources

Materials	Source	Processing method	Cell types/*In vivo*	Relevant outputs	Reference
Animal origin					
Collagen	Calf	Hydrogel with BMSCs encapsulated	Implanted subcutaneously in rabbits	Induced *in vivo* BMSCs chondrogenesis, without GFs addition	[20]
	Bovine skin	Electrospinning	hBMSCs	Support MSCs growth without compromising their osteogenic differentiation capability	[17]
	Wistar rat tails	Self-assembled collagen matrix	Neural cells	Neurite outgrowth within the 3-D matrix	[19]
	Bovine	Lyophilization	Oral epithelial cells and fibroblasts	Fibroblasts synthesized new ECM where the average collagen fiber diameter was close to that of native oral mucosa	[18]
Fibronectin	Human plasma	Polyester aminolysis followed by Fn coupling to electrospun PLLC NFM	Porcine esophageal epithelial cells	Promoted epithelium regeneration	[28]
	Human	Injectable hydrogel composed by fibrin and fibronectin	Knife-cut cavity in the rat spinal cord	Gel integrated with the host spinal cord tissue, and supported robust growth of axons	[26]
	Human serum	Covalent immobilization onto chitosan membranes	SaOs-2 human osteoblast-like cells	Fn improved the biological performance of the chitosan membrane	[27]

(Continued)

TABLE 8.1 *(Continued)*

Materials	Source	Processing method	Cell types/*In vivo*	Relevant outputs	Reference
Hyaluronic acid / Hyaluran	Lactic acid bacteria	Wetspinning method	Rabbit articular chondrocytes	Cartilage-like ECM formation with strong collagen type II expression	[31]
	Sodium hyaluronate	Hydrogel	Human intra-articular injection	Symptomatic benefit which persisted for 6 months	[34]
	Acrylated hyaluronic acid	Hydrogel	hMSCs for rat calvarial defect regeneration	HA-based hydrogel can be used for cell and growth factor carriers for tissue regeneration.	[32]
Fibrin	Human fibrinogen	Gels	Dorsal root ganglia (DRG) neurite extension	Cells responded to their underlying substrate and attempted the necessary processes to remodel the surrounding ECM.	[41]
	Tisseel VH fibrin sealant	Fibrin glue	Cells in fibrin glue injected into the infarcted area of the rat coronary artery	Fibrin glue increases cell transplant survival, decreases infarct size, and increases blood flow to ischemic myocardium.	[42]
	PEGylated fibrin	Patch	Porcine MSCs	Increases MSC viability and causes phenotypic changes in MSCs consistent with endothelial cells.	[43]
Chondroitin sulfate	Shark cartilage	Hydrogel	Chondral defects in rabbits	Postponed OA by improving the biomechanical and histological properties of knee joints	[56]
	Bovine tendon	Cross-linked scaffolds	Allogenic skin fibroblasts implanted on the dorsum of Sprague-Dawley rats	Successful repair of full-thickness defects on skin of rats	[55]

Heparan sulfate	Bovine kidney	Cross-linked to collagenous matrices	Subcutaneous implantation in rats	Promoted angiogenesis *in vivo* and the formation of new tissue	[63]
		Immobilized on electrospun PCL nanofiber meshes	Scaffolds seeded with hMSCs precommitted to osteogenic lineage and implanted subcutaneously into the dorsum of nude rats	Survival of the implanted human cells was affected by the host response to the implant regardless of the presence of HS	[64]
Dermatan sulfate	Recombinant	Gel with collagen	Human coronary artery smooth muscle cells	Modulated collagen structure and tissue stiffness, aid cellular adhesion	[58]
	Porcine intestinal mucosa	PEG-containing alginate/chitosan/dermatan sulfate microspheres	Mouse fibroblasts (NIH 3T3 cell line)	Significant stimulation of cell proliferation *in vitro*	[59]
Silk	*Bombyx mori* silk cocoons	Films	Human and rabbit corneal fibroblast (hCF)	hCF remained viable, proliferated throughout the construct, and initial native matrix was produced	[70]
		Porous microstructures in blends of silk fibroin and hyaluran	Bone marrow MSCs	Silk fibroin/hyaluronan scaffolds resulted in more efficient tissue formation compared with plain silk scaffolds	[68]
		Fibroin microparticles for BMP-2 delivery	C2C12 cells and in a rat ectopic model	No significant cytotoxicity and evidence of ectopic bone formation	[71]
Chitosan	Chitin (medium molecular weight)	Insulin-loaded chitosan particle-aggregated scaffolds	ATDC5 cell line	Effective promotion of chondrogenic differentiation	[81]
	Chitin (75–80% deacetylation)	Cross-linking and freeze-drying	Human venous fibroblasts	Enhanced cell adhesion, proliferation, and viability	[80]
	Chitin	*Unidirectional freezing* method followed by freeze-drying	Sciatic nerve defect in rats	Nerve regeneration and functional recovery equivalent to that of an autograft	[79]

(Continued)

TABLE 8.1 *(Continued)*

Materials	Source	Processing method	Cell types/*In vivo*	Relevant outputs	Reference
Plant origin					
Starch	Corn starch/ethylene vinyl alcohol coated with a biomimetic calcium phosphate corn starch and cellulose acetate	Extrusion with blowing agent	Rat femur model	All materials exhibited a favorable bony response rapidly forming initial *connective tissue* seen around all scaffolds	[101]
	Corn starch/ polycaprolactone	Fiber bonding	Bovine articular chondrocytes	*In vitro* assessment ofcartilaginous markers	[105]
		Fiber bonding	Endothelial cells	ECs maintain a normal phenotype and genotype on SPCL fiber-mesh scaffolds	[107]
Algae origin					
Alginate	Alginate–heparin gel releasing bFGF	Heparin and alginate covalently cross-linked with ethylenediamine	Subcutaneous implantation	Cellular infiltration and angiogenesis were shown to occur inthe matrix for the 2 weeks of implantation	[114]
	Chitosan–alginate hydrogel	Thermally induced phase separation followed by freeze-drying	*In vitro* assessment of cartilage markers with HTB-94 cells	Cells maintained their round-shape morphology and express cartilage-specific markers	[120]
	Calcium phosphate cement paste combined with alginate hydrogel microbeads		hUCMSCs encapsulated	hUCMSCs differentiated into the osteogenic phenotype and the cell construct presented load-bearing capability	[122]

Carrageenan	Red marine algae	Alginate hydrogels with TGF-β1 by ionic cross-linking	Human ASCs	Carrageenan hydrogel withTGF-β1 incorporated enhanced the cartilage differentiation of hASCs	[128]
	Red marine algae	Alginate hydrogels by ionic cross-linking	hASCs hNCs ATDC5 cell line	The hydrogel supported all types of chondrogenic related cells.	[129]
Ulvan	Green algae, *Ulva lactuca*	3-D porous structures by freeze-drying and cross-linking	L929	Cytotoxic evaluation showed that the structures were not toxic to cells.	[137]
Microbial origin					
Cellulose	Bacterial cellulose membranes were supplied from Fibrocel®	Membranes with hydroxyapatite incorporated	Noncritical bone defects in rat tibiae	Bone formation after 4 weeks	[144]
	Gluconacetobacter xylinus	Liophylization	Primary bovine chondrocytes	Cartilage	[141]
Gellan gum	Gelzan®, from *Sphingomonas elodea*	Hydrogel obtained by ionic cross-linking with CaCl2	*In vitro* cytotoxic evaluation with L929 cell line *In vitro* encapsulation with hNCs	Hydrogel revealed to be noncytotoxic hNCs kept viable and maintain its phenotype	[147]
		Hydrogel obtained by ionic cross-linking with CaCl2	Human articular chondrocytes implantation in nude mice	Hydrogels support the growth and ECM deposition of human articular chondrocytes when implanted *in vivo*	[148]
		Ionic- and photocross-linked methacrylated gellan gum hydrogels	*In vitro* evaluation with fibroblast cells (L929 cells) and human intervertebral disk (hIVD) cells *In vivo* subcutaneous implantation in rats	*In vitro* results showed that both types of cells were viable up to 21 days of culturing; *in vivo* implantation evidenced no necrosis, calcification, and acute inflammatory reaction	[152]

(Continued)

TABLE 8.1 *(Continued)*

Materials	Source	Processing method	Cell types/*In vivo*	Relevant outputs	Reference
Pullulan	Pullulan from *Aerobasidium pullulans*	3-D porous scaffolds by freeze-drying	Rat MSCs	Rat MSCs proliferate and maintain its characteristics	
	Pullulan from *Aerobasidium pullulans* Nanohydroxyapatite	3-D porous scaffods by freeze-drying	Human MSCs Two different *in vivo* models: (a) nonosseous (subcutaneously in rats and intramuscularly in goats), (b) osseous (femoral defects in femur condyle of rat and goat mandibles)	*In vitro* studies induced the formation of multicellular aggregates and expression of early and late bone-specific markers of human MSCs in the medium without osteogenic factors; *in vivo* studies showed that the scaffolds induced a highly mineralized tissue in the three models, whatever the site of implantation	
Dextran	*Leuconostoc mesenteroides*	Methacrylated dextran (dex-MA) and hydroxyethyl-methacrylated dextran (dex-lactate-HEMA)	Human skin fibroblasts (cell line PK84)	*In vitro* studies showed that both microspheres and hydrogels did not exert cytotoxic effects over the cells.	
		Dextran-tyramine (Dex-tyr) hydrogels by enzymatic cross-linking reaction with platelet lysate	Human bone marrow MSCs	Dex-tyr hydrogels with platelet lysate incorporated promoted proliferation and triggered chondrogenic differentiation of MSCs.	[173]
		Dextran hydrogels prepared by cross-linking with BMP-2 incorporated	*In vivo* subcutaneous implantation in rats	Hydrogels, when combined with extracted bovine BMP to enhance the bone formation, was evaluated in a rat ectopic model.	[175]

Fn, fibronectin; NFM, nanofiber meshes; OA, osteoarthritis; PEG, polyethylene glycol.

making these variants promising candidates to be used as bioactive surfaces [25]. Moreover, with the objective of mimicking normal epithelium regeneration in a synthetic scaffold, *in vitro*, poly (L-lactide-co-caprolactone) (PLLC) was processed into a nanofibrous scaffold using electrospinning technology, and fibronectin was engrafted onto the scaffold fibers. This scaffold was found to greatly promote the epithelium regeneration [28].

8.2.1.3 Hyaluronic Acid Hyaluronan, also named HA or hyaluronate has been extracted from bovine vitreous humor, rooster combs, or umbilical cords [29]. HA is an important polysaccharide naturally present in the human body and it has various biological roles in human tissues. It is a GAG component of connective tissue, synovial fluid, and in the vitreous humor of the eye [29]. In cartilage, for example, HA organizes the ECM into a strong structure via assembling the large PG aggrecan [30]. It is a highly biocompatible polysaccharide and is often used for biomedical applications including cartilage [31], bone [32], or viscous supplementation in joint diseases [33]. HA injection is a common treatment for osteoarthritis [34], as HA molecules can elicit pro- and anti-inflammatory responses depending in its molecular weight. HA also plays important roles in protein adhesion and provides attachment sites between the chondrocytes and the ECM of articular cartilage [35].

HA-based TE strategies have been used extensively in clinic. For example, the Hyalograft®C (Anika Therapuetics, Bedford, MA) membrane has been used to repair cartilage defects. In a study using this membrane and the technique of matrix-induced chondrocyte implantation (MACI), patients were followed after implantation during 2 years, by magnetic resonance imaging [36]. The postoperative observations showed dynamic processes in cartilage repair over time, with positive effects. Another report of a 5-year follow-up of the MACI technique also shown good results: 8 out of 11 patients rated the function of their knees as much better or better than before the surgery [37].

8.2.1.4 Fibrin Fibrin is not a regular component of the ECM but it exists as a temporary matrix that will be replaced by ECM [38]. Fibrin is the result of fibrinogen polymerization in the presence of thrombin [39]. Its main biological functions are blood clotting, fibrinolysis, cellular and matrix interactions, inflammation, and wound healing [40]. Fibrin is generally used in form of gels [41], glue [42], or patches [43]. Fibrin hydrogels for cartilage TE are usually seeded with chondrocytes to produce composite constructs cultured *in vitro* or *in vivo* to generate cartilaginous matrix [44]. The major disadvantages of fibrin hydrogels include the poor mechanical properties and rapid enzyme-catalyzed degradation [30].

Recently, a 3-D fibrin-based patch was developed for cardiac TE applications. Authors have successfully engineered a human cardiac tissue patch starting from human embryonic stem cell (hESC)-derived cardiomyocytes. The structural and functional properties of these patches provide the closest *in vitro* approximation of native human heart tissue [45].

8.2.1.5 Proteoglycans PGs, including CS, DS, HS, and KS, comprise a core protein to which GAG chains are covalently linked, and are an important structural family of macromolecules found in the ECM [46]. They are one of the major classes of natural compounds found in the ECM that convey important properties to cells, connective tissues, and basal membranes [47]. Cell-surface PGs and ECM are involved in cell signaling, proliferation, adhesion, and motility [48]. PGs have been applied in TE mostly in scaffolds (matrigels and collagen–CS matrices), where PGs or their GAG chains are incorporated into the scaffold to promote cell growth, tissue remodeling, and intracellular signaling [48].

CS has been isolated from natural sources such as bovine cartilage [49], chicken cartilage [50], sturgeon [51], shark, and ray [52]. Both CS and DS chains have interesting functions in the development of the central nervous system, wound repair, infection, growth factor signaling, morphogenesis, and cell division [53]. It is normal to perform a CS cross-linking treatment with other polymers such as gelatin or collagen to increase the stability of the materials [54,55]. A study was performed to evaluate the efficiency of intra-articular injection of CS carried by a hydrogel in the treatment of chondral defects in adult rabbit models [56]. The biomechanical and histological properties of the repaired cartilage were both improved by this hydrogel [56].

DS can be isolated from marine species such as ray skin and the body of clams [57]. DS can act as a stabilizer, cofactor, and/or coreceptor for growth factors, cytokines and chemokines, a regulator for enzyme activity, or as a signaling molecule in response to cellular damage such as wound or infection [53]. Recombinant DS peptidoglycosaminoglycans have been studied for TE, showing that they are able to modulate the structure of other biomaterials, for example, collagen, thus aiding in the cellular adhesion [58]. Likewise, DS incorporated into alginate/chitosan microspheres, helped to stimulate cell proliferation [59].

HS is also a member of the GAG family and its structure is related to that of heparin [60]. HS is usually used to describe heparin-like by-products of the industrial preparation of heparin from animal tissues such as bovine fetal rib growth plate [61]. HS is found in all animal tissues as a PG, in which two or three HS chains are attached in close proximity to cell surface or ECM proteins [62]. HS functions are related to interactions between ECM components essential for the control of cell proliferation, differentiation, adhesion, and migration [60]. HS is normally used for cross-linking with other biomaterials [63] or immobilized in a scaffold [64]. HS conjugated with chitosan has been used for preparing nanoparticles for osteogenic differentiation induction, with promising results [65].

KS is another type of GAG that comprises different forms present in diverse tissues, including articular cartilage, reproductive tissue, and neural tissue [66]. It has been isolated from bovine cornea [46].

8.2.1.6 Silk Silk is the building element of many arthropod nets, cocoons, and prey traps [67]. Silks are a family of structural proteins that are biocompatible, degradable, mechanically superior, and that can be chemically modified for a vast range of biomedical applications [68,69] such as cornea [70] or bone [71].

Fibroin is one of the two main kinds of silkworm silk protein, and has been investigated for biomedical applications. It presents high biocompatibility, biodegradability, limited inflammatory response, and excellent mechanical properties

[67,69]. Silk polymers have a side chain that can be used for binding to globular proteins and growth factors, which has been exploited to produce a platform for precursor cell differentiation [14].

The efficiency of a silk fibroin-based biodegradable material to generate small diameter grafts was evaluated. The graft was implanted in rat abdominal aorta, and it showed excellent long-term patency and optimal mechanical properties. Endothelial and smooth muscle cells migrated into the graft after implantation and organized into endothelial and medial layers, generating a vascular-like structure [72].

8.2.1.7 Chitosan Chitosan is a natural polymer that has interesting properties for TE applications. It is biodegradable, biocompatible, and has many structural similarities to GAGs [73]. Chitosan is obtained from chitin, the second most abundant polysaccharide (after cellulose), which is the major element of the shells of many crustaceans, such as shrimps or crabs [73]. Chitosan is obtained by a de-N-deacetylation (DD) of chitin [73]. This process is usually made in alkaline conditions, resulting into chitosans with different degrees of DD and, thus, different molecular weights [74]. The source of chitosan and the chemical modification may influence both parameters. The DD has been shown to be important for cell biocompatibility, attachment, and growth [75]. One of the interesting properties of chitosan is its ability to be processed into porous structures, which support cell growth and ECM deposition [76]. Additionally, chitosan is reported to degrade *in vivo* mainly by enzymatic hydrolysis [74,77]. Lysozyme, one enzyme present in the human body, has been used to enhance the formation of pores *in situ*, in chitosan scaffolds coated with calcium phosphate [78].

Chitosan-based scaffolds have been studied for many TE applications, such as nerve regeneration [79] or cartilage repair. In this tissue, chitosan-based scaffolds promoted cell proliferation [80], differentiation [81], and metabolic activity [82], as well as mesenchymal stem cell (MSC) chondrogenesis *in vitro* [83] and *in vivo* [84]. Chitosan has been explored for preparing hydrogels, namely a very promising gene-activated chitosan–gelatin matrix [85]. The referred matrices were capable of releasing, in a controlled fashion, transforming growth factor beta 1 (TGF-β1), and promoted chondrocyte proliferation. When seeded with human articular chondrocytes (hACs), this injectable hydrogel was able to regenerate and repair a lesion made in bovine articular cartilage, showing the great potential of this novel cell delivery system for cartilage TE [86].

8.2.2 Plant Origin

8.2.2.1 Cellulose Cellulose is the most abundant naturally occurring polymer of glucose, found as the main constituent of plants and natural fibers such as cotton and linen. Cellulose is also produced by some bacteria, *Acetobacter xylinun*, being the most well studied [87]. *Gluconacetobacter xylinus*, another type of bacteria, is one of the most efficient producers of cellulose [88]. Both types of celluloses (plant and bacterial) are chemically identical, a polymer of β-D-glucose units, which accounts for its high crystallinity and its insolubility in water and other common solvents. Cellulose is degraded by cellulases existent in bacteria and fungi, not present in the human body. It is also to some extent degraded by hydrolysis, although highly resistant, due to its compact structure.

Plant-derived cellulose has not been explored much in the biomedical field, since the human body is not able to degrade it. One of the few reports consists of the development of a thin film using a layer-by-layer technique, combining anionic rod-like cellulose nanocrystals with chitosan [89].

8.2.2.2 Starch Starch is the main carbohydrate reservoir of higher plants, found in storage organs, such as grains of maize and rice or the tubers of cassava and potatoes, present in human diet. Increasingly, starch is also used as a renewable raw material, as a source of energy after conversion to ethanol, and for many different industrial applications [90]. Starch is a relatively simple polymer composed of amylose (20–30%) and amylopectin (70–80%). Amylose is a linear polysaccharide of several thousand units of glucose linked together by $(1 \rightarrow 4)$ bonds. Amylopectin, consist mainly of $(1 \rightarrow 4)$ linked glucose residues, is a branched molecule with $(1 \rightarrow 6)$ linkages at every 25–30 glucose units distance [91].

Different relative weight percentages of both molecules are present in different types of starch, exhibiting significant differences in terms of properties [92]. For instance, for most cereal starches, the relative weight percentages are 18–33% of amylose and 72–82% of amylopectin. Starch is a semicrystalline, with a degree of crystallinity between 15% and 45% [91].

The original application of starch was in the food industry, as a food additive. In its native form, it is used as thickener, when heated in water, the helices within the amylopectin of starch melt and the granule starts to swell, increasing the viscosity of the solution [93]. Starch is also used in the paper industry, where it is used as a filler to give the strength to the final paper sheet. It is also used in paper coatings to improve quality [90]. Starch is enzymatically degraded by α-amylase, which exists in human blood serum [94].

In the biomedical field, Reis *et al.* [95] have developed extensive work concerning the investigation of several blends of corn starch with synthetic polymers namely: polyethylene-vinyl alcohol (SEVA-C); cellulose acetate (SCA); polycaprolactone (PCL), and polylactic acid (SPLA). These starch blends showed to be biocompatible both *in vitro* [96,97] and *in vivo* [98]. Starch blends can be processed by thermoplastic techniques into different shapes, such as microspheres [99], 3-D porous scaffolds [100–102], and hydrogels [103]. Starch microspheres were used to release growth factors to enhance osteogenic differentiation [99]. Osteoblast like cells adhered and proliferated in SEVA-C scaffolds processed by injection molding technology [104]. Furthermore, the same materials and also its composite with hydroxyapatite, showed to be biocompatible *in vitro*, as well as *in vivo* in an intramuscular and intracortical model in goats [96]. Another study using the same composition of SEVA-C coated with a biomimetic calcium phosphate layer, and SCA scaffolds produced by extrusion with blowing agents implanted in femurs of rats, showed bone formation [101].

Starch combined with polycaprolactone (SPCL) scaffolds, cultured with bovine articular chondrocytes, showed enhanced cartilaginous matrix [105,106]. This same blend also evidenced good results with bone marrow stromal cells

(BMSCs) differentiated into osteogenic lineage [100]. It has also demonstrated that endothelial cells (ECs) grown on SPCL scaffolds display an adequate phenotype and genotype, which indicates the angiogenic potential of these scaffolds [107,108].

SPCL scaffolds combining SPCL micro- and PCL nanomotifs, respectively produced by rapid prototyping (RP) and electrospinning techniques showed to be suitable for osteoblasts like cells [102].

8.2.3 Algae Origin

8.2.3.1 Alginate Alginates are polysaccharides isolated from brown algae such as *Laminaria hyperborea* and *Lessonia* [109]. Alginate gels are used in several applications, ranging from the food industry to pharmaceutical manufacturing [110].

In the biomedical field, alginate hydrogels have been used for drug delivery. anti-cancer drugs, such as daunomycin [111], or the combination delivery of several drugs, such as methotrexate, doxorubicin, and mitoxantrone, using different release methods [112], has been achieved. Growth factors were also incorporated in alginate hydrogels [113–116]. For instance, alginate beads with vascular endothelial growth factor (VEGF) incorporated were shown to stimulate ECs *in vitro*, being three to five times more potent than the same mass of VEGF added directly to the culture [113]. In another study, alginate and heparin were covalently cross-linked with ethylenediamine, to produce matrices that released basic fibroblast growth factor (bFGF). These constructs were implanted subcutaneously in the dorsal area of rats and were able to promote angiogenesis [114]. A similar approach consisted of incorporating poly (lactic-co-glycolic acid) microspheres capable of controlling the release of bFGF into a porous alginate [115]. The released bFGF induced the proliferation of cardiac fibroblasts *in vitro*, and when implanted in rat peritoneum, increased the number of penetrating capillaries [115]. In another strategy, a nanofiber mesh tube was combined with a peptide-modified alginate hydrogel impregnated with human recombinant bone morphogenetic protein (rhBMP-2). This construct, which released this growth factor *in situ*, was implanted in a rat segmental bone defect and shown to regenerate bone [116].

The combination of chitosan and alginate to produce structures to be used in the biomedical field has been extensively explored [117–120]. Chitosan–alginate polyelectrolyte complex membranes were evaluated as potential wound-dressing materials with *in vitro* culture of human fibroblasts and *in vivo* compared with conventional gauze dressing, leading to an accelerated healing of incision wounds in a rat model [117]. Chitosan–alginate scaffolds fabricated by thermally induced phase separation followed by freeze-drying were studied using HTB-94 chondrosarcoma cell line [120]. These scaffolds were shown to promote cell proliferation and enhance phenotype expression of HTB-94 chondrocytes [120]. Alginate-based chitosan hybrid polymer fibers showed improved adhesion capacity with chondrocytes in comparison with alginate polymer fibers [118].

Alginate has also been proposed for bone regeneration strategies [116,121,122], such as the previously described nanofiber mesh tube with the alginate hydrogel inside with rhBMP-2 to release this growth factor. Other reports use alginate hydrogels as injectable systems for bone regeneration. A modified alginate hydrogel with an arginine-glycine-aspartic acid (RGD)-containing peptide promoted osteoblast adhesion and spreading. Primary rat calvarial osteoblast transplantation revealed increased bone formation at 16 and 24 weeks with RGD-modified alginate compared with unmodified alginate [121]. Calcium phosphate cement paste was combined with hydrogel microbeads to encapsulate human umbilical cord mesenchymal stem cells (hUCMSCs), which in turn differentiated down the osteogenic pathway [122].

8.2.3.2 Carrageenan Carrageenan is a sulfated polysaccharide extracted from red marine algae, *Rhodophyceae*. It is comprised of a linear backbone built up by β-D-galactose smd 3,6-anhydro-α-D-galactose partially sulfated [123]. Carrageenans are divided into three families, accordingly with the number and position of sulfate groups: ι, κ, λ corresponding to mono-, di- and tri-sulfate groups. The first two are gel-forming systems and the third is a thickening agent that is most commonly used in the biomedical field [47].

κ-Carrageenan hydrogel beads were produced by an ionotropic gelation method. Platelet-derived growth factor (PDGF-BB) was incorporated into these gels and the constructs were evaluated for bone TE [124]. Grenha *et al.* described the preparation of chitosan and carrageenan nanoparticles obtained by ionic complexation for drug delivery [125]. Porous scaffolds composed of gelatin and κ-carrageenan produced by freeze-drying followed by chemical cross-linking were subcutaneously implanted in Wistar rats, showing that the scaffold was biodegradable and biocompatible with low antigenicity [126].

Different formulations and processing parameters of alginate and carrageenan hydrogels in the form of beads and fibers were studied to determine the best conditions required to achieve the best behavior in terms of mechanical stability, functionality, and cell viability, and both types of morphologies showed similar results [127]. Carrageenan-based hydrogels were used to encapsulate both ATDC5 cell line and human adipose-derived stem cells (hASCs), and TGF-β1 to check about the chondrogenic enhancement potential of this strategy [128].

Hydrogels of κ-carrageenan were used to encapsulate hASCs, human nasal chondrocytes (hNCs), or ATDC5 cells for cartilage regeneration, and were demonstrated to be a good support for all cell types [129]. The same authors also described the use of carrageenan hydrogels for the delivery of stem cells obtained from adipose tissue differentiated into the chondrogenic phenotype [130]. They analyzed the mechanical properties of the hydrogels with encapsulated cells and observed an increase in stiffness and viscoelastic properties with increasing time in culture with chondrogenic medium [130]. Mihaila *et al.* reported a photocross-linkable methacrylated carrageenan with controllable compressive moduli, swelling ratios, and pore size distributions [131]. Moreover, by micromolding approaches, spatially controlled geometries and cell distribution patterns were obtained enabling the development of cell-material platforms to be applied to a broad range of TE strategies [131].

8.2.3.3 Ulvan Ulvan is a sulfated polysaccharide present in the cell wall of *Ulvaceae* genera algae, and it is mostly composed of rhamnose, glucuronic acid, iduronic acid, and xylose [132]. One notable characteristic of ulvan is the presence of rare sugars within its backbone rhamnose, glucuronic acid, iduronic acid, and xylose rhamnose. The presence of iduronic acid is unusual as this sugar residue has never been identified in algal polysaccharides and it is an important constituent of mammalian GAGs, including heparin and CS [133].

This polysaccharide has not been explored as extensively as the ones previously described. Only recently, researchers started to explore its use in biomedical field. Different extraction methods have been exploited, such as hot-water extraction [134] and hot water with previous extraction of dried algae with dichloromethane and acetone, which removed most of the lipids [135].

Ulvan hydrogels produced from ulvan macromers containing unsaturated groups sensitive to ultraviolet (UV) photopolymerization showed to be stable under physiological conditions, in opposition of hydrogels obtained by UV cross-linking of glycidyl methacrylate conjugated ulvan [136]. 3-D ulvan porous structures were fabricated by the chemical cross-linking of ulvan with 1,4-butanediol diglycidyl ether and freeze-drying. These structures were evaluated *in vitro* with L929 cell line for biomedical applications and showed promising results [137].

8.2.4 Microbial Origin

8.2.4.1 Bacterial Cellulose The most widely used application of bacterial cellulose (BC) in medicine has been a membrane in the treatment of renal failure. Worldwide trends in the use of membranes for the treatment of chronic renal failure indicate a move away from cellulose-based membranes in favor of synthetic membranes [138]. Other biomedical applications of BC are in wound healing as a topical cover. XCell® and Biofill® are cellulose-based products already available in the market [139]. Another application of BC is in the form of tubes for blood vessel replacement, although not reached to humans, animal studies show very promising results in this field [140]. BC has also been investigated for producing scaffolds for cartilage applications [141,142]. For bone TE, scaffolds of BC with ceramics [143,144] have shown good results in this field.

8.2.4.2 Gellan Gum Gellan gum (GG) is a high linear molecular weight bacterial exopolysaccharide obtained from *Sphingomonas elodea.* It is composed of tetrasacchride of (1→4)-L-rhamnose-α(1→3)-D-glucose-b(1→4)-D-glucuronic acid-b(1→4)-D-glucose as a repeating unit. In its native form, two acyl substituents, D-acetate and D-glycerate, are present. The high-acyl form produces transparent soft elastic and flexible gels that are resistant to heat and acid, whereas the low-acyl form produces firm, nonelastic brittle gels [145]. The first report about GG hydrogel production reported the cross-linking of the gel by just adding culture medium [146].

GG hydrogels present properties that can be used in the context of cartilage regeneration. Oliveira *et al.* used this hydrogel to encapsulate hNCs [147]. These authors also studied GG hydrogels in combination with hACs, and implanted these constructs subcutaneously in nude mice and the hydrogels were capable of supporting the growth and ECM deposition of human articular chondrocytes [148]. GG hydrogels were also tested for their ability to be used as injectable systems able to deliver and maintain chondrocytes by *in situ* gelation, and support cell viability and production of ECM [149]. These hydrogels coupled with adipose tissue derived autologous cells were able to regenerate rabbit full-thickness articular cartilage defects [150]

Methacrylated GG hydrogels, obtained either by ionic or photocross-linking were tested for intervertebral disc regeneration [151,152]. They showed to be biocompatible *in vitro* using mouse lung fibroblast cells and human intervertebral disk cells and when subcutaneously implanted [152]. GG hydrogels were also reinforced with biocompatible and biodegradable GG microparticles [153]. Recently, PCL nanofibers produced by electrospinning were sprayed with a GG solution, creating a nanofiber-reinforced GG gel with enhanced mechanical properties and with structures mimicking the native nucleus pulposus (interior region of intervertebral disk) [154].

8.2.4.3 Pullulan Pullulan is a linear polysaccharide obtained from the fermentation of the yeast *Aureobasidium pullulans*. The backbone is formed by glycosidic linkages of $\alpha(1 \rightarrow 6)$ D-glucopyranose and $\alpha(1 \rightarrow 4)$ D-glucopyranose units in a 1 : 2 ratio. It has numerous applications; in the food and beverage industry as filler, in pharmaceuticals as a coating agent, and in manufacturing and electronics due to its ability to form films and fiber. The use of pullulan in the biomedical field is increasing contemporarily due to its nontoxic, nonimmunogenic, biocompatible, and inert nature [155].

Pullulan is highly water soluble hence, it is used as a carrier for drugs and helps in controlled release in plasma. For these purposes, hydrophobized pullulan is used as a drug delivery carrier. In one example, pullulan was used as a carrier for anticancer drugs [156]. Hydrogel nanoparticles of pullulan have recently been prepared in order to develop a carrier system for gene transfer into mammalian cells [157]. Pullulan microspheres have also been grafted with poly(N-isopropylacrylamide-co-acrylamide) to produce pH- and temperature-sensitive microspheres [158].

Pullulan and dextran scaffolds produced by freeze-drying and cross-linked with sodium trimetaphosphate, showed to be a suitable substrate for culture of rat MSCs [159].

Scaffolds composed of pullulan and dextran, supplemented with nanocrystalline hydroxyapatite particles combined with human BMSCs without osteogenic factors promoted bone formation in three different *in vivo* models [160], showing that these scaffolds are suitable for bone TE applications.

8.2.4.4 Dextran Dextrans (Dex) are bacterial extracellular polysaccharides, synthesized from sucrose by beneficial lactic acid bacteria, such as *Leuconostoc mesenteroides* and *Lactobacillus brevis*, and also by the dental plaque-forming species *Streptococcus mutans* [161]. These polysaccharides are composed of $\alpha(1\rightarrow6)$-linked D-glucopyranosyl backbone modified with small side chains of D-glucose branches with $\alpha(1\rightarrow2)$, $\alpha(1\rightarrow3)$, and $\alpha(1\rightarrow4)$-linkage.

Dextrans are particularly used in the pharmaceutical field for drug and protein delivery [162]. Dextrans are used to increase the longevity of therapeutic agents in circulation, which is achieved mainly through relatively longer blood half-lives of high molecular weight of dextran conjugates of therapeutic agents, compared with the intact drug or protein [161]. Methacrylated dextran (dex-MA) and lactate-hydroxyethyl methacrylated dextran (dex-lactate-HEMA) with different cross-links were used to release recombinant human interleukin-2 (IL-2), showing that the release of this immune system molecule can be modulated with the amount of water content and cross-linking density [163]. Microparticles of dex-lactate-HEMA showed to have dual potential to deliver proteins and DNA. These microparticles were surrounded by a membrane permeable to water, but impermeable to both entrapped drugs and gel degradation products, and released their contents with the swelling pressure [164]. Dextran microspheres were also tested for nasal drug delivery systems, with no immune response of nasal mucosa [165]. The biocompatibility of dex-MA and dex-lactate-HEMA hydrogels was evaluated by subcutaneous implantation in rats for up to 6 weeks and also the relationship between *in vitro* and *in vivo* degradation profiles. Both types of hydrogels showed to be biocompatible with no evidence of necrosis, immunocitoxicity, or damage to muscle tissue. *In vitro* and *in vivo* degradation were comparable [166].

Functionalized dextran-derived hydrogels retained rhBMP-2 in a variable manner depending on their functionalization ratio. These dextran hydrogels combined with bovine BMP in a rat ectopic model showed enhanced bone formation [167]. These same hydrogels also have the ability to bind and release recombinant human TGF-β1 [168]. Dextran hydrogels produced by enzymatic synthesis [169], with immobilized RGD peptide and microencapsulated VEGF, were used to encapsulate hESCs to modulate the vascular differentiation of these cells [170]. Dextran hydrogels formed by enzymatic cross-linking of dextran–tyramine conjugates [171] have also been seeded and cultured with chondrocytes to be used for cartilage regeneration [172]. Furthermore, platelet lysate was incorporated into these hydrogels, in the polymer solution prior to gelation and enzymatic cross-linking to determine the effect of platelet lysate on MSC behavior [173]. The addition of this supplement in the hydrogel did not affect the mechanical properties or the porosity and promoted proliferation and chondrogenic differentiation of MSCs [173]. Dex-MA macroporous and interconnected scaffolds were produced using polyethylene glycol (PEG) [174]. The presence of PEG in the solution allowed forming different types of structures from a microporous gel to a macroporous gel or macroporous interconnected structures [174].

8.3 CONCLUSIONS

Natural ECM component materials have been extensively used in the development of biomaterials for TERM. Among the animal origin biomaterials, collagen has been by far the most used for clinical applications in the regeneration of skin or cartilage. HA, fibrin, or silk are other animal origin biomaterials that have reached the clinic and reported promising results. Plants are another source of biomaterials, although plant-derived materials have not been as extensively studied for TERM purposes, with plant cellulose being the less studied due to its inherent difficulties. Starch-based biomaterials have been explored with a few research groups, showing remarkable results for bone TE. Algae-derived biomaterials have been used mainly as cell, protein, and drug delivery vehicles and a support matrix for TE. The sulfate groups of these biomaterials modulate cell behavior in a tissue regeneration context, which is an opportunity to exploit the clinical potential of marine-origin polysaccharides. Alginate is the best studied, but ulvan has been exploited recently and presents properties that may be very useful for TERM approaches. Microbial-origin materials also present interesting characteristics for TERM applications. From those, GG is the less exploited, although it presents very attractive properties for different fields.

Animal-origin materials present several advantages over the other natural origins, but their intrinsic variability and some ethical concerns can limit their widespread use for TERM approaches. The clinical application of plant-, algae-, and microbial-derived biomaterials will constitute a long and challenging road, since the regulatory context of medical devices and advanced therapy medicinal products is extremely demanding, but certainly, the rich array of unique properties of this class of materials will ensure them an important role in that very demanding context.

Acknowledgments

The authors thank the Portuguese Foundation for Science and Technology (FCT) for the post-doc grants of Albino Martins (SFRH/BPD/70669/2010), Marta Alves da Silva (SFRH/BPD/73322/2010), and Ana Costa-Pinto (SFRH/BPD/90332/2012), as well as for the OsteoGraphy (PTDC/EME-MFE/2008) and MaxBone (PTDC/SAU-ENB/115179/2009) projects.

REFERENCES

[1] Badylak, S.F. *Biomaterials* **2007**, *28*, 3587.
[2] Muiznieks, L.D., Keeley, F.W. *Biochimica et Biophysica Acta* **2013**, *1832*, 866.
[3] Langer, R., Vacanti, J.P. *Science* **1993**, *260*, 920.
[4] Lutolf, M.P., Hubbell, J.A. *Nature Biotechnology* **2005**, *23*, 47.
[5] Hutmacher, D.W., Schantz, J.T., Lam, C.X., Tan, K.C., Lim, T.C. *Journal of Tissue Engineering and Regenerative Medicine* **2007**, *1*, 245.
[6] Frantz, C., Stewart, K.M., Weaver, V.M. *Journal of Cell Science* **2010**, *123*, 4195; Alberts, B., Johnson, A., Lewis, J., Raff, M., Roberts, K., Walter, P. *Molecular Biology of the Cell*; Garland Science, London, UK: 2002.
[7] Bosman, F.T., Stamenkovic, I. *The Journal of Pathology* **2003**, *200*, 423.
[8] Rosso, F., Giordano, A., Barbarisi, M., Barbarisi, A. *Journal of Cellular Physiology* **2004**, *199*, 174.
[9] Zagris, N. *Micron* **2001**, *32*, 427.
[10] Zhang, Y.Z., Lim, C.T., Ramakrishna, S., Huang, Z.M. *Journal Of Materials Science-Materials In Medicine* **2005**, *16*, 933.
[11] Yang, S.F., Leong, K.F., Du, Z.H., Chua, C.K. *Tissue Engineering* **2001**, *7*, 679.
[12] Salgado, A.J., Coutinho, O.P., Reis, R.L. *Macromolecular Bioscience* **2004**, *4*, 743.
[13] Freyman, T.M., Yannas, I.V., Gibson, L.J. *Progress in Materials Science* **2001**, *46*, 273.
[14] George, A., Ravindran, S. *Nano Today* **2010**, *5*, 254.
[15] Khadka, D.B., Haynie, D.T. *Nanomedicine: Nanotechnology, Biology and Medicine* **2012**, *8*, 1242.

[16] Meena, C., Mengi, S.A., Deshpande, S.G. *Proceedings of the Indian Academy of Sciences, Chemical Sciences* **1999**, *111*, 319.
[17] Shih, Y.R.V., Chen, C.N., Tsai, S.W., Wang, Y.J., Lee, O.K. *Stem Cells* **2006**, *24*, 2391.
[18] Kinikoglu, B., Auxenfans, C., Pierrillas, P., Justin, V., Breton, P., Burillon, C., Hasirci, V., Damour, O. *Biomaterials* **2009**, *30*, 6418.
[19] Labour, M.N., Banc, A., Tourrette, A., Cunin, F., Verdier, J.M., Devoisselle, J.M., Marcilhac, A., Belamie, E. *Acta Biomaterialia* **2012**, *8*, 3302.
[20] Zheng, L., Fan, H.S., Sun, J., Chen, X.N., Wang, G., Zhang, L., Fan, Y.J., Zhang, X.D. *Journal of Biomedical Materials Research Part A* **2010**, *93A*, 783.
[21] Meimandi-Parizi, A., Oryan, A., Moshiri, A. *Journal of Biomedical Science* **2013**, *20*, doi: 10.1186/1423-0127-20-28.
[22] Gille, J., Behrens, P., Volpi, P., de Girolamo, L., Reiss, E., Zoch, W., Anders, S. *Archives of Orthopaedic and Trauma Surgery* **2013**, *133*, 87.
[23] Faralli, J.A., Schwinn, M.K., Gonzalez, J.M., Filla, M.S., Peters, D.M. *Experimental Eye Research* **2009**, *88*, 689.
[24] Koide, A., Bailey, C.W., Huang, X.L., Koide, S. *Journal of Molecular Biology* **1998**, *284*, 1141.
[25] Roy, D.C., Hocking, D.C. *Tissue Engineering. Part A* **2013**, *19*, 558.
[26] King, V.R., Alovskaya, A., Wei, D.Y.T., Brown, R.A., Priestley, J.V. *Biomaterials* **2010**, *31*, 4447.
[27] Custodio, C.A., Alves, C.M., Reis, R.L., Mano, J.F. *Journal of Tissue Engineering and Regenerative Medicine* **2010**, *4*, 316.
[28] Zhu, Y.B., Leong, M.F., Ong, W.F., Chan-Park, M.B., Chian, K.S. *Biomaterials* **2007**, *28*, 861.
[29] Almond, A. *Cellular and Molecular Life Sciences* **2007**, *64*, 1591.
[30] Zhao, W., Jin, X., Cong, Y., Liu, Y.Y., Fu, J. *Journal of Chemical Technology and Biotechnology* **2013**, *88*, 327.
[31] Yamane, S., Iwasaki, N., Majima, T., Funakoshi, T., Masuko, T., Harada, K., Minami, A., Monde, K., Nishimura, S. *Biomaterials* **2005**, *26*, 611.
[32] Kim, J., Kim, I.S., Cho, T.H., Lee, K.B., Hwang, S.J., Tae, G., Noh, I., Lee, S.H., Park, Y., Sun, K. *Biomaterials* **2007**, *28*, 1830.
[33] Fu, L.L.K., Maffulli, N., Chan, K.M. *Clinical Rheumatology* **2001**, *20*, 98.
[34] Huskisson, E.C., Donnelly, S. *Rheumatology* **1999**, *38*, 602.
[35] Yoo, H.S., Lee, E.A., Yoon, J.J., Park, T.G. *Biomaterials* **2005**, *26*, 1925.
[36] Trattnig, S., Pinker, K., Krestan, C., Plank, C., Millington, S., Marlovits, S. *European Journal of Radiology* **2006**, *57*, 9.
[37] Behrens, P., Bitter, T., Kurz, B., Russlies, M. *Knee* **2006**, *13*, 194.
[38] Mano, J.F., Silva, G.A., Azevedo, H.S., Malafaya, P.B., Sousa, R.A., Silva, S.S., Boesel, L.F., Oliveira, J.M., Santos, T.C., Marques, A.P., Neves, N.M., Reis, R.L. *Journal of the Royal Society, Interface* **2007**, *4*, 999.
[39] Janmey, P.A., Winer, J.P., Weisel, J.W. *Journal of the Royal Society, Interface* **2009**, *6*, 1.
[40] Mosesson, M.W., Siebenlist, K.R., Meh, D.A. *Annals of the New York Academy of Sciences* **2001**, *936*, 11.
[41] Man, A.J., Davis, H.E., Itoh, A., Leach, J.K., Bannerman, P. *Tissue Engineering. Part A* **2011**, *17*, 2931.
[42] Christman, K.L., Vardanian, A.J., Fang, Q.Z., Sievers, R.E., Fok, H.H., Lee, R.J. *Journal of the American College of Cardiology* **2004**, *44*, 654.
[43] Zhang, G., Wang, X.H., Wang, Z.L., Zhang, J.Y., Suggs, L. *Tissue Engineering* **2006**, *12*, 9.

[44] Bensaid, W., Triffitt, J.T., Blanchat, C., Oudina, K., Sedel, L., Petite, H. *Biomaterials* **2003**, *24*, 2497.
[45] Zhang, D.H., Shadrin, I.Y., Lam, J., Xian, H.Q., Snodgrass, H.R., Bursac, N. *Biomaterials* **2013**, *34*, 5813.
[46] Weyers, A., Yang, B., Solakyildirim, K., Yee, V., Li, L.Y., Zhang, F.M., Linhardt, R.J. *The FEBS Journal* **2013**, *280*, 2285.
[47] Silva, T.H., Alves, A., Ferreira, B.M., Oliveira, J.M., Reys, L.L., Ferreira, R.J.F., Sousa, R.A., Silva, S.S., Mano, J.F., Reis, R.L. *International Materials Reviews* **2012**, *57*, 276.
[48] Ferdous, Z., Grande-Allen, K.J. *Tissue Engineering* **2007**, *13*, 1893.
[49] Lamari, F.N., Theocharis, A.D., Asimakopoulou, A.P., Malavaki, C.J., Karamanos, N.K. *Biomedical Chromatography* **2006**, *20*, 539.
[50] Nakano, T., Pietrasik, Z., Ozimek, L., Betti, M. *Process Biochemistry* **2012**, *47*, 1909.
[51] Zhao, T., Zhou, Y., Mao, G.H., Zou, Y., Zhao, J.L., Bai, S.Q., Yang, L.Q., Wu, X.Y. *Journal of the Science of Food and Agriculture* **2013**, 93–1633.
[52] Garnjanagoonchorn, W., Wongekalak, L., Engkagul, A. *Chemical Engineering and Processing* **2007**, *46*, 465.
[53] Thelin, M.A., Bartolini, B., Axelsson, J., Gustafsson, R., Tykesson, E., Pera, E., Oldberg, A., Maccarana, M., Malmstrom, A. *The FEBS Journal* **2013**, *280*, 2431.
[54] Lynn, A.K., Best, S.M., Cameron, R.E., Harley, B.A., Yannas, I.V., Gibson, L.J., Bonfield, W. *Journal of Biomedical Materials Research Part A* **2010**, *92A*, 1057.
[55] Wang, W.H., Zhang, M., Lu, W., Zhang, X.J., Ma, D.D., Rong, X.M., Yu, C.Y., Jin, Y. *Tissue Engineering. Part C: Methods* **2010**, *16*, 269.
[56] Hui, J.H., Chan, S.W., Li, J., Goh, J.C., Li, L., Ren, X.F., Lee, E.H. *Journal of Molecular Histology* **2007**, *38*, 483.
[57] Volpi, N., Maccari, F. *Glycobiology* **2009**, *19*, 356.
[58] Paderi, J.E., Sistiabudi, R., Ivanisevic, A., Panitch, A. *Tissue Engineering. Part A* **2009**, *15*, 2991.
[59] Wen, Y.H., Grondahl, L., Gallego, M.R., Jorgensen, L., Moller, E.H., Nielsen, H.M. *Biomacromolecules* **2012**, *13*, 905.
[60] Stringer, S.E., Gallagher, J.T. *The International Journal of Biochemistry & Cell Biology* **1997**, *29*, 709.
[61] Govindraj, P., West, L., Koob, T.J., Neame, P., Doege, K., Hassell, J.R. *The Journal of Biological Chemistry* **2002**, *277*, 19461.
[62] Iozzo, R.V. *Annual Review of Biochemistry* **1998**, *67*, 609.
[63] Pieper, J.S., Hafmans, T., van Wachem, P.B., van Luyn, M.J.A., Brouwer, L.A., Veerkamp, J.H., van Kuppevelt, T.H. *Journal of Biomedical Materials Research* **2002**, *62*, 185.
[64] Luong-Van, E., Grondahl, L., Song, S.J., Nurcombe, V., Cool, S. *Journal of Molecular Histology* **2007**, *38*, 459.
[65] Hempel, U., Moller, S., Noack, C., Hintze, V., Scharnweber, D., Schnabelrauch, M., Dieter, P. *Acta Biomaterialia* **2012**, *8*, 4064.
[66] Funderburgh, J.L. *IUBMB Life* **2002**, *54*, 187.
[67] Foo, C.W.P., Kaplan, D.L. *Advanced Drug Delivery Reviews* **2002**, *54*, 1131.
[68] Garcia-Fuentes, M., Meinel, A.J., Hilbe, M., Meinel, L., Merkle, H.P. *Biomaterials* **2009**, *30*, 5068.
[69] Vepari, C., Kaplan, D.L. *Progress in Polymer Science* **2007**, *32*, 991.
[70] Lawrence, B.D., Marchant, J.K., Pindrus, M.A., Omenetto, F.G., Kaplan, D.L. *Biomaterials* **2009**, *30*, 1299.

[71] Bessa, P.C., Balmayor, E.R., Hartinger, J., Zanoni, G., Dopler, D., Meinl, A., Banerjee, A., Casal, M., Redl, H., Reis, R.L., van Griensven, M. *Tissue Engineering. Part C: Methods* **2010**, *16*, 937.
[72] Enomoto, S., Sumi, M., Kajimoto, K., Nakazawa, Y., Takahashi, R., Takabayashi, C., Asakura, T., Sata, M. *Journal of Vascular Surgery* **2010**, *51*, 155.
[73] Brown, C.D., Hoffman, A.L. in *Methods of Tissue Engineering* (Eds: A. Atala, R. Lanza), Academic Press, San Diego, CA: 2002, p. 565.
[74] Tuzlakoglu, K., Reis, R.L. in *Natural-Based Polymers for Biomedical Applications* (Ed.: R.L. Reis), Woodhead Publishing Limited and CRC Press LLC, Cambridge, UK: 2008, p. 357.
[75] Prasitsilp, M., Jenwithisuk, R., Kongsuwan, K., Damrongchai, N., Watts, P. *Journal of Materials Science* **2000**, *11*, 773.
[76] Alves da Silva, M.L., Crawford, A., Mundy, J.M., Correlo, V.M., Sol, P., Bhattacharya, M., Hatton, P.V., Reis, R.L., Neves, N.M. *Acta Biomaterialia* **2010**, *6*, 1149.
[77] Kim, I.Y., Seo, S.J., Moon, H.S., Yoo, M.K., Park, I.Y., Kim, B.C., Cho, C.S. *Biotechnology Advances* **2008**, *26*, 1.
[78] Martins, A.M., Pereira, R.C., Leonor, I.B., Azevedo, H.S., Reis, R.L. *Acta Biomaterialia* **2009**, *5*, 3328.
[79] Hu, X.Y., Huang, J.H., Ye, Z.X., Xia, L., Li, M., Lv, B.C., Shen, X.F., Luo, Z.J. *Tissue Engineering. Part A* **2009**, *15*, 3297.
[80] Zhu, C.H., Fan, D.D., Ma, X.X., Xue, W.J., Yu, Y.Y., Luo, Y.N., Liu, B.W., Chen, L. *Journal of Bioactive and Compatible Polymers* **2009**, *24*, 560.
[81] Malafaya, P.B., Oliveira, J.T., Reis, R.L. *Tissue Engineering. Part A* **2010**, *16*, 735.
[82] Griffon, D.J., Sedighi, M.R., Schaeffer, D.V., Eurell, J.A., Johnson, A.L. *Acta Biomaterialia* **2006**, *2*, 313.
[83] Ragetly, G.R., Griffon, D.J., Lee, H.B., Fredericks, L.P., Gordon-Evans, W., Chung, Y.S. *Acta Biomaterialia* **2010**, *6*, 1430.
[84] Mrugala, D., Bony, C., Neves, N., Caillot, L., Fabre, S., Moukoko, D., Jorgensen, C., Noel, D. *Annals of the Rheumatic Diseases* **2008**, *67*, 288.
[85] Guo, T., Zhao, J., Chang, J., Ding, Z., Hong, H., Chen, J., Zhang, J. *Biomaterials* **2006**, *27*, 1095.
[86] Pereira, R.C., Scaranari, M., Castagnola, P., Grandizio, M., Azevedo, H.S., Reis, R.L., Cancedda, R., Gentili, C. *Journal of Tissue Engineering and Regenerative Medicine* **2009**, *3*, 97.
[87] Jung, J.Y., Khan, T., Park, J.K., Chang, H.N. *Korean Journal of Chemical Engineering* **2007**, *24*, 265.
[88] Klemm, D., Heublein, B., Fink, H.-P., Bohn, A. *Angewandte Chemie International Edition* **2005**, *44*, 3358.
[89] de Mesquita, J.P., Donnici, C.L., Pereira, F.V. *Biomacromolecules* **2010**, *11*, 473.
[90] Röper, H. *Starch—Stärke* **2002**, *54*, 89.
[91] Wang, T.L., Bogracheva, T.Y., Hedley, C.L. *Journal of Experimental Botany* **1998**, *49*, 481.
[92] Buléon, A., Colonna, P., Planchot, V., Ball, S. *International Journal of Biological Macromolecules* **1998**, *23*, 85.
[93] Jobling, S. *Current Opinion in Plant Biology* **2004**, *7*, 210.
[94] Azevedo, H.S., Gama, F.M., Reis, R.L. *Biomacromolecules* **2003**, *4*, 1703.
[95] Reis, R.L., Cunha, A.M. *Journal of Materials Science. Materials in Medicine* **1995**, *6*, 786; Reis, R.L., Cunha, A.M., Allan, P.S., Bevis, M.J. *Journal of Polymers for Advanced Technologies* **1996**, *7*, 784; Gomes, M.E., Ribeiro, A.S., Malafaya, P.B., Reis, R.L., Cunha, A.M. *Biomaterials* **2001**, *22*, 883.

[96] Mendes, S.C., Reis, R.L., Bovell, Y.P., Cunha, A.M., van Blitterswijk, C.A., de Bruijn, J.D. *Biomaterials* **2001**, *22*, 2057.
[97] Marques, A.P., Reis, R.L., Hunt, J.A. *Biomaterials* **2002**, *23*, 1471.
[98] Santos, T.C., Marques, A.P., Horing, B., Martins, A.R., Tuzlakoglu, K., Castro, A.G., van Griensven, M., Reis, R.L. *Acta Biomaterialia* **2010**, *6*, 4314.
[99] Silva, G.A., Coutinho, O.P., Ducheyne, P., Shapiro, I.M., Reis, R.L. *Tissue Engineering* **2007**, *13*, 1259.
[100] Gomes, M.E., Holtorf, H.L., Reis, R.L., Mikos, A.G. *Tissue Engineering* **2006**, *12*, 801.
[101] Salgado, A.J., Coutinho, O.P., Reis, R.L., Davies, J.E. *Journal of Biomedical Materials Research. Part A* **2007**, *80A*, 983.
[102] Martins, A., Chung, S., Pedro, A.J., Sousa, R.A., Marques, A.P., Reis, R.L., Neves, N.M. *Journal of Tissue Engineering and Regenerative Medicine* **2009**, *3*, 37.
[103] Elvira, C., Mano, J.F., San Roman, J., Reis, R.L. *Biomaterials* **2002**, *23*, 1955.
[104] Gomes, M.E., Reis, R.L., Cunha, A.M., Blitterswijk, C.A., de Bruijn, J.D. *Biomaterials* **2001**, *22*, 1911.
[105] Oliveira, J.T., Crawford, A., Mundy, J.M., Moreira, A.R., Gomes, M.E., Hatton, P.V., Reis, R.L. *Journal of Materials Science. Materials in Medicine* **2007**, *18*, 295.
[106] da Silva, M.A., Crawford, A., Mundy, J., Martins, A., Araujo, J.V., Hatton, P.V., Reis, R.L., Neves, N.M. *Tissue Engineering. Part A* **2009**, *15*, 377.
[107] Santos, M.I., Fuchs, S., Gomes, M.E., Unger, R.E., Reis, R.L., Kirkpatrick, C.J. *Biomaterials* **2007**, *28*, 240.
[108] Santos, M.I., Tuzlakoglu, K., Fuchs, S., Gomes, M.E., Peters, K., Unger, R.E., Piskin, E., Reis, R.L., Kirkpatrick, C.J. *Biomaterials* **2008**, *29*, 4306.
[109] Lee, K.Y., Mooney, D.J. *Progress in Polymer Science* **2012**, *37*, 106.
[110] Augst, A.D., Kong, H.J., Mooney, D.J. *Macromolecular Bioscience* **2006**, *6*, 623.
[111] Bouhadir, K.H., Kruger, G.M., Lee, K.Y., Mooney, D.J. *Journal of Pharmaceutical Sciences* **2000**, *89*, 910.
[112] Bouhadir, K.H., Alsberg, E., Mooney, D.J. *Biomaterials* **2001**, *22*, 2625.
[113] Peters, M.C., Isenberg, B.C., Rowley, J.A., Mooney, D.J. *Journal of Biomaterials Science. Polymer Edition* **1998**, *9*, 1267.
[114] Tanihara, M., Suzuki, Y., Yamamoto, E., Noguchi, A., Mizushima, Y. *Journal of Biomedical Materials Research* **2001**, *56*, 216.
[115] Perets, A., Baruch, Y., Weisbuch, F., Shoshany, G., Neufeld, G., Cohen, S. *Journal of Biomedical Materials Research. Part A* **2003**, *65A*, 489.
[116] Kolambkar, Y.M., Dupont, K.M., Boerckel, J.D., Huebsch, N., Mooney, D.J., Hutmacher, D.W., Guldberg, R.E. *Biomaterials* **2011**, *32*, 65.
[117] Wang, L.H., Khor, E., Wee, A., Lim, L.Y. *Journal of Biomedical Materials Research* **2002**, *63*, 610.
[118] Iwasaki, N., Yamane, S.T., Majima, T., Kasahara, Y., Minami, A., Harada, K., Nonaka, S., Maekawa, N., Tamura, H., Tokura, S., Shiono, M., Monde, K., Nishimura, S.I. *Biomacromolecules* **2004**, *5*, 828.
[119] Knill, C.J., Kennedy, J.F., Mistry, J., Miraftab, M., Smart, G., Groocock, M.R., Williams, H.J. *Carbohydrate Polymers* **2004**, *55*, 65;Li, Z.S., Ramay, H.R., Hauch, K.D., Xiao, D.M., Zhang, M.Q. *Biomaterials* **2005**, *26*, 3919.
[120] Li, Z.S., Zhang, M.Q. *Journal of Biomedical Materials Research. Part A* **2005**, *75A*, 485.
[121] Alsberg, E., Anderson, K.W., Albeiruti, A., Franceschi, R.T., Mooney, D.J. *Journal of Dental Research* **2001**, *80*, 2025.
[122] Zhao, L.A., Weir, M.D., Xu, H.H.K. *Biomaterials* **2010**, *31*, 6502.
[123] Bartkowiak, A., Hunkeler, D. *Colloids and Surfaces B: Biointerfaces* **2001**, *21*, 285.

[124] Santo, V.E., Frias, A.M., Carida, M., Cancedda, R., Gomes, M.E., Mano, J.F., Reis, R.L. *Biomacromolecules* **2009**, *10*, 1392.

[125] Grenha, A., Gomes, M.E., Rodrigues, M., Santo, V.E., Mano, J.F., Neves, N.M., Reis, R.L. *Journal of Biomedical Materials Research. Part A* **2010**, *92A*, 1265.

[126] Lim, Y.M., Gwon, H.J., Choi, J.H., Shin, J., Nho, Y.C., Jeong, S.I., Chong, M.S., Lee, Y.M., Kwon, I.K., Kim, S.E. *Macromolecular Research* **2010**, *18*, 29.

[127] Popa, E.G., Gomes, M.E., Reis, R.L. *Biomacromolecules* **2011**, *12*, 3952.

[128] Rocha, P.M., Santo, V.E., Gomes, M.E., Reis, R.L., Mano, J.F. *Journal of Bioactive and Compatible Polymers* **2011**, *26*, 493.

[129] Popa, E., Reis, R., Gomes, M. *Biotechnology and Applied Biochemistry* **2012**, *59*, 132.

[130] Popa, E.G., Caridade, S.G., Mano, J.F., Reis, R.L., Gomes, M.E. *Journal of Tissue Engineering and Regenerative Medicine* **2013**, doi: 10.1002/term.1683.

[131] Mihaila, S.M., Gaharwar, A.K., Reis, R.L., Marques, A.P., Gomes, M.E., Khademhosseini, A. *Advanced Healthcare Materials* **2013**, *2*, 895.

[132] Ray, B., Lahaye, M. *Carbohydrate Research* **1995**, *274*, 313.

[133] Alves, A., Sousa, R.A., Reis, R.L. *Journal of Applied Phycology* **2013**, *25*, 407.

[134] Paradossi, G., Cavalieri, F., Pizzoferrato, L., Liquori, A.M. *International Journal of Biological Macromolecules* **1999**, *25*, 309.

[135] Alves, A., Caridade, S.G., Mano, J.F., Sousa, R.A., Reis, R.L. *Carbohydrate Research* **2010**, *345*, 2194;Costa, C., Alves, A., Pinto, P.R., Sousa, R.A., da Silva, E.A.B., Reis, R.L., Rodrigues, A.E. *Carbohydrate Polymers* **2012**, *88*, 537.

[136] Morelli, A., Chiellini, F. *Macromolecular Chemistry and Physics* **2010**, *211*, 821.

[137] Alves, A., Sousa, R.A., Reis, R.L. *Journal of Biomedical Materials Research. Part A* **2013**, *101A*, 998.

[138] Grassmann, A., Gioberge, S., Moeller, S., Brown, G. *Nephrology Dialysis Transplantation* **2005**, *20*, 2587.

[139] Czaja, W., Krystynowicz, A., Bielecki, S., Brown, R.M., Jr. *Biomaterials* **2006**, *27*, 145.

[140] Klemm, D., Schumann, D., Udhardt, U., Marsch, S. *Progress in Polymer Science* **2001**, *26*, 1561;Schumann, D., Wippermann, J., Klemm, D., Kramer, F., Koth, D., Kosmehl, H., Wahlers, T., Salehi-Gelani, S. *Cellulose* **2009**, *16*, 877.

[141] Svensson, A., Nicklasson, E., Harrah, T., Panilaitis, B., Kaplan, D.L., Brittberg, M., Gatenholm, P. *Biomaterials* **2005**, *26*, 419.

[142] Muller, F.A., Muller, L., Hofmann, I., Greil, P., Wenzel, M.M., Staudenmaier, R. *Biomaterials* **2006**, *27*, 3955.

[143] Jiang, L.Y., Li, Y.B., Wang, X.J., Zhang, L., Wen, J.Q., Gong, M. *Carbohydrate Polymers* **2008**, *74*, 680.

[144] Saska, S., Barud, H.S., Gaspar, A.M., Marchetto, R., Ribeiro, S.J., Messaddeq, Y. *International Journal of Biomaterials* **2011**, *2011*, doi: 10.1155/2011/175362.

[145] Cascone, M.G., Barbani, N., Giusti, C.C.P., Ciardelli, G., Lazzeri, L. *Journal of Biomaterials Science, Polymer Edition* **2001**, *12*, 267.

[146] Smith, A.M., Shelton, R.M., Perrie, Y., Harris, J.J. *Journal of Biomaterials Applications* **2007**, *22*, 241.

[147] Oliveira, J.T., Martins, L., Picciochi, R., Malafaya, P.B., Sousa, R.A., Neves, N.M., Mano, J.F., Reis, R.L. *Journal of Biomedical Materials Research. Part A* **2010**, *93A*, 852.

[148] Oliveira, J.T., Santos, T.C., Martins, L., Silva, M.A., Marques, A.P., Castro, A.G., Neves, N.M., Reis, R.L. *Journal of Tissue Engineering and Regenerative Medicine* **2009**, *3*, 493.

[149] Oliveira, J.T., Santos, T.C., Martins, L., Picciochi, R., Marques, A.P., Castro, A.G., Neves, N.M., Mano, J.F., Reis, R.L. *Tissue Engineering. Part A* **2010**, *16*, 343.

[150] Oliveira, J.T., Gardel, L.S., Rada, T., Martins, L., Gomes, M.E., Reis, R.L. *Journal of Orthopaedic Research* **2010**, *28*, 1193.
[151] Silva-Correia, J., Oliveira, J.M., Caridade, S.G., Oliveira, J.T., Sousa, R.A., Mano, J.F., Reis, R.L. *Journal of Tissue Engineering and Regenerative Medicine* **2011**, *5*, E97.
[152] Silva-Correia, J., Zavan, B., Vindigni, V., Silva, T.H., Oliveira, J.M., Abatangelo, G., Reis, R.L. *Advanced Healthcare Materials* **2013**, *2*, 568.
[153] Pereira, D.R., Silva-Correia, J., Caridade, S.G., Oliveira, J.T., Sousa, R.A., Salgado, A.J., Oliveira, J.M., Mano, J.F., Sousa, N., Reis, R.L. *Tissue Engineering. Part C: Methods* **2011**, *17*, 961.
[154] Thorvaldsson, A., Silva-Correia, J., Oliveira, J.M., Reis, R.L., Gatenholm, P., Walkenstrom, P. *Journal of Applied Polymer Science* **2013**, *128*, 1158.
[155] Prajapati, V.D., Jani, G.K., Khanda, S.M. *Carbohydrate Polymers* **2013**, *95*, 540.
[156] Na, K., Bae, Y. *Pharmaceutical Research* **2002**, *19*, 681.
[157] Gupta, M., Gupta, A.K. *Journal of Controlled Release* **2004**, *99*, 157.
[158] Fundueanu, G., Constantin, M., Ascenzi, P. *Biomaterials* **2008**, *29*, 2767.
[159] Autissier, A., Le Visage, C., Pouzet, C., Chaubet, F., Letourneur, D. *Acta Biomaterialia* **2010**, *6*, 3640.
[160] Fricain, J.C., Schlaubitz, S., Le Visage, C., Arnault, I., Derkaoui, S.M., Siadous, R., Catros, S., Lalande, C., Bareille, R., Renard, M., Fabre, T., Cornet, S., Durand, M., Leonard, A., Sahraoui, N., Letourneur, D., Amedee, J. *Biomaterials* **2013**, *34*, 2947.
[161] Larsen, C. *Advanced Drug Delivery Reviews* **1989**, *3*, 103.
[162] Mehvar, R. *Journal of Controlled Release* **2000**, *69*, 1.
[163] Cadee, J.A., de Groot, C.J., Jiskoot, W., den Otter, W., Hennink, W.E. *Journal of Controlled Release* **2002**, *78*, 1.
[164] De Geest, B.G., Stubbe, B.G., Jonas, A.M., Van Thienen, T., Hinrichs, W.L.J., Demeester, J., De Smedt, S.C. *Biomacromolecules* **2005**, *7*, 373.
[165] Pereswetoff-Morath, L. *Advanced Drug Delivery Reviews* **1998**, *29*, 185.
[166] Cadée, J.A., van Luyn, M.J.A., Brouwer, L.A., Plantinga, J.A., van Wachem, P.B., de Groot, C.J., den Otter, W., Hennink, W.E. *Journal of Biomedical Materials Research* **2000**, *50*, 397.
[167] Maiga-Revel, O., Chaubet, F., Jozefonvicz, J. *Carbohydrate Polymers* **1997**, *32*, 89.
[168] Maire, M., Logeart-Avramoglou, D., Degat, M.-C., Chaubet, F. *Biomaterials* **2005**, *26*, 1771.
[169] Ferreira, L., Gil, M.H., Dordick, J.S. *Biomaterials* **2002**, *23*, 3957.
[170] Ferreira, L.S., Gerecht, S., Fuller, J., Shieh, H.F., Vunjak-Novakovic, G., Langer, R. *Biomaterials* **2007**, *28*, 2706.
[171] Jin, R., Hiemstra, C., Zhong, Z.Y., Feijen, J. *Biomaterials* **2007**, *28*, 2791.
[172] Jin, R., Teixeira, L.S.M., Dijkstra, P.J., Zhong, Z.Y., van Blitterswijk, C.A., Karperien, M., Feijen, J. *Tissue Engineering. Part A* **2010**, *16*, 2429.
[173] Teixeira, L.S.M., Leijten, J.C.H., Wennink, J.W.H., Chatterjea, A.G., Feijen, J., van Blitterswijk, C.A., Dijkstra, P.J., Karperien, M. *Biomaterials* **2012**, *33*, 3651.
[174] Levesque, S.G., Lim, R.M., Shoichet, M.S. *Biomaterials* **2005**, *26*, 7436.
[175] Maire, M., Chaubet, F., Mary, P., Blanchat, C., Meunier, A., Logeart-Avramoglou, D. *Biomaterials* **2005**, *26*, 5085.

PART II

BIO-INSPIRED TISSUE ENGINEERING

CHAPTER 9

BIO-INSPIRED DESIGN OF SKIN REPLACEMENT THERAPIES

DENNIS P. ORGILL

Division of Plastic Surgery, Brigham and Women's Hospital, Harvard Medical School, Boston, MA, USA

9.1 INTRODUCTION

As the largest organ of the body, the skin provides critical barrier and protective functions. At a basic level, the skin can be described as a bilayer material with a top layer termed the epidermis: an ectodermal derivative that produces a barrier to fluid flux and bacterial invasion. The main cell population is made up of keratinocytes that divide near the dermis, proliferate, and undergo apoptosis with necrotic cells forming a robust barrier known as the *stratum corneum*. The epidermis is avascular, but has a rich network of nerve endings. It contains melanocytes, which provide color, and Langerhans cells, which provide immune function.

The lower, thicker layer of the skin is the dermal layer, composed primarily of collagen, glycosaminoglycans, and elastin. It is rich in fibroblasts and has both a superficial and deep plexus of blood vessels that are connected by perforating vessels. The dermis is strong, flexible, and provides substantial protection from external trauma. Between the epidermis and the dermis is the basement membrane, which provides a convoluted structure with inverting cones of epidermis invaginating into the dermis. In addition, skin adnexal glands, including hair follicles and sebaceous glands that are epidermal derivatives, also invaginate into the epidermis. Epidermal cells that line these structures have the capacity to divide, proliferate, and reconstitute the epidermis.

Due to burns, frostbite, trauma, malignancy, infection, and congenital defects, humans can present with various degrees of skin loss that can be problematic. For small injuries, there is a well-developed system of wound healing that works through a sequential series of events including hemostasis, inflammation, proliferation, and remodeling. For deeper injuries that reach the dermis, the healing response results in wound contraction and the formation of scar tissue. In cases in which wounds

Bio-inspired Materials for Biomedical Engineering, First Edition. Edited by Anthony B. Brennan and Chelsea M. Kirschner.

heal under tension and/or patients have a genetic predisposition, healing can produce heavy scars, which can result in hypertrophic scars or keloids.

9.2 BIO-INSPIRATION OF SKIN REPLACEMENT THERAPY

9.2.1 Observations

9.2.1.1 The Epidermis Has the Capacity to Regenerate Superficial burns, dermabrasion, cosmetic chemical peels, and laser therapy can damage nearly the entire epidermis. The regeneration that occurs often leaves a better cosmetic result than prior to injury. The capacity for stem cells at the base of the epidermis and along the hair follicles is substantial.

9.2.1.2 Only the Superficial Dermis Has the Capacity to Regenerate Dunkin [1] performed an important study in which an incision of variable depth was made in the forearms of human volunteers. In instances when the incision was less than 0.56 mm, or one-third of the thickness of the dermis, no visible scar was present 3 months after injury. It is well known among surgeons that those patients with superficial burns that can heal within 21 days generally do not form a substantial scar. In contrast, those that heal beyond 21 days have a high likelihood of developing heavy scarring.

9.2.1.3 The Mammalian Response to Deep Dermal Injury Is Wound Contraction and Scarring Unlike amphibians, which have the capacity to regenerate entire limbs, mammals respond to injury through the twin processes of scarring and wound contraction. There may have been an evolutionary advantage in developing mechanisms for faster wound closure to avoid complications of infection.

9.2.1.4 Structural Dermal Elements Can Block Wound Contraction Skin grafts are known to block wound contraction [2] with grafts that have a higher percentage of dermal elements, resulting in less wound contraction. This effect is likely due to the innate structure of dermal macromolecules, such as collagen. Erhlich compared wound healing of experimental frostbite injuries, which do not denature macromolecules, with those of burns, which do denature macromolecules, and found that the burns healed through wound contraction, whereas the frostbite injuries healed with very little contraction, preserving the dermal elements [3].

9.2.1.5 Skin Elements Have the Capacity to Self-Organize Through a variety of manipulations, the skin has a remarkable ability to reorganize with epidermal elements tending to migrate towards the wound surface.

9.3 BIOMIMETIC SOLUTIONS

There are many interesting aspects of skin anatomy, immunology, and physiology that have inspired several products that are clinically used today. In this chapter, the author suggests aspects of skin structure and biology that may have inspired the inventors of each of the three technologies that have since led to the design of specific products on the market today. There are many other skin substitute technologies that are either available clinically or in some stage of preclinical development throughout the world. These three examples introduce the concepts of bio-inspired skin by illustrating how the investigators prioritized skin function in their design.

9.3.1 Epidermal Replacement: Epicell® (Genzyme Tissue Repair)

Perhaps the most critical function of skin for human survival is the capacity of the epidermis to form an adequate barrier to water loss and bacterial invasion. Howard Green and James Rhinwald (formerly at MIT and now at Harvard Medical School) were pioneers in cell culture and described a method to culture keratinocytes into a multilayered construct. Critical to their success was the observation that keratinocytes produced more rapidly in culture when placed on top of a feeder layer of fibroblasts [4]. They subsequently developed a method to produce sheets of multilayered keratinocytes that could be used clinically [5,6]. These were initially used on large body surface area burn victims after debridement. In several cases, this technology was attributed to saving these patients' lives. Many treated patients had burns over 90% of their body, with very little area from which to harvest conventional skin grafts. The capacity to grow large quantities of skin from a small biopsy proved to be very beneficial in achieving initial closure of these very difficult cases. The long-term results of these cases showed that often, these constructs of keratinocyte sheets used in isolation led to unstable skin. Surgeons have found that these constructs work best when used on large flat anterior surfaces and when used with a dermal replacement, such as cadaver allograft. Cuono described a technique where cadaver allograft would be placed following eschar excision of the burn and covered with cadaver skin allografts for about 2 weeks [7]. The epidermis would be removed through a process of dermabrasion and the cultured keratinocytes would be placed onto the dermal constructs. This technique led to a better take of the cultured cells and improved long-term stability of the skin.

9.3.2 Treatment of Chronic Wounds: Apligraf® (Organogenesis)

Eugene Bell, working in the Biology Department at MIT, was also an expert in cell culture. He realized the importance of the interaction of the epidermis with the underlying dermis and, thus, felt that skin replacement constructs could be improved if they were composed of both dermal and epidermal analogs. He made the important discovery that fibroblasts in collagen gels would induce contraction of these gels *in vitro* [8]. On top of the fibroblast populated lattice, cultured keratinocytes could be placed to form a multilayered result. The initial clinical studies in burn victims never received much attention from surgeons, and at some point, these constructs were trialed in chronic wounds. In addition, to simplify manufacturing issues, the construct switched to using allogenic neonatal foreskin as a source material. Initially, this product was thought to *take* (become vascularized by the host) on wounds, but extensive studies have shown that cells from this product do not integrate into the

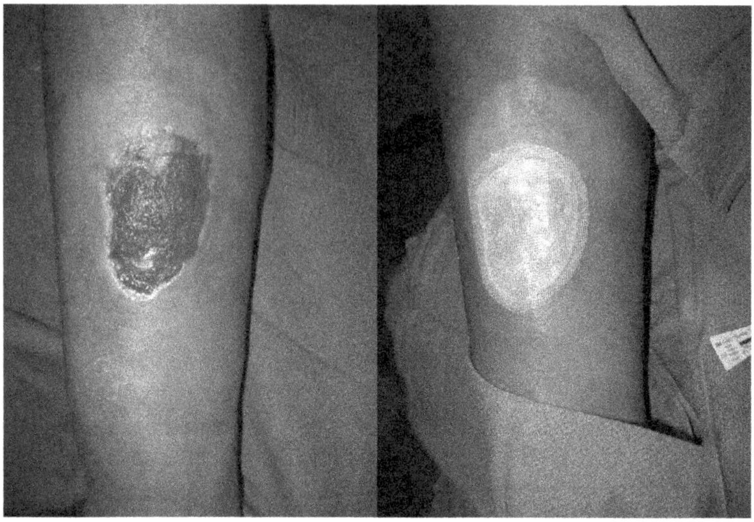

FIGURE 9.1 Application of allogenic product to leg wound. (Courtesy of Lauren R. Bayer, PA-C.) *(See insert for color representation of the figure.)*

host. Nevertheless, these constructs have been used successfully in chronic wounds (Figure 9.1). The current thought is that the cells derived from neonatal foreskins provide a large amount of growth factors that facilitate wound closure.

9.3.3 Dermal Regeneration: Integra®

Ioannis Yannas, while working in the Mechanical Engineering Department at MIT, teamed up with John Burke from Harvard Medical School to design a scaffold for dermal replacement. Dr. Yannas, an expert in polymers, had done basic work on collagen and realized that the rich chemistry and complexity of these molecules had enormous potential as degradable biomaterial. Collagen is the major protein component within skin and has significant mechanical properties. Altering physicochemical parameters such as cross-link density, polymer orientation, crystallinity, and the degree of banding can alter these properties significantly. When type I collagen is exposed to acid at pH 3, it becomes swollen and loses its banding, which also reduces its clotting abilities. Preclinical studies showed optimal ranges for pore size and degradation of the matrix. This work led to a polymeric scaffold that allows cell ingrowth from the wound surface that will replace and eventually degrade the scaffold. Currently, the clinical material is made of type I collagen derived from bovine Achilles tendon, which is processed to remove telopeptides and coprecipitated with chondroitin-6-sulfate and then freeze-dried to form a highly porous solid. A silicone elastomer is then applied as a coating to the entire surface.

The current clinical material, Integra® is applied to burns and other debrided wounds. For small areas of coverage, keratinocytes can migrate into the edge of the

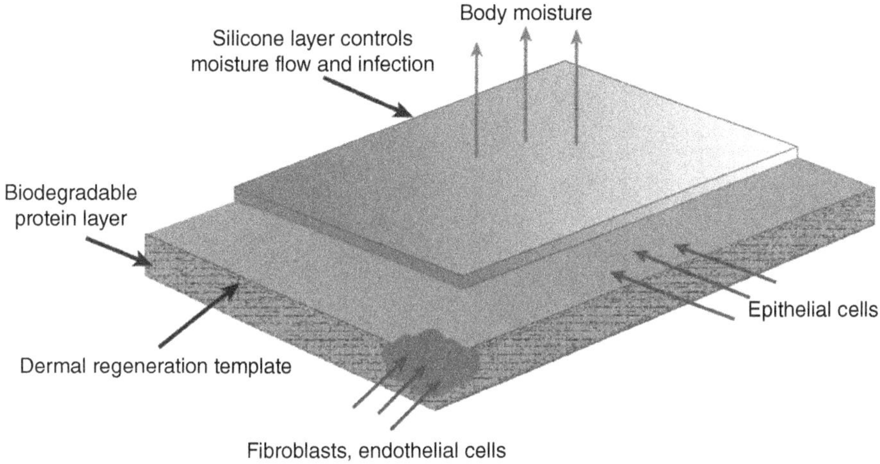

FIGURE 9.2 Schematic of the bilayer device. (Reprinted from Yannas, I.V., Orgill, D.P., Burke, J.F. Template for skin regeneration. *Plastic and Reconstructive Surgery 127* Suppl 1, 60S–70S, 2011 with permission from Wolters Kluwer Health. Modified from original Yannas, I.V., Burke, J.F., Orgill, D.P., Skrabut, E.M. Wound tissue can utilize a polymeric template to synthesize a functional extension of skin. *Science* **1982**; *215*, 174–176 with permission from the American Association for the Advancement of Science.)

wound beneath the silicone elastomer (Figure 9.2), which is spontaneously ejected from the scaffold. In larger wounds, the silicone elastomer can be removed and a very thin skin graft applied. This allows minimal donor site morbidity and can achieve excellent color and texture match of the skin (Figure 9.3).

These scaffolds were originally designed for burns and have been shown in clinical trials to perform well when carefully applied [9,10]. They are more easily infected than conventional skin grafts, but if carefully monitored and treated with topical antibiotics, can have satisfactory results with low complication rates. The elasticity and lack of scarring that occurs with this type of skin has been encouraging. Its use has expanded to treat diabetic foot wounds and complex wounds, including those with small areas of exposed bone or tendon [11]. We have found it particularly useful in treating scalp wounds with exposed calvarium. By burring the bone down to the diploic space and applying the scaffold for 6–8 weeks, it slowly revascularizes, and then a very thin skin graft is applied. The result is a much simpler surgical procedure than conventional scalp flaps or free tissue transfer [12].

Experimentally, cells can be added to the scaffold to obtain nearly complete regeneration of skin within 14 days [13]. In these cases, cells were seeded by a centrifugation process and was found to self-organize into a bilayer skin construct with many features comparable with normal skin. In particular, the dermal architecture appears to be more woven than that of scar tissue, which tends to have unidirectional short parallel fibers. This procedure allows a single-stage reconstruction of skin that could be highly beneficial if properly developed for patients.

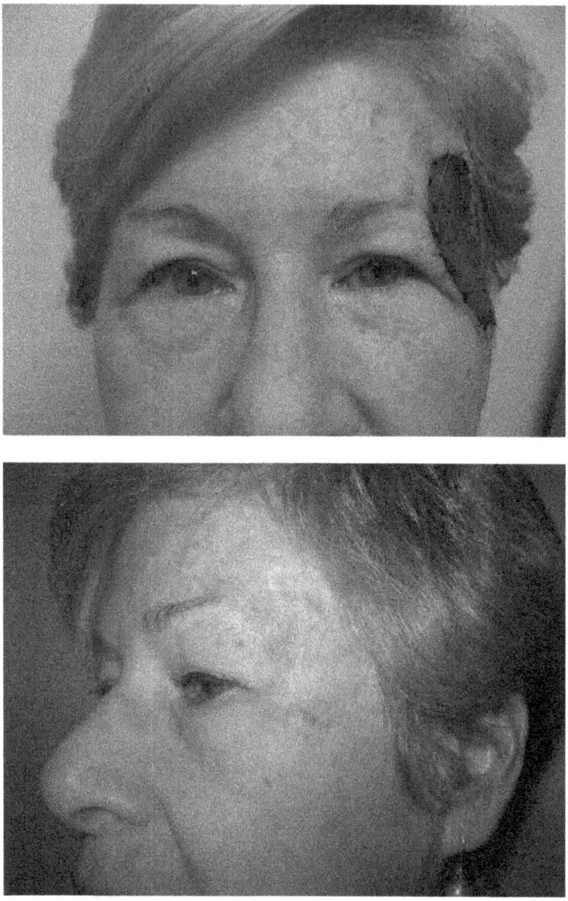

FIGURE 9.3 A 65-year-old woman treated with resection and application of a dermal regeneration template. The silicone was removed at 1 month and a thick skin graft was taken from the upper arm. At 1 year, there is an excellent color match of the skin. (Reprinted from Yannas, I.V., Orgill, D.P., Burke, J.F. Template for skin regeneration. *Plastic and Reconstructive Surgery* 127 Suppl 1, 60S–70S, 2011 with permission from Wolters Kluwer Health.) *(See insert for color representation of the figure.)*

9.4 DISCUSSION

The skin is a complex organ that has major variation in structure and function even in adjacent areas. For example, the glabrous skin on the palm and soles differs significantly from the skin on the dorsum of these structures. On the face, the skin on the nose is very different from forehead skin, which differs from that around the mouth. Our goal in skin replacement therapy should be to develop the capacity to regenerate skin that is an exact replica of what is removed. To date, we do not have the full capacity to regenerate adnexal structures such as hair follicles and sweat glands. Much work needs to be done to clarify the sensory reinnervation that occurs into these various constructs, as well as the development of immune function. As molecular biologic methods are being applied, more are being learned about the regenerative capacity of these structures. Through hair micrografts, it is possible to do reproducible transfer of hair follicles to other areas of the body. Navsaria used hair follicles placed through the silicone of Integra® to resurface wounds with keratinocytes [14].

The future is likely to involve some combination of specific stem cells and scaffolds. Several investigators have already shown the ability of stem cells to facilitate wound closure. The exact mechanism of improvement remains uncertain. There is some evidence that in certain cases, cells can engraft and proliferate [13]. In other cases, it may be that there are specific biomolecules within stem cell extracts that facilitate growth and regeneration. As more are learned about stem cells, it is likely that a combination of specific stem cells with an appropriately designed scaffold may provide a reproducible method to design a highly specific regenerative template for skin replacement therapy.

REFERENCES

[1] Dunkin, C.S., Pleat, J.M., Gillespie, P.H., Tyler, M.P., Roberts, A.H., McGrouther, D.A. *Plastic and Reconstructive Surgery* **2007**, *119*, 1722.

[2] Yannas, I.V., Lee, E., Orgill, D.P., Skrabut, E.M., Murphy, G.F. *Proceedings of the National Academy of Sciences of the United States of America* **1989**, *86*, 933.

[3] Ehrlich, H.P. *Progress in Clinical and Biological Research* **1988**, *266*, 243.

[4] Rheinwald, J.G., Green, H. *Cell* **1975**, *6*, 331.

[5] Gallico, G.G., 3rd, O'Connor, N.E., Compton, C.C., Kehinde, O., Green, H. *The New England Journal of Medicine* **1984**, *311*, 448.

[6] O'Connor, N., Mulliken, J., Banks-Schlegel, S., Kehide, O., Green, H. *Lancet* **1981**, *1*, 75.

[7] Cuono, C.B., Langdon, R., Birchall, N., Barttelbort, S., McGuire, J. *Plastic and Reconstructive Surgery* **1987**, *80*, 626.

[8] Bell, E., Ivarsson, B., Merrill, C. *Proceedings of the National Academy of Sciences of the United States of America* **1979**, *76*, 1274.

[9] Burke, J.F., Yannas, I.V., Quinby, W.C., Jr., Bondoc, C.C., Jung, W.K. *Annals of Surgery* **1981**, *194*, 413.

[10] Heimbach, D., Luterman, A., Burke, J., Cram, A., Herndon, D., Hunt, J., Jordan, M., McManus, W., Solem, L., Warden, G., Zawacki, B. *Annals of Surgery* **1988**, *208*, 313.

[11] Iorio, M.L., Shuck, J., Attinger, C.E. *Plastic and Reconstructive Surgery* **2012**, *130*, 232S.

[12] Corradino, B., Di Lorenzo, S., Leto Barone, A.A., Maresi, E., Moschella, F. *Journal of Plastic, Reconstructive & Aesthetic Surgery* **2010**, *63*, e245.

[13] Yannas, I.V., Burke, J.F., Orgill, D.P., Skrabut, E.M. *Science* **1982**, *215*, 174.

[14] Navsaria, H.A., Ojeh, N.O., Moiemen, N., Griffiths, M.A., Frame, J.D. *Plastic and Reconstructive Surgery* **2004**, *113*, 978.

CHAPTER 10

EPITHELIAL ENGINEERING: FROM SHEETS TO BRANCHED TUBES

HYE YOUNG KIM and CELESTE M. NELSON

Departments of Chemical and Biological Engineering and Molecular Biology, Princeton University, Princeton, NJ, USA

10.1 INTRODUCTION

The epithelium is one of the major tissue types in the body, commonly arranged as either a single layer or multiple layers of cells [1,2]. Epithelial cells are polarized within the epithelium such that their apical surfaces face fluid-filled spaces, such as the inside of the tube, while their basal surfaces are juxtaposed to the extracellular environment (Figure 10.1A). Cohesion within this polarized epithelium is mediated by intercellular junctions, including tight junctions and desmosomes, thus enabling the tissue to form a contiguous sheet, which can fold and form a tubular structure. Notably, epithelia cover the surfaces of the body and line hollow tissues where they function to protect or enclose organs.

Construction of the basic structure and function of the epithelium begins during development, when populations of epithelial cells undergo dramatic changes in shape, including collective migration, bending, folding, or assembling a tube, which subsequently extends or branches to build a mature organ (Figure 10.1B–D). Here, we introduce examples of epithelial morphogenesis that occur in nature which can provide useful insights into how epithelial tissues sculpt themselves into various shapes. These morphogenetic behaviors illuminate gaps in our knowledge of regulatory cues, including the biochemical and mechanical signals that are present within the substratum, the soluble factors that direct cellular behaviors, and the intracellular machinery that drives epithelial shape changes. We also highlight engineering strategies that can be used to dissect the effects of the individual regulatory components by mimicking external environments and various stimuli during epithelial sheet migration, folding, tubulogenesis, and branching.

Bio-inspired Materials for Biomedical Engineering, First Edition. Edited by Anthony B. Brennan and Chelsea M. Kirschner.

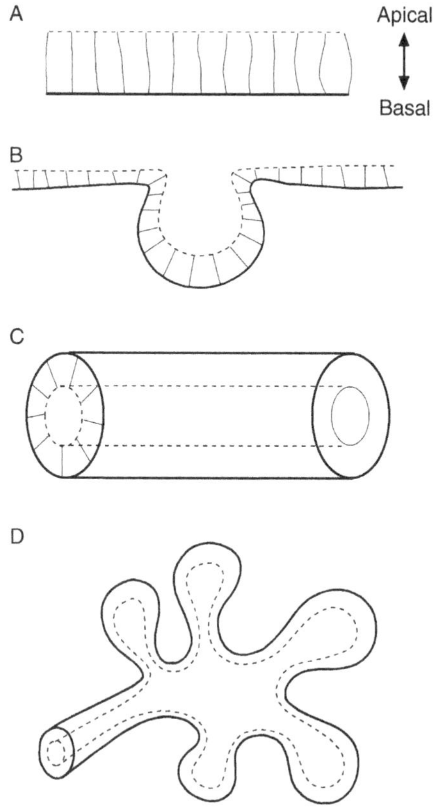

FIGURE 10.1 Epithelial morphogenesis. Schematic of various epithelial shapes. (A) Tightly connected epithelial sheet with distinct apical (dotted line) and basal (filled line) polarity. (B) Folded epithelium. (C) Epithelial tube. (D) Branching epithelium.

10.2 INSPIRATION FROM THE BIOLOGY OF EPITHELIAL MORPHOGENESIS

10.2.1 Collective Migration of Epithelial Sheets

The epithelium moves collectively to cover the whole embryo during the early stages of development, forms the internal lining of organs at later stages, and even fills holes during wound healing in the mature organism. These concerted epithelial movements are observed in many developmental processes, and tracking labeled cells within a moving epithelial sheet has provided significant insight into collective cell behaviors. For example, the animal–vegetal axis is the first one to form in the vertebrate embryo, with an animal pole that represents the future ectoderm and mesoderm and a vegetal pole that represents the future endoderm. The simple epithelium of the enveloping layer (EVL) of the zebrafish embryo moves toward the vegetal pole from the animal side, and eventually covers the entire embryo during

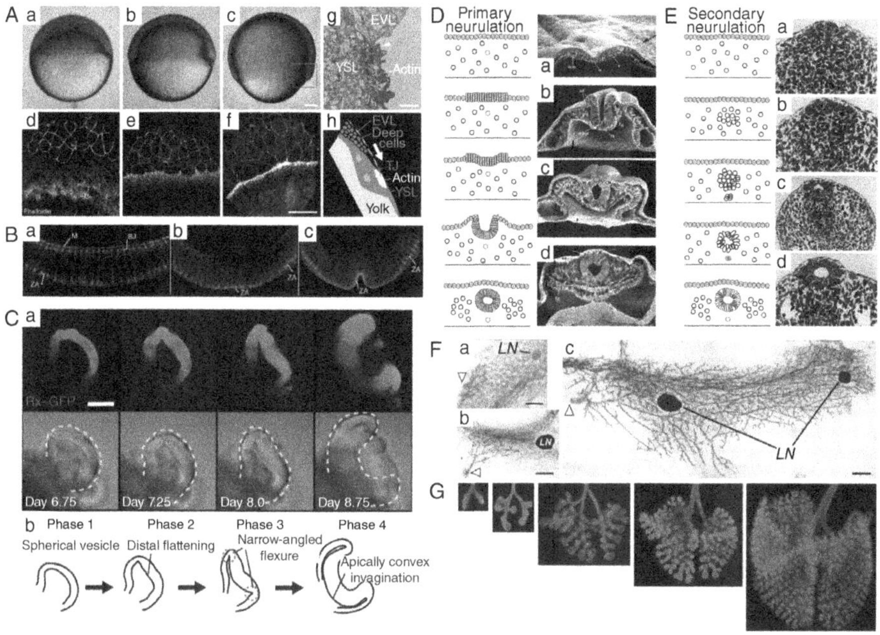

FIGURE 10.2 Epithelial morphogenesis during development. (A) Epithelial movement within the enveloping layer (EVL) of the zebrafish embryo. Note that the monolayer of cells moves over the surface of the embryo. (B) Epithelial folding during gastrulation of *Drosophila*. Shown are the adherens junctions (zonula adherens [ZA]), the basal junctions (BJ), and myosin (M). Note the bending of the cells as the ventral furrow forms. (C) Multiple steps of folding, relaxing, and growth of epithelium during optic cup morphogenesis. Shown are Rx-GFP-labeled retinal anlagen, which first appear as the optical vesicle. (D) Neural tube formation during primary neurulation in the chicken embryo. (E) Secondary neurulation in the mouse embryo. (F) Branching morphogenesis of the mouse mammary gland. (G) Branching morphogenesis of the embryonic mouse lung. (Adapted from References 5, 24, 32, 37, 51, 53, and 78.)

a process known as epiboly (Figure 10.2A) [3]. These animal-to-vegetal movements of the epithelium are guided by underlying microtubule arrays that extend in parallel to the direction of migration [4], which is facilitated by pulling through contractions of the actin cytoskeleton and cell shape changes at the vegetal margin of the tissue (Figure 10.2A(d–f)) [5,6].

Epithelial sheets are usually accompanied by an underlying mesenchyme and extracellular matrix (ECM). Evidence of increased fibronectin assembly under the migrating epithelium [7] suggests a possible role for the ECM to act as a directional cue, which has been shown in some model systems [8,9]. Whereas the ECM has been thought to provide a guidance cue for migrating cells, live imaging has revealed that the ECM in the embryo is not fixed in place. Instead, analysis of the movements of ECM and epithelial cells suggest concerted motions [10], such that the cells move as flowing streams that merge at the primitive streak in avian embryos [11,12]. In the abdomen of *Drosophila*, the epidermal cells divide and then migrate in an oriented direction until the whole abdomen is covered [13], which differs from the oriented cell division that often controls epithelial morphogenesis. These examples suggest a major role for cell migration in embryonic development.

In addition to these morphogenetic movements, epithelial sheets migrate collectively to repair injured tissues. Unlike adult tissues, embryonic wounds do not induce an inflammatory response and heal rapidly without a scar [14]. Careful examination of the wound healing process in the epidermis of the wing bud of the embryonic chicken indicates that a contiguous F-actin cable, called the actin purse string, runs around the smooth margin of the wound and plays a major role in closing the wounded area [15]. In contrast, wound healing of the epithelium in *Xenopus* embryos requires the contraction of deep cells and epithelial protrusions, in addition to the actin purse string [16]. Epithelial cells can thus exhibit similar collective movements that are driven by disparate underlying physical mechanisms.

10.2.2 Folding of Epithelial Sheets

As epithelial tissues migrate to their final positions, these flat sheets begin to bend and fold to build additional dimensions into the embryo: by adding another germ layer during gastrulation, making hollow tubular structures during neurulation, or forming a spherical cup during eye development (Figure 10.2B–E). Tissue bending is an active process that has been proposed to use a number of mechanisms, including localized changes in cell shape driven by actomyosin contractions at the apical side of the epithelium, known as apical constriction (reviewed in Reference 17), differential cell proliferation, or differential positioning of adherens junctions [18]. These changes at localized regions of the mechanically linked sheet can cause it to bend. During gastrulation, apical constriction causes the epithelium to fold and, thereby, induces the tissue that covers the surface of the embryo to invaginate and penetrate inside to form an additional germ layer [19–22]. A similar mechanism of apical contractions drives gastrulation of the *Drosophila* embryo (Figure 10.2B) [23,24], where the dynamics of tissue folding has been best characterized for the formation of the ventral furrow, which internalizes to form a tube and marks the onset of gastrulation [25,26]. Live imaging and quantitative analysis of actin and myosin revealed that pulsed actomyosin contractions at the apical cortex pull the sites of adherens junctions inwards [26]. These contractions are stabilized incrementally between cycles through a ratchet-like mechanism, which constricts the apical side of a strip of ventral cells that fold into the embryo [26]. The polarized constrictions along the length of the cells of the ventral furrow result from the tissue-level tension integrated along the anterior–posterior axis [25].

In addition to this one-step folding of the epithelial sheet, further bending can arise at defined *hinge points*, where cell wedging takes place to provide additional nodes to fold. During primary neurulation in the mouse and chick (Figure 10.2D), the neural epithelium begins to form one acute bend (at the medial hinge point [MHP]) along the midline, resulting in a V-shaped groove, and later develops two additional dorsolateral hinge points (DLHPs), where subsequent inward bending occurs [27–30]. Additionally, the medial hinge cells have longer cell cycle times as compared with the others, which also enhances the folding of the neural epithelium [31].

More complex folding events have been uncovered during the morphogenesis of the optic cup using embryonic stem cell-derived neuroepithelial tissue (Figure 10.2C). Four distinct steps of morphogenesis were found to be controlled by local epithelial properties, driven by actomyosin contraction and subsequent tissue stiffening and tissue growth (Figure 10.2C) [32]. First, the monolayered epithelium bulges out from tissue aggregates with intense active myosin accumulation along the apical surface. Subsequently, during the second phase, the distal end flattens as the cells differentially contract, producing a flexible distal end and stiff lateral sides. During the third phase, cells in the lateral edge make a hinge point that causes inward buckling of the flattened distal epithelium. Invagination continues in a cell proliferation-dependent manner during the fourth and final phase [32]. Despite differences in underlying cellular mechanism, all epithelial folding results from these kinds of spatial patterns in cell behaviors.

10.2.3 Tubulogenesis

Epithelial tubes are essential functional units in many organs, acting as pipes that are used to transport the gases and liquids of the body. Epithelial tubes form with a distinct polarity in which the apical surface of the tissue faces the inside of a hollow lumen (Figure 10.1C). Several organs, including the neural tube, lung, kidney, and mammary gland, develop from simple epithelial tubes and their extensions, connections, and elaborations enable the construction of complex architectures. Tubular structures can be generated by two mechanisms: folding of the polarized epithelial sheet or by eliminating cells or producing a hollow lumen from an unpolarized cell mass. Neural tube formation is a well-known example, in which an epithelial tube is made by folding a polarized sheet via mechanisms described in the previous section. Once the epithelium folds, the lateral edges of the tissue must fuse together to form a closed tube (Figure 10.2D). To induce this open-ended epithelium to seal, the cells rearrange and change shape as they intercalate with each other in a direction perpendicular to the length of the tube under the control of the planar cell polarity (PCP) signaling pathway [33–36]; the elongated and actively moving cells cause the whole tissue to converge at the midline. In contrast, during secondary neurulation, the posterior neural tube forms from mesenchymal cells that condense into a solid rod and then transform into an epithelial tube (Figure 10.2E) [37,38]. Posterior neural tube formation in the avian embryo has especially distinguishable phases: formation of a medullary cord by dorsal cell aggregation, differentiation of central and peripheral cells, cavitation of multiple lumens, and coalescence of these into a single larger lumen [38]. A similar mechanism for tubulogenesis has been observed during formation of the gut in the zebrafish embryo [39].

In addition, tube formation during development of the salivary gland in *Drosophila* begins from a population of specified cells called the placode, which sequentially internalize through cell shape changes [40], rearrangement, and directed cell migration (for review, see References 41 and 42). Though assembly of the tubular structure is driven by various strategies, it requires dramatic epithelial cell movements and cell shape changes.

10.2.4 Branching Morphogenesis

The simple epithelial tubes described above often elongate, expand, and bifurcate at the ends to create a three-dimensional (3-D) tree-like structure via branching morphogenesis (Figure 10.1D), which generates a variety of organs including the mammary gland, salivary gland, kidney, and lungs. Each organ uses a unique process to build the complicated network of branched epithelial tubes [43]. Branching morphogenesis of the vertebrate submandibular gland (SMG) is especially well characterized. The spherical epithelial bud first forms shallow clefts along its surface that deepen and divide to generate multiple buds. The formation of these clefts is closely associated with fibronectin assembly [44–46], such that epithelial cells adjacent to fibronectin convert their cell–cell adhesions to cell–matrix adhesions as they communicate through a Rho kinase-mediated integrin and actomyosin network [47–49].

Unlike other branched organs, the mammary gland elaborates into a ramified structure during puberty in response to hormonal stimulation (Figure 10.2F) [50,51]. In the mouse, the ends of mammary ducts form bulbous structures called terminal end buds (TEBs), which consist of multiple cell types including layers of epithelium. The TEBs are thought to extend into the surrounding stroma by collective cell rearrangements without extending cellular protrusions [52]. The extended TEBs undergo a further branching process of elongation and bifurcation that is closely related to the behaviors of myoepithelial cells that tightly surround the budding epithelium.

While the branching morphogenesis of both salivary and mammary glands is designed to acquire an increased surface area necessary for their function, the branching process in the lung adds a hierarchical component via highly organized iterative morphogenetic steps. Careful examination of the temporal sequence of airway morphogenesis indicates a stereotyped branching process (Figure 10.2G) in which the epithelium uses only three routines: domain branching, planar bifurcation, and orthogonal bifurcation [53]. Repetition of these branching routines in a specific order enables construction of the highly organized and yet evenly spaced epithelium that is suitable for the lung, which resides in the limited volume of the chest cavity.

10.3 ENGINEERING APPROACHES TO MIMIC EPITHELIAL MORPHOGENESIS

Observing and analyzing epithelial morphogenesis in model organisms allows us to understand how epithelial sheets move and shape into tubes, and even sculpt themselves into intricate branched structures. However, it is challenging to dissect individual factors that regulate epithelial morphogenesis in whole embryos or intact tissues, which are exposed simultaneously to multiple biochemical and mechanical cues. Here we describe current engineering approaches that were inspired from epithelial morphogenesis *in vivo*, and used to uncover additional processes involved in building various shapes of epithelial tissues.

10.3.1 Making a Sheet

Epithelial morphogenesis and wound healing require regulatory mechanisms to direct the movements of epithelial sheets. These mechanisms require that cells interpret geometric and mechanical cues in the substratum, in addition to signals from neighboring cells within the tissue. A number of engineering approaches have been developed to investigate factors that contribute to collective behaviors within epithelial sheets, including quantifying the direction and speed of sheet movements and relative behaviors of cells located at different positions within the tissue.

The most well-characterized factor that controls morphogenesis of epithelial sheets is the geometry of the underlying substratum, from which cells interpret signals for motility, proliferation, and shape [54]. A simple microcontact printing technique using a microfabricated elastomeric stamp has been used to provide defined adhesive geometries for monolayers of cells. Culturing monolayers on different widths of adhesive substratum (Figure 10.3A) revealed distinct modes of collective cell behaviors. Cells migrating over narrow substrata (20 μm) use a contraction–relaxation type of motility, whereas those migrating on wider substrata (400 μm) move continuously in one direction (Figure 10.3A) [55]. Geometric control within epithelial sheets has also been investigated by exposing free regions in the substratum, thus mimicking the geometry of a wound. Removing various shapes and sizes of elastomeric barriers from within intact epithelial sheets introduces an open space without scarring, and this introduction of available substratum triggers cellular movement (Figure 10.3B). This approach has been used to test effects of wound shapes as well as to identify major cellular behaviors that are required for epithelial sheets to fill open spaces [56–58]. Additional geometric control has been obtained from using photoswitchable surfaces. For example, an initially nonadhesive substratum (polyethylene glycol [PEG] conjugated with 2-nitrobenzyl groups) can be switched to an adhesive substratum after exposure to ultraviolet (UV) light, which cleaves the photoremovable PEG from the surface, thus exposing the underlying adhesive glass (Figure 10.3C). In this way, adhesive regions are first patterned precisely using a photomask for initial tissue formation, and UV exposure is used to change the surface chemistry, increase the area of adhesive substratum, and thereby induce expansion of the already formed epithelial tissue [59]. Because of the fine geometric control and switchable chemistry of the substratum, this approach is useful for studying how the dynamics of expansion of epithelial tissues vary depending on the curvature of the boundary [59].

In addition to geometric cues from the substratum, the mechanical compliance of the external environment is known to have a significant influence on cellular behaviors *in vivo*, as well as in culture [60]. Developing embryos change their stiffness as they age with about 10-fold difference in tissue stiffness across the

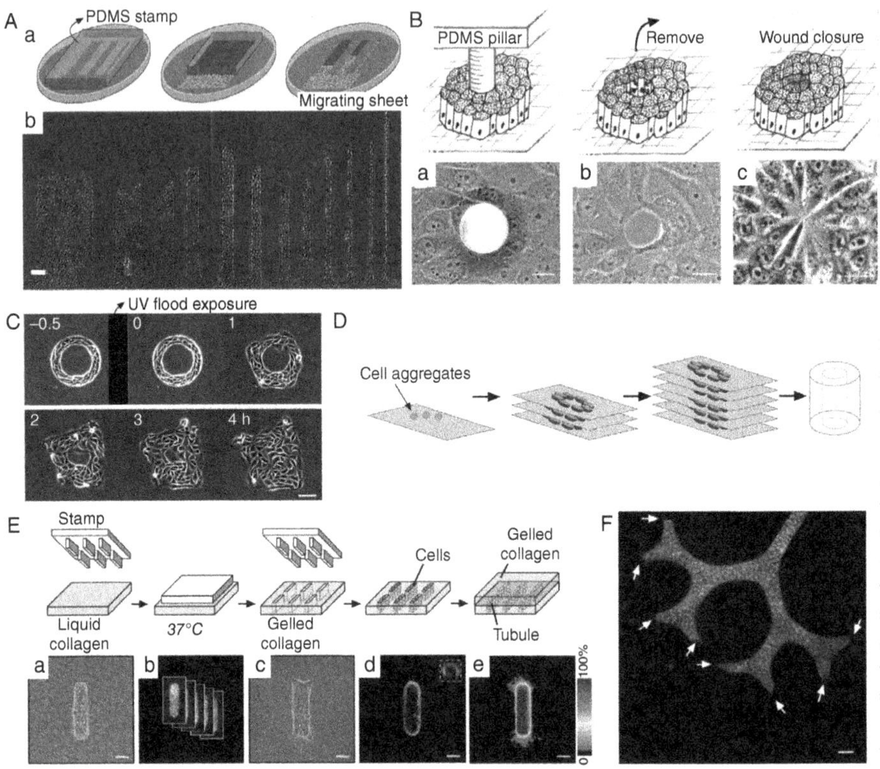

FIGURE 10.3 Engineering approaches that mimic epithelial morphogenesis. (A) Micropatterned adhesive substratum for investigating epithelial sheet migration. (B) Mimicking wounded epithelium using PDMS pillars. (C) Switchable substratum for expansion of epithelial sheets. (D) 3-D printing for constructing a biological tube. (E,F) 3-D micropatterned tubes for investigating branching morphogenesis. (Adapted from References 55, 57, 59, 71, and 72.) *(See insert for color representation of the figure.)*

various germ layers [61], suggesting natural heterogeneity within the mechanical environments encountered by migrating epithelial sheets during morphogenesis. A common approach to test the effects of the mechanical environment on epithelial sheet migration is to use polymeric substrata such as polyacrylamide (PAA) or polydimethylsiloxane (PDMS), in which tuning the relative concentration of cross-linker modulates the stiffness [62,63]. When epithelial tissues are cultured on substrata of varying stiffness in wound healing assays, the epithelial cells increase their collective migration speed, persistence, and directionality, such that they coordinate movements over stiff substrata [64]. Moreover, substratum stiffness affects the transmission of directional cues by establishing cell polarity via a gradient of myosin II [64]. Subsequent approaches should emphasize understanding how epithelial tissues integrate biochemical and mechanical signals that induce coordinated sheet movements, even in a 3-D environment.

10.3.2 Folding a Sheet

Epithelia often bend and fold during organ development (Figure 10.2B–E). These sequential processes are akin to using origami to make 3-D structures from two-dimensional (2-D) sheets of paper. During epithelial morphogenesis, groups of cells within specific regions of the epithelium generate folding forces using actomyosin contractions. As a consequence, one strategy to mimic epithelial folding would be to stimulate the major force-producing machinery (i.e., actomyosin network) at desired locations within the epithelial sheet. Several approaches have been proposed to induce localized cell contractions within an epithelium, including electrical stimulation, laser activation, and nanoperfusion of soluble factors [65]. These external stimuli can induce localized cell contractions within an epithelial sheet, but have only been tested for transient induction of contractions. Repetitive application or increasing the period of stimulus may induce prolonged cellular forces that are enough to bend or fold the targeted regions of the epithelium, thus, possibly mimicking epithelial folding within native tissues as well as enabling study of the required downstream contributors.

10.3.3 Making a Tube

Tubes serve as the most basic functional unit for many conducting organs in the body. In nature, epithelial tubulogenesis results from either cellular aggregates or sheets that simultaneously form a polarized inner cavity, or from folding sheets to make a tubular structure (Figure 10.2D,E). Borrowing similar concepts from epithelial tube morphogenesis, Mardin–Darby canine kidney (MDCK) epithelial cells were embedded in 3-D collagen gels and stimulated with growth factors [66]. The 3-D cellular aggregates formed hollow cysts, which extended to make lumen-containing tubes [67,68]. The steps of this self-assembly during tubulogenesis are similar to those observed in posterior neural tube formation, and therefore can provide useful insight into the mechanisms regulating collective cellular behaviors.

In addition to self-assembling tubes, several approaches have been developed to engineer biological tubes a priori. Tubes can be built from epithelial sheets that are wrapped around a hollow cylinder multiple times; such physically fabricated tubes have been used as replacements for the aorta [69]. Alternatively, biological tubes can be constructed using 3-D bioprinting techniques. Using a frame comprised of a thin sheet or small rods (Figure 10.3D), small cell aggregates can be printed at chosen locations within the 3-D space. The printed cell aggregates act as building blocks and fuse to form a contiguous tissue that mimics the shape of a tube [70,71]. Nevertheless, it is challenging to construct a tube that is physically stable, yet biologically functional to replace the native tissue. Multiple strategies used to assemble similar shaped epithelial tubes during development (Figure 10.2D,E) may provide essential clues to enhance current engineering approaches.

10.3.4 Making a Branch

Branching morphogenesis is a complex and intricate process that is thought to require precise spatiotemporal control of growth factors and reciprocal interactions between the branching epithelium and its surrounding mesenchyme. To dissect the directional cues that instruct branching sites along an epithelial tube, 3-D micropatterned tissues have been constructed in different shapes using micromolding of collagen cavities (Figure 10.3E,F) [72]. Using this approach, hundreds of tubules were microfabricated by introducing geometrically patterned cavities into collagen gels that were filled with mammary epithelial cells and used to quantify the pattern of branch initiation. These simplified 3-D microscale tissues revealed that the initial geometry of the epithelial tube provides an important cue for determining sites of branching by influencing the concentration profile of morphogens secreted from the epithelium itself. Besides geometric and biochemical cues, the mechanical properties of the microenvironment also influence the branching pattern. The microfabricated mammary epithelial tubules were found to branch from regions within the tissue exposed to high mechanical stress [73], which are accompanied by physical changes in the surrounding collagen gels [74].

In addition to the formation of new branches from a preexisting tube, branching processes involve interactions between neighboring mesenchymal cells, as well as other branching epithelia. 3-D epithelial tissues sandwiched between two layers of ECM gels (matrigel and collagen) [75,76] permitted formation of tubules that connect to neighboring tissues. These tubular networks were found to be initiated and maintained by traction forces generated by cells through collagenous ECM, indicating that mechanical feedback of epithelial tissues can be another guidance cue for branching morphogenesis [74,76].

10.4 CONCLUSION

Epithelial tissue lines all internal organs and covers the surfaces of the body. The status and function of the epithelium are closely related to many clinical conditions, from life-threatening cancers to small abrasions on the skin. Decades of effort in developing tissue replacements have resulted in commercial products that enhance wound healing processes [77], and brighten the future of tissue engineering approaches for regenerative medicine. However, this is a small step forward if we consider the range of epithelial tissues that line different organs with distinct shapes and functions.

The exploration of tissue behaviors within 3-D environments similar to those found *in vivo* has a short history compared with traditional 2-D cell culture. Due to the increased complexity of the 3-D setting, tissue engineers often aim to construct a tissue that is physically analogous to that found *in vivo*. Surprisingly little attention has been paid to the integration of the physical and biochemical mechanisms observed to regulate naturally evolving tissue morphogenesis. We believe it is critical to understand the epithelial morphogenesis that occurs in developing embryos; epithelial sheet migration, folding, tubulogenesis, and branching can be harnessed as discrete steps in the engineering of tissues *ex vivo*. Despite obscure regulatory cues in morphogenesis and limited engineering strategies, rapidly developing technologies and integration of expertise in developmental biology and engineering may lead to significant progress in the field of epithelial tissue engineering.

Acknowledgments

Work from the authors' laboratory was supported in part by grants from the NIH (GM083997 and HL110335), the David & Lucile Packard Foundation, the Alfred P. Sloan Foundation, the Camille & Henry Dreyfus Foundation, and Susan G. Komen for the Cure. C.M.N. holds a Career Award at the Scientific Interface from the Burroughs Wellcome Fund.

REFERENCES

[1] Hay, E.D. *Acta Anatomica* **1995**, *154*, 8.
[2] St Johnston, D., Ahringer, J. *Cell* **2010**, *141*, 757.
[3] Warga, R.M., Kimmel, C.B. *Development* **1990**, *108*, 569.
[4] Solnica-Krezel, L., Driever, W. *Development* **1994**, *120*, 2443.
[5] Köppen, M., Fernández, B.G., Carvalho, L., Jacinto, A., Heisenberg, C.-P. *Development* **2006**, *133*, 2671.
[6] Yu, J.A., Foley, F.C., Amack, J.D., Turner, C.E. *Developmental Biology* **2011**, *349*, 225.
[7] Latimer, A., Jessen, J.R. *Matrix Biology* **2010**, *29*, 89.
[8] Goto, T., Davidson, L., Asashima, M., Keller, R. *Current Biology* **2005**, *15*, 787.
[9] Davidson, L.A., Marsden, M., Keller, R., DeSimone, D.W. *Current Biology* **2006**, *16*, 833.
[10] Zamir, E.A., Rongish, B.J., Little, C.D. *PLoS Biology* **2008**, *6*, e247.
[11] Cui, C., Yang, X., Chuai, M., Glazier, J.A., Weijer, C.J. *Developmental Biology* **2005**, *284*, 37.
[12] Chuai, M., Zeng, W., Yang, X., Boychenko, V., Glazier, J.A., Weijer, C.J. *Developmental Biology* **2006**, *296*, 137.
[13] Bischoff, M., Cseresnyés, Z. *Development* **2009**, *136*, 2403.
[14] Bullard, K.M., Longaker, M.T., Lorenz, H.P. *World Journal of Surgery* **2003**, *27*, 54.
[15] Martin, P., Lewis, J. *Nature* **1992**, *360*, 179.
[16] Davidson, L.A., Ezin, A.M., Keller, R. *Cell Motility and the Cytoskeleton* **2002**, *53*, 163.
[17] Sawyer, J.M., Harrell, J.R., Shemer, G., Sullivan-Brown, J., Roh-Johnson, M., Goldstein, B. *Developmental Biology* **2010**, *341*, 5.
[18] Wang, Y.-C., Khan, Z., Kaschube, M., Wieschaus, E.F. *Nature* **2012**, *484*, 390.
[19] Lee, J.-Y., Harland, R.M. *Developmental Biology* **2007**, *311*, 40.
[20] Davidson, L.A., Koehl, M.A., Keller, R., Oster, G.F. *Development* **1995**, *121*, 2005.
[21] Beane, W.S., Gross, J.M., McClay, D.R. *Developmental Biology* **2006**, *292*, 213.
[22] Sherrard, K., Robin, F., Lemaire, P., Munro, E. *Current Biology* **2010**, *20*, 1499.
[23] Sweeton, D., Parks, S., Costa, M., Wieschaus, E. *Development* **1991**, *112*, 775.
[24] Dawes-Hoang, R.E., Parmar, K.M., Christiansen, A.E., Phelps, C.B., Brand, A.H., Wieschaus, E.F. *Development* **2005**, *132*, 4165.
[25] Martin, A.C., Gelbart, M., Fernandez-Gonzalez, R., Kaschube, M., Wieschaus, E.F. *The Journal of Cell Biology* **2010**, *188*, 735.
[26] Martin, A.C., Kaschube, M., Wieschaus, E.F. *Nature* **2009**, *457*, 495.
[27] Lawson, A., Anderson, H., Schoenwolf, G.C. *The Anatomical Record* **2001**, *262*, 153.
[28] Smith, J.L., Schoenwolf, G.C., Quan, J. *The Journal of Comparative Neurology* **1994**, *342*, 144.
[29] Schoenwolf, G.C., Franks, M.V. *Developmental Biology* **1984**, *105*, 257.
[30] Shum, A.W., Copp, A. *Anatomy and Embryology* **1996**, *194*, 65.

[31] Smith, J.L., Schoenwolf, G.C. *Cell and Tissue Research* **1988**, *252*, 491.
[32] Eiraku, M., Takata, N., Ishibashi, H., Kawada, M., Sakakura, E., Okuda, S., Sekiguchi, K., Adachi, T., Sasai, Y. *Nature* **2011**, *472*, 51.
[33] Nishimura, T., Honda, H., Takeichi, M. *Cell* **2012**, *149*, 1084.
[34] Wallingford, J.B., Harland, R.M. *Development* **2002**, *129*, 5815.
[35] Kim, H.Y., Davidson, L.A. *Journal of Cell Science* **2011**, *124*, 635.
[36] Davidson, L.A., Keller, R.E. *Development* **1999**, *126*, 4547.
[37] Schoenwolf, G.C. *American Journal of Anatomy* **1984**, *169*, 361.
[38] Schoenwolf, G.C., Delongo, J. *American Journal of Anatomy* **1980**, *158*, 43.
[39] Bagnat, M., Cheung, I.D., Mostov, K.E., Stainier, D.Y.R. *Nature Cell Biology* **2007**, *9*, 954.
[40] Myat, M.M., Andrew, D.J. *Development* **2000**, *127*, 679.
[41] Maruyama, R., Andrew, D.J. *Developmental Dynamics* **2012**, *241*, 119.
[42] Abrams, E.W., Vining, M.S., Andrew, D.J. *Trends in Cell Biology* **2003**, *13*, 247.
[43] Kim, H.Y., Nelson, C.M. *Organogenesis* **2012**, *8*, 56.
[44] Onodera, T., Sakai, T., Hsu, J.C.-F., Matsumoto, K., Chiorini, J.A., Yamada, K.M. *Science* **2010**, *329*, 562.
[45] Larsen, M., Wei, C., Yamada, K.M. *Journal of Cell Science* **2006**, *119*, 3376.
[46] Sakai, T., Larsen, M., Yamada, K.M. *Nature* **2003**, *423*, 876.
[47] Daley, W.P., Gervais, E.M., Centanni, S.W., Gulfo, K.M., Nelson, D.A., Larsen, M. *Development* **2012**, *139*, 411.
[48] Daley, W.P., Kohn, J.M., Larsen, M. *Developmental Dynamics* **2011**, *240*, 2069.
[49] Daley, W.P., Gulfo, K.M., Sequeira, S.J., Larsen, M. *Developmental Biology* **2009**, *336*, 169.
[50] Gjorevski, N., Nelson, C.M. *Nature Reviews. Molecular Cell Biology* **2011**, *12*, 581.
[51] Sternlicht, M.D. *Breast Cancer Research* **2006**, *8*, 201.
[52] Ewald, A.J., Brenot, A., Duong, M., Chan, B.S., Werb, Z. *Developmental Cell* **2008**, *14*, 570.
[53] Metzger, R.J., Klein, O.D., Martin, G.R., Krasnow, M.A. *Nature* **2008**, *453*, 745.
[54] Théry, M. *Journal of Cell Science* **2010**, *123*, 4201.
[55] Vedula, S.R.K., Leong, M.C., Lai, T.L., Hersen, P., Kabla, A.J., Lim, C.T., Ladoux, B. *Proceedings of the National Academy of Sciences* **2012**, *109*, 12974.
[56] Poujade, M., Grasland-Mongrain, E., Hertzog, A., Jouanneau, J., Chavrier, P., Ladoux, B., Buguin, A., Silberzan, P. *Proceedings of the National Academy of Sciences* **2007**, *104*, 15988.
[57] Anon, E., Serra-Picamal, X., Hersen, P., Gauthier, N.C., Sheetz, M.P., Trepat, X., Ladoux, B. *Proceedings of the National Academy of Sciences* **2012**, *109*, 10891.
[58] Klarlund, J.K. *Proceedings of the National Academy of Sciences* **2012**, *109*, 15799.
[59] Rolli, C.G., Nakayama, H., Yamaguchi, K., Spatz, J.P., Kemkemer, R., Nakanishi, J. *Biomaterials* **2012**, *33*, 2409.
[60] Discher, D.E., Janmey, P., Wang, Y.-L. *Science* **2005**, *310*, 1139.
[61] Zhou, J., Kim, H.Y., Davidson, L.A. *Development* **2009**, *136*, 677.
[62] Lo, C.M., Wang, H.B., Dembo, M., Wang, Y.L. *Biophysical Journal* **2000**, *79*, 144.
[63] Balaban, N.Q., Schwarz, U.S., Riveline, D., Goichberg, P., Tzur, G., Sabanay, I., Mahalu, D., Safran, S., Bershadsky, A., Addadi, L., Geiger, B. *Nature Cell Biology* **2001**, *3*, 466.
[64] Ng, M.R., Besser, A., Danuser, G., Brugge, J.S. *The Journal of Cell Biology* **2012**, *199*, 545.
[65] Joshi, S.D., von Dassow, M., Davidson, L.A. *Experimental Cell Research* **2010**, *316*, 103.
[66] Montesano, R., Schaller, G., Orci, L. *Cell* **1991**, *66*, 697.

[67] Pollack, A.L., Barth, A.I., Altschuler, Y., Nelson, W.J., Mostov, K.E. *Journal of Cell Biology* **1997**, *137*, 1651.
[68] Pollack, A.L., Runyan, R.B., Mostov, K.E. *Developmental Biology* **1998**, *204*, 64.
[69] Sekine, H., Shimizu, T., Yang, J., Kobayashi, E., Okano, T. *Circulation* **2006**, *114*, I.
[70] Norotte, C., Marga, F.S., Niklason, L.E., Forgacs, G. *Biomaterials* **2009**, *30*, 5910.
[71] Mironov, V., Boland, T., Trusk, T., Forgacs, G., Markwald, R.R. *Trends in Biotechnology* **2003**, *21*, 157.
[72] Nelson, C.M., VanDuijn, M.M., Inman, J.L., Fletcher, D.A., Bissell, M.J. *Science* **2006**, *314*, 298.
[73] Gjorevski, N., Nelson, C.M. *Integrative Biology* **2010**, *2*, 424.
[74] Gjorevski, N., Nelson, C.M. *Biophysical Journal* **2012**, *103*, 152.
[75] Lee, G.Y., Kenny, P.A., Lee, E.H., Bissell, M.J. *Nature Methods* **2007**, *4*, 359.
[76] Guo, C.-L., Ouyang, M., Yu, J.-Y., Maslov, J., Price, A., Shen, C.-Y. *Proceedings of the National Academy of Sciences* **2012**, *109*, 5576.
[77] Shevchenko, R.V., James, S.L., James, S.E. *Journal of the Royal Society, Interface* **2010**, *7*, 229.
[78] Colas, J.-F., Schoenwolf, G.C. *Developmental Dynamics* **2001**, *221*, 117.

CHAPTER 11

A BIOMIMETIC APPROACH TOWARD THE FABRICATION OF EPITHELIAL-LIKE TISSUE

HONGJUN WANG and MENG XU

Department of Chemistry, Chemical Biology and Biomedical Engineering, Stevens Institute of Technology, Hoboken, NJ, USA

11.1 INTRODUCTION

The skin is the largest organ and covers the entire exterior surface of the body. It plays a crucial role in maintaining the homeostasis of internal tissues and organs, keeping the balance of moisture and temperature. The skin also serves as an efficacious barrier against hostile attack from the external environment and protects the body from mechanical, osmotic, and thermal damage. The skin contains three primary layers: epidermis, dermis, and subcutaneous fat. The epidermis is waterproof and acts as a barrier to infection. The dermis provides support to the epidermis and accommodates various appendages of the skin. The subcutaneous fat serves as padding and insulation for the body, in addition to storing energy and supplying oxygen via the blood vessels. With constant exposure to the external environment, damage to the skin cannot be avoided, and in the worst case, it can threaten the human life. The skin has a remarkable ability to regenerate in the case of injury to the epidermal layer or the superficial region of the dermis; however, when the injury is deep, especially with a complete devastation of the hair follicles, the damaged skin cannot spontaneously regenerate. In this regard, massive deep skin loss can lead to acute physiologic imbalance and ultimately, cause significant disability or even death. It becomes essential to properly close the wounded area with an effective dressing or with skin grafts as soon as possible to prevent fluid loss and potential infection and to facilitate wound healing with skin regeneration.

Skin wound healing is a complex process that involves the migration and proliferation of keratinocytes and fibroblasts and the synthesis and remodeling of

Bio-inspired Materials for Biomedical Engineering, First Edition. Edited by Anthony B. Brennan and Chelsea M. Kirschner.

new extracellular matrix (ECM). During this process, the cellular functions are synergistically regulated by various growth factors and cytokines, which are upregulated by the wounded environment [1] via corresponding cell membrane–bound receptors. To help the closure of hard-to-heal, deep wounds, several treatment options are available. Among them, split-thickness autologous skin grafting remains the gold standard. However, the limited availability of autografts along with the morbidity and pain associated with tissue harvesting sites can be a significant barrier for its application. The next best alternative to autografts is decellularized allografts, which provide a natural three-dimensional (3-D) ECM structure and basement membrane and are known for their role in promoting normal wound healing [2]. However, prolonged time is required for cells to migrate into an acellular matrix from surrounding tissues [3–5], which can significantly delay new tissue formation and reepithelialization. In recognition of the challenges of autografts and allografts, efforts have been made to create skin-like constructs via a tissue engineering approach. Currently, several tissue-engineered skin grafts are available for wound repair. Despite successful demonstration of their potential utility in wound repair, current commercial tissue-engineered grafts suffer from the following: (1) prolonged culture time to create the grafts (typically 2–3 weeks) [6], (2) low rate of acceptance by the host (40–60%, not clinically acceptable) [7], (3) poor handleability due to fragile mechanical durability, and (4) high cost associated with the use of skin grafts (e.g., annual medical cost of using Apligraf® [Organogenesis, Canton, MA] for an individual patient is approximately $20,000). Chronic wound patients frequently suffer from diseases such as diabetes and obesity [8]. Hence, it remains highly desirable to fabricate functional skin grafts for regenerating damaged skin tissues with full restoration of the lost biological functions, but at a low cost and within a clinically acceptable time windows (e.g., less than 2 weeks).

Nanofibrous matrices prepared from electrospinning can maximally recapture the dimension and morphology of native ECM and support the adhesion and growth of various cells, showing great potential in the field of skin tissue engineering. This chapter highlights how various nanofibrous matrices are prepared, and specific elaboration is made on the fabrication of nanofibrous scaffold for skin tissue regeneration.

11.2 SKIN ECM AND ITS FUNCTION

Skin is comprised of several different cell types. Keratinocytes are the most common cell type in the epidermis and form the surface barrier layer. Melanocytes are found in the epidermis but mainly reside in the basal layer, providing color to the skin. Fibroblasts are the major cellular components of dermal layer, which provides strength and resilience to the skin (Figure 11.1). The interactions between skin cells and their surroundings, that is, the ECM, play a crucial role in regulating the normal functions of the skin. For example, epidermal keratinocytes rest on a thin, ultrafine fibrous membrane called the basement membrane, mainly composed of types IV and VII collagen and laminins. The basement membrane is the fusion of two laminae, the basal lamina and the reticular lamina. The primary function of the basement

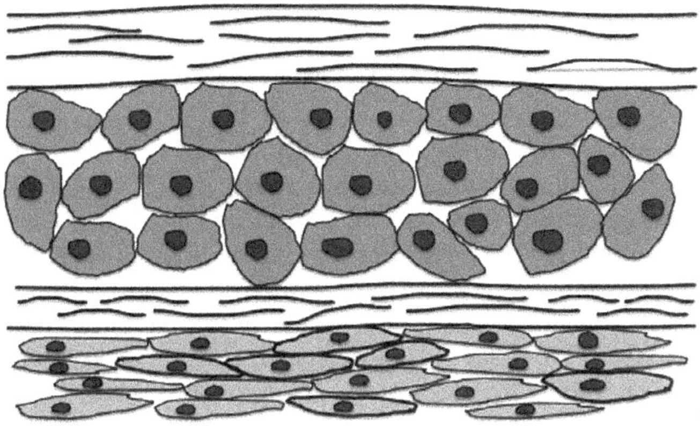

FIGURE 11.1 Schematic illustration of the structure of natural skin. Keratinocytes are the most common cells in the epidermis and fibroblasts are the major cellular components of dermis. Compared with the fibroblasts that are embedded in fibrous ECM, keratinocytes rest on a thin ultrafine fibrous membrane called the basement membrane.

membrane is to anchor the epithelium to the dermis. It also acts as a mechanical barrier, preventing malignant cells from invading the deeper tissues. Since the basement membrane proteins have been found to accelerate differentiation of endothelial cells, it is also essential for angiogenesis. The basement membrane not only allows the anchorage of keratinocytes, but also regulates their phenotype via the interaction with cell membrane–bound integrins. Unlike the epidermis, in the dermis, dermal cells (e.g., fibroblasts) are embedded in a 3-D fibrous composite ECM consisting primarily of collagen (notably, type I collagen) and elastin fibers embedded in an amorphous matrix of mucopolysaccarides. The collagen is arranged in a planar array of highly undulated, wavy, and meandering fibers in all directions, thus allowing the skin to stretch but preventing overstretching. The collagen and elastin fibers may be cross-linked with ground substance and between themselves to provide mechanical stability to the skin [9].

11.3 SKIN TISSUE ENGINEERING AND SCAFFOLD DESIGN

The primary purpose of skin tissue engineering is to provide a temporary barrier, a dermal matrix, or a transfer mechanism to facilitate the regeneration of damaged skin by using synthetic or natural materials. On complete healing, these materials should eventually disappear from the healed skin, by way of biodegradation or replacement by newly formed tissue. It is of great benefit to combine the wound healing process and the cellular elements of living tissue with sophisticated biomaterials to produce living skin equivalents with sufficient size and desirable functions. Recently, a great deal of attention has been paid to the utilization of fibrous scaffolds for tissue formation mainly due to the dimensional and morphological similarity of the fibers to native ECM, which in turn could regulate cell attachment, migration, proliferation, and differentiation. Various fabrication techniques have been explored to address some potential challenges associated with the fabrication of complex and multifunctional fibrous substrates, including the creation of large pores to facilitate cell infiltration, controlled formation of nanometer-sized fibers, and incorporation of the spatial distribution of specific proteins for cell attachment and remodeling. The fibrous structure fabricated by these techniques can determine the shape and mechanical performance of tissues and provide the cells with instructive external cues to guide tissue formation. It also provides physical supports for cells to attach and to grow. To appropriately maintain cell phenotype during skin tissue engineering, it is desirable for the scaffold to recapitulate the major features of native ECM of skin on a multilevel scale, from the composition, morphology, and topography to spatial organization. We believe that nanofibrous scaffolds will be an ideal substrate to support skin tissue formation due to their similarity to ECM fibers in both dimension and morphology. Indeed, the advantages of nanofibrous scaffolds in promoting skin cell growth and maintaining proper phenotype have been demonstrated in a number of studies [10–13].

In the search for appropriate substrates to support the formation of skin-like structures, diverse techniques have been explored and used to fabricate scaffolds with nano- and microscale features to mimic the composition and structure of normal skin. Great progress has been made, and several of these products are commercially available to heal wounds as a temporary dressing or permanent substitutes, such as Alloderm® (LifeCell, Branchburg, NJ), Oasis® (Healthpoint, Fort Worth, TX), Integra® Dermal Regeneration Template (Integra Life Sciences, South Plainfield, NJ), and Biobrane® (Bertek Pharmaceuticals, Sugarland, TX). These scaffolds, in sponge, fibrous, or gel format, are produced from different polymers that support skin tissue formation. Salt leaching, emulsion freeze-drying, high-pressure gas expansion, phase separation, and electrospinning are common techniques to fabricate porous scaffolds for skin tissue formation [14]; these are further elaborated below. In addition, various parameters of biomimetic nanofibrous scaffolds that regulate tissue formation are discussed and their utilization for creating skin grafts is highlighted.

11.3.1 Particle Leaching Technique

The particle leaching method is one of the well-established techniques for fabricating porous scaffolds. This technique is based on a biodegradable polymer mixed with a salt in an organic solvent and then poured in a mold for evaporating the solvent. A structure with cellular pores can be obtained after the dissolution of the salt particle or porogen. Acceptable dimensional shrinkage may occur after salt leaching, but the structural integrity of scaffolds with microsized pores will be well maintained. By changing the size of salt particulates and the salt/polymer ratio, the pore size and porosity parameters of the scaffold can be controlled. Various porogens, such as salts, carbohydrates, and polymers can be used to produce porous materials. Water-soluble polymers of poly(ethylene oxide) (PEO) or gelatin can also be incorporated with some water-insoluble or slow-degrading polymers to produce artificial skin grafts [15]. Lee *et al.* [16] studied porous gelatin scaffolds that were prepared using a salt leaching method and compared them with scaffolds fabricated using a freeze-drying method for gelatin-containing artificial skin. The results showed that the scaffolds prepared from the salt leaching method had larger pore structures than that prepared by freeze-drying. After 1 week of *in vitro* culturing, fibroblasts showed a good affinity to the scaffold and were mainly distributed on the surface of the macropores. Sponges formed using the freeze-drying method had a dense surface that did not allow cells to penetrate into the inner areas. The mechanical strength and the rate of biodegradation were easily modulated by the addition of salt. It was found that the artificial dermis rather than the acellular sponge improved the reepithelialization on a full-thickness skin defect [16]. However, residual salts remaining in the scaffolds, rough morphologies of surface transferred from salt, irregularly shaped pores, and their poor interconnectivity pose challenges for cell seeding and culture [17].

11.3.2 Emulsion Freeze-Drying

Freeze-drying is one of the most extensively used methods to produce scaffolds with porosity greater than 90% [15]. The pore size depends on the growth rate of ice crystals during the freeze-drying process. In this method, an emulsion of a polymer with organic solvent and water is quickly frozen to set the pore structure before the two phases separate. Then the scaffold is freeze-dried to eliminate the residual solvents. After the removal of the frozen solvent, the remaining spaces become pores. This yields a fine porous scaffold compatible with cell cultures in tissue engineering. By adjusting the polymer concentration, using different solvents, or varying the cooling rate, phase separation could occur via different mechanisms, resulting in scaffolds with various morphologies. Although freeze-drying can prevent the disintegration of porous structure, it is so time- and energy-consuming that the whole scaffold fabrication process becomes inefficient and economically uncompetitive. Another problem encountered in the application of freeze-drying to the preparation of scaffolds is the occurrence of surface skin. During the freeze-drying stage, if the temperature is not kept sufficiently low, the polymer matrix is not rigid enough to resist the interfacial tension caused by the evaporation of the solvent. Freeze-dried porous scaffolds made from chitosan have been used to fabricate skin grafts; in particular, those collagen/chitosan composite scaffolds cross-linked by glutaraldehyde (GA) showed great potential as dermal equivalents with enhanced biostability and good biocompatibility [18].

11.3.3 High-Pressure Gas Expansion Methods

High-pressure gas expansion methods were developed to fabricate macroporous sponges from synthetic biodegradable polymers for potential use in tissue engineering without the use of organic solvents [19]. In this method, a biodegradable polymer was saturated with CO_2 by exposure to high-pressure CO_2 gas at room temperature in a pellet mold. Then the pressure was reduced to atmospheric levels rapidly to decrease the solubility of the gas in the polymer. This created a thermodynamic instability for the CO_2 dissolved in the polymer disks, and resulted in the nucleation and growth of gas cells within the polymer matrix. By using this technique, open pores can be obtained within the polymer sponges. The porosity of the sponges could be controlled by the perform production technique, in which the disks preformed with compression molding or solvent casting were used to create pores with a certain size, then polymers such as polyglycolic acid (PGA) and poly(lactic-co-glycolic) acid (PLGA) were mixed in the disks to form foams with pore structure. Also, the porosity can be controlled by mixing crystalline and amorphous polymers. Fiber-reinforced foams could also be produced by placing polymer fibers within the polymer matrix before CO_2 gas processing [20]. This matrix exhibits enhanced mechanical properties and can be utilized to form 3-D skin tissues. Also, this fabrication method may provide an ideal system for drug and/or growth factor incorporation into polymers used as skin tissue engineering matrices.

11.3.4 Phase Separation Method

The phase separation technique is based on thermodynamic demixing of a homogeneous polymer–solvent solution into a polymer-rich phase and a polymer-poor phase, usually by either exposure of the solution to another immiscible solvent or cooling the solution below a binodal solubility curve [21]. This cooling, known as thermally induced phase separation (TIPS), uses thermal energy as a latent solvent to induce phase separation [15]. The quenched polymer solution below the freezing point of the solvent is subsequently freeze-dried to produce porous structure. One of the advantages of this technique is that various porous structures can be easily obtained by adjusting various thermodynamic and kinetic parameters. The TIPS technique has been used commercially to produce microporous membranes for filtration, but the pore size of the resultant membrane is too small to be applied for cell seeding, which requires a pore diameter at least above 100 μM, with an open cellular morphology. The TIPS technique has been previously reported for the preparation of porous polylactic acid (PLA) scaffolds intended to use as tissue scaffolds. Although a wide array of microporous (1–10 μm) isotropic morphologies were reported in response to a slight change in the TIPS parameters, whether they have macroporous and open cellular structure throughout the matrix was not discussed [21]. Large pore size and an open porous structure are critical parameters for cell seeding and neovascularization when implanted *in vivo*. When compared with the prefabricated solid form, which requires additional and complicated processing, the solution-based phase separation technique is relatively simple and easy to manipulate. However, there have been limited studies on the selection of the porogen phase that can be properly available for the processing of biopolymer scaffolds.

11.3.5 Electrospinning Technique

Compared with the approaches mentioned earlier, electrospinning has recently received a tremendous amount of attention due to its low setup cost, easy operation, high production rate, good reproducibility, and the capability to fabricate fiber scaffolds from versatile materials. In addition, the nanofiber scaffolds have a large surface area to interact with cells and varying porosities and interfiber pore sizes controlling cellular infiltration. The electrospinning device contains a spinneret, a

TABLE 11.1 Synthetic and Natural Polymers for Electrospun Nanofibers for Skin Grafts

Polymer	Solvent	Concentration	Application	Reference
Poly (ε-caprolactone)/ Collagen (type I) blend	HFIP	10%	Dermal substitute	[98]
Poly (ε-caprolactone)/ gelatin blend	TFE	10%	Dermal substitute	[51]
Polyurethanes (PU)	Mixture of THF and DMF	25%	Wound dressing	[118]
Chitosan/Gelatin blend	TFA or mixture of TFA and DCM	4–7%	Skin regeneration	[119]
Polyvinyl alcohol (PVA)/chitosan blend	Distilled water	9%	Wound dressing	[120]
Poly(lactic-co-glycolic acid) (PLGA)/collagen (type I) blend	HFIP	PLGA/col (50:50 w/w)	Wound dressing	[121]
Collagen(type I)	Hexafluoro-2-propanol	10%	Skin substitute	[122]
Silk/PEO	HFIP	7.5%	Wound dressing	[49]
Gelatin	TFE	10%	Dermal substitute	[123]

HFIP, 1,1,1,3,3,3-hexafluoro-2-propanol; DCM, dichloromethane; DMF, dimethylformamide; THF, tetrahydrofuran; TFE, 2,2,2-trifluoroethanol; TFA, trifluoroacetic acid.

syringe pump, a high voltage supply, and a grounded conductive surface for fiber collection. Nanofibers fabricated by electrospinning closely mimic the architecture of native skin tissue at the nanometer scale. These nanofibrous matrices are ideal for skin tissue engineering applications due to their large surface area and distinct properties such as low density, high porosity, controllable pore size, and exceptional mechanical properties [22,23]. There are several parameters that can affect the electrospinning process and determine the physical properties of the obtained nanofibrous matrices [24–27]; these include polymer molecular weight, polymer solution properties, voltage, flow rate, distance between spinneret and collector (working distance), motion of the grounded collector, and ambient conditions (temperature, humidity, and air velocity). A variety of materials including synthetic and natural polymers (e.g., PLGA, poly-L-lactide (PLLA), polycaprolactone (PCL), poly(ethylene oxide terephthalate) (PEOT)-poly(butylene terephthalate) (PBT), collagen, chitosan, or blends of these materials) [28–33] have been successfully electrospun into micro- or nanosized fibers with diameters ranging from 50 to 1000 nm or greater [34,35]. Several synthetic and natural polymers that are used in electrospinning nanofibers for skin grafts have been briefly summarized in Table 11.1.

The surfaces of most electrospun fibers are smooth with a solid cross-section. Efforts have been made to create secondary unique structures in the electrospun fibers, for example, core/shell composite fibers [36,37], tubular fibers [38], multichannel tubular structure [39], and porous fibers [40] to achieve novel properties and specific functionalities, such as the incorporation of drug molecules for controlled release [41] and promotion of cell anchorage [42] to the surface. Many potential applications can be identified with these structures. To fabricate either tubular or core/shell composite fibers, the coaxial electrospinning technique has been developed. By using a spinneret consisting of two coaxial capillaries with different diameters, two solutions can be accommodated for coelectrospinning. Inspired by this setup, a spinneret with multiple capillaries embedded in a plastic syringe at three vertices of an equilateral triangle was fabricated for multifluidic compound jet electrospinning [43].

The electrospinning technique also offers the opportunity to control the thickness and composition of the nanofibrous matrices [34,44–46]. Extensive efforts have been made to explore the potential utilization of electrospun nanofibers for tissue regeneration. Electrospun nanofiber scaffolds showed great advantages in designing biomimetic skin graft as a result of their morphological and dimensional similarity to the native skin ECM fibers [47] and facilitate the incorporation of various biomolecules including growth factors [48,49]. It has been demonstrated in a number of studies that nanofibrous scaffolds can promote cell growth and maintain proper phenotype [10–13]. All these unique properties of nanofibers have led to the effort to directly apply nanofibers for wound repair as a wound dressing/matrix or as scaffolds for skin tissue engineering applications [50–52]. There are several critical parameters that can influence skin tissue formation when using the electrospun nanofibrous matrices as skin grafts.

11.3.5.1 Pore Size Interpore size, the size of the pores that are created by nanofibers, is one of the parameters that can influence skin tissue formation. The control of interfiber pore size is a critical issue especially in the case of cell infiltration and tissue ingrowth. The relationship between pore size and cell migration has been investigated and several results indicated that the cells could infiltrate into electrospun nanofiber mats with the average pore size of only 1.5 μm, which is much smaller than the minimum pore size of 10 μm assumed for cell infiltration in other scaffolds [53]. A possible explanation for this finding may result from the ability of cells to push aside individual fibers in the electrospun nonwoven mats. Lowery *et al.* [54] found that cells in the fibrous scaffolds with pores larger than 300 μm are no longer able to effectively bridge the interfiber gaps. Conversely, for those scaffolds with the pore sizes less than 1 μm, cell infiltration can be completely prevented (cell barrier) [54]. Based on these findings, an optimal pore size of 6–20 μm has been proposed for effective cell infiltration, which corresponds to approximately the average diameter of suspended cells. However, for most electrospun nanofiber meshes, the pore size is less than 5 μm, smaller than the cell size, which constrains the cells penetrating into the meshes [55]. A large number of studies have also looked into the effect of porosity and pore size on cell proliferation and infiltration by varying the fiber diameter [56–68]. However, the alteration of fiber diameter may lead to various unintended cellular responses, for example, cell proliferation and differentiation [54]. Obviously, a key challenge to systematically investigate the correlation between pore size and cell infiltration and cell proliferation is how to precisely control the pore size independent of other structural parameters, such as fiber diameter and porosity. With current electrospinning setups, this is very difficult to achieve especially considering the interdependency among fiber diameter, pore size, and porosity.

The opportunity to increase the pore size by manipulating nanofiber diameter is very limited. Regarding this, attempts have been made to improve cell infiltration by using enzyme-degradable natural polymers [69], or coelectrospinning with sacrificing nanofibers, which will be removed afterwards to generate large pores [70]. Recent methods used in increasing pore size of electrospun nanofibers include salt leaching [14,71,72], incorporating solid crystals in the fibers on the collection device [73,74], wet electrospinning on a bath collector [75,76], combining nanofibers and microfibers [56,77,78], laser/ultraviolet (UV) irradiation [79,80], and controlling deposition of nanofibers using an electric field [81–83].

11.3.5.2 Fiber Diameter The influence of fiber diameter on proliferation and migration of cells on electrospun fiber meshes has been extensively investigated. The results showed variations that may have resulted from the fact that not all the studies have taken the potential contribution of porosity into consideration. Among various studies, the most appealing findings were made with electrospun PCL fiber scaffolds, in which the fiber diameter ranged from 0.1 to 1.6 μm with uniform pore sizes (above 50 μm) [57,84,85]. The initial proliferation of cells on the scaffold surface and the subsequent migration into the scaffold with all diameters was observed via confocal microscopy [84]. Cell proliferation rates on the scaffold increased with decreased fiber diameter of submicron fibers. An exception to this trend was observed on beaded fibers, which had a strong detrimental effect on cell morphology and proliferation. For microfibers with diameters larger than 1 μm, cell proliferation rates increased with fiber diameter. This trend was confirmed and extended to even larger fiber diameters on PCL microfibers by other studies [54,58,85]. A possible explanation to these results given by Chen *et al.* [57] was that the specific surface area is dramatically increased as the fiber diameter of those submicron fibers decrease, which is critical for the fiber to adsorb protein from serum for cell attachment and proliferation, as cell attachment and proliferation by submicron fibers can only be observed in the presence of serum. When the fiber diameters were below 0.7 μm, electrospun PCL fibers undergo stretch-induced molecular orientation, resulting in higher tensile strength, altered crystallinity, and preferential presentation of chemical motifs at the fiber surface. A study by Christopherson *et al.* [86] further corroborates the influence of protein adsorption. An improved proliferation of neural stem cells on poly(ether sulfone) (PES) nanofibers was observed compared with microfibers but only in the presence of fibroblast growth factor (FGF)-2. In summary, fiber diameter indeed affects cell proliferation rate independent of pore size. For scaffolds prepared from PCL, submicron fibers show superior protein absorption capacity when diameter is below a certain number, which results in improved cell attachment and proliferation. For other materials, submicron fibers showed no difference or decreased cell proliferation in comparison with microfibers.

11.3.5.3 Fiber Spatial Arrangement Typically, electrospun nanofibers are collected on a grounded stationary flat surface, on which the fibers exhibit a random arrangement, however, by using various collectors, such as collecting on a grounded rotator, parallel electrodes, rotating disks, or a square wire loop, a particular orientation can be achieved (Figure 11.2). Studies have shown that aligned nanofibers can guide the cell adhesion and spatial arrangement of intracellular cytoskeletal proteins, and as a result, lead to the elongation of cells along the fiber orientation. From the results of several studies, aligned nanofibers offer improved mechanical stability and degradation properties for specific applications. The effect of nanofiber spatial arrangement on fibroblast behaviors has been studied [1,31,87] and the arrangement patterns influence the alignment and migration of fibroblasts and adipose stromal cells [88,89]. It was found that fibroblasts showed a lower adhesion affinity to aligned collagen nanofibers but a higher proliferation rate compared with those on the random ones [90]. Although the exact mechanism for how fiber

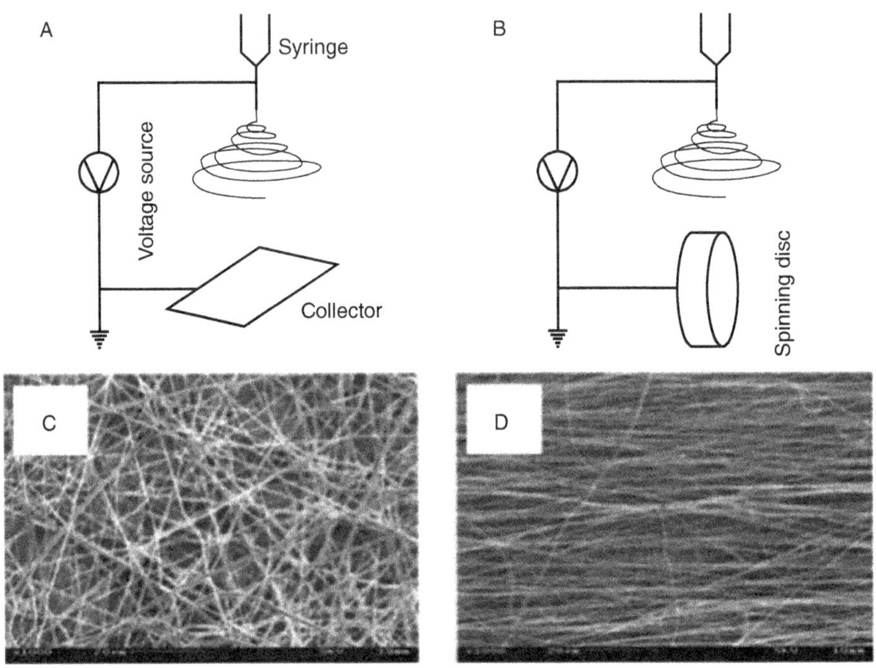

FIGURE 11.2 (A) Schematic depiction of the electrospinning process to obtain random nanofiber meshes. A charged solution is drawn from the tip and the residual random fibers collect on a grounded stationary plate. (B) A spinning disk technique is commonly employed to create aligned electrospun fibers. (C) Random nanofiber meshes collected on a grounded plate. (D) Aligned nanofiber meshes prepared utilizing the spinning disk.

orientation induces fibroblastic responses remains elusive, it is believed that integrins on cell membranes play a dominant role. We have recently found that aligned PCL/collagen nanofibers promote fibroblast migration and alignment by activating the integrin β_1 pathway [91]. Meanwhile, the activation of integrin β_1 also induces the fibroblast-to-myofibroblast differentiation, relevant to wound healing. Also, in another research from our group, we found that the spatial arrangement of PCL/collagen nanofiber scaffolds can regulate the wound healing related behaviors of human adipose stromal cells (hASCs). From the results, elongated cell morphology, higher proliferation, and faster migration rates were observed for hASCs cultured on the aligned nanofibers, showing that hASCs could detect the nanofiber spatial arrangement and then distinctively respond. The expression of ECM-related genes in hASCs revealed higher synthesis capacity for critical ECM molecules including tropoelastin, collagen I, and matrix metalloproteinase (MMP)-1 on the aligned nanofibers. Integrins α_5, β_1, β_3, β_6, and transforming growth factor (TGF)-β_1 were differentially regulated by PCL/collagen nanofiber arrangements. The aligned fiber orientation upregulated the gene expression of integrins β_1 and β_3 as early as day 1 and the upregulation remained for the rest of the experimental times, while integrin α_5 expression was downregulated for all the times investigated. And it was found that TGF-β_1 was highly upregulated by aligned nanofibers. With the increasing interest in using nanofibers for wound repair, the mechanistic understanding of nanofiber-induced cellular responses, like adhesion, spreading, migration, proliferation, and differentiation [92–94], becomes highly desirable.

11.4 BIOMIMETIC APPROACH TOWARD THE FORMATION OF EPITHELIAL-LIKE TISSUE USING ELECTROSPUN NANOFIBERS

11.4.1 Incorporation of ECM Components into Electrospun Nanofibers

To better support the formation of skin-like tissue, it is crucial for the skin cells to experience a microenvironment that is as similar to their native one as much as possible. In this regard, a biomimetic scaffold should mimic the key features of natural ECM of skin to facilitate cell recruitment/seeding, adhesion, proliferation, differentiation, and skin tissue formation. Multiple strategies have been evolving to incorporate this biomimetic concept into scaffold design. One of the logical approaches is to prepare scaffolds out of the materials with similar composition to natural skin. As a matter of fact, many natural ECM proteins have been used for wound repair and skin tissue regeneration, for example, collagen, a main ECM component of skin, has been extensively used in fabricating wound dressings and tissue engineered skin equivalents such as Apligraf®. However, pure electrospun collagen nanofibers normally have poor mechanical properties and degrade rapidly, making them unsuitable for a long-term application unless the fibers are further chemically cross-linked [95]. However, chemical modification often leads to the loss of active sites for cell adhesion. To better maintain the biological activity and achieve mechanical stability, composite collagen nanofibers containing other materials that can enhance mechanical strength would be a practical option [96]. Powell and Boyce [97] studied the correlation of mechanical strength and biological affinity of such composite nanofibers with various PCL/collagen ratios and found that the addition of PCL to collagen (as little as 10%) was sufficient to achieve a good balance between mechanical strength and biological activity. In our group, we have demonstrated that the PCL/collagen nanofibers containing 25% (w/w) type I collagen are effective at promoting the adhesion and proliferation of skin dermal fibroblasts and keratinocytes [98]. Besides PCL/collagen nanofibers, other composite fibers have also been evaluated for skin applications. For example, PLGA/dextran nanofibers were tested for their ability to support dermal fibroblasts and favorable fiber/fibroblast interactions were observed with the resemblance of a dermal-like architecture. Similarly, PLGA/collagen, PLLA/gelatin, and polyvinyl alcohol (PVA)/chitosan nanofibers have been particularly studied for their potential application for skin tissue engineering [28,99].

11.4.2 On-Site Layer-by-Layer Cell Assembly Approach for 3-D Tissue Formation

Studies show that multifunctional electrospun nanofibers have many superior advantages in regulating cell functions. However, the difficulty for cells to infiltrate through the micrometer-sized interfiber spaces is one of the key problems associated with the use of electrospun nanofibers as tissue engineered scaffolds. By using fast-degrading natural polymers or increasing the pore size by spinning thicker fibers, cell infiltration could be improved [100]. However, the intrinsic nutrition gradient from the exterior to the interior of the scaffold preferentially retains the cells on the periphery of the scaffold, while limiting cell viability in the center of the construct. In addition, many tissues in the body are rather thick, like muscle or fat. To mimic these tissues requires an optimal pore size for the transport of nutrient and metabolic waste through thick tissue. Although thick nonwoven electrospun fiber mats (>1 mm in thickness) have been utilized to form thicker tissues [101,102], the reduced pore size in such mats block cell–cell communication between layers. In response to this particular challenge, together with the spatial manipulation of cell arrangement and fiber arrangement, more innovative approaches are needed to incorporate the cells into fiber meshes. Srouji *et al.* [103] examined the possibility of forming 3-D constructs using nanofiber mats seeded with cells, in which human bone marrow-derived mesenchymal stem cells (hMSCs) were cultured on nonwoven 1 : 1 PCL/collagen fiber mats (100–200 μm in thickness) for 24 hours and then stacked into 3-D structure and cultured under perfusion. A similar approach was also taken to form cardiac tissue [104]. Although cell penetration is not a big issue in this study as a result of large pores (>50 μm), however, it is a tremendous challenge for the operator to stack ultrafine nanofibers together into a 3-D structure as the cell–nanofiber mats are very fragile.

In this regard, Yang *et al.* have developed an on-site layer-by-layer cell assembly approach for 3-D tissue formation [98]. Briefly, a thin layer (5–10 μm) of 3 : 1 PCL/collagen electrospun nanofibers was collected on the medium surface, and then normal human fibroblasts (used as model cells) were evenly seeded upon the fiber

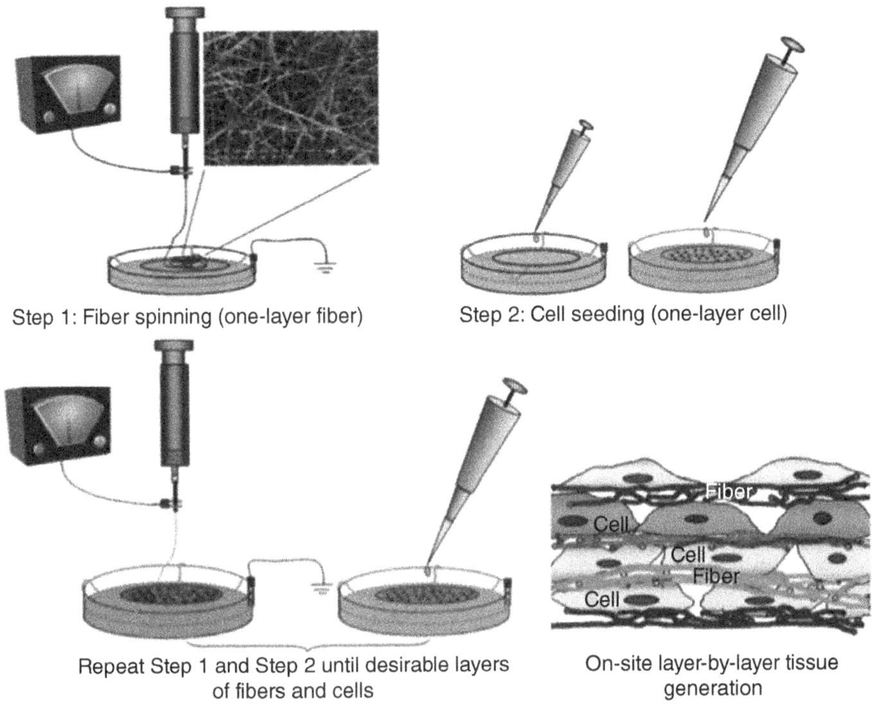

FIGURE 11.3 A schematic illustration of the on-site layer-by-layer cell assembly while electrospinning. As indicated by different colors, both fiber and cell layers can be varied during the cell assembly to create a customized final 3-D construct according to the design. (Reprinted from Reference 98, with permission from Mary Ann Liebert, Inc. Publishers).

surface. By repeating these steps, a 3-D multilayer fibroblast/nanofiber structure can be obtained right on the site of electrospinning, and the cells are included throughout the constructs without concern about later cell infiltration, which is similar to *in vivo* tissues where cells were embedded in ECM fibers (Figure 11.3). During the layering of the cells, each fiber layer would be less than a cell thickness, and cells adhere to adjacent layers similar to *in vivo* circumstances and deposit ECM components to hold the layers together. Since cell seeding and nanofiber collection take place on the surface of cell culture medium under a sterile condition, cell viability was well maintained and no damage was observed to the cells after the assembly. Another advantage of this system is the flexibility to vary cell density and cell type for each cell layer and the composition and thickness for each nanofiber layer during this layer-by-layer tissue rebuilding. With the opportunity to precisely control the composition of fiber layers, the fiber layer thickness, fiber diameter, and fiber orientation, as well as inclusion of bioactive molecules into the fibers, it is possible to create a specific 3-D microenvironment for individual cells within the same cell–material construct. This cell layering approach not only yields a well-controlled initial cell distribution across the constructs, but also has the potential to form a uniform tissue based on our observation with the culture of osteoblast/nanofiber and fibroblast/

nanofiber layered constructs. Additionally, this approach offers a way to create multilayered constructs with distinct spatial arrangement of various types of cells and has a high potential for bilayered skin tissue formation. It is necessary to mention that layer separation was observed in the prepared cell/fiber 3-D constructs, but it became invisible within 3 days after the extended culture. A compelling approach proposed to circumvent the aforementioned cell seeding/infiltration problem is to spray cells onto the electrospun nanofibers simultaneously with electrospinning, which has been successfully demonstrated by Stankus *et al.* Using this technique, the investigators achieved microintegration of smooth muscle cells into a biodegradable, elastomeric poly (ester urethane) urea (PEUU) fiber matrix [105]. Cell electrospraying shows promises in bringing cells inside the fiber meshes, yet with some drawbacks, including less control of cell distribution as well as the potential cell damage from high voltage and toxic organic solvents during cell electrospraying.

11.4.3 Formation of 3-D Epithelial-Like Tissues

The successful formation of cell/fiber structures may lead to the formation of 3-D tissue. Following this 3-D cell layering approach, efforts have been made to fabricate tissue-engineered dermal substitutes or bilayered skin equivalents. For example, culture of 10-layer fibroblast–PCL/collagen fiber constructs assembled for 3 and 7 days could result in the formation of dermal-like tissues. In brief, human dermal fibroblasts (passage 3–4) and 1 : 1 collagen/PCL nanofibers were alternately layer-by-layer assembled into a 3-D construct on the fibroblast-culture medium surface. In total, about 10 layers of cells/nanofibers were obtained. The seeded cell density for each layer (3 cm in diameter) was 1×10^5 cells in 1-mL medium. Uniform cell distribution across the assembled construct was observed by staining the cross-sections of the cell/nanofiber construct cultured overnight with 4',6-diamidino-2-phenylindole (DAPI) for nuclei. Results showed a homogeneous, 3-D distribution of cells throughout the full thickness of the construct. After assembly, the cell/nanofiber constructs were continuously cultured in fibroblast culture medium. The cross-sections of the cultured constructs were examined under a fluorescent microscope. It was found that the cell/fiber construct became compact over a prolonged culture. No clear layers could be recognized after the culture for 3 days. Cross-sections of the constructs cultured for 7 days were stained with hematoxylin and eosin (H&E). It was found that fibroblasts showed an elongated spindle shape uniformly distributed among fibers. The tissue morphology was very similar to native dermis. Based on this observation, it is possible for us to fabricate dermal-like substitutes within 7 days. The exploration of forming epidermal/dermal bilayer skin grafts was similarly made by layer-by-layer assembly of fibroblasts and keratinocytes together with PCL/collagen nanofibers into a multilayer cell/fiber structure (18 layers of fibroblast/fiber in the lower part and two layers of keratinocyte/fiber on the top) and then culturing for additional days. Cell seeding density for fibroblasts was 1×10^5 cells/layer and for keratinocytes was 1×10^6 cells/layer. The average assembly time for one 20-layer cell/fiber construct is about 30 minutes. After culturing in a coculture media (3 : 1 dulbecco's modified eagle medium/F (DMEM/F)-12 containing 5% fetal calf serum (FCS), 1% penicillin and streptomycin, 5 μg/mL

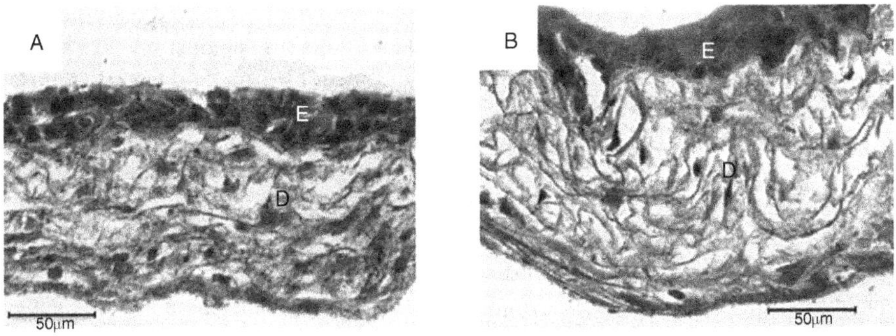

FIGURE 11.4 H&E-stained cross-sections of bilayer skin tissues composed of epidermal (E) and dermal (D) layers and formed by culturing L-b-L assembled cell/fiber constructs for 3 days (A) and 7 days (B). Green broken line outlines the border between E and D. (Reprinted from Reference 98, with permission from Mary Ann Liebert, Inc. Publishers.) *(See insert for color representation of the figure.)*

insulin, 0.4 μg/mL hydrocortisone, 1 μM isoproterenol, and 1 mM ascorbic acid 2-phosphate) for 3 days, a bilayer skin structure (epidermal and dermal layer) was clearly seen on the H&E-stained cross-sections (Figure 11.4). The seeded keratinocytes remained on the surface and formed a continuous epidermal layer, and fibroblasts retained in the lower part with a uniform distribution. A tight binding between the epidermal layer and dermal layer was observed. At this time, two layers of keratinocytes still could be recognized by careful examination, but it became invisible by day 7. The bilayer structure remained the same even after culture for 7 days, except that both the epidermal layer and dermal layer became thicker (about 50% increase for both layers). It is necessary to mention that the formed bilayer skin substitutes have good mechanical strength (~4 MPa for the ultimate tensile stress), which is comparable with native skin as well [106]. Following the cell layering approach as described earlier, we have successfully fabricated both dermal-like or bilayered skin-like constructs by further culturing the assembled constructs [98]. This approach has a great potential for creating large size grafts. Overall, the nanofiber-enabled cell layering allows for the creation of dermal-like tissue for skin grafts and is a one-step approach, which provides the cells with a biomimetic growing microenvironment.

11.5 FUTURE PERSPECTIVE AND CHALLENGE

The simple and easy setup of electrospinning has initiated a wide-ranging effort to fabricate fibers from various materials with diameters in nanometers/micrometers, or with a secondary structure like nanopores or nanochannels within the fibers. With the help of biomimetic nanofibers, it becomes possible to rapidly form 3-D substitutes containing both epidermal and dermal layers within a week. However, native skin shows much more complexity than any of current skin substitutes. Regarding this, part of the ongoing effort is to create some of the appendage structures such as hair follicles in skin substitutes [107]. An ideal graft should restore the normal skin functions such as the barrier formation, pigmentory defense against UV irradiation, thermoregulation, and mechanical and esthetic functions. It is hard to restore all of the functions with the existing skin substitutes. Using autologous skin cells for skin grafts has great potential, however, they are not available off the shelf for the immediate treatment of burns or trauma and their use is not feasible for large-scale commercialization. In the search for an alternative cell source to autologous skin cells, adipose-derived stromal cells (ASCs) have drawn particular attention because they are easy to access, minimally invasive, have the capacity to self-renew, and large quantities can be harvested at a time [108,109]. There is increasing evidence that ASCs promote the proliferation and migration of fibroblasts, stimulate angiogenesis, and accelerate reepithelialization from the wound edge [110–115]. However, more clinical results are still needed to demonstrate their effectiveness in wound healing *in vivo*. Currently, there are no good alternative sources for keratinocytes. Exploring the way to engineer such cells will be the key to successfully developing tissue-engineered skin substitutes.

Nanofibers demonstrate great effectiveness in supporting skin cells to form 3-D substitutes; however, there is still a need to further investigate the interactions between fiber composition and morphology and cells for functional tissue formation. Comprehensive studies are necessary to further optimize the nanofiber parameters (e.g., diameter, pore size, spatial arrangement, composition) for a faster and better wound healing [54,57,89,95,116,117]. Considering the fact that many other molecules, in addition to collagen, such as growth factors are involved in cell signaling in the native skin ECM, it may be also necessary to release various signaling molecules, differentiation factors, and protein domains engineered to facilitate cell migration and adhesion [117] in further studies. With all the further insightful understanding, it may allow us to create skin substitutes ready for transplantation right after assembly without further culture in the near future.

Additionally, regarding the use of tissue-engineered skin grafts for clinical wound repair, some skin-related biological events such as the cell–cell and cell–ECM interactions, barrier function of skin, the wound healing process, angiogenesis, regulation of pigmentation, and skin contraction, can be studied by using skin grafts as models. Tissue-engineered skin grafts also can be used in the understanding of skin diseases such as melanoma invasion, psoriasis and skin blistering disorders, and drug and cosmetic product evaluation. As envisioned, there will be an exponential increase in use of skin grafts as a 3-D testing model, in which physiological interactions among different skin cells that cannot occur in conventional monolayer cultures are possible.

11.6 CONCLUSION

Electrospinning offers a great opportunity to produce a variety of fibers from a spectrum of materials, which results in the formation of epithelial-like scaffolds with diverse functionality and tunable spatial distribution of bioactive molecules. Together with the ability to incorporate cells directly into a multilayered fiber structure, it is possible to formulate a customized microenvironment specific for fabricating tissue-engineered skin grafts within 7 days. Further optimization of the nanofiber composition, especially the formulation of the customized microenvironment specific for each type of skin cell, will significantly facilitate skin tissue formation and, as a consequence, drastically decrease production time. The development of nanofiber-assisted cell layering following a layer-by-layer deposition technique or coelectrospraying has enabled 3-D tissue formation. In summary, electrospun nanofibers provide an avenue to fabricate tissue-engineered skin with complex, bio-inspired hierarchical architecture as well as an *in vitro* physiologically relevant platform for studying cell–cell or cell–matrix interactions.

REFERENCES

[1] Venugopal, J.R., Zhang, Y., Ramakrishna, S. *Artificial Organs* **2006**, *30*, 440–446.
[2] Wainwright, D.J. *Burns* **1995**, *21*, 243–248.
[3] Clark, R.A. *The American Journal of the Medical Sciences* **1993**, *306*, 42–48.
[4] McClain, S.A., Simon, M., Jones, E., Nandi, A., Gailit, J.O., Tonnesen, M.G., Newman, D., Clark, R.A. *American Journal of Pathology* **1996**, *149*, 14.
[5] Clark, R. *Dermatologic Clinics* **1993**, *11*, 20.
[6] Wang, H.J., *et al. Tissue Engineering* **2004**, *10*, 1054–1064.
[7] Bello, Y.M., Falabella, A.F., Eaglstein, W.H. *American Journal of Clinical Dermatology* **2001**, *2*, 305–313.
[8] Schonfeld, W.H., *et al. Wound Repair and Regeneration* **2000**, *8*, 251–257.
[9] Abrams, G.A., *et al. Cell and Tissue Research* **2000**, *299*, 39–46.
[10] Li, W., *et al. Journal of Biomechanics* **2007**, *40*, 1686–1693.
[11] Min, B. *Biomaterials* **2004**, *25*, 1289–1297.
[12] Ji, Y., *et al. Biomaterials* **2006**, *27*, 3782–3792.
[13] Chua, K., *et al. Biomaterials* **2005**, *26*, 2537–2547.
[14] Mikos, A.G., *et al. Polymer* **1994**, *35*, 1068–1077.
[15] Garg, T., *et al. Critical Reviews in Therapeutic Drug Carrier Systems* **2012**, *29*, 1–63.
[16] Lee, S.B., *et al. Biomaterials* **2005**, *26*, 1961–1968.
[17] Chen, G., Ushida, T., Tateishi, T. *Journal of Biomedical Materials Research* **2000**, *51*, 273–279.
[18] Ma, L., *et al. Biomaterials* **2003**, *24*, 4833–4841.
[19] Harris, L.D., Kim, B.S., Mooney, D.J. *Journal of Biomedical Materials Research* **1998**, *42*, 396–402.
[20] Mooney, D.J., *et al. Biomaterials* **1996**, *17*, 1417–1422.
[21] Nam, Y.S., Park, T.G. *Journal of Biomedical Materials Research* **1999**, *47*, 8–17.
[22] Bhattarai, S.R., *et al. Biomaterials* **2004**, *25*, 2595–2602.
[23] Ma, Z., *et al. Tissue Engineering* **2005**, *11*, 101–109.
[24] Shin, Y. *Polymer* **2001**, *42*, 9955–9967.
[25] Zong, X., *et al. Polymer* **2002**, *43*, 4403–4412.
[26] Fridrikh, S.V., *et al. Physical Review Letters* **2003**, *90*, 144502.
[27] Dhirendra, S., Katti, K.W.R., Ko, F.K., Laurencin, C.T. *Journal of Biomedical Materials Research* **2004**, *70B*, 11.
[28] Matthews, J.A., *et al. Biomacromolecules* **2002**, *3*, 232–238.
[29] Li, D., *et al. Nano Letters* **2005**, *5*, 913–916.

[30] Yuge, L., *et al. Stem Cells and Development* **2006**, *15*, 921–929.
[31] Zhang, Y.Z., *et al. Biomacromolecules* **2005**, *6*, 2583–2589.
[32] Moroni, L., *et al. Biomaterials* **2006**, *27*, 4911–4922.
[33] Geng, X., Kwon, O.H., Jang, J. *Biomaterials* **2005**, *26*, 5427–5432.
[34] Chun, D.H.R.A.I. *Nanotechnology* **1996**, *7*, 8.
[35] Shin, Y. *Applied Physics Letters* **2001**, *78*, 1149.
[36] Zhang, Y., *et al. Journal of Materials Science. Materials in Medicine* **2005**, *16*, 933–946.
[37] He, W., *et al. Biomaterials* **2005**, *26*, 7606–7615.
[38] Dosunmu, O.O., *et al. Nanotechnology* **2006**, *17*, 1123–1127.
[39] Zhao, Y., Cao, X., Jiang, L. *Journal of the American Chemical Society* **2007**, *129*, 764–765.
[40] McCann, J.T., Marquez, M., Xia, Y. *Journal of the American Chemical Society* **2006**, *128*, 1436–1437.
[41] Huang, Z.M., *et al. Journal of Biomedical Materials Research. Part A* **2006**, *77*, 169–179.
[42] Yashiki, S., *et al. Journal of Bioscience and Bioengineering* **2001**, *92*, 385–388.
[43] Zhao, Y., Cao, X., Jiang, L. *Journal of the American Chemical Society* **2007**, *129*, 764–765.
[44] Doshi, J., Reneker, D.H. *Journal of Electrostatics* **1995**, *35*, 151–160.
[45] Dzenis, Y. *Science* **2004**, *304*, 1917–1919.
[46] Krishnan, J., *et al. Journal of Nanoscience and Nanotechnology* **2004**, *4*, 52–65.
[47] Barnes, C.P., *et al. Advanced Drug Delivery Reviews* **2007**, *59*, 1413–1433.
[48] Casper, C.L., *et al. Biomacromolecules* **2005**, *6*, 1998–2007.
[49] Schneider, A., *et al. Acta Biomaterialia* **2009**, *5*, 2570–2578.
[50] Venugopal, J., Ramakrishna, S. *Applied Biochemistry and Biotechnology* **2005**, *125*, 147–158.
[51] Chong, E.J., *et al. Acta Biomaterialia* **2007**, *3*, 321–330.
[52] Kumbar, S.G., *et al. Biomedical Materials* **2008**, *3*, 034002.
[53] Boland, E.D., *et al. Journal of Biomedical Materials Research Part B: Applied Biomaterials* **2004**, *71B*, 144–152.
[54] Lowery, J.L., DattaG, N., Rutledge, C. *Biomaterials* **2010**, *31*, 491–504.
[55] Karageorgiou, V., Kaplan, D. *Biomaterials* **2005**, *26*, 5474–5491.
[56] Pham, Q.P., Sharma, U., Mikos, A.G. *Biomacromolecules* **2006**, *7*, 2796–2805.
[57] Chen, M., *et al. Tissue Engineering* **2007**, *13*, 579–587.
[58] Soliman, S., *et al. Acta Biomaterialia* **2010**, *6*, 1227–1237.
[59] Li, W.-J., *et al. Journal of Biomedical Materials Research* **2002**, *60*, 613–621.
[60] He, W., *et al. Tissue Engineering* **2005**, *11*, 1574–1588.
[61] Kang, X., *et al. Biomaterials* **2007**, *28*, 450–458.
[62] Seth, D., McCullen, D.R.S., Roberts, W.A., Clarke, L.I., Bernacki, S.H., Gorga, R.E., Loboa, E.G. *International Journal of Nanomedicine* **2007**, *2*, 11.
[63] Lee, S.J., *et al. Biomaterials* **2008**, *29*, 2891–2898.
[64] Del Gaudio, C., *et al. Journal of Biomedical Materials Research Part A* **2009**, *89A*, 1028–1039.
[65] Venugopal, J.R., *et al. Artificial Organs* **2008**, *32*, 388–397.
[66] Heydarkhan-Hagvall, S., *et al. Biomaterials* **2008**, *29*, 2907–2914.
[67] Schenke-Layland, K., *et al. Biomaterials* **2009**, *30*, 4665–4675.
[68] Gui-Bo, Y., *et al. Journal of Biomedical Materials Research Part A* **2010**, *93A*, 158–163.
[69] Puppi, D., *et al. Progress in Polymer Science* **2010**, *35*, 403–440.

[70] Phipps, M.C., *et al. Biomaterials* **2012**, *33*, 524–534.
[71] Lee, Y.H., *et al. Biomaterials* **2005**, *26*, 3165–3172.
[72] Nam, J., *et al. Tissue Engineering* **2007**, *13*, 2249–2257.
[73] Simonet, M., *et al. Polymer Engineering & Science* **2007**, *47*, 2020–2026.
[74] Leong, M.F., *et al. Journal of Biomedical Materials Research Part A* **2009**, *91A*, 231–240.
[75] Ki, C.S., *et al. Journal of Applied Polymer Science* **2007**, *106*, 3922–3928.
[76] Yokoyama, Y., *et al. Materials Letters* **2009**, *63*, 754–756.
[77] Kwon, I.K., Kidoaki, S., Matsuda, T. *Biomaterials* **2005**, *26*, 3929–3939.
[78] Shim, I.K., *et al. Journal of Biomedical Materials Research. Part A* **2009**, *90*, 595–602.
[79] Sundararaghavan, H.G., MetterJ, R.B., Burdick, A. *Macromolecular Bioscience* **2010**, *10*, 265–270.
[80] Yixiang, D., *et al. Tissue Engineering Part A* **2008**, *14*, 1321–1329.
[81] Zhang, D., Chang, J. *Nano Letters* **2008**, *8*, 3283–3287.
[82] Zhang, K., *et al. Biomedical Materials* **2009**, *4*, 035004.
[83] Vaquette, C., Cooper-White, J.J. *Acta Biomaterialia* **2011**, *7*, 2544–2557.
[84] Chen, M., Patra, P.K., Warner, S.B., Bhowmick, S. *Biophysical Reviews and Letters* **2006**, *1*, 26.
[85] Chen, M., *et al. Journal of Tissue Engineering and Regenerative Medicine* **2009**, *3*, 269–279.
[86] Christopherson, G.T., Song, H., Mao, H.-Q. *Biomaterials* **2009**, *30*, 556–564.
[87] Dalby, M.J., *et al. Tissue Engineering* **2002**, *8*, 1099–1108.
[88] Xie, J., *et al. ACS Nano* **2010**, *4*, 5027–5036.
[89] Fu, X., Wang, H. *Tissue Engineering Part A* **2011**, *18*, 631–642.
[90] Zhong, S., *et al. Journal of Biomedical Materials Research Part A* **2006**, *79A*, 456–463.
[91] Huang, C., Fu, X., Liu, J., Qi, Y., Li, S., Wang, H. *Biomaterials* **2011**, *33*, 1791–1800.
[92] Webster, T.J., *et al. Biomaterials* **2001**, *22*, 1327–1333.
[93] Eisenbarth, E., *et al. Biomaterials* **1996**, *17*, 1399–1403.
[94] Dalby, M.J., *et al. Tissue Engineering* **2002**, *8*, 453–467.
[95] Rho, K.S., *et al. Biomaterials* **2006**, *27*, 1452–1461.
[96] Yang, X., Ogbolu, K., Wang, H. *Journal of Experimental Nanoscience* **2008**, *3*, 329–345.
[97] Powell, H.M., Boyce, S.T. *Tissue Engineering Part A* **2009**, *15*, 2177–2187.
[98] Yang, X., Shah, J.D., Wang, H. *Tissue Engineering Part A* **2009**, *15*, 945–956.
[99] Noh, H.K., *et al. Biomaterials* **2006**, *27*, 3934–3944.
[100] Ekaputra, A.K., *et al. Biomacromolecules* **2008**, *9*, 2097–2103.
[101] Kidoaki, S., Kwon, I.K., Matsuda, T. *Biomaterials* **2005**, *26*, 37–46.
[102] Park, S.H., *et al. Acta Biomaterialia* **2008**, *4*, 1198–1207.
[103] Srouji, S., *et al. Journal of Materials Science: Materials in Medicine* **2008**, *19*, 1249–1255.
[104] Ishii, O., *et al. The Journal of Thoracic and Cardiovascular Surgery* **2005**, *130*, 1358–1363.
[105] Stankus, J.J., *et al. Biomaterials* **2006**, *27*, 735–744.
[106] Edwards, C., Marks, R. *Clinics in Dermatology* **1995**, *13*, 375–380.
[107] Mahjour, S.B., Ghaffarpasand, F., Wang, H. *Tissue Engineering Part B: Reviews* **2011**, *18*, 15–23.
[108] Tran, T.T., Kahn, C.R. *Nature Reviews Endocrinology* **2010**, *6*, 195–213.
[109] Branski, L.K., *et al. Burns* **2009**, *35*, 171–180.

[110] Kim, W.-S., *et al. Journal of Dermatological Science* **2007**, *48*, 15–24.
[111] Park, B.-S., *et al. Dermatologic Surgery* **2008**, *34*, 1323–1326.
[112] Rajashekhar, G., *et al. Stem Cells* **2008**, *26*, 2674–2681.
[113] Altman, A.M., *et al. Stem Cells* **2009**, *27*, 250–258.
[114] Tsuji, W., *et al. Tissue Engineering Part A* **2008**, *15*, 83–93.
[115] Wang, H.-J., *et al. Tissue Engineering* **2004**, *10*, 1054–1064.
[116] Bashur, C.A., Dahlgren, L.A., Goldstein, A.S. *Biomaterials* **2006**, *27*, 5681–5688.
[117] Patel, S., *et al. Nano Letters* **2007**, *7*, 2122–2128.
[118] Zeugolis, D.I., *et al.*, *Biomaterials* **2008**, *29*, 2293–2305.
[119] Dhandayuthapani, B., *et al.*, *J. Biomed. Mater. Res. Part B Appl. Biomater.* **2010**, *94*, 264–272.
[120] Zhang, Y., *et al.*, *Biomaterials* **2008**, *29*, 4314–4322.
[121] Ngiam, M., *et al. Bone* **2009**, *45*, 4–16.
[122] Powell, H.M., *et al.*, *Biomaterials* **2008**, *29*, 834–843.
[123] Zhang, Y.Z., *et al.*, *Nanotechnology* **2006**, *17*, 901–908.

CHAPTER 12

NANO- AND MICROSTRUCTURED ECM AND BIOMIMETIC SCAFFOLDS FOR CARDIAC TISSUE ENGINEERING

QUENTIN JALLERAT and JOHN M. SZYMANSKI

Department of Biomedical Engineering, Carnegie Mellon University, Pittsburgh, PA, USA

ADAM W. FEINBERG

Department of Biomedical Engineering, Department of Materials Science and Engineering, Carnegie Mellon University, Pittsburgh, PA, USA

12.1 INTRODUCTION

Cardiovascular disease (CVD) has become the principal cause of death in the developed world, causing one in three US deaths in 2009 (787,931 total) [1]. The most common form of CVD is coronary heart disease, where the coronary vasculature supplying the heart tissue with oxygen and nutrients becomes occluded. When starved of oxygen for too long, it causes a heart attack, or myocardial infarction (MI), where the cardiomyocytes (CMs) die and are unable to regenerate. MI was responsible for one in six US deaths in 2009, and associated health care costs amounted to $195.2 billion [1]. This demonstrates the increasing burden of CVD on the society. Moreover, patient morbidity leads to a considerable decrease in their standard of living. To address this issue, significant basic and clinical research efforts are focused on the development of strategies to restore cardiac function.

Therapies to treat MI and heart failure face unique challenges, chief among them is the absence of natural regeneration in adult cardiac tissue. CM proliferation, important during development, nearly disappears after birth [2]. Instead, postnatal myocardial growth is mostly due to CM maturation and enlargement, termed

Bio-inspired Materials for Biomedical Engineering, First Edition. Edited by Anthony B. Brennan and Chelsea M. Kirschner.

hypertrophy, which can bring a 20-fold increase in cell size [3]. Recent findings suggest that there is a basal level of CM proliferation even in adults, but fewer than 1% are replaced every year, meaning more than 50% of CMs are sustained throughout our lifetime [4]. After injury, the percentage of CM proliferation near the infarct site rises to 3%, but it is insufficient to naturally recover cardiac function [5]. Damaged myocardium is then further remodeled during the acute inflammatory response and is eventually replaced by hypoxic and noncontractile scar tissue. Therapeutic approaches aim to recover full or partial cardiac function, either by inducing the formation of viable cardiac tissue or by compensating for the damaged tissue with medical devices.

With disease progression or recurrent MI, patients eventually suffer heart failure, characterized by the inability to pump enough blood to the body. A heart transplant is the gold standard for treatment, and while it is estimated that 40,000 patients could benefit, there are only about 2200 donor hearts available each year [6]. To address the urgent need for circulatory assist, devices such as total artificial hearts and left ventricular assist devices have been developed [7]. These mechanical pumps work short term, but have serious limitations as they can trigger thrombosis, leading to blood clots that circulate to the brain or the lungs, block small arteries, and induce stroke or embolism. Furthermore, excessive shear stress within these mechanical assist devices can damage the blood cells, causing hemolysis. Thus, survival is improved only for ~5 years, with these devices used predominantly for bridge to transplantation instead of destination therapy for heart failure [8]. More recently, stem cells have been used in an attempt to regenerate and repair damaged heart tissue [9–13]. Stem cell therapy can effectively modulate the effects of the acute inflammatory response after a heart attack via the release of paracrine factors that reduce CM death and preserve functional myocardium that would otherwise be remodeled to form scar tissue [11]. However, long-term outcomes are hindered by the inability to recover the bulk of cardiac muscle lost during MI. More than 90% of stem cells injected into the infarct die after a week in the hostile environment with little to no new muscle formation or functional integration [11]. Additionally, exogenous cells introduced into the complex cardiac environment have the potential to create arrhythmia due to inadequate electromechanical coupling [14].

Cardiac tissue engineering has emerged as a promising solution to address these challenges and produce functional cardiac tissue. While approaches vary, typically, polymer scaffolds are engineered to provide physical and chemical cues that guide the behavior of cardiac cells and instruct them to proliferate, differentiate, and eventually assemble into a viable tissue. The primary goal is to develop a construct with dense, anisotropic cardiac muscle that has electromechanical function on par with ventricular myocardium. The ultimate tissue engineering application is to repair an MI in a patient, by either replacing the damaged tissue with an *in vitro*-engineered cardiac muscle graft or implanting a scaffold *in vivo* that promotes endogenous repair. Additionally, engineered cardiac muscle has great utility as an *in vitro* model system of cardiac function, enabling researchers to test the effects of drugs for safety and toxicity, as well as discover new compounds to treat a range of CVD. To inform scaffold design, cardiac tissue engineering draws from cardiac anatomy and physiology, advances in stem cell biology and, most importantly, from a better understanding of the heart's native scaffold, the extracellular matrix (ECM). The ECM is essential to cardiac structure and function and it helps provide and transduce the physical and chemical cues required by cardiac cells to maintain tissue homeostasis. For these reasons, researchers have developed a range of technologies to engineer scaffolds that mimic the cardiac ECM in order to achieve the production of functional cardiac constructs. Cardiac tissue engineering, by using bio-inspired scaffolds to induce tissue growth, has the potential to treat CVD in millions of patients every year.

In this chapter, we provide an overview of the heart and the biomimetic approaches used to engineer cardiac muscle tissue. We first describe the unique structure and function of cardiac muscle (myocardium), discuss how this dictates scaffold design, and then define the benchmarks used to evaluate the performance of engineered cardiac tissue. Next, we present the different techniques researchers have developed to fabricate tissue-engineered cardiac scaffolds that mimic ECM cues observed in the heart. For each case, we discuss the relevant physical, mechanical and/or chemical properties and the particular advantages and limitations of the approach. Finally, we discuss the persistent challenges to engineering functional cardiac tissue and the future directions of the field.

12.2 STRUCTURE AND FUNCTION OF THE MYOCARDIUM

12.2.1 Multiscale Hierarchy of the Contractile Apparatus

The myocardium features a complex hierarchical organization that spans from the molecular to tissue scales, encompassing eight orders of spatial magnitude (Figure 12.1). At the nanometer scale, contractile forces rely on the interaction of molecular motors composed of actin and myosin filaments organized into overlapping bands that form the basic contractile unit termed the sarcomere. Sarcomeres assemble into cytoskeletal filaments termed myofibrils that span the entire CM cell body, and myofibrils within CMs bundle in parallel to form aligned contractile structures. The 2–4 billion CMs in the heart are cylindrical in shape, with a length to width aspect ratio of 7 : 1, and are longitudinally connected end to end by specialized structures, termed intercalated disks, that mechanically and electrically couple the cells into multicellular, contractile myofibers [3,15]. At the tissue scale, the heart consists of lamellar-like layers of aligned cardiac myofibers wrapped around the heart to form the walls of the ventricles and the atria. In the left ventricle, which sends the blood to the body through the systemic circulation, myofiber orientation varies linearly throughout the thickness of the ventricular wall (Figure 12.2). This architecture is responsible for translating the uniaxial contraction of individual cells into an actual reduction in the volume of the heart chambers during a pumping cycle. Thus, it is the coordinated action of billions of actin–myosin molecular motors at the nanometer scale, each generating piconewton forces, that produces tissue-scale forces up to 10 mN/mm^2 of myocardium [16,17]. Engineered cardiac constructs need to reproduce the complex hierarchical structure of the myocardium in order to provide clinically relevant contractile forces.

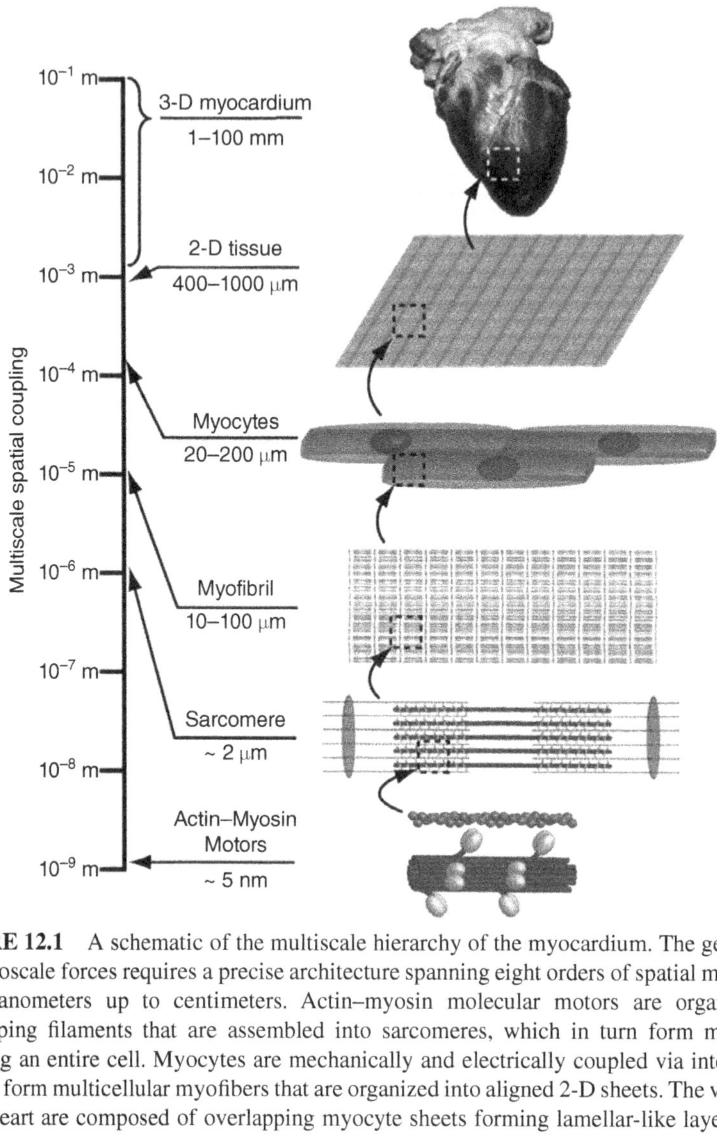

FIGURE 12.1 A schematic of the multiscale hierarchy of the myocardium. The generation of macroscale forces requires a precise architecture spanning eight orders of spatial magnitude from nanometers up to centimeters. Actin–myosin molecular motors are organized as overlapping filaments that are assembled into sarcomeres, which in turn form myofibrils spanning an entire cell. Myocytes are mechanically and electrically coupled via intercalated disks to form multicellular myofibers that are organized into aligned 2-D sheets. The ventricles in the heart are composed of overlapping myocyte sheets forming lamellar-like layers.

12.2.2 Mechanical Anisotropy

The complex arrangement of aligned myofibers gives the myocardium highly anisotropic mechanical properties. For example, the native myocardium of the left ventricle of a rat has a Young's modulus of 157 ± 84 kPa parallel and 84 ± 8 kPa perpendicular to the myofibers (Figure 12.3), with similar values reported for human and porcine hearts [18–22]. The importance of the mechanical properties and anisotropy is illustrated by the profound alterations observed in disease states. In a rat

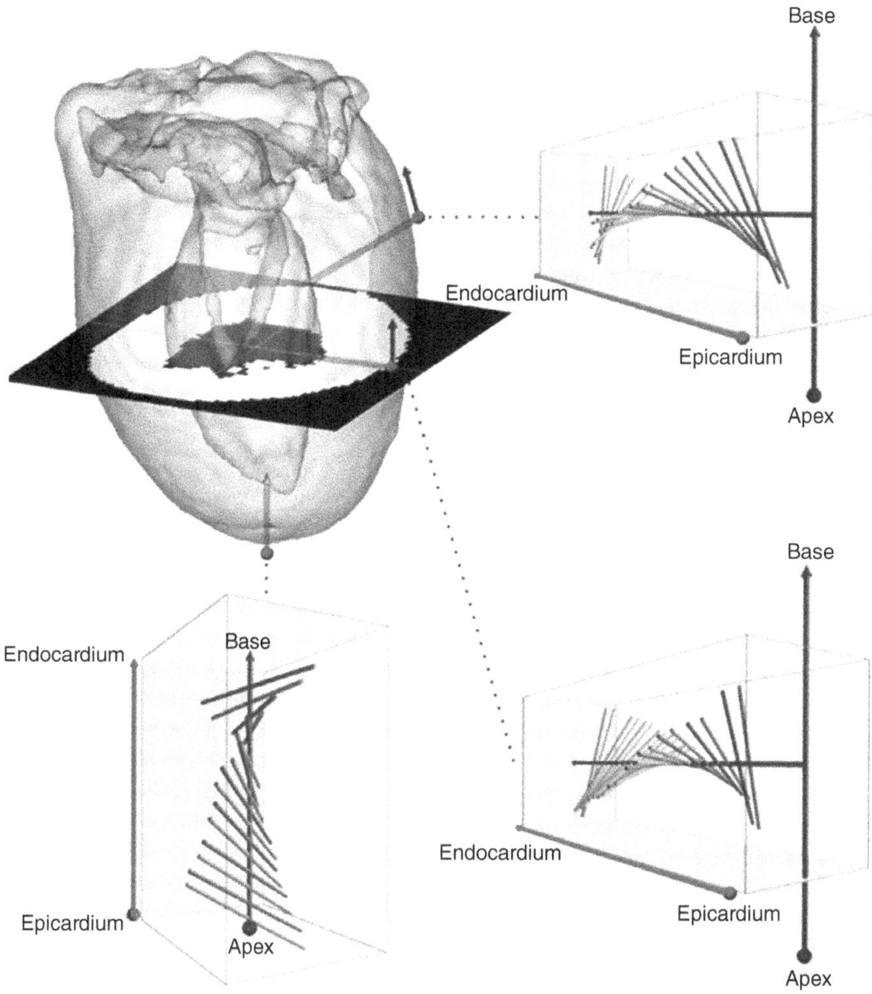

FIGURE 12.2 Myofiber orientation in a rat heart. The orientation of the myofibers was reconstructed by fitting a generalized helicoid model to an MRI dataset. The schematic of the heart shows the three areas of observation (red penetrating arrows) at the base, equator, and apex (clockwise from top right). At each location, the orientation of the myofibers through the myocardium, from endocardium (innermost layer) to the epicardium (outermost) is shown. Myofiber orientation was reconstructed (blue rods) by interpolating orientations obtained from MRI data (red rods). MRI, magnetic resonance imaging. (Reprinted with permission from Reference 141.)

model of MI, ventricular stiffness increased twofold at 6 weeks after the MI, while the scar tissue within the MI became completely isotropic in mechanical properties [23]. *In vitro*, CMs cultured on substrates stiffer than normal myocardium have demonstrated a decreased beating frequency, further suggesting that ventricular stiffening can adversely affect CM contraction and cardiac function [24]. This indicates that in order to achieve therapeutic success, scaffolds for cardiac tissue

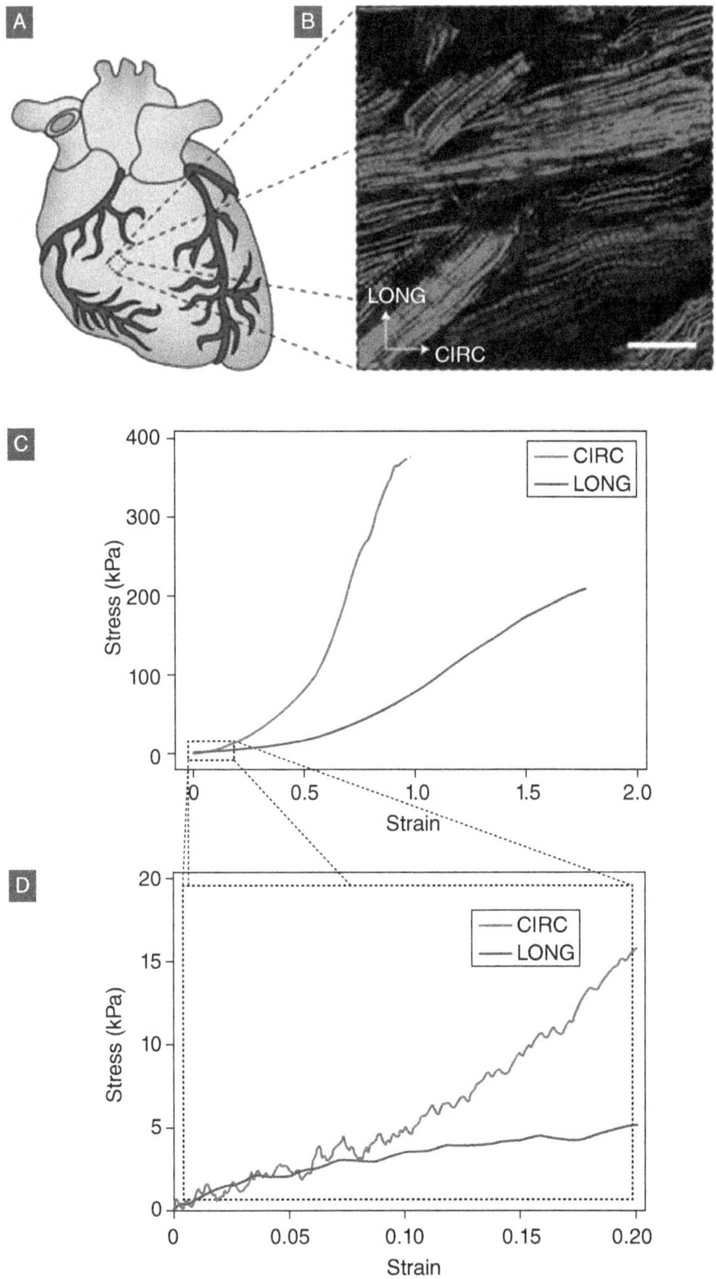

FIGURE 12.3 Mechanical and structural characteristics of the myocardium. The typical arrangement of aligned myofibers gives the myocardium highly anisotropic tensile properties. (A) A schematic representation of a mammalian heart. (B) Confocal microscopy image of the right ventricular myocardium of an adult rat showing the oriented myofibers, labeled for F-actin (green) and cell nuclei (blue). (C,D) Uniaxial tensile stress–strain plots of right ventricular myocardium along the circumferential and longitudinal direction illustrates the mechanical anisotropy (C, full range to demonstrate failure properties; D, physiologic regime). Scale bar in (B) is 50 μm. CIRC and LONG stand for circumferential and longitudinal axes, respectively. (Reprinted with permission from Reference 18.) *(See insert for color representation of the figure.)*

engineering must match the unique mechanical properties of the native myocardium. Specifically, CMs contract ~10% during the cardiac cycle in order to empty the ventricle and eject blood. Achieving this deformation in an engineered cardiac tissue requires that the effective elastic modulus of the combined cells and scaffold be closely matched to the native myocardium.

12.2.3 Innervation and the Conduction System

The heart has a complex control system to regulate heart rate and the contraction cycle, and like every muscle, the heart is densely innervated. Heart rate and cardiac output are influenced by a network of afferent and efferent neurons that extend to the brain and the spinal cord, as well as interconnected neurons within the heart [25]. However, direct control of heart rate originates in the sinoatrial (SA) node, a group of cells in the right atria that initiate each contraction of the heart by regularly and autonomously depolarizing (Figure 12.4) [26,27]. Electrical signals generated in the SA node propagate as an action potential through the myocardium via the gap junctions, which are intercellular ion channels that transmit electrical potentials via Ca^{2+} flux. The arrangement of CMs in aligned myofibers accounts for the anisotropic electrical properties, with faster action potential propagation longitudinally than

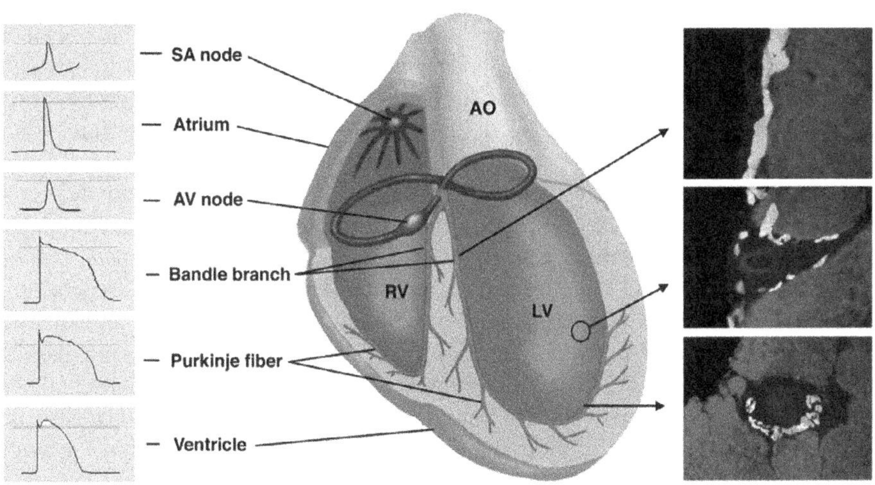

FIGURE 12.4 The cardiac conduction system. (A) Representative action potentials for components of the conduction system in the chick heart illustrated in (B). The action potentials originating in the SA node undergo several transformations as they travel through the atria, the AV node, and finally the Purkinje fibers and the ventricles. Panels on the right show the Purkinje fibers (green) within the myocardium (red) at different locations in the ventricles. (C) Subendocardial Purkinje fibers. (D) Branch point from subendocardial Purkinje fibers. (E) Intramural Purkinje fibers. AO, aorta; AV, atrioventricular; LV, left ventricle; RV, right ventricle; SA, sinoatrial. (Reproduced with permission from Reference 26.) *(See insert for color representation of the figure.)*

transverse [28]. From the atria, the action potential reaches the atrioventricular (AV) node, which connects to a ring of CMs that loops between the ventricles and the atria. The AV node slows conduction by introducing a delay to ensure enough time for the atria to empty into the ventricles. The action potential is then transmitted to the ventricular conduction system, which comprises the Purkinje fibers. The specialized CMs of the Purkinje fibers have higher conduction speed in order to rapidly distribute the action potential throughout the ventricles so that they can contract synchronously. Together, the nerves, nodes, and ventricular conduction system coordinate and regulate contraction to meet systemic needs. However, disease states can disrupt normal contraction and cause an arrhythmia, which can be fatal. Loss of synchrony, whether due to genetic defect or tissue remodeling after MI, prevents coordinated contraction of the ventricles, thus reducing or eliminating blood flow [25,29,30]. For cardiac tissue engineering, grafts intended to replace infarcted scar tissue must integrate successfully with the native neuronal and conduction systems and synchronize with the viable myocardium in order to improve cardiac function. Failure to do so will create a pro-arrhythmic interface between the graft and the patient's myocardium. It remains to be determined how difficult it is to achieve this integration, as there are no clinical trials of engineered tissue grafts that have yet been conducted in humans.

12.2.4 Vascularization

The CMs in the heart are constantly contracting, creating a high metabolic demand that requires a large supply of oxygen and nutrients. To meet this need, the heart is the most vascularized organ in the body with capillaries spaced $\sim$20 μm, which maximizes mass transport to and from the CMs [31]. The coronary vasculature forms an extremely intricate vessel network branching throughout the myocardium from the $\sim$6 mm diameter coronary artery to the $\sim$10 μm diameter capillaries (Figure 12.5). The high capillary density means each CM is in contact with endothelial cells from at least one capillary [32]. While endothelial cells line the capillaries, larger vessels are supported by pericytes and smooth muscle cells that line the outer perimeter and maintain vessel integrity and vasoactivity. These vascular cells represent a significant fraction of the total number of cells in the heart and have an essential role in regulating CM function. In the context of cardiac tissue engineering, tissues must have sufficient CM density to generate contractile force and high capillary density to provide adequate nutrient transport in three-dimensional (3-D) constructs. For example, cell density of more than 5×10^5 cells/mm^3 is observed in the adult rat myocardium, of which $\sim$17% are cardiac vascular cells (endothelial cells and pericytes) [33,34]. The cardiac endothelium, in particular, can regulate cardiac function via paracrine signaling between endothelial cells and CMs [35,36]. For example, when cultured with endothelial cells *in vitro*, CMs showed improved survival rate and contractile performance [37]. In cardiac tissue engineering, it is important to account for the coronary vascular system in terms of nutrient mass transport to highly metabolic CMs, as well as specific cardiac endothelium–CM signaling. Recapitulating both aspects is likely necessary in order to achieve clinically relevant CM density and function.

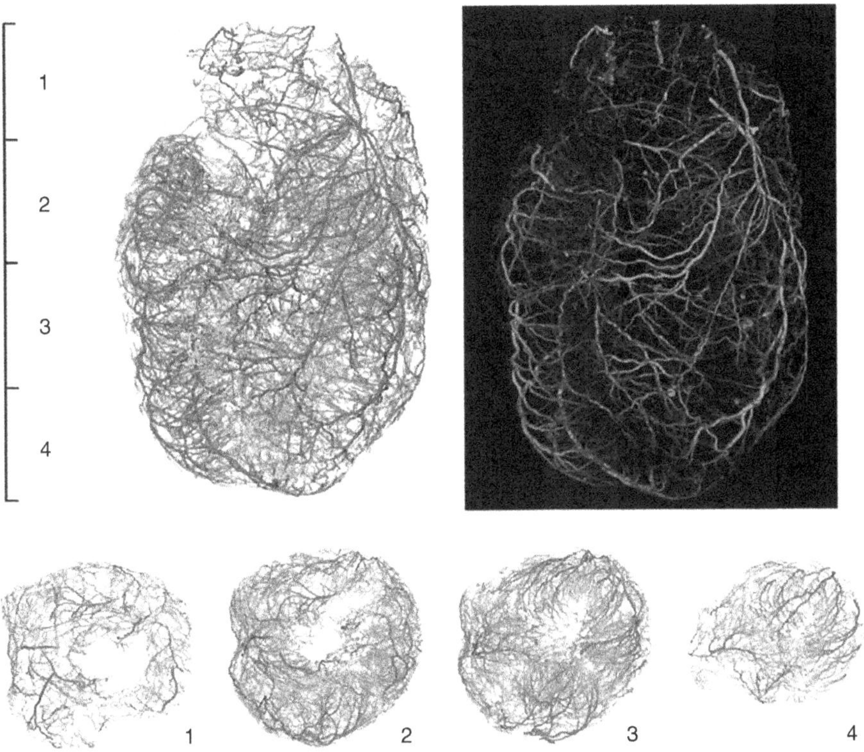

FIGURE 12.5 The whole vasculature of an adult rat heart was reconstructed (top left) from micro-CT data (top right). Transverse sections obtained in four planes (below) show the penetrating network of capillaries. The color of the rendered vessels corresponds to the intensity of the voxels in the original dataset. (Reprinted with permission from Reference 142.) *(See insert for color representation of the figure.)*

12.2.5 Extracellular Matrix

The cardiac ECM is a 3-D network of fibrillar proteins and glycosaminoglycans (GAGs) that provides structural support and serves as an essential substrate for physical and chemical signaling. Each cell type in the heart contributes in some way to the synthesis, assembly, and remodeling of the ECM in their local environment. However, cardiac fibroblasts are the primary cell type involved in the maintenance of the ECM and represent 64% of the total number of cells in the adult rat heart, though they account for only ~18% of the cardiac volume [33,38]. Cells interact with the ECM through integrin receptors on their membrane that can bind to the ECM and form a mechanical linkage to the cytoskeleton within the cell. Through this binding interaction, cells can exert mechanical forces on the ECM to migrate through their environment or to manipulate and assemble the surrounding ECM. In order to remodel the ECM, cells also release specialized enzymes, the matrix metalloproteinases (MMPs), which can cleave specific ECM components [39].

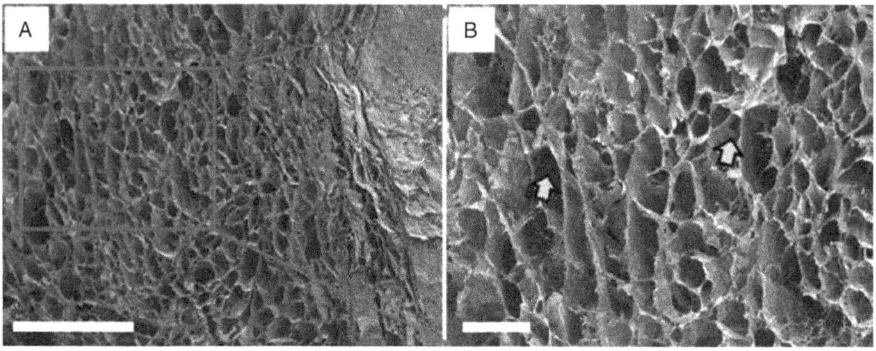

FIGURE 12.6 Scanning electron microscopy (SEM) images of porcine myocardium after decellularization using the detergent sodium dodecyl sulfate (SDS). (A) Cross-section view shows the porous topography. Scale bar is 400 μm. (B) At higher magnification, the intricate network of interconnecting pores (yellow arrows) throughout the ECM is visible. Scale bar is 100 μm. (Adapted with permission from Reference 20.)

There are a large number of proteins within the cardiac ECM that contribute to its structural, mechanical, and chemical properties. The first element of the ECM interacting directly with the CMs is the basement membrane, a thin protein layer composed mostly of laminin, fibronectin, collagen type IV, and perlecan, a basement membrane-specific heparan sulfate [40,41]. The basement membrane acts as a selective barrier for soluble factors, as well as a linkage to the structural fibers of the extracellular environment. Beyond the basement membrane, collagens type I and III provide tensile strength and maintain the shape of the heart while also distributing the contractile forces to the whole organ [42–44]. They form a complex weave that surrounds bundles of adjacent CMs and long coiled strands aligned with the main myofiber orientation [45]. Each CM is tethered to the collagen matrix by costameres, protein assemblies that link the sarcomeres to integrin receptors to the ECM, which prevent slippage or injury under excessive loading [44,46]. Additionally, the myocardium contains the structural fiber elastin, a cross-linked, flexible protein that is highly expressed in blood vessel walls and contributes to the heart's elasticity [20]. Structurally, the cardiac ECM has characteristics of the fibrillar and laminar components, featuring an interconnected network of small pores, organized very tightly around cardiac cells. For example, the average pore diameter was 21.4 μm in decellularized porcine myocardium (Figure 12.6) [20]. The fibrillar proteins also exhibit a wide range of diameters, ranging from ∼1 μm for elastin, from 10 nm to 10 μm for collagen type I, and from 5 nm to 1 μm for fibronectin [47–49].

The importance of the ECM in normal cardiac function is highlighted by the maladaptive changes that occur in structure and composition with disease and aging. For example, the ECM collagen content and the diameter of the collagen fibers increase with age. This contributes to loss of elasticity (i.e., increase in stiffness) and is associated with a higher risk of CVD [42]. Furthermore, abnormal turnover of ECM proteins such as collagens, fibronectins, and laminins due to increased MMP activity is observed in many cardiomyopathies [50]. The ECM is also important in wound healing in the heart. During the acute phase of MI, the normal cardiac ECM is replaced by a provisional fibrin-based matrix. In the chronic phase, a dense, collagen type I-rich scar tissue is formed that is stiffer and has lost the mechanical anisotropy of the healthy tissue, which further hinders cardiac contraction [23,51]. Thus, the cardiac ECM is a key model system for designing cardiac tissue engineering scaffolds, and further highlights the need to recapitulate properties of the healthy ECM in order to produce viable cardiac constructs.

12.3 BIO-INSPIRED DESIGN REQUIREMENTS OF CARDIAC TISSUE ENGINEERING SCAFFOLDS

The structure and function of the myocardium as highlighted in Section 12.1 forms the basis from which we can implement the bio-inspired design of cardiac tissue engineering scaffolds. While the myocardium contains complex contractile, electrical, and vascular networks, it is the cardiac ECM that integrates these together into a functional tissue system. Thus, it is the cardiac ECM that many researchers look to as a design template for building biomimetic scaffolds. Significant research has focused on the structural, mechanical, and biochemical characteristics of the ECM and methods to recapitulate a subset of these properties using synthetic or naturally derived polymer scaffolds.

Porosity is a key component of scaffolds because CMs need to be able to infiltrate and couple together into a dense network to form functional myocardium, requiring a network of interconnected pores. In scaffolds that cannot be easily remodeled by the cells, the absolute minimum size for pores is the size of a cell, which based on measurements of the decellularized myocardium is $\sim$20 μm [20]. Because the myocardium is cell dense, the scaffold must have high porosity or be able to be resorbed or remodeled by the cells to achieve a high porosity. Based on the ECM density in the heart, this requires a scaffold porosity of >90% to sustain high cell density, ensure homogenous cell seeding, and support oxygen and nutrients mass transport essential to the function of cardiac tissue. In the case of low porosity or very small pore size, specialized techniques such as the use of medium perfusion, are required for efficient cell seeding and oxygen and nutrients delivery to the bulk of the scaffold.

The cardiac ECM is composed of protein fibers such as collagen type I and III, fibronectin, and elastin, which are organized in parallel with the myofibers and provide alignment cues to the constituent CMs. It is important to recapitulate the structural characteristics of these fibers because CMs, like all cells, respond to physical structures in the nanometer to micrometer range via a contact guidance effect. The key physical attributes of ECM fibers are the diameter (typically ranging from 5 nm to 10 μm), spatial density, and 3-D orientation. Thus, scaffolds with fibers or fiber-like features that mimic the structure of the cardiac ECM can be effective for the alignment of CMs and other cardiac cells into anisotropic tissues.

In the myocardium, cardiac cells are tightly connected to the ECM via integrin receptors, and this adhesive binding is necessary for survival and function. For example, disruption of fibronectin binding, observed in mice without the integrin $\alpha_5\beta_1$, is embryonically lethal [52]. The specific ECM protein composition and organization is also important, as fibronectin, which is expressed at higher levels during cardiac development, has been shown to foster cardiac progenitor cell differentiation into CMs [53]. Cardiac tissue engineering scaffolds composed of ECM proteins typically have intrinsic ligands for integrin receptors to bind cells, comparable with the native ECM. However, synthetic materials, polysaccharides, and other non-ECM protein materials must be modified in some way to add cell adhesive motifs. Typically, this is achieved by grafting or adsorbing polypeptides or whole ECM proteins (such as fibronectin) to the materials as either a surface or bulk modification.

In the heart, cells remodel the cardiac ECM continuously to allow for myofiber formation, angiogenesis, and high cell density. For example, collagens are completely replaced every 160–240 days and noncollagen proteins are replaced 10 times as fast [54]. It is essential that cardiac tissue engineering scaffolds degrade in a controlled manner in order for cells to organize and maintain their own ECM. Scaffolds made from native components of the ECM can readily be remodeled by cells because they already express the appropriate enzymes, such as MMPs. For synthetic materials, remodeling can be achieved by using biodegradable polymers that cleave by hydrolysis and/or integrating enzyme-cleavable domains, comparable with the MMP-degradable sites in ECM proteins. Moreover, the kinetics of scaffold degradation should be coordinated with ECM production and remodeling to match the rate of muscle formation and maintain structural integrity throughout the process.

Finally, scaffolds need to integrate with the electrical and mechanical properties of the cardiac environment. For example, constructs must exhibit high fatigue resistance by reversibly withstanding mechanical strains of $\sim$13% over a large number of contraction–relaxation cycles, millions if grafted in the actual heart [55,56]. Scaffolds also need to match the mechanical compliance and anisotropy of the myocardium to promote CM alignment and contractility. Based on rat and human myocardium, the Young's modulus should be in the range of 10 kPa to 1 MPa and be higher in the myofiber direction than transverse. The CMs must also be aligned uniaxially within each lamellar layer, for example 50% of the CMs in the rat right ventricle are aligned within $\pm 20°$ of the mean alignment direction [18]. Furthermore, scaffolds must not negatively impact the electrical properties of cardiac tissue, in particular, it must have similar conduction velocity and anisotropy to native myocardium and the impedance should be low to avoid hindering contraction and causing necrotic/hypoxic myocardium [57].

12.4 APPROACHES TO FABRICATING ECM BIOMIMETIC SCAFFOLDS

12.4.1 Porous Scaffolds

Fabricating scaffolds with an interconnected network of pores is the most straightforward way to obtain a thick 3-D structure that can be seeded with cells throughout and sustain high cell density. For cardiac tissue engineering, porogen leaching and lyophilization are the methods that have produced the best results. In porogen leaching, a polymer solution is mixed with a porogen (e.g., salts or sugars) and polymerized. The porogen is then dissolved using a solvent, which creates voids within the polymer matrix. By controlling the size and density of porogen particles, scaffolds can be engineered with pore sizes ranging from one to hundreds of micrometers and porosities of up to 95% of the scaffold volume [58]. For example, Radisic *et al.* used sodium chloride (NaCl) particles as porogens to fabricate poly(glycerol sebacate) (PGS) scaffolds with interconnected pores between 75 and 150 μm in diameter and 90% porosity (Figure 12.7A) [59]. These scaffolds allowed seeding of CMs and cardiac fibroblasts at high density *in vitro* and when implanted in a mice model of MI, acellular porous PGS scaffolds were readily invaded by host cells and supported the formation of new blood vessels [59,60]. Porogen leaching also can be adapted to provide more control over the macroscale architecture of the scaffold. For example, Madden *et al.* engineered porous scaffolds that combined regularly spaced polycarbonate (PC) rods and poly(methyl methacrylate) (PMMA) beads in order to create an array of parallel channels in the poly(2-hydroxyethyl methacrylate-co-methacrylic acid) (pHEMA-co-MAA) hydrogel scaffold (Figure 12.7B) [61]. The result was a high-porosity scaffold with an array of 60-μm wide channels in place of the rods that fostered CM bundle formation and 30-μm pores in place of the beads that improved neovascularization in an infarct model.

Lyophilization, or freeze-drying, is another method for creating porous scaffolds, and is most commonly used for naturally derived biopolymers. To do this, a polymer dissolved in water is frozen and then exposed to near vacuum to sublimate the ice crystals that formed during the freezing process, leaving behind pores. This technique can be used with polysaccharides such as alginate to produce scaffolds with tunable pore size and mechanical properties (Figure 12.7C) [62]. Furthermore, polysaccharide scaffolds can be chemically modified with cell-adhesive proteins found in the ECM. For example, alginate scaffolds with ~100-μm pores fabricated by lyophilization were modified with the integrin-binding peptide arginine–glycine–aspartic acid (RGD), and heparin-binding protein (HBP) [63]. Culturing CMs on the alginate–RGD–HBP scaffolds resulted in increased expression of sarcomeric α-actinin, indicating increased muscle volume compared with CMs cultured on nonmodified alginate scaffolds. Additionally, lyophilization can be used to engineer scaffolds with anisotropic pore structures by controlling thermal gradients during the freezing process. For example, when a collagen–GAG solution is frozen in a Teflon cylinder between two copper plates, the difference in thermal conductivity between copper and Teflon creates a unidirectional pore structure. After freeze-drying, these scaffolds had elongated pores up to 243 μm in diameter, with an aspect ratio of up to 12.6 : 1, which promoted alignment and beating of the human CM cell line HL-1 *in vitro* (Figure 12.7D) [64].

As described, porogen leaching and lyophilization are techniques commonly used to fabricate scaffolds with controllable porosity. In general, scaffolds are produced with isotropic microstructure, although recent work has demonstrated that a degree of anisotropy can be engineered into the scaffolds using more advanced techniques. While promising results have been obtained with these methods, it

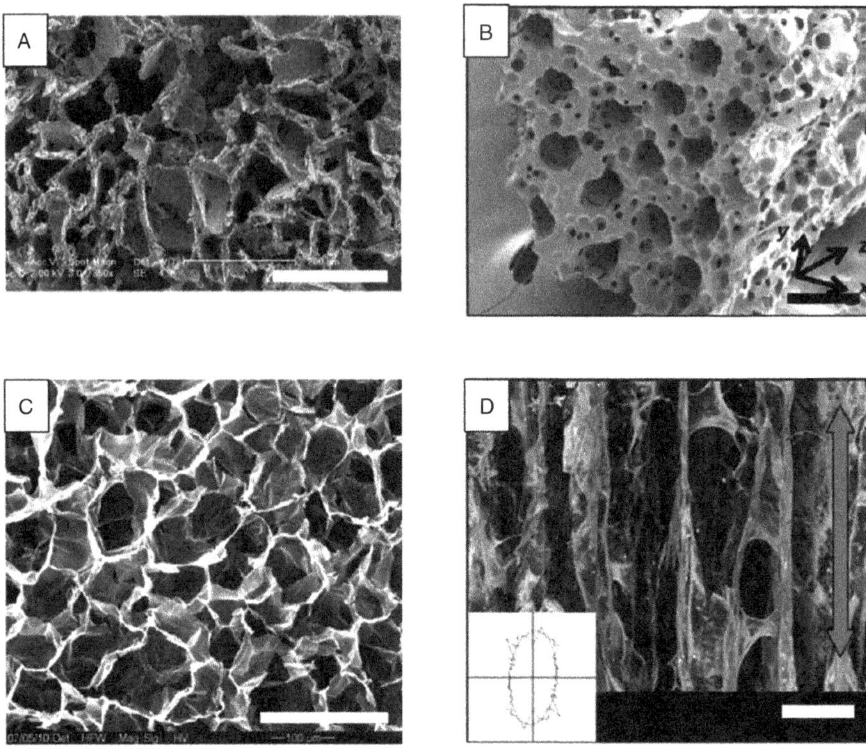

FIGURE 12.7 SEM images of porous scaffolds for cardiac tissue engineering. (A) PGS scaffolds were created by salt leaching. NaCl particles were incorporated in PGS polymer solution during polymerization then dissolved in water to leave 75- to 150-μm pores. Scanning electron micrographs reveal an extensive network of interconnected pores. Scale bar is 200 μm. (B) pHEMA-co-MAA hydrogel scaffold were fabricated using an array of rods PC as well as PMMA beads. After dissolution of the porogen, the network of interconnected 30-μm pores promoted angiogenesis, while the 60-μm diameter channels induced the formation of myocyte bundles. Scale bar is 100 μm. (C) Alginate sponges were made by freeze-drying a solution of cross-linked alginate. Scaffolds with 97-μm pores were modified with binding peptides to improve cell adhesion. Scale bar is 200 μm. (D) Anisotropic collagen–GAG scaffolds were fabricated using freeze-drying. To obtain elongated pores, the solution was frozen in a Teflon cylinder between two copper plates. The arrow marks the scaffold's axis. Insert shows the best fit ellipse to the average pore shape. Scale bar is 200 μm. (Adapted with permission respectively from References 59, 61, 63, and 143.)

remains to be seen whether the high degree of anisotropy seen in the native myocardium can be replicated. Furthermore, these techniques provide limited control of scaffold architecture at the micrometer and nanometer scales. Whether it is necessary to engineer the scaffold to mimic the ECM at this scale is still an active area research, but should be considered as a potential limitation in the context of cardiac tissue engineering.

12.4.2 Micro- and Nanofiber Scaffolds

Major components of the cardiac ECM, including collagen type I and III, are found in the form of fibers of varying diameter and length. Engineering scaffolds with fibrillar topography is an essential approach to mimic the cardiac ECM because fiber diameter has been shown to affect cell attachment, migration, and spreading [65]. Electrospinning is the most common technique used to produce fibrillar scaffolds for cardiac tissue engineering using a wide range of synthetic and natural materials, while rotary jet spinning and surface-initiated assembly are two newer techniques that offer additional control over scaffold properties.

Electrospinning employs a strong electrical field to draw a fiber from an electrically charged polymer solution ejected through a needle onto a grounded collector (see Reference 66 for an in-depth review). The relative simplicity of electrospinning has fostered the development of multiple approaches to control scaffold composition, fiber size, and architecture. Stationary collectors can be used to generate isotropic scaffolds while rotating collectors such as a disks or mandrels can be used to generate aligned, anisotropic scaffolds [67–69]. For example, Shin *et al.* used a stationary collector to generate an isotropic scaffold consisting of 250-nm diameter polycaprolactone fibers that supported myofibril formation and contractility of cultured CMs [70]. To generate anisotropy, Zong *et al.* generated an isotropic scaffold which consists of 1-μm diameter poly(L-lactide) (PLA) fibers (Figure 12.8A), which were uniaxially stretched after electrospinning and promoted alignment of cultured CMs (Figure 12.8B) [71]. The electrospinning process can also be tailored to control scaffold mechanical properties. For example, Amoroso *et al.* used a rotating mandrel that was translating at varying speed along its longitudinal axis to control the number of overlapping fibers in the scaffold and mimic the bending stiffness of a heart valve [69]. The composition of the polymer solution used in electrospinning also has a direct effect on fiber diameter and the electrical and mechanical properties. In one example, increasing the ratio of poly(aniline) to gelatin solution reduced the diameter of electrospun fibers from 803 to 61 nm, while increasing the tensile modulus from 499 to 1384 MPa [72]. The scaffolds demonstrated significantly more proliferation of cardiac myoblasts compared with glass or tissue culture-treated plastic substrates. Poly(aniline) is also a conductive polymer and thus can be used to modify the electrical properties of engineered scaffolds. For example, rat CMs cultured on electrospun poly(L-lysine)-poly(aniline) scaffolds supported greater CM viability when the cells were electrically stimulated to contract [73]. Electrospinning is also capable of creating scaffolds with a range of fiber diameters nanometer to micrometer scales, similar to the cardiac ECM where diameters range from 5-nm fibronectin fibrils to 10-μm collagen type I fibers. To achieve this, multiple polymer solutions can be coelectrospun and collected on the same surface to obtain a range of fiber characteristics. For example, 50- to 500-μm diameter poly(lactic-co-glycolic acid) fibers were electrospun together with 2- to 4-μm diameter fibrin fibers to create multiscale scaffolds (Figure 12.8C,D), which promoted expression of the CM markers sarcomeric α-actinin, troponin, and tropomyosin in mesenchymal stem cells [74].

Electrospinning is a popular technique to create micro- and nanofiber scaffolds that mimic the size of fibers found in the cardiac ECM. However, there are a number

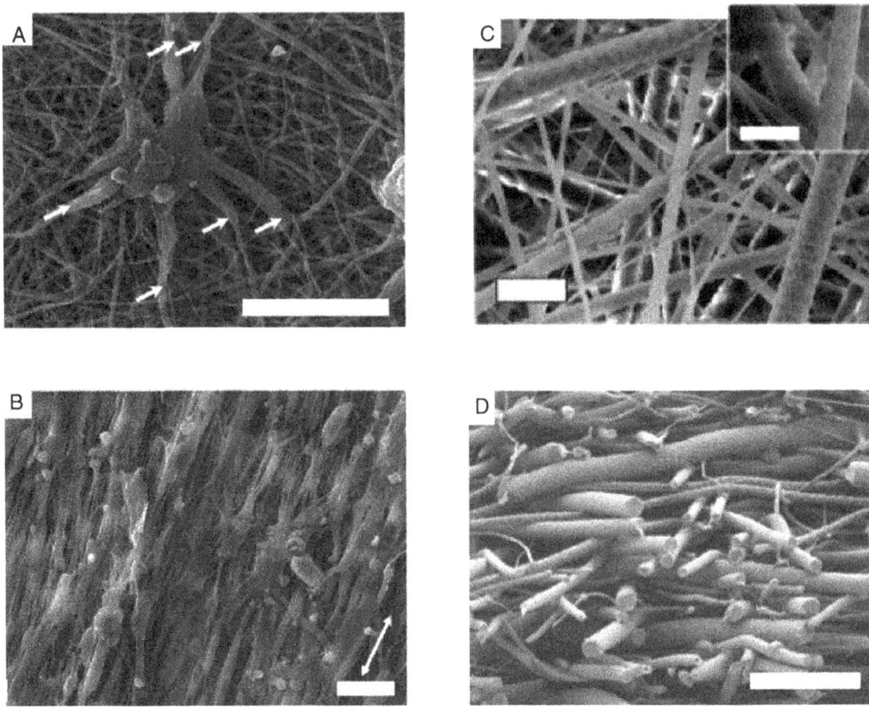

FIGURE 12.8 Electrospinning of anisotropic and multiscale scaffolds. Adult rat CMs were seeded on electrospun scaffolds of PLA that were (A) isotropic or (B) anisotropic due to being uniaxially stretched. Arrows in (A) show the filopodia-like structure that the cells create to spread on the scaffold while the arrow in (B) indicates the main fiber orientation that CMs follow. SEM images, scale bars are 40 μm. Another application of electrospinning is the fabrication of multiscale scaffolds. (C) Two solutions of polycaprolactone of different concentrations were electrospun simultaneously on the same collector to produce fibers with two mean diameters, 3.3 μm and 0.6 μm. Scale bars are 10 μm and 5 μm (insert). (D) SEM cross-sections show the ECM-like range of fiber diameters. Scale bar is 40 μm. (Adapted with permission from References 71 and 144.)

of limitations that are important to consider. First, electrospinning does not produce a specific fiber diameter; rather it produces scaffolds with a distribution of fiber diameters. Whether this is an issue depends on a number of factors, including the sensitivity of cells of interest to fiber morphology. Second, electrospinning produces scaffolds that are typically dense fiber mats, which results in relatively low porosities and pore sizes that are smaller than cells, limiting infiltration. One approach to solve this issue has been the incorporation of electrospun poly(glycolic acid) fibers within a freeze-dried collagen–GAG scaffold, which increased the mechanical stiffness and improved the ability to culture cardiac stem cells [75]. Third, electrospinning of synthetic polymers can mimic the size of ECM fibers, but not the cell-adhesive motifs that bind integrins, in these cases it is necessary to coat the fibers with an

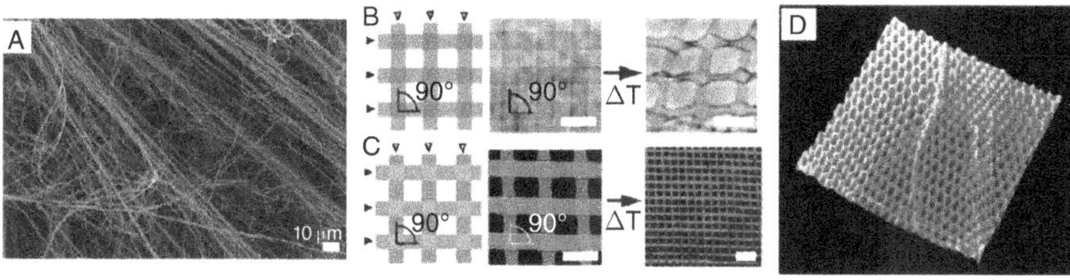

FIGURE 12.9 Alternatives to electrospinning to create micro- and nanofiber scaffolds. Rotary jet spinning uses a high-speed rotating spindle to draw fibers from synthetic and natural materials. (A) SEM images of gelatin fibers show a high degree of alignment. Surface-initiated assembly is a technique that mimics cell-mediated assembly and provides control of the scaffold nano- to macroscale structure and composition. (B) Schematic (left) and optical phase image (right) of two patterns of 20-μm width by 20-μm spacing fibronectin lines microcontact-printed orthogonally onto PIPAAm. After some time (ΔT) in cooling water, the mesh termed nanofabric is released and maintains its shape. (C) The same pattern was created with fibronectin (green) and laminin (red) and was released as a bicomponent nanofabric (right) showing that SIA can be used to control the architecture and composition of biomimetic ECM nanofabrics. (D) Three-dimensional, false-colored rendering of a fibronectin mesh with 20-μm wide elliptical holes observed by scanning electron microscopy. The nanofabric shows fishnet-like ripples. Scale bars are 40 μm in B and C and 100 μm for the released bicomponent nanofabrics. X, Y axes are 360 μm in D. (Adapted with permission from References 77 and 78.) *(See insert for color representation of the figure.)*

ECM protein or related cell-adhesive peptide. Finally, electrospinning uses strong electrical fields that make it incompatible with some materials, and some polymers will not form fibers under the spinning conditions. This has led to the development of alternative fiber spinning techniques that can produce nano- and microfibers with similar diameters, but from a different range of materials. For example, rotary jet spinning (RJS) uses a high-speed rotating spindle to create strong centrifugal forces that push a polymer solution through an integrated nozzle [76]. The spun fibers such as PLA are assembled inside a cylindrical collector with a high degree of anisotropy and have been shown to support CM attachment and uniaxial alignment [77]. RJS works with synthetic polymers such as PLA, polyaorylic acid, and polyethylene glycol (PEG) and natural biopolymers such as gelatin (Figure 12.9A), and represents one of the alternative fiber spinning techniques being developed to address some of the limitation of electrospinning.

While the cardiac ECM contains a variety of nano- and microfibers, cells do not assemble these using fiber spinning techniques such as those typically used for manmade materials. Rather, ECM protein fibers are polymerized using a range of receptor-mediated and enzymatically driven processes. For example, fibronectin fibers are assembled on the cell membrane using integrin receptors in a process termed fibrillogenesis [49]. The integrin receptors bind fibronectin dimers and actin-based contraction unfolds the dimers and exposes cryptic fibrillogenesis sites that bind adjacent fibronectin dimers. Certain ECM protein fibers such as fibronectin, laminin, and collagen type IV cannot be fabricated using most known fiber spinning techniques, but mimicking the way cell assembles these ECM fibers may be an alternative approach. Recently, Feinberg *et al.* developed a technique termed surface-initiated assembly (SIA) that enables the fabrication of ECM protein nanofibers with control of the nano- to macroscale fiber organization and composition [78]. In SIA, fibronectin is adsorbed to polydimethylsiloxane (PDMS) where it unfolds due to hydrophobic surface interactions. The unfolded fibronectin with exposed, cryptic fibrillogenesis domains is then transferred to a sacrificial surface, which can be dissolved to release the assembled fibronectin fibers. This process mimics cell-mediated assembly of the ECM, where integrins bound to the cytoskeleton are used to unfold fibronectin dimers to expose their self-binding (fibrillogenesis) sites [48]. SIA can create freestanding, nanometer-thick ECM protein nanofibers (Figure 12.9B) and nanofabrics (Figure 12.9D), using single or multiple ECM proteins such as fibronectin, laminin, fibrinogen, and collagens type I and IV. A unique aspect of SIA is its ability to combine multiple ECM fibers into the same scaffold, such as laminin and fibronectin (Figure 12.9C), which should enable the engineering of scaffolds that more closely mimic the structure and composition of the native cardiac ECM. This approach has been used to engineer anisotropic cardiac constructs such as CM threads and sheets; however, the current limitation is how to extend this technology to 3-D engineered cardiac tissues.

12.4.3 Synthetic and Naturally Derived Hydrogels

Naturally derived hydrogels are widely used in cardiac tissue engineering because they are relatively simple to fabricate and possess many of the properties of the native cardiac ECM, including a fibrillar network morphology, integral cell-adhesive binding sites, high water content, and high porosity [79]. For example, collagen type I can be polymerized into a hydrogel at physiologic temperatures and has been used to engineer cardiac tissue with a high density of CMs and good electrical coupling [80]. Matrigel is another example, consisting primarily of a mixture of the basement membrane proteins laminin, collagen type IV, and entactin. It can be polymerized as a hydrogel with embedded CMs to engineer 3-D cardiac tissues with high cell density and contractile properties [81–83]. Multicomponent ECM gels have also been derived from several tissues such as porcine bladder and more recently from porcine myocardium (Figure 12.10A), which have been decellularized, lyophilized, and then digested enzymatically to extract the ECM components into the solution [84,85]. These ECM-based hydrogels can be polymerized under physiologic conditions due to the high collagen type I content and closely match the ECM protein composition of the native myocardium, though GAGs, fibronectin, and elastin may be lost [20,86–88]. Culturing CMs in these myocardium-derived ECM gels has shown enhanced cell migration and differentiation compared with regular collagen type I hydrogels [89,90]. These gels have also been injected *in vivo* to improve postinfarct recovery and have been shown to promote angiogenesis by binding growth factors more effectively than pure collagen type I hydrogels [91].

Natural hydrogels also present the unique advantage of being easily degraded and remodeled by the cells, which facilitates cell self-organization into 3-D constructs. For example, Zimmermann *et al.* have seeded CMs in ring-shaped

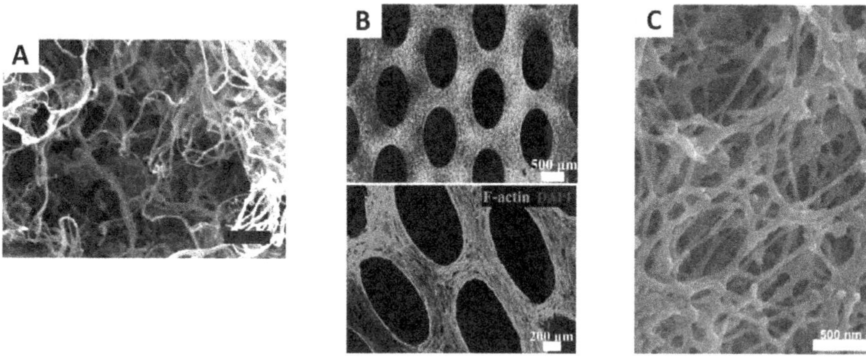

FIGURE 12.10 Examples of hydrogels for cardiac tissue engineering. Myocardial ECM gels can be obtained by decellularization, lyophilization, and enzymatic digestion. (A) The solubilized ECM components gel under physiological conditions into fibrous multicomponent hydrogels with ECM-like structure revealed by SEM. Scale bar is 1 μm. (B) Fibrin–matrigel hydrogels cast around an array of micropillars are remodeled by myocytes to form a contractile cardiac construct with local anisotropy. Scale bars are 500 μm (top) and 200 μm (bottom). (C) Synthetic polypeptides are designed to self-assemble into nanofibrous hydrogels and to mimic VEGF to induce angiogenesis. (Adapted with permission from References 89, 94, and 100.) *(See insert for color representation of the figure.)*

collagen–matrigel hydrogels cast around Teflon cylinders. Over time, the embedded cells pull and compact the surrounding ECM gel, causing the construct to contract around the inner cylinder and increase circumferential tension. This mechanical strain induced CM alignment and the formation of 3-D contractile, anisotropic cardiac tissue, termed engineered heart tissue [92]. Related work has demonstrated the formation of functional engineered heart tissue using additional ECM protein gels such as fibrin [93]. Furthermore, CMs and cardiac fibroblasts have been combined to create anisotropic cardiac tissue using sutures as anchor points to form self-organized cardiac fibers and using fibrin–matrigel cast around microfabricated post arrays to produce anisotropic cardiac sheets (Figure 12.10B) [94,95].

There are many advantages to natural hydrogels, but there are also limitations that need to be considered. First, the ECM proteins must be obtained from an animal or cell source, which introduces batch-to-batch variability in properties and an increase in cost. Furthermore, the animal source of these ECM proteins has the potential to induce an unwanted inflammatory immune response when implanted *in vivo*. Second, these hydrogels are typically isotropic in microstructure, which limits their ability to engineer anisotropic cardiac tissue. As mentioned, cell-generated compaction of these gels can be used to induce alignment, but engineering anisotropy into the scaffold at the time of fabrication may be advantageous. Third, the ability of cells to degrade and remodel the ECM hydrogels can be a problem if it is degraded too rapidly and the mechanical properties deteriorate before adequate muscle tissue has formed. Thus, the cell and ECM density, cross-link density, and susceptibility to MMP degradation must be carefully tuned to optimized cardiac muscle formation.

Synthetic hydrogels offer a number of advantages over natural hydrogels because the chemical structure, molecular weight, cross-link density, water content, hydrolytic stability, and cross-linking chemistry can be readily tuned. Furthermore, recent advances have enabled the integration of custom polypeptides that functionalize synthetic hydrogels with biologically active cell binding and enzymatic degradation sites. For example, the fibronectin-derived adhesion ligand amino acid sequence arginine–glycine–aspartic acid–serine–proline (RGDSP) integrated into PEG hydrogels improved viability of the HL-1 CM cell line [96]. In another example, Kraehenbuehl *et al.* synthesized a bioactive PEG hydrogel featuring adhesion ligands for the integrins $\alpha_v\beta_3$ and $\alpha_5\beta_1$, as well as an MMP substrate site, which promoted CM survival and angiogenesis when injected together with endothelial cells into an infarct [97]. A similar study showed that PEG hydrogels with MMP degradability and integrin-binding RGDSP polypeptides could improve cardiac progenitor cell differentiation [98]. A different approach is the design of synthetic polypeptides that self-assemble into nanofibrous hydrogels under specific stimuli, similar to the way ECM proteins assemble into fibers. For example, polypeptides can be delivered as an injectable solution that polymerizes *in situ* and they can be modified to deliver growth factors such as pro-angiogenic vascular endothelial growth factor (VEGF) [97]. Guo *et al.* successfully added a heparin-binding domain sequence to the peptide RADA16 (its amino acid sequence is arginine–alanine–aspartic acid–alanine, repeated four times) to trigger a sustained and continuous release of VEGF, mimicking a natural release profile [99]. In another example, Webber *et al.* designed a polypeptide with a sequence mimicking VEGF that fostered angiogenesis *in vivo* (Figure 12.10C) [100]. Self-assembling polypeptides have also been modified to bind other growth factors such as the platelet-derived growth factor (PDGF), which is essential during cardiovascular development. When injected *in vivo*, such self-assembling polypeptides demonstrated a protective effect on CMs near the infarct site and induced angiogenesis [101].

These synthetic hydrogels are an area of active research and are continually improving, however, there are limitations that need to be considered. First, there is generally poor microstructural control of anisotropy, similar to natural hydrogels, and thus most of these hydrogels are polymerized as isotropic systems. Although the cost of synthesizing synthetic hydrogels is usually relatively inexpensive, certain monomers can be expensive and the integrated polypeptides used to create biofunctionality add significant cost to the system. Furthermore, even though biological motifs such as integrin ligands and MMP degradation sites are added, this represents only a tiny fraction of functional binding sites within intact ECM proteins. For example, growth factor binding sites and mechanosensitive changes in protein conformation are typically absent.

12.4.4 Nano- and Microfabricated Scaffolds

Nano- and microfabrication techniques for cardiac tissue engineering enable the formation of scaffolds with anisotropic structure on the same scale as fibers in the native cardiac ECM. In 2-D, sheets of cardiac muscle have been engineered using photolithographic techniques to generate microscale surface topography, nanoscale

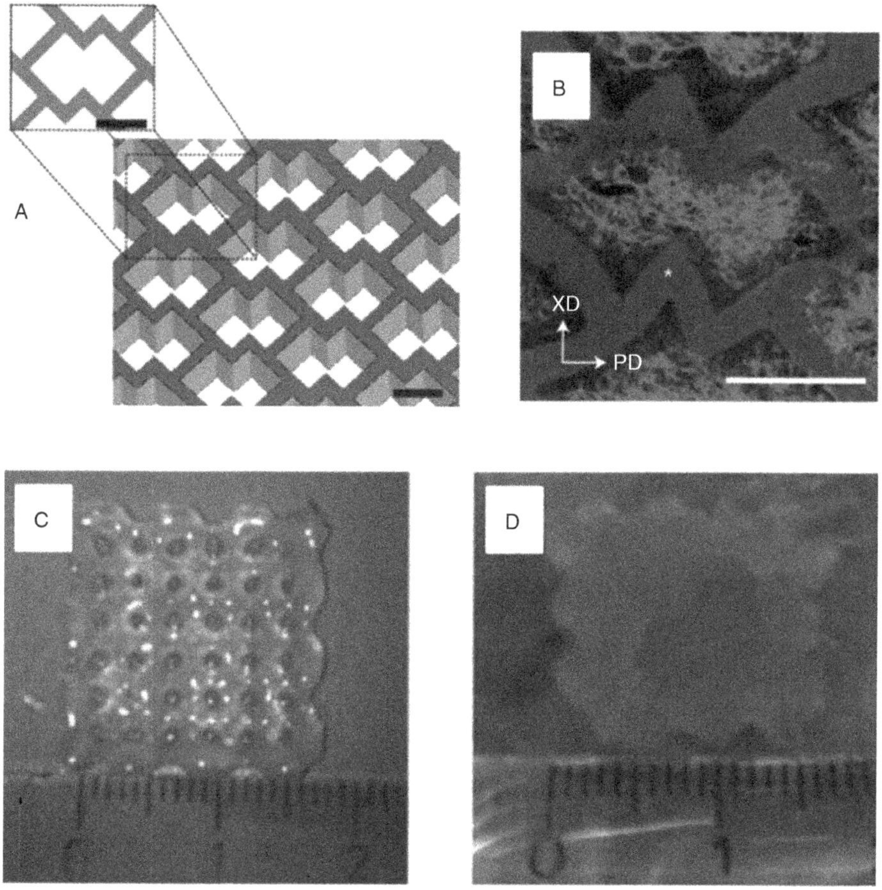

FIGURE 12.11 Microfabricated scaffolds for cardiac tissue engineering. (A) Design of PGS scaffolds laser-cut to create an accordion-like honeycomb structure. (B) The structural anisotropy of the scaffolds guided myocyte alignment, observed by immunostaining for F-actin (green) and nuclei (blue). 3-D printing can fabricate scaffolds layer by layer, with control over the microarchitecture. 2×2 cm 3-D-printed scaffolds with (C) or without (D) $\sim$1 mm pores can be seeded with cardiac progenitors. Scale bars are 200 μm in A and B, and the ruler is in centimeters in C and D. (Adapted with permission from References 18 and 107.)

surface topography, and micropatterned fibronectin lines on the cell culture substrates [102–105]. These approaches are effective at cardiac tissue engineering in 2-D, but are planar fabrication techniques and thus difficult to translate to 3-D. To address these limitations, two techniques traditionally applied with industrial plastics and metals, laser cutting, and 3-D printing, are now being used to fabricate tissue engineering scaffolds. Laser cutting is a subtractive method that uses a focused, high-powered laser to cut and ablate material. For example, accordion-like honeycombs were cut into PGS scaffolds to create an anisotropic structure that guided uniaxial CM alignment *in vitro* (Figure 12.11A,B) [18]. Furthermore, the constructs had anisotropic mechanical properties with a tensile modulus of 32 kPa parallel and 19 kPa perpendicular to the myofiber direction, comparable with the native rat myocardium (54 kPa and 20 kPa, respectively). Researchers have also used laser cutting to create blood vessel-like parallel channels in PGS sponges to increase medium perfusion and improve CM viability [106]. However, laser cutting works best for making vertical cuts on planar materials, limiting the complexity of 3-D scaffold architectures that can be engineered, and resolution is typically greater than the diameter of most ECM fibers, for example, 100 μm. Another approach that is advancing rapidly is 3-D printing, that is, solid-free form fabrication, which is an additive technique that deposits material layer by layer to create scaffolds with control of the 3-D architecture. In one of the first cardiac tissue engineering applications, alginate gels were printed as 2 × 2 cm meshes in simple square lattice pattern and seeded with CM progenitor cells that attached and maintained viability (Figure 12.11C,D) [107]. Miller *et al.* also demonstrated that 3-D printing can be used to create microvascular networks, which though achieved in the context of engineered liver tissue, could be readily adapted to create a perfusable engineered cardiac construct [108]. However, currently the resolution of 3-D printing technology using hydrogels is limited to hundreds of micrometers and improves to ~50 μm using thermoplastics such as PLA. Given the demonstrated effectiveness of nano- and microscale features in cardiac tissue engineering, improvements in 3-D printing will be required for it to become a viable scaffold fabrication technique. Fortunately, this is a rapidly developing area of research and there are already 3-D printers that can achieve micrometer scale resolution and will likely be adapted to use biomaterials suitable for cardiac tissue engineering in the near future.

12.4.5 Cell-Generated ECM Scaffolds

The scaffold fabrication techniques described so far can mimic various aspects of the cardiac ECM, but they do not replicate the actual composition and structure. To address this, researchers have tried to minimize the number of artificial cues and instead, induce cardiac cells *in vitro* to synthesize and assemble their own ECM and self-organize into a tissue. These cell-generated ECM scaffolds, a process also termed scaffold-free tissue engineering, can produce CM-dense tissues with tightly integrated ECM composed of the major protein components including collagens, fibronectin, and laminin. For example, Okano and coworkers pioneered cell sheet engineering by developing a thermoresponsive cell culture dish based on a poly(N-isopropylacrylamide) (PIPAAm) grafted surface [109]. CMs cultured on these substrates formed dense, 2-D confluent monolayers that, when released, maintained structural integrity with robust cell–cell junctions and an interconnected ECM. Anisotropic CM sheets have been engineered by modification of the grafted PIPAAm surface (Figure 12.12) and microcontact printing of fibronectin lines on the PIPAAm surface [110,111]. The transition from 2-D sheets to 3-D is achieved by stacking cell sheets to form multilayered constructs. Up to four layers have been demonstrated to couple together and beat synchronously, and when engrafted on an infarct *in vivo*, improved heart function and cell survival compared with cell injection [112,113]. However, these cell sheet scaffolds are limited to ~100 μm in thickness due to the

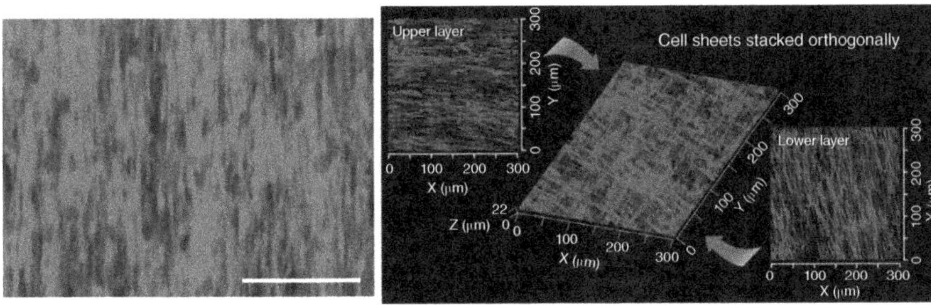

FIGURE 12.12 Cell sheet engineering is a scaffold-free approach to cardiac tissue engineering. Human fibroblasts were seeded on an anisotropic PIPAAm layer. After release, the cell sheets produced their own anisotropic ECM, revealed by observation of highly aligned collagen type I fibers (green, left). Aligned cell sheets can be stacked at different angles to create multilayer constructs (right) with F-actin (red) and nuclei (blue). Scale bar is 100 μm. (Adapted with permission from Reference 110.) *(See insert for color representation of the figure.)*

lack of microvasculature and, thus, thick 3-D scaffolds suitable for grafting in the human heart are still in development. Another cell-generated ECM scaffold approach, in this case designed to create larger 3-D scaffolds, uses suspension culture on rotating orbital shakers to generate a cell mass that assembles a 3-D ECM and dense CM tissue. Stevens *et al.* demonstrated the ability to combine CMs, endothelial cells, and fibroblasts to create a 3-D vascularized, engineered cardiac tissue that functionally integrated in the rat heart when implanted [114]. While this approach did not produce anisotropic muscle tissue, it does demonstrate the potential to use cell-generated ECM to engineer larger 3-D vascularized constructs. In total, these examples of cell-generated ECM constructs demonstrate that many aspects of the native cardiac ECM can be replicated by inducing cardiac cells to make their own ECM. As these techniques develop further, it is likely that the structural and functional properties of these constructs will be improved.

12.4.6 Decellularized ECM Scaffolds

Rather than mimicking the native cardiac ECM, recent work has focused on using the cardiac ECM itself as a tissue engineering scaffold. In 2008, Ott *et al.* demonstrated that the rat heart can be decellularized to produce an ECM scaffold that retains the structure and most of the protein composition of the native cardiac ECM and will support CM integration and muscle formation [133]. To do this, whole rat hearts were perfusion decellularized through the coronary vasculature using the detergent SDS to produce a collagen type I scaffold with lesser amounts of laminin, collagen type IV, and other components (Figure 12.13). Endothelial cells seeded into the vasculature could repopulate the capillary network, while CMs injected into the myocardium could repopulate sections of the ventricle and spontaneously contract. This has served as an important proof of concept that mimicking the cardiac ECM in detail, in this case by using the cardiac ECM itself, can enable the engineering of

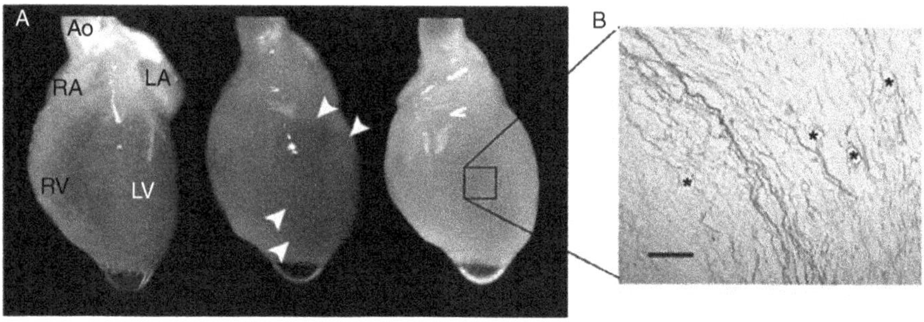

FIGURE 12.13 Decellularization of whole rat hearts. (A) Photographs of cadaveric rat hearts before, during and after perfusion of SDS detergent over 12 hours (1% SDS in deionized water, 77.4 mmHg, 20°C); Ao, aorta; LA, left atrium; LV, left ventricle; RA, right atrium; RV, right ventricle. The RV, then the atria and the LV are cleared of cellular material, rendering the heart translucent. (B) H&E staining of decellularized heart showing leftover matrix and the complete absence of cells. Scale bar is 200 μm. The technique maintains large vasculature conduits (black asterisks). H&E, hematoxylin and eosin. (Adapted with permission from Reference 133.)

functional, vascularized cardiac tissues. However, there are limitations, as the recellularized hearts generated only 2% of the contractile force of normal adult rat hearts, highlighting the challenge of repopulating whole decellularized hearts with the constituent cells. For example, the extreme density and small pore size of the ECM in the myocardium (~20 μm) inhibits cell infiltration and homogenous cell seeding throughout the construct, leading to inadequate CM density [20]. Of course, many research groups are actively working to address these challenges (see Reference 116 for an in-depth review). In an alternative approach, Godier-Furnémont *et al.* fabricated a composite graft made of decellularized human cardiac ECM sheets held together into a 3-D construct using fibrin glue seeded with human mesenchymal progenitor cells [87]. When implanted *in vivo* in a mouse MI model, this composite graft improved cardiac output, demonstrating the potential regenerative cardiac therapy. Current research is focused (a) on improving scaffolds for cardiac tissue engineering through characterization and optimization of the decellularization process to better preserve ECM structure and (b) on expanding the techniques beyond the rat heart to clinically translatable porcine and cadaveric human hearts [20,115–118].

12.5 PERSISTENT CHALLENGES

12.5.1 Cell Sources for Cardiomyocytes

Cardiac tissue engineering requires cardiac cells, and CMs are the most difficult cell type to obtain in sufficient numbers. Human CMs are terminally differentiated and cannot be expanded in culture to generate more CMs. Furthermore, human adult CMs lack the necessary plasticity to attach and integrate into a tissue engineering scaffold, and instead, will rapidly die. Thus, human CMs with developmental plasticity must be derived from some type of stem cell population. Adult cardiac stem cells and cardiac progenitor cells have been identified by the expression of the transcription factors c-KIT and Sca-1, are self-renewing, and can differentiate into CMs, smooth muscle cells, and endothelial cells [119,120]. In the infarcted area, they produce new muscles as well as new capillaries that link to the existing coronary vasculature. Furthermore, they can be consistently isolated for culture *in vitro* using the cardiosphere technique [121]. However, their role *in vivo* and their ability to reliably deliver large number of CMs for cardiac tissue engineering is still controversial, and to date has not been proven as a viable source for large numbers of new CMs [3]. Human embryonic stem cells (hESCs) can be differentiated into mature cardiac cells including CMs, cardiac fibroblasts, and endothelial cells [122,123]. However, while useful for proof of concept, hESCs are allogeneic and, thus, an engineered cardiac construct used *in vivo* would require immunosuppression, similar to a heart transplant. Furthermore, ethical concerns and the cost associated with harvesting cells from human embryos have led to the development of alternative approaches. Human induced pluripotent stem cells (hIPSCs) developed by Yamanaka and coworkers enable patient-specific adult cells to be reverted to a pluripotent state and then differentiated to nearly any cell type, including CMs [124–127]. This is a potentially unlimited source of CMs and other cardiac cells without the ethical concerns and immune rejection issues of hESCs. However, CMs derived from ESCs or hIPSCs are embryonic in phenotype (i.e., immature) and have the potential to form teratomas. Recent efforts have sought to circumvent the pluripotent stem cell state by the direct reprogramming, or transdifferentiation, of fibroblasts into CMs, but have not yet been tested for cardiac tissue engineering [128–130]. What this means is that human CMs can be differentiated from hESCs and hIPSCs in large numbers suitable for cardiac tissue engineering, however, significant work remains in optimizing the differentiation process to obtain specific CM types, standardizing the process to obtain consistent results, and ensuring cells are terminally differentiated, all of which are necessary for these cells to be clinically translatable in an engineered tissue.

12.5.2 Vascularization

Vascularization and nutrient mass transport is a persistent challenge in the field of tissue engineering as a whole and particularly important in cardiac tissue engineering due to the high metabolic activity of muscle tissue. To address this issue, researchers have developed multiple approaches, such as bio-inspired materials and scaffolds that deliver VEGF to increase vasculogenesis when implanted *in vivo* [99,100,131,132]. Additionally, CMs and endothelial cells can be cocultured in cell sheets, synthetic scaffolds, hydrogels, and acellular ECM scaffolds to promote neovessel formation [21,81,133–137]. Scaffolds can also be engineered with channels to promote capillary network formation or pores for vascular invasion from the host when implanted [61,138]. 3-D printing and soft lithography have also been used to engineer 3-D vascular networks [108,139,140]. All of these methods are useful tools to investigate the mechanisms behind the formation and function of microvascular networks. However, these approaches have limitations, and specifically for cardiac tissue engineering, cannot generate the high-density, parallel capillary networks found in the myocardium. New advances in the engineering of microvascular networks are needed to engineer dense, anisotropic cardiac tissues with embedded microvascular networks that can provide sufficient nutrient supply.

12.6 THE FUTURE OF CARDIAC TISSUE ENGINEERING

In this chapter, we have described the structure and function of the heart and a range of techniques to mimic the native cardiac ECM for the design and fabrication of cardiac tissue engineering scaffolds. Mimicking the cardiac ECM, whether composition, structure, or mechanical properties, provides clear benefits and results in engineered cardiac tissue, with improved electromechanical function. While we have organized these various fabrication approaches into distinct sections, it should be realized that many researchers are combining multiple techniques in order to engineer next-generation scaffolds with better performance. As the field evolves, we anticipate that new techniques will enable improved engineering of the cellular environment from the nanometer to macroscales in order to instruct cardiac cells to form cardiac tissue. For example, design inspiration may be drawn from cardiac morphogenesis during embryonic development, which in humans is the only period when there is significant CM proliferation and muscle formation. Particularly, the ECM cues that guide this process could be integrated into cardiac tissue engineering scaffolds to promote regeneration in adults. Ultimately, as cardiac tissue engineering research evolves and matures, recovering from heart failure should no longer require a transplant. Instead, treating an MI or heart failure should be possible by implanting an engineered cardiac graft to replace damaged tissue or to serve as a scaffold to enhance endogenous cardiac repair.

REFERENCES

[1] Go, A.S., Mozaffarian, D., Roger, V.L., Benjamin, E.J., Berry, J.D., Borden, W.B., Bravata, D.M., Dai, S., Ford, E.S., Fox, C.S., Franco, S., Fullerton, H.J., Gillespie, C., Hailpern, S.M., Heit, J.A., Howard, V.J., Huffman, M.D., Kissela, B.M., Kittner, S.J., Lackland, D.T., Lichtman, J.H., Lisabeth, L.D., Magid, D., Marcus, G.M., Marelli, A., Matchar, D.B., McGuire, D.K., Mohler, E.R., Moy, C.S., Mussolino, M.E., Nichol, G., Paynter, N.P., Schreiner, P.J., Sorlie, P.D., Stein, J., Turan, T.N., Virani, S.S., Wong, N.D., Woo, D., Turner, M.B. *Circulation* **2013**, *127*, e6–e245.

[2] Sedmera, D., Thompson, R.P. *Dev. Dyn.* **2011**, *240*, 1322–1334.

[3] Laflamme, M.A., Murry, C.E. *Nature* **2011**, *473*, 326–335.

[4] Bergmann, O., Bhardwaj, R.D., Bernard, S., Zdunek, S., Barnabé-Heider, F., Walsh, S., Zupicich, J., Alkass, K., Buchholz, B.A., Druid, H., Jovinge, S., Frisén, J. *Science* **2009**, *324*, 98–102.

[5] Senyo, S.E., Steinhauser, M.L., Pizzimenti, C.L., Yang, V.K., Cai, L., Wang, M., Wu, T.-D., Guerquin-Kern, J.-L., Lechene, C.P., Lee, R.T. *Nature* **2013**, *493*, 433–436.

[6] Jahanmir, S., Hunsberger, A.Z., Heshmat, H., Tomaszewski, M.J., Walton, J.F., Weiss, W.J., Lukic, B., Pae, W.E., Zapanta, C.M., Khalapyan, T.Z. *Artif. Organs* **2008**, *32*, 366–375.
[7] Joyce, L.D., DeVries, W.C., Hastings, W.L., Olsen, D.B., Jarvik, R.K., Kolff, W.J. *Trans. Am. Soc. Artif. Intern. Organs* **1983**, *29*, 81–87.
[8] Gray, N.A., Selzman, C.H. *Am. Heart J.* **2006**, *152*, 4–10.
[9] Davani, S., Deschaseaux, F., Chalmers, D., Tiberghien, P., Kantelip, J.-P. *Cardiovasc. Res.* **2005**, *65*, 305–316.
[10] Passier, R., Van Laake, L.W., Mummery, C.L. *Nature* **2008**, *453*, 322–329.
[11] Segers, V.F.M., Lee, R.T. *Nature* **2008**, *451*, 937–942.
[12] Forrester, J.S., Price, M.J., Makkar, R.R. *Circulation* **2003**, *108*, 1139–1145.
[13] Clifford, D.M., Fisher, S.A., Brunskill, S.J., Doree, C., Mathur, A., Watt, S., Martin-Rendon, E. *Cochrane Database Syst. Rev.* **2012**, (2), CD006536.
[14] Menasché, P. *Circulation* **2009**, *119*, 2735–2740.
[15] Gerdes, A.M., Capasso, J.M. *J. Mol. Cell. Cardiol.* **1995**, *27*, 849–856.
[16] Bershitsky, S.Y., Tsaturyan, A.K. *Biophys. J.* **1995**, *69*, 1011–1021.
[17] Janssen, P.M., Lehnart, S.E., Prestle, J., Hasenfuss, G. *J. Mol. Cell. Cardiol.* **1999**, *31*, 1419–1427.
[18] Engelmayr, G.C., Cheng, M., Bettinger, C.J., Borenstein, J.T., Langer, R., Freed, L.E. *Nat. Mater.* **2008**, *7*, 1003–1010.
[19] Chen, Q.-Z., Harding, S.E., Ali, N.N., Lyon, A.R., Boccaccini, A.R. *Mater. Sci. Eng. R Rep.* **2008**, *59*, 1–37.
[20] Wang, B., Tedder, M.E., Perez, C.E., Wang, G., de Jongh Curry, A.L., To, F., Elder, S.H., Williams, L.N., Simionescu, D.T., Liao, J. *J. Mater. Sci. Mater. Med.* **2012**, *23*, 1835–1847.
[21] Wang, B., Borazjani, A., Tahai, M., Curry, A.L.D.J., Simionescu, D.T., Guan, J., To, F., Elder, S.H., Liao, J. *J. Biomed. Mater. Res. A* **2010**, *94*, 1100–1110.
[22] Sarig, U., Au-yeung, G.C.T., Wang, Y., Bronshtein, T., Dahan, N., Boey, F.Y.C., Venkatraman, S.S., Machluf, M. *Tissue Eng. A* **2012**, *18*, 2125–2137.
[23] Fomovsky, G.M., Holmes, J.W. *Am. J. Physiol. Heart Circ. Physiol.* **2010**, *298*, H221–H228.
[24] Engler, A.J., Carag-Krieger, C., Johnson, C.P., Raab, M., Tang, H.-Y., Speicher, D.W., Sanger, J.W., Sanger, J.M., Discher, D.E. *J. Cell Sci.* **2008**, *121*, 3794–3802.
[25] Armour, J.A. *Am. J. Physiol. Regul. Integr. Comp. Physiol.* **2004**, *287*, R262–R271.
[26] Pennisi, D.J., Rentschler, S., Gourdie, R.G., Fishman, G.I., Mikawa, T. *Int. J. Dev. Biol.* **2002**, *46*, 765–775.
[27] Moorman, A.F.M., de Jong, F., Denyn, M.M.F.J., Lamers, W.H. *Circ. Res.* **1998**, *82*, 629–644.
[28] Young, R.J., Panfilov, A.V. *Proc. Natl. Acad. Sci. U.S.A.* **2010**, *107*, 15063–15068.
[29] Nogami, A. *Pacing Clin. Electrophysiol.* **2011**, *34*, 624–650.
[30] Kimura, K., Ieda, M., Fukuda, K. *Circ. Res.* **2012**, *110*, 325–336.
[31] Stoker, M.E., Gerdes, A.M., May, J.F. *Anat. Rec.* **1982**, *202*, 187–191.
[32] Kaneko, N., Matsuda, R., Toda, M., Shimamoto, K. *Am. J. Physiol. Heart Circ. Physiol.* **2011**, *300*, H754–H761.
[33] Banerjee, I., Fuseler, J.W., Price, R.L., Borg, T.K., Baudino, T.A. *Am. J. Physiol. Heart Circ. Physiol.* **2007**, *293*, H1883–H1891.
[34] Mandarim-De-Lacerda, C.A., Meirelles Pereira, L.M. *Pathobiology* **2008**, *68*, 36–42.
[35] Brutsaert, D.L. *Physiol. Rev.* **2003**, *83*, 59–115.
[36] Ramaciotti, C., Sharkey, A., McClellan, G., Winegrad, S. *Proc. Natl. Acad. Sci. U.S.A.* **1992**, *89*, 4033–4036.

[37] Narmoneva, D.A., Vukmirovic, R., Davis, M.E., Kamm, R.D., Lee, R.T. *Circulation* **2004**, *110*, 962–968.
[38] Vliegen, H.W., Van der Laarse, A., Cornelisse, C.J., Eulderink, F. *Eur. Heart J.* **1991**, *12*, 488–494.
[39] Spinale, F.G. *Physiol. Rev.* **2007**, *87*, 1285–1342.
[40] Farhadian, F., Contard, F., Sabri, A., Samuel, J.L., Rappaport, L. *Cardiovasc. Res.* **1996**, *32*, 433–442.
[41] Lipke, D.W., McCarthy, K.J., Elton, T.S., Arcot, S.S., Oparil, S., Couchman, J.R. *Hypertension* **1993**, *22*, 743–753.
[42] Gazoti Debessa, C.R., Mesiano Maifrino, L.B., Rodrigues de Souza, R. *Mech. Ageing Dev.* **2001**, *122*, 1049–1058.
[43] Eghbali, M., Weber, K.T. *Mol. Cell. Biochem.* **1990**, *96*, 1–14.
[44] Weber, K.T. *J. Am. Coll. Cardiol.* **1989**, *13*, 1637–1652.
[45] Pope, A.J., Sands, G.B., Smaill, B.H., LeGrice, I.J. *Am. J. Physiol. Heart Circ. Physiol.* **2008**, *295*, H1243–H1252.
[46] Samarel, A.M. *Am. J. Physiol. Heart Circ. Physiol.* **2005**, *289*, H2291–H2301.
[47] Gerson, C.J., Elkins, R.C., Goldstein, S., Heacox, A.E. *Cryobiology* **2012**, *64*, 33–42.
[48] Mao, Y., Schwarzbauer, J.E. *Matrix Biol.* **2005**, *24*, 389–399.
[49] Singh, P., Carraher, C., Schwarzbauer, J.E. *Annu. Rev. Cell Dev. Biol.* **2010**, *26*, 397–419.
[50] Ertl, G., Frantz, S. *Cardiovasc. Res.* **2005**, *66*, 22–32.
[51] Liehn, E.A., Postea, O., Curaj, A., Marx, N. *J. Am. Coll. Cardiol.* **2011**, *58*, 2357–2362.
[52] Yang, J.T., Rayburn, H., Hynes, R.O. *Development* **1993**, *119*, 1093–1105.
[53] Schenke-Layland, K., Nsair, A., Van Handel, B., Angelis, E., Gluck, J.M., Votteler, M., Goldhaber, J.I., Mikkola, H.K., Kahn, M., Maclellan, W.R. *Biomaterials* **2011**, *32*, 2748–2756.
[54] Jugdutt, B.I. *Circulation* **2003**, *108*, 1395–1403.
[55] Spotnitz, H.M. *J. Thorac. Cardiovasc. Surg.* **2000**, *119*, 1053–1077.
[56] Buckberg, G., Hoffman, J.I.E., Mahajan, A., Saleh, S., Coghlan, C. *Circulation* **2008**, *118*, 2571–2587.
[57] Schwartzman, D., Chang, I., Michele, J.J., Mirotznik, M.S., Foster, K.R. *J. Interv. Card. Electrophysiol.* **1999**, *3*, 213–224.
[58] Stella, J.A., Amore, A.D., Wagner, W.R., Sacks, M.S. *Acta Biomater.* **2010**, *6*, 2365–2381.
[59] Radisic, M., Park, H., Martens, T.P., Salazar-Lazaro, J.E., Geng, W., Wang, Y., Langer, R., Freed, L.E., Vunjak-Novakovic, G. *J. Biomed. Mater. Res. A* **2008**, *86*, 713–724.
[60] Radisic, M., Park, H., Chen, F., Salazar-Lazzaro, J.E., Wang, Y., Dennis, R., Langer, R., Freed, L.E., Vunjak-Novakovic, G. *Tissue Eng.* **2006**, *12*, 2077–2091.
[61] Madden, L.R., Mortisen, D.J., Sussman, E.M., Dupras, S.K., Fugate, J.A., Cuy, J.L., Hauch, K.D., Laflamme, M.A., Murry, C.E., Ratner, B.D. *Proc. Natl. Acad. Sci. U.S.A.* **2010**, *107*, 15211–15216.
[62] Shapiro, L., Cohen, S. *Biomaterials* **1997**, *18*, 583–590.
[63] Sapir, Y., Kryukov, O., Cohen, S. *Biomaterials* **2011**, *32*, 1838–1847.
[64] Gonnerman, E.A., Kelkhoff, D.O., McGregor, L.M., Harley, B.A.C. *Biomaterials* **2012**, *33*, 8812–8821.
[65] Kumbar, S.G., James, R., Nukavarapu, S.P., Laurencin, C.T. *Biomed. Mater.* **2008**, *3*, 034002.
[66] Teo, W.E., Ramakrishna, S. *Nanotechnology* **2006**, *17*, R89–R106.

[67] Kai, D., Prabhakaran, M.P., Jin, G., Ramakrishna, S. *J. Biomed. Mater. Res. B Appl. Biomater.* **2011**, *98B*, 379–386.

[68] Ricotti, L., Polini, A., Genchi, G.G., Ciofani, G., Iandolo, D., Mattoli, V., Menciassi, A., Dario, P., Pisignano, D. *Conf. Proc. IEEE Eng. Med. Biol. Soc.* **2011**, *2011*, 3597–3600.

[69] Amoroso, N.J., D'Amore, A, Hong, Y., Rivera, C.P., Sacks, M.S., Wagner, W.R. *Acta Biomater.* **2012**, *8*, 4268–4277.

[70] Shin, M., Ishii, O., Sueda, T., Vacanti, J.P. *Biomaterials* **2004**, *25*, 3717–3723.

[71] Zong, X., Bien, H., Chung, C.-Y., Yin, L., Fang, D., Hsiao, B.S., Chu, B., Entcheva, E. *Biomaterials* **2005**, *26*, 5330–5338.

[72] Li, M., Guo, Y., Wei, Y., MacDiarmid, A.G., Lelkes, P.I. *Biomaterials* **2006**, *27*, 2705–2715.

[73] Fernandes, E., Zucolotto, V., De Queiroz, A. *J. Macromol. Sci. Pure Appl. Chem.* **2010**, *47*, 1203–1207.

[74] Sreerekha, P.R., Menon, D., Nair, S.V., Chennazhi, K.P. *Tissue Eng. A.* **2013**, *19*, 849–859.

[75] Hosseinkhani, H., Hosseinkhani, M., Hattori, S., Matsuoka, R., Kawaguchi, N. *J. Biomed. Mater. Res. A* **2010**, *94*, 1–8.

[76] Mellado, P., McIlwee, H.A., Badrossamay, M.R., Goss, J.A., Mahadevan, L., Kit Parker, K. *Appl. Phys. Lett.* **2011**, *99*, 203107.

[77] Badrossamay, M.R., McIlwee, H.A., Goss, J.A., Parker, K.K. *Nano Lett.* **2010**, *10*, 2257–2261.

[78] Feinberg, A.W., Parker, K.K. *Nano Lett.* **2010**, *10*, 2184–2191.

[79] Li, Z., Guan, J. *Polymers* **2011**, *3*, 740–761.

[80] Yost, M.J., Baicu, C.F., Stonerock, C.E., Goodwin, R.L., Price, R.L., Davis, J.M., Evans, H., Watson, P.D., Gore, C.M., Sweet, J., Creech, L., Zile, M.R., Terracio, L. *Tissue Eng.* **2004**, *10*, 273–284.

[81] Leung, B.M., Sefton, M.V. *Tissue Eng. A.* **2010**, *16*, 3207–3218.

[82] Sondergaard, C.S., Mathews, G., Wang, L., Jeffreys, A., Sahota, A., Wood, M., Ripplinger, C.M., Si, M.-S. *Ann. Thorac. Surg.* **2012**, *94*, 1241–1249.

[83] Dengler, J., Song, H., Thavandiran, N., Massé, S., Wood, G.A., Nanthakumar, K., Zandstra, P.W., Radisic, M. *Biotechnol. Bioeng.* **2011**, *108*, 704–719.

[84] Freytes, D.O., Martin, J., Velankar, S.S., Lee, A.S., Badylak, S.F. *Biomaterials* **2008**, *29*, 1630–1637.

[85] Duan, Y., Liu, Z., O'Neill, J., Wan, L.Q., Freytes, D.O., Vunjak-Novakovic, G. *J. Cardiovasc. Transl. Res.* **2011**, *4*, 605–615.

[86] Cigliano, A., Gandaglia, A., Lepedda, A.J., Zinellu, E., Naso, F., Gastaldello, A., Aguiari, P., De Muro, P., Gerosa, G., Spina, M., Formato, M. *Biochem. Res. Int.* **2012**, *2012*, 979351.

[87] Godier-Furnémont, A.F.G., Martens, T.P., Koeckert, M.S., Wan, L., Parks, J., Arai, K., Zhang, G., Hudson, B., Homma, S., Vunjak-Novakovic, G. *Proc. Natl. Acad. Sci. U.S.A.* **2011**, *108*, 7974–7979.

[88] Johnson, T.D., Lin, S.Y., Christman, K.L. *Nanotechnology* **2011**, *22*, 494015.

[89] Singelyn, J.M., DeQuach, J.A., Seif-Naraghi, S.B., Littlefield, R.B., Schup-Magoffin, P.J., Christman, K.L. *Biomaterials* **2009**, *30*, 5409–5416.

[90] French, K.M., Boopathy, A.V., Dequach, J.A., Chingozha, L., Lu, H., Christman, K.L., Davis, M.E. *Acta Biomater.* **2012**, *8*, 4357–4364.

[91] Seif-Naraghi, S.B., Horn, D., Schup-Magoffin, P.J., Christman, K.L. *Acta Biomater.* **2012**, *8*, 3695–3703.

[92] Zimmermann, W.-H. *Circ. Res.* **2001**, *90*, 223–230.

[93] Schaaf, S., Shibamiya, A., Mewe, M., Eder, A., Stöhr, A., Hirt, M.N., Rau, T., Zimmermann, W.-H., Conradi, L., Eschenhagen, T., Hansen, A. *PLoS ONE* **2011**, *6*, e26397.
[94] Zhang, D., Shadrin, I.Y., Lam, J., Xian, H.-Q., Snodgrass, H.R., Bursac, N. *Biomaterials* **2013**, *34*, 5813–5820.
[95] Baar, K., Birla, R., Boluyt, M.O., Borschel, G.H., Arruda, E.M., Dennis, R.G. *FASEB J.* **2005**, *19*, 275–277.
[96] Jongpaiboonkit, L., King, W.J., Lyons, G.E., Paguirigan, A.L., Warrick, J.W., Beebe, D.J., Murphy, W.L. *Biomaterials* **2008**, *29*, 3346–3356.
[97] Kraehenbuehl, T.P., Ferreira, L.S., Hayward, A.M., Nahrendorf, M., van der Vlies, A.J., Vasile, E., Weissleder, R., Langer, R., Hubbell, J.A. *Biomaterials* **2011**, *32*, 1102–1109.
[98] Kraehenbuehl, T.P., Zammaretti, P., Van der Vlies, A.J., Schoenmakers, R.G., Lutolf, M.P., Jaconi, M.E., Hubbell, J.A. *Biomaterials* **2008**, *29*, 2757–2766.
[99] Guo, H., Cui, G., Yang, J., Wang, C., Zhu, J., Zhang, L., Jiang, J., Shao, S. *Biochem. Biophys. Res. Commun.* **2012**, *424*, 105–111.
[100] Webber, M.J., Tongers, J., Newcomb, C.J., Marquardt, K.-T., Bauersachs, J., Losordo, D.W., Stupp, S.I. *Proc. Natl. Acad. Sci. U.S.A.* **2011**, *108*, 13438–13443.
[101] Hsieh, P.C.H., Davis, M.E., Gannon, J., MacGillivray, C., Lee, R.T. *J. Clin. Invest.* **2006**, *116*, 237.
[102] Chung, C., Bien, H., Sobie, E.A., Dasari, V., McKinnon, D., Rosati, B., Entcheva, E. *FASEB J.* **2011**, *25*, 851–862.
[103] Kim, D.-H., Lipke, E.A., Kim, P., Cheong, R., Thompson, S., Delannoy, M., Suh, K.-Y., Tung, L., Levchenko, A. *Proc. Natl. Acad. Sci. U.S.A.* **2010**, *107*, 565–570.
[104] Feinberg, A.W., Feigel, A., Shevkoplyas, S.S., Sheehy, S., Whitesides, G.M., Parker, K.K. *Science* **2007**, *317*, 1366–1370.
[105] Feinberg, A.W., Alford, P.W., Jin, H., Ripplinger, C.M., Werdich, A.A., Sheehy, S.P., Grosberg, A., Parker, K.K. *Biomaterials* **2012**, *33*, 5732–5741.
[106] Radisic, M., Deen, W., Langer, R., Vunjak-Novakovic, G. *Am. J. Physiol. Heart Circ. Physiol.* **2005**, *288*, H1278–H1289.
[107] Gaetani, R., Doevendans, P.A., Metz, C.H.G., Alblas, J., Messina, E., Giacomello, A., Sluijter, J.P.G. *Biomaterials* **2012**, *33*, 1782–1790.
[108] Miller, J.S., Stevens, K.R., Yang, M.T., Baker, B.M., Nguyen, D.-H.T., Cohen, D.M., Toro, E., Chen, A.A., Galie, P.A., Yu, X., Chaturvedi, R., Bhatia, S.N., Chen, C.S. *Nat. Mater.* **2012**, *11*, 768–774.
[109] Shimizu, T., Yamato, M., Kikuchi, A., Okano, T. *Tissue Eng.* **2001**, *7*, 141–151.
[110] Takahashi, H., Nakayama, M., Shimizu, T., Yamato, M., Okano, T. *Biomaterials* **2011**, *32*, 8830–8838.
[111] Williams, C., Xie, A.W., Yamato, M., Okano, T., Wong, J.Y. *Biomaterials* **2011**, *32*, 5625–5632.
[112] Sekine, H., Shimizu, T., Dobashi, I., Matsuura, K., Hagiwara, N., Takahashi, M., Kobayashi, E., Yamato, M., Okano, T. *Tissue Eng. A.* **2011**, *17*, 2973–2980.
[113] Shimizu, T., Okano, T. *Biomaterials* **2003**, *24*, 2309–2316.
[114] Stevens, K.R., Kreutziger, K.L., Dupras, S.K., Korte, F.S., Regnier, M., Muskheli, V., Nourse, M.B., Bendixen, K., Reinecke, H., Murry, C.E. *Proc. Natl. Acad. Sci. U.S.A.* **2009**, *106*, 16568–16573.
[115] Weymann, A., Loganathan, S., Takahashi, H., Schies, C., Claus, B., Hirschberg, K., Soós, P., Korkmaz, S., Schmack, B., Karck, M., Szabó, G. *Circ. J.* **2011**, *75*, 852–860.
[116] Badylak, S.F., Taylor, D., Uygun, K. *Annu. Rev. Biomed. Eng.* **2011**, *13*, 27–53.
[117] Akhyari, P., Aubin, H., Gwanmesia, P., Barth, M., Hoffmann, S., Huelsmann, J., Preuss, K., Lichtenberg, A. *Tissue Eng. C Methods* **2011**, *17*, 915–926.

[118] Crapo, P.M., Gilbert, T.W., Badylak, S.F. *Biomaterials* **2011**, *32*, 3233–3243.
[119] Beltrami, A.P., Barlucchi, L., Torella, D., Baker, M., Limana, F., Chimenti, S., Kasahara, H., Rota, M., Musso, E., Urbanek, K., Leri, A., Kajstura, J., Nadal-Ginard, B., Anversa, P. *Cell* **2003**, *114*, 763–776.
[120] Oh, H., Bradfute, S.B., Gallardo, T.D., Nakamura, T., Gaussin, V., Mishina, Y., Pocius, J., Michael, L.H., Behringer, R.R., Garry, D.J., Entman, M.L., Schneider, M.D. *Proc. Natl. Acad. Sci. U.S.A.* **2003**, *100*, 12313–12318.
[121] Davis, D.R., Zhang, Y., Smith, R.R., Cheng, K., Terrovitis, J., Malliaras, K., Li, T.-S., White, A., Makkar, R., Marbán, E. *PLoS ONE* **2009**, *4*, e7195.
[122] Chien, K.R., Domian, I.J., Parker, K.K. *Science* **2008**, *322*, 1494–1497.
[123] Caspi, O., Lesman, A., Basevitch, Y., Gepstein, A., Arbel, G., Habib, I.H.M., Gepstein, L., Levenberg, S. *Circ. Res.* **2007**, *100*, 263–272.
[124] Takahashi, K., Yamanaka, S. *Cell* **2006**, *126*, 663–676.
[125] Takahashi, K., Tanabe, K., Ohnuki, M., Narita, M., Ichisaka, T., Tomoda, K., Yamanaka, S. *Cell* **2007**, *131*, 861–872.
[126] Gai, H., Leung, E.L.-H., Costantino, P.D., Aguila, J.R., Nguyen, D.M., Fink, L.M., Ward, D.C., Ma, Y. *Cell Biol. Int.* **2009**, *33*, 1184–1193.
[127] Zhang, J., Wilson, G.F., Soerens, A.G., Koonce, C.H., Yu, J., Palecek, S.P., Thomson, J.A., Kamp, T.J. *Circ. Res.* **2009**, *104*, e30–e41.
[128] Qian, L., Huang, Y., Spencer, C.I., Foley, A., Vedantham, V., Liu, L., Conway, S.J., Fu, J., Srivastava, D. *Nature* **2012**, *485*, 593–598.
[129] Song, K., Nam, Y.-J., Luo, X., Qi, X., Tan, W., Huang, G.N., Acharya, A., Smith, C.L., Tallquist, M.D., Neilson, E.G., Hill, J.A., Bassel-Duby, R., Olson, E.N. *Nature* **2012**, *485*, 599–604.
[130] Teunissen, B.E.J., Smeets, P.J.H., Willemsen, P.H.M., De Windt, L.J., Van der Vusse, G.J., Van Bilsen, M. *Cardiovasc. Res.* **2007**, *75*, 519–529.
[131] Lin, Y.-D., Luo, C.-Y., Hu, Y.-N., Yeh, M.-L., Hsueh, Y.-C., Chang, M.-Y., Tsai, D.-C., Wang, J.-N., Tang, M.-J., Wei, E.I.H., Springer, M.L., Hsieh, P.C.H. *Sci. Transl. Med.* **2012**, *4*, 146ra109.
[132] Ruhrberg, C., Gerhardt, H., Golding, M., Watson, R., Ioannidou, S., Fujisawa, H., Betsholtz, C., Shima, D.T. *Genes Dev.* **2002**, *16*, 2684–2698.
[133] Ott, H.C., Matthiesen, T.S., Goh, S.-K., Black, L.D., Kren, S.M., Netoff, T.I., Taylor, D.A. *Nat. Med.* **2008**, *14*, 213–221.
[134] Bel, A., Planat-Bernard, V., Saito, A., Bonnevie, L., Bellamy, V., Sabbah, L., Bellabas, L., Brinon, B., Vanneaux, V., Pradeau, P., Peyrard, S., Larghero, J., Pouly, J., Binder, P., Garcia, S., Shimizu, T., Sawa, Y., Okano, T., Bruneval, P., Desnos, M., Hagège, A.A., Casteilla, L., Pucéat, M., Menasché, P. *Circulation* **2010**, *122*, S118–S123.
[135] Sekine, H., Shimizu, T., Hobo, K., Sekiya, S., Yang, J., Yamato, M., Kurosawa, H., Kobayashi, E., Okano, T. *Circulation* **2008**, *118*, S145–S152.
[136] Masumoto, H., Matsuo, T., Yamamizu, K., Uosaki, H., Narazaki, G., Katayama, S., Marui, A., Shimizu, T., Ikeda, T., Okano, T., Sakata, R., Yamashita, J.K. *Stem Cells* **2012**, *30*, 1196–1205.
[137] Mo, X., Xu, C., Kotaki, M., Ramakrishna, S. *Biomaterials* **2004**, *25*, 1883–1890.
[138] Maidhof, R., Marsano, A., Lee, E.J., Vunjak-Novakovic, G. *Biotechnol. Prog.* **2010**, *26*, 565–572.
[139] Wu, W., DeConinck, A., Lewis, J.A. *Adv. Mater.* **2011**, *23*, H178–H183.
[140] Zheng, Y., Chen, J., Craven, M., Choi, N.W., Totorica, S., Diaz-Santana, A., Kermani, P., Hempstead, B., Fischbach-Teschl, C., López, J.A., Stroock, A.D. *Proc. Natl. Acad. Sci. U.S.A.* **2012**, *109*, 9342–9347.
[141] Savadjiev, P., Strijkers, G.J., Bakermans, A.J., Piuze, E., Zucker, S.W., Siddiqi, K. *Proc. Natl. Acad. Sci. U.S.A.* **2012**, *109*, 9248–9253.

[142] Lee, J., Beighley, P., Ritman, E., Smith, N. *Med. Image Anal.* **2007**, *11*, 630–647.
[143] O'Brien, F., Harley, B., Yannas, I., Gibson, L. *Biomaterials* **2004**, *25*, 1077e86.
[144] Soliman, S., Pagliari, S., Rinaldi, A., Forte, G., Fiaccavento, R., Pagliari, F., Franzese, O., Minieri, M., Di Nardo, P., Licoccia, S., Traversa, E. *Acta Biomater.* **2010**, *6*, 1227–1237.

CHAPTER 13

STRATEGIES AND CHALLENGES FOR BIO-INSPIRED CARDIOVASCULAR BIOMATERIALS

ELAINE L. LEE

Department of Biomedical Engineering, Boston University, Boston, MA, USA

JOYCE Y. WONG

Department of Biomedical Engineering, Boston University, Boston, MA, USA; Department of Materials Science and Engineering, Boston University, Boston, MA, USA

13.1 NEED FOR CARDIOVASCULAR BIOMATERIALS

13.1.1 Cardiovascular Tissue Regeneration

Harnessing the cell's capacity to renew, revive, or restore is a hallmark of regenerative medicine. Unfortunately, the human heart and blood vessels have little true regenerative capacity, especially when injured. Despite many improvements to biomaterials and major advances in stem cell biology, nature's sophisticated hierarchical organization and adaptability is still the benchmark achievement that modern biomedical engineering technologies seek to replicate. Bio-inspired cardiovascular tissue engineering strives to apply structure–function relationships in the development of biological substitutes to restore, maintain, or improve tissue function [1]. We review here the current techniques for bio-inspired cardiovascular materials and the milestones we as a field have yet to reach.

13.1.2 Unmet Clinical Need: Incidence and Prevalence of Cardiovascular Disease

In its 2013 update, the American Heart Association estimated that 83.6 million American adults (more than one in three) have one or more types of cardiovascular disease (CVD), which encompasses high blood pressure and coronary heart disease (myocardial infarction [MI], angina pectoris, heart failure, stroke, and congenital heart defects) [2]. A global health problem, CVD impacts all races, genders, ages,

Bio-inspired Materials for Biomedical Engineering, First Edition. Edited by Anthony B. Brennan and Chelsea M. Kirschner.

and socioeconomic groups, with an average of one death every 40 seconds in the United States alone. Atherosclerosis—the underlying systemic disease in which inflammation results in fatty plaques and scar tissue build up along artery walls—leads to many coronary events (e.g., MI, stroke); an estimated 397,000 coronary artery bypass graft (CABG) procedures were performed in 2010 alone.

In addition to CVD that can be managed with preventive measures, between 4 and 10 per 1000 live births in the United States will also present with congenital heart defects (i.e., structural problems resulting from abnormal formation of the heart or major blood vessels) [3,4]. Although some minor defects may resolve spontaneously, an estimated 800,000 American adults were thought to be living with a congenital heart defect in 2000, some whose disease went undetected during childhood [5]. Pediatric patients with major malformations are usually treated surgically, but tissue availability is limited and the current clinical options do not grow with the child. Health expenditures for CVD were estimated at $312.6 billion for 2009 [2]; by comparison, the estimated cost of all cancer and benign neoplasms was $228 billion in 2008.

13.1.3 Need for Tissue-Engineered Solutions

MI can lead to myocardial necrosis or loss of viable myocardium that is thinned or fails to contract [6]. The infarction can take weeks to months to heal, but loss of terminally differentiated cardiomyocytes (often 1 billion or more) [7,8], the remodeling process, and resulting scar tissue may limit function [7]. Moreover, the environment of the stressed heart may lead to maladaptive remodeling that can lead to arrhythmogenesis, heart dilatation, wall thinning, and decreased cardiac output [9–12]. Although heart transplantation remains the most viable option for end-stage heart failure patients, donor hearts remain limited (3.6 per 1000 deaths in 2010) [13]. Ventricular assist devices and other mechanical circulatory support have a projected duration of 5 years, but still have associated thrombogenic issues. Although allografts avoid thromboembolisms, immunogenicity has led to degenerative changes (e.g., calcification) and eventual graft loss within 8–10 years [14]. Because autologous cell sources avoid rejection issues and the need for immunosuppressive medication [15,16], autologous myoblasts and other multipotent adult stem cells have been implanted clinically, but arrhythmias and exacerbating inflammation remain long-term concerns [17]. Although direct cell injection of embryonic stem cells (ESCs) has shown some improvement to cardiac function, approximately 90% of cells are lost in circulation [18], and impure cell populations still carry risk of teratoma formation [8,19]. Thus, tissue-engineered myocardial patches aim for (a) targeted delivery of pure cell populations, (b) integration with three-dimensional (3-D) structure and mechanical compliance with fatigue resistance, (c) synchronous electrical integration of viable cells, and (d) rapid vascularization to support cell survival.

For vascular grafts, autologous vessels are most desirable as blood vessel substitutes in coronary bypass procedures and other small-caliber blood vessel replacements (<6 mm). However, because of trauma, disease, or prior surgeries, suitable autologous vessels (e.g., saphenous vein or internal mammary artery) are unavailable in >10% of patients [20,21]. Success with cryopreserved cadaveric grafts, umbilical vein grafts, and arterial allografts has also been limited [22–26]. Although effective for large vessel replacements with high flow and low resistance, synthetic materials (e.g., Dacron, expanded polytetrafluoroethylene [ePTFE]) display suboptimal behavior and have high risk of thrombosis in smaller vessel replacements [27–29]. Additionally, vascular grafts have had problems with intimal hyperplasia, accelerated atherosclerosis, and compliance and diameter mismatches at the anastomoses [30–34]. Ideal tissue-engineered vascular graft substitutes require (a) proper mechanical properties, such as burst pressure and viscoelastic properties [35]; (b) contractile response or vascoactivity [36,37]; (c) nonthrombogenicity [37]; and (d) responsiveness to postimplantation remodeling by the host tissue [38].

In addition to the clinical need for cardiovascular replacements, tissue engineering offers strategic advantages as model systems to investigate specific biologic questions [39]. The ability to manipulate composition, geometry, and microstructure in a controlled environment that can be monitored makes tissue-engineered structures (e.g., organ-on-a-chip) more ideal than biologic explants in the study of cell–matrix interactions, cell and matrix mechanics, and mechanobiology. Furthermore, in combination with the technological advances in computational modeling, more complex questions can be addressed in understanding the pathology of CVD or in tissue model development for drug screening, to which we refer to other more in-depth reviews [40–44]. Computational modeling that predicts and validates experimental results could shed new light on structure–property–performance relationships. Likewise, we refer to other reviews for *in vivo* experimental models that can be used to assess tissue-engineered cardiovascular constructs [45–49].

13.2 STRUCTURE EQUALS FUNCTION: FOCUS ON STRATEGIES THAT INTRODUCE HIERARCHICAL ORGANIZATION

13.2.1 Tissue Engineering Tools

Although the immediate clinical goal is to restore functionality (e.g., porcine valve replacements in pediatric patients despite lack of native leaflet trilayered structure), the long-term goal of tissue engineering is to manipulate cells and biomaterials to develop constructs that recapitulate native tissue composition and structure to guide postimplantation remodeling, growth, and self-maintenance. Generally, taking a bottom-up assembly approach and building materials up molecule by molecule to form supramolecular architectures and integrating intrinsic (e.g., gene manipulation) and extrinsic (e.g., mechanical conditioning) cues, *tissue engineering seeks to incorporate hierarchical organization into the tissue structure* using (Figure 13.1):

1. *Scaffolds*, which provide the foundational architecture and mechanical properties, and the inactive structure on which cells act
2. *Bioreactors and bioactive molecules*, which provide short-term active exogenous forces (i.e., biochemical and biophysical cues) to which the cells must react

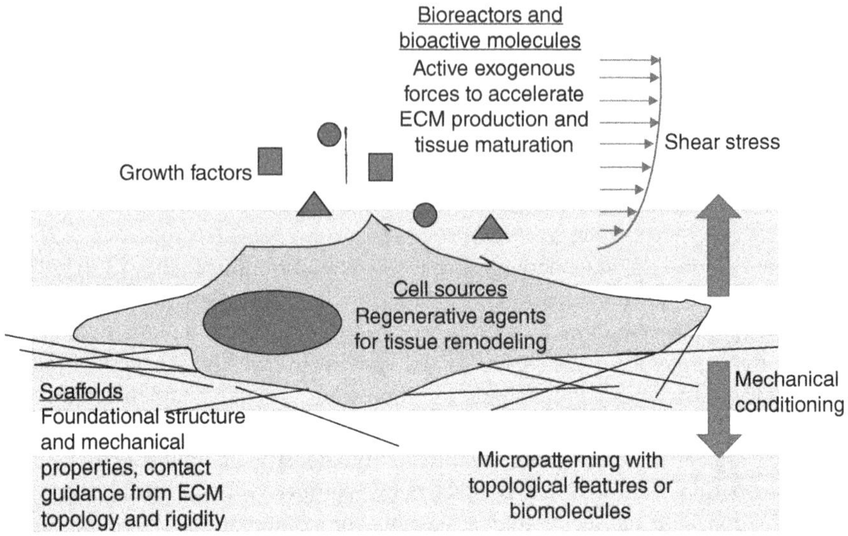

FIGURE 13.1 Schematic showing tissue engineering tools for controlling hierarchical structure, comprising of scaffolding techniques, bioreactors and biomolecules, and cell source manipulations.

3. *Cell sources*, of which the specific cell phenotype and its subsequent extracellular matrix (ECM) remodeling capabilities provide both short- and long-term regenerative agents.

Although at one point, material scientists sought to make inert biomaterials for implantation, as we continue to learn more about the biologic response to implants (e.g., inflammation, thrombus formation), the new paradigm in tissue engineering is to create more complex, active biomaterials that can interact with cells, bioactive molecules, and its microenvironment to initiate an appropriate host response [50]. Instead of avoiding inflammation, we may now engineer the physical characteristics of biomaterials to promote cell invasion and a modulated inflammatory or immune response to elicit desired remodeling (e.g., enzyme degradation, monocyte activation). We may also modify the surfaces with therapeutic agents (e.g., drug delivery, immobilized growth factor) to elicit a specific response.

In constructing structural tissues *in vitro*, tissue engineers must lay enough of a structural template for cells to remodel the injured structure at the target site before achieving compositional and functional integration with the surrounding environment [39,51,52]. The choice of scaffold must maintain its mechanical integrity as the cells entrapped within or invading the scaffold produce enough ECM to rebuild the compromised structure, a process that usually takes days. Once structural remodeling has taken place, compositional remodeling can occur, in which the scaffolding can be broken down and replaced simultaneously by cell-produced ECM, a process that can take weeks to months. Table 13.1 lists some examples of parameters to consider when developing tissue-engineered constructs.

TABLE 13.1 Some Design and Remodeling Parameters to Consider before and after Implantation [39,63,73,76]

	What parameters can be controlled before implantation?	What parameters must be accounted for after implantation?
Scaffold	• Biopolymer material selection mechanical properties: density, modulus, strength, viscoelasticity, anisotropy • Biopolymer material processing to modulate mechanical properties: ◦ Bulk material dip coating or compression molding and particulate leaching: pore size, geometry ◦ Electrospinning fibers: number, pore size, diameter, entanglement/branching frequency, network and fibril stiffness, intra- and interfiber cross-linking ◦ Solvent casting and particulate leaching films: polymer size, distribution, pore size, lamination for stacking ◦ Micropatterning with soft lithography: topological features, length scale • Biopolymer surface modification to increase cell adhesion: passivation, hydrophobicity • Nutrient availability: diffusional limitations determined by construct thickness	• Biopolymer: enzymatic triggers, degradation sites, degradation product toxicity and removal, immune response, inflammatory response • Cell secretion: extracellular matrix secretion for scaffolding
Bioreactors and bioactive molecules	• Medium composition: serum, supplemental growth factors that affect cell proliferation, ECM production, ECM degradation, and so on • Biochemical cues (functionalizing via scaffold surface modification): drug loading, immobilized growth factors, or ligands • Biophysical cues: mechanical preconditioning, electrical stimulation	• Biophysical cues: mechanical stresses (static vs. cyclic loading, mechanical stretching vs. hydrodynamic shear), electrical stimulation/integration • Nutrient delivery: perfusion, convective transport of oxygen, nutrients, and waste
Cell sources	• Cell type: species, age, passage number, culture conditions • Cell concentration: release of autocrine/paracrine factors related to construct thickness • Cell modifications: genetic manipulation (e.g., proteins for imaging)	• Cell phenotype: long-term maintenance, de-/trans-/differentiation, cell adhesion, ECM secretion, release of autocrine/paracrine factors • Nutrient delivery: oxygen transfer via microvasculature requires proper cell type

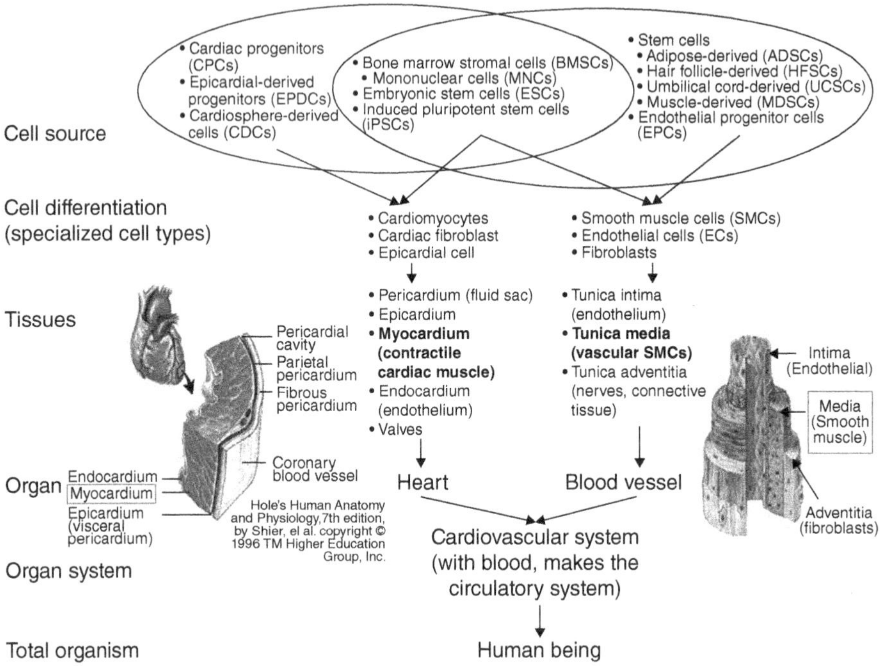

FIGURE 13.2 Hierarchical organization of the cardiovascular system in the human body. This chapter focuses on the tissue engineering of the myocardium and the tunica media [8,205,256,257].

13.2.2 Hierarchical Structure of the Heart and Blood Vessels

The basic hierarchical organization in the heart and blood vessels is seen in Figure 13.2. The heart is composed of striated muscle organized into anisotropic laminae that wrap around ventricular cavities [53,54]. Three major cell types comprise cardiac tissue: cardiomyocytes, fibroblasts, and endothelial cells (ECs) [55,56]. Cardiomyocytes form the contractile myocardium layer—the most difficult layer to replicate—at approximately 2 cm thick; high oxygen and nutrient demand is a factor in constructing viable tissue. Several nonmuscular structures are embedded within these layers: neural and vascular networks, collagen fibrils, and fibroblasts [53,57–60]. To maximize peak force generation, structural organization and cytoskeletal and sarcomere alignment are critically important [53,61]. Fibroblasts provide ECM proteins and paracrine factors [55], while ECs line vasculature to prevent thrombus formation [56].

Blood vessels have three layers—tunica intima, media (contractile), and adventitia—that vary in thickness according to the size and type of vessel [62,63]. In the tunica intima, ECs align and orient in the axial direction in the lumen, in contact with blood flow and attached to the basement membrane connective tissue, to regulate vasomotor tone and solute diffusion. The media layer is composed mostly of sheets or bundles of smooth muscle cells (SMCs, containing actin and myosin filaments) oriented in the circumferential direction to resist pulsatile flow. SMCs can operate under either a contractile or a synthetic (promotes ECM secretion) phenotype [64,65]. The media is reinforced by elastin embedded between small amounts of collagen fibers, oriented in the axial direction, with the ratio of elastin to collagen varying from 0.25 to 1.5, depending on the artery type [66]. The wavy, undulating structure of elastin straightens under loads at physiologic pressures; collagen fibers straighten in $<10\%$ of fibers, but become load bearing above normal physiologic pressures; and distension is limited by stiffness as collagen exhibits nonlinear viscoelastic behavior [67]. The adventitia is a collagenous ECM layer of fibroblasts and nerves that serves mostly to give rigidity to the vessel. Tissue engineering focuses on the media, as this layer contributes mechanical strength and elasticity to the vessel, followed by the vasoactive intima.

The most commonly found collagen in cardiovascular structures, *type I collagen* forms the biomechanical scaffold on which cells attach [62,68]. Collagen has a Young's modulus that ranges 5–10 MPa and provides the strength to resist failure at high pressures [66]; elastin has a Young's modulus of 0.4–1 MPa and provides reversible extensibility, with a stretch ratio of 1.6, and fibers can maintain elastic properties up to 140% of extension [69]. However, despite its assumed necessity for viscoelasticity, elastin is often absent in engineered grafts because elastin is minimally expressed in adults [70] and because the purification process is labor-intensive [71]. Additionally, because tropoelastin (i.e., soluble monomeric form of elastin) must aggregate on the cell surface and be cross-linked via helper proteins, recreating the complex temporal, spatial, and structural interactions to form functional elastic fibers remains a significant challenge [67,72].

13.3 TISSUE ENGINEERING APPROACHES TO CARDIOVASCULAR BIOMATERIALS

13.3.1 Scaffolds: Foundational Architecture and Mechanical Properties

Scaffolds provide the initial 3-D pore structure for cells to adhere, migrate, proliferate, differentiate, secrete ECM, and apoptose [73]. Over the long term, scaffolds can also release biochemical signals to maintain cell phenotype, induce differentiation, and influence metabolism. The scaffolding composition and its subsequent cellular remodeling can affect the cell shape, cytoskeleton, elastic modulus [74], and mechanotransduction of cells [75]. Thus, the design of the scaffolding material itself plays a significant role in controlling a tissue-engineered construct's architecture and mechanical properties as cells act upon scaffolding structures to self-organize its ECM.

Current tissue-engineered cardiovascular constructs often integrate unsuccessfully because grafts breakdown from either (a) failure from the repetitiveness or magnitude of the physiologic loads or (b) lack of oxygen and nutrient transport to implanted cells [76]. Transport by diffusion can only support four to seven cell layers (approximately 100–200 μm thick) [77–81]. Efforts of using scaffolds that recreate native cardiovascular tissue architecture and mechanical properties fall into one of the following categories: synthetic polymers, natural ECM, and decellularized matrix. Examples of each biomaterial and its corresponding advantages and disadvantages are listed in Table 13.2.

Isotropic porous scaffolds have inadequate mechanical strength for load-bearing tissues like the heart and blood vessels. Structural anisotropy is necessary for directing blood flow, as well as for excitation propagation in cardiac muscle; however, engineering constructs with aligned cells has been challenging. The methods in which biomaterials are fabricated or processed can influence the resulting anisotropy (e.g., forming fibers or sheets), as well as cell adhesion (e.g., passivating biomaterials or changing hydrophobicity) and other mechanical properties [82]. We focus our discussion on methods of introducing anisotropy via *topographical cues using contact guidance principles* using (a) soft lithography on the nano- and microscale, (b) electrospinning to encourage self-assembling tissues, and (c) decellularization to preserve existing ECM topological and molecular cues of tissues.

To introduce hierarchical structure to tissue-engineered constructs, the topography of the actual scaffold provides physical *contact guidance* cues to recapitulate the spatial cues of the microstructure in the native environment. Tissue engineers often use *soft lithography* techniques to easily create and inexpensively reproduce various elastomeric stamps, molds, and masks (see reviews elsewhere) to introduce physical topographies into constructs (Figure 13.3) [83–85]. Using soft lithography to mold topographical substrates (usually polydimethylsiloxane [PDMS]) that mimic physical native features, contact guidance has been shown to be an important factor in inducing vascularization in biomaterials, even without using growth factors [39]. Host cells have been able to penetrate membranes with pore sizes ranging from 0.8 to 8 μm [86], while studies found the optimum pore size was approximately 60 μm [87,88]. Elastomeric scaffolds have been able to recreate the aligned cardiac cell structures and physiological mechanical properties using accordion-like honeycomb-shaped pores; however, despite synchronous contraction of cardiac cells, macroscopic in-plane compression was not achieved [89–91]. Reducing the strut widths can increase scaffold compressibility, although the resulting increase in pore size must also be balanced with submicron-sized structures to guide cell alignment without compromising compressibility overall. Although contractile strength studies of engineered cardiac muscle tissues are rare, a few studies have found that cellular alignment (i.e., anisotropically aligned in comparison with isotropic) in thin films of cardiomyocytes increases contractile strength $>1000\%$ [53] and can increase twitch force up to 180% [61]. Peak systolic stresses in these anisotropically aligned films can reach >10 kPa when stretched at 0.5–5.0 Hz, in comparison with isotropic films that only reached peak systolic stresses around 1 kPa at 0.5 Hz and did not fully relax during the contractile cycle when stretched at >1.0 Hz [53].

Hydrogels, which can have different cross-linking densities and compositions to control substrate stiffness that in turn influences cell morphology and phenotype, encompasses a large portion of the literature on cardiovascular biomaterials; however, applying spatial cues to hydrogels still requires micropatterning to apply biomolecules (discussed in a later section) or lithography techniques (e.g., applying masks during photopolymerization or using molds for shaping). For an in-depth review

TABLE 13.2 Overview of Biomaterials Used for Scaffolding Structures for Cardiovascular Tissue Engineering (Adapted from References 63, 76, 93, and 200)

Material	Examples	Advantages	Disadvantages
Synthetic polymers	• Poly(glycolic acid) (PGA)[a] [199,261] • Poly(lactic acid) (PLA)[a] [262] • Poly(ε-caprolactone) (PCL) [263,264] • Poly(hydroxylalkanoate) (PHA) [265] • Poly(trimethylene carbonate) (PTMC) [266] • Poly(glycerol sebacate) (PGS) [91] • Polyurethane	• Cheap and readily available • Tunable physical and chemical properties via processing • Tunable bulk and surface modifications for controlling topological features and patterns	• Toxicity of degradation products • Suboptimal cell attachment and proliferation requires functionalizing surfaces • Mechanical properties of native tissues not yet optimized for tissue engineering
Extracellular matrix	• Collagen [117] • Elastin [267,268] • Gelatine • Fibrinogen • Hyaluronic acid • Chitosan • Alginate • Silk fibroin, recombinant spider silk • Small intestinal submucosa (SIS)	• Good cell attachment • Good cell signaling • Components found in cardiac muscle and blood vessels	• Mechanically weak • Expensive, especially purification of materials
Decellularized tissue	• Allogeneic [96–98] • Xenogeneic [99–102]	• Good biocompatibility • Mechanical properties of native tissue preserved (for vessels) [63]	• Difficult and nonuniform cell seeding • Poor cell migration because of small porous size of ECM structure • Laborious cleaning procedures • High mechanical stiffness because of ECM compaction (for the heart) [108,110]

[a]Indicates Food and Drug Administration (FDA) approval.

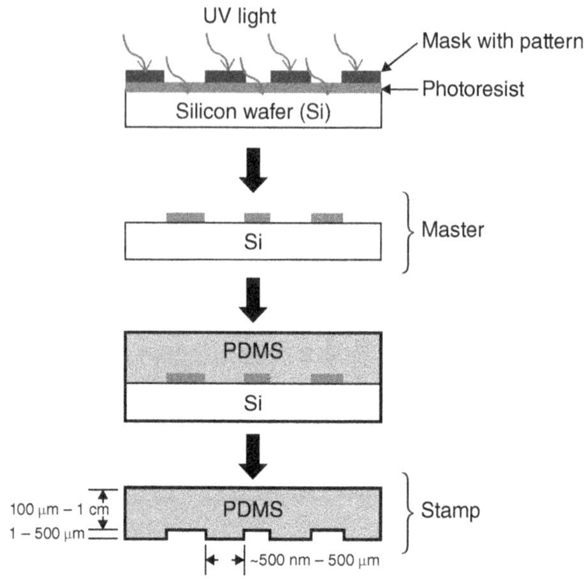

FIGURE 13.3 Schematic of making a master PDMS stamp with precise spatial cues using soft lithography. (Adapted from Kane *et al.* [85].)

specific to spatiotemporal control of hydrogels, we refer to the recent publication from Kharkar and colleagues [92].

To compensate for mechanical strength while still accommodating contractility, *electrospun nanofibers* can recreate topological cues on the same length scale as native ECM to encourage cellular self-assembly along 3-D fiber-like elements [63,93,94]. Electrospinning allows many parameters to be manipulated, including fiber orientation, density, and length; moreover, synthetic polymers, natural ECM, and polymer blends can be electrospun. Suspended nonwoven unidirectional nanofibers packed at 20- to 30-μm spacing provided enough guidance and structural support to grow a continuous anisotropic tissue using cardiomyocytes; tissue stripes and gaps in the tissue formed instead at greater packing distances [93]. Similarly, SMCs orient along fiber lengths on electrospun fabrics [95].

Decellularization uses combinations of mechanical means, detergents, enzyme inhibitors, and buffers to treat natural tissues, usually allogeneic [96–98] or xenogeneic [99–102] in origin. Cells are stripped to (a) passivate antigenic epitopes and (b) leave behind natural ECM for its biocompatibility and mechanical properties [14]. As shown in biaxial mechanical tests separating tissue layers consisting of mostly collagen fibers under tension, anisotropic materials can be separated longitudinally and isotropic materials can be separated circumferentially; however, the processing methods for removing cellular material can alter fiber alignment and matrix integrity, thus changing the mechanical behavior of the resulting scaffolds [103,104]. In addition, decellularized tissue scaffolds can be reseeded with autologous differentiated cells to further reduce immunogenicity [97,105]. However, the tight matrix structure usually makes cell migration difficult during recellularization [14,100], and the treatment process itself may cause structural alterations [103,106,107]. Acellularized myocardial scaffolds have shown stiffer mechanical behavior that may be restored with recellularization [108–110]. For vascular tissue engineering, human umbilical veins [111,112], coronary arteries [113], urinary bladder [114], and small intestinal submucosa (SIS) [21,115] have been used as decellularized scaffolds, with similar difficulties in cell infiltration. Clinical use of decellularized tissues remains limited, with efforts focused on *in vitro* or accelerated *in situ* with autologous cells [14,112]. A methods comparison of whole organ decellularization [112] and the structure and function of intact decellularized tissues [103] have been reviewed in-depth elsewhere.

13.3.2 Bioreactors and Bioactive Molecules: Active Exogenous Forces

Although tissue engineers have made great strides forward, culture times for many current constructs simply are still too long ($\geq$1 month) [116–119], which increases risk of contamination, or the cell may change phenotype [65,120,121] or dedifferentiate [118,122–126]. Inappropriate or absent stimulations may cause cells to become disorganized and apoptose [63]. Tissue engineers try to simulate native physiologic conditions to elicit desired cellular remodeling responses *in vitro*, and adding an exogenous cell signal to a construct encourages accelerated ECM production, maturation, and integration of engineered tissues [39,127–130], which generally fall into three categories:

1. *Biochemical Cues.* Scaffolding surfaces can be functionalized with bioactive or bioadhesive molecules with methods such as *microcontact printing* (μCP). Additionally, the culture environment can be supplemented with growth factors and other chemokines.
2. *Electrical Fields.* Cardiac contractions are propagated by electrical signals, to which any grafts must synchronize, or arrhythmias may occur. Electric fields have also been naturally detected in skin wounds, damaged tissues, and vasculature, which may be useful in directing migration of cells for repair and could be exploited for tissue engineering constructs [131].
3. *Mechanical Conditioning (Bioreactors).* Cardiovascular cells are subjected to constant mechanical forces (e.g., cyclic deformation, pulsatile flow, fluid shear) that are necessary to direct blood flow. However, injured tissues may have undesired stress environments that negatively impact remodeling (e.g., cardiac dilatation, atherosclerotic lesions), such as disarranging cells essential for effective contractions, or releasing (or lack thereof) autocrine and paracrine factors [132–135].

The number of studies dedicated to determining the optimal combination of *growth factors* and other bioactive molecules to produce desired behavior (e.g., increase ECM secretion, increase yield of preferentially differentiated cells) is

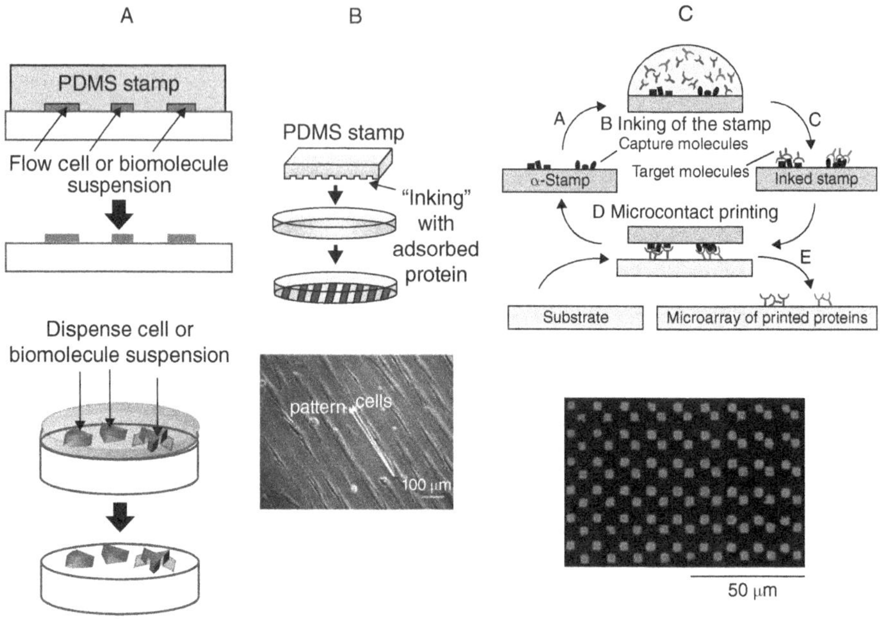

FIGURE 13.4 Schematic of soft lithography techniques used to create micropatterning on substrates using (A) blocking methods and solution dispensing with microchannels or stencils, (B) microcontact printing of adsorbed proteins using conformal contact, and (C) affinity contact printing using immobilized ligands for conformal contact printing of target biomolecules. (Panel A adapted from Park and Shuler [136]; panel B adapted from Williams *et al.* [143,258]; panel C from Renault *et al.* [141].) *(See insert for color representation of the figure.)*

abundant. Tissue engineers also modify scaffolds to control drug release, immobilize angiogenic growth factors and other ligands, or engineer chemoattractant gradients to encourage cell adhesion and migration. We focus our discussion here on *microfabrication*, which can be used to *micropattern* scaffolds and substrates with bioadhesive molecules or ECM proteins that allow for temporal or spatial control of bioactive molecules to increase alignment or improve other hierarchical structures (Figure 13.4) [136]. Biomolecules can be applied by either restricting or stamping the area intended to be activated.

Soft lithography techniques can be used to create *microfluidic channels or stencils* for restricting the delivery of bioactive molecules and cells (Figure 13.4A) [136]. Masks with microchannels or cutouts are formed using PDMS, which are then self-sealed on top of the substrate. Suspensions of cells or biomolecules can then be (a) streamed through the microchannels using laminar flow or (b) added directly to the exposed stencil cutout areas for attachment before removing the mask. The masks allow for spatial control, but diffusion of small molecules and leakage of cells can occur using these blocking methods.

μCP allows patterned shapes that resemble native architectures to be created, giving precise means to apply spatial cell adhesion cues to surfaces (Figure 13.4B)

[85,137]. A precisely patterned stamp (usually PDMS) is generated using soft lithography techniques that can then be *inked* or coated with bioactive molecules or ECM proteins and applied to a surface using conformal contact. Because PDMS is hydrophobic and DNA and other biomolecules are hydrophilic, the stamp can undergo oxygen plasma treatment, chemical treatment (usually silanes), or adsorption of polar molecules (e.g., positively-charged dendrimers) to allow homogenous inking with hydrophilic molecules [138]. Resolution and feature definition by μCP are limited to the mechanical properties of the stamp, and any deformities the stamp (e.g., buckling) may undergo during stamping may also be observed on the stamped substrate. Hybrid copolymers that allow for stiffer molds may be used for creating features on the nanoscale or with higher relief aspect ratios [139]. Additionally, most stamps only ink one molecule at a time, although creating a stamp with a microfluidic network to allow simultaneous inking of several molecules has been suggested [140]. Similarly, affinity and other supramolecular interactions (e.g., target and ligand molecules) can be used in *affinity contact printing* (αCP) (Figure 13.4C), but with the added advantage of reversible interactions through the complementary ligand units, which are usually immobilized on the stamp, so that stamps can be reused [138,140,141]. αCP allows for simultaneous capture of different molecules (e.g., DNA, antibodies) from a heterogeneous solution and easy inking for repeated printing, but the interaction between the target molecule and the ligand must be weaker than the interaction between the ligand and the stamp. μCP is more commonly used for cardiovascular applications, such as patterning fibronectin on substrates to recreate the aligned striations of cardiac muscle [142] or SMCs for vascular patches [143].

Electric fields can also be used to increase the structural organization of cardiovascular cells in tissue-engineered constructs. Applying synchronous contractions to cardiomyocytes induces preferred structural alignment in direction of applied electric field, which directly translates to higher average conduction velocity in comparison with nonstimulated tissues [144]. Cardiomyocyte alignment reduces the excitation threshold voltage and increases the amplitude of contraction, as aligned cells were shown to propagate signal along the longitudinal direction faster than in the transverse direction [53,145]. Additionally, contractile protein (troponin I, myosin heavy chain, sarcomeric α-actin) and gap junction protein connexin-43 (Cx43) expression increased in aligned cells [146,147]. Distribution of Cx43 expression has been shown to change significantly in cardiac diseases that lead to arrhythmias [148–150].

Although obvious for cardiac tissue, applying electrical fields also has potential in vascular tissue engineering, as endothelial progenitor cells (EPCs) have responded similarly with elongation and alignment [131]. Inhibiting the vascular endothelial growth factor (VEGF) receptor completely stopped electric field-induced migration of EPCs. Similarly, pulsatile electrical stimulation has been used in scaffolding made from the electroresponsive polymer polyacrylic acid to allow cells to infiltrate the scaffolding and induce alignment of SMCs [151]. To a lesser extent, *magnetic fields* have also been applied to hydrogels embedded with magnetic wires to induce collagen orientation in vascular grafts [152] or changes to cell morphology [153].

In addition to improving construct size, cellularity, and molecular composition [39,154–156], *bioreactors* can be used to enhance cellular anisotropy and mechanical behavior under specific hydrodynamic conditions [77] and physical stimuli, such as dynamic compression and cyclic stretch [157–160]. Multiple bioreactor devices have been made that apply tensile loading, compressive loading, and convective flow to study biologic responses and engineer tissues, which are described in detail elsewhere [161–163]. Because compressive systems are generally nondirectional in applying hydrostatic pressure, the focus on applying physiologic mechanical stimulation to bring structural hierarchy to cardiovascular tissue engineering constructs has been on cyclic tension, fluid shear, and hydrodynamic forces.

The application of uniaxial *cyclic tension* to cardiovascular cells (cardiomyocytes, ECs, SMCs) has shown repeatedly that the cells self-organize and elongate in an orientation perpendicular to the stretch axis in the absence of a scaffold [127,164–171]. This alignment may allow for coordinating wave-like contraction propagation and for absorbing deforming forces from contraction. As with electric fields, applying stretch has shown increased anisotropic distribution of Cx43 [166,172]. Furthermore, uniaxial cyclic tension applied longitudinally to an ECM scaffold with active cellular remodeling will further align fibers, especially collagen, along the stretching direction [173–176]. Again, the evidence indicates that scaffolds provide important architectural foundation cues via contact guidance, as cells and mechanical stimuli work synergistically to remodel the matrix.

Rotary suspension cultures and spinner flasks have been used to apply *fluid shear* and *hydrodynamic forces* on pluripotent stem cells to (a) control aggregation into embryoid bodies (EBs) and (b) accelerate differentiation into cardiovascular cells [177], especially cardiomyocytes [178–186] and ECs [187–191]. Suspension cultures allow for size-controllable scale-up production of EBs over tedious methods like hanging drop formation and other static suspension cultures that produce heterogeneous EB sizes and, thus, heterogeneous differentiation and low yields [192–194]. Fluid shear has also been used in maturation of larger 3-D constructs, controlling the size of scaffold-free cardiac patch constructions [195,196], discussed in the section below.

Convective flow from *perfusion bioreactors* has been used to homogenize initial cell seeding densities and precondition vascular grafts using pulsatile flow [39,63,116,119,197–199]. Because the functionality of ECs correlates with shear stress, engineered vessels are often flowed under physiologic pulsatile conditions to line the lumens with ECs. The physiologic flow conditions also modulate collagen remodeling in tissue-engineered blood vessels [198]. Additionally, constructs cultured under pulsatile flow had tissues that were twice as thick as those under constant flow, and resulting burst strengths were an order of magnitude higher for small-caliber grafts seeded with various cell types [116,119]. Increased alignment of cells and collagen were also observed in constructs under pulsatile flow for over a month, while constructs under constant flow also demonstrated dedifferentiated cell phenotypes.

Although exogenous cues greatly enhance tissue maturation, contact guidance using topographical cues, as discussed in the previous section, has still been demonstrated in numerous studies to have a greater effect on cellular alignment cues.

For example, in studies comparing the interactive effects of contact guidance and electrical stimulation, cardiomyocytes and cardiac fibroblasts preferentially oriented along topographical alignment cues, even when electric fields were applied perpendicular to the grooves [200–202]. When electrical stimulation fields were applied parallel to topographically aligned cells, the effect was additive in elongating cells. Additionally, smaller spatial cues (e.g., 1- to 10-μm grooves) have a greater effect on cell elongation, as focal adhesion protein vinculin was observed to be localized along the grooves [201], and larger grooves may yield less functional tissues as gap junction formation between cells for cellular communication is impeded [203]. A similar study comparing the interactive effects of contact guidance and mechanical conditioning found similar results, in which mechanical stretch was applied to cells aligned perpendicular or parallel to the stretch direction in micro-grooves [204]. Unconstrained cells that are mechanically stimulated normally orient perpendicular to the direction of stretch; cells previously aligned within grooves perpendicular to the applied strain showed a significant increase in the strength of cellular alignment. However, cells constrained within microgrooves parallel to the strain direction tended to stay aligned in the groove direction, suggesting that the microtopographical cue has a greater effect than mechanical stimulation. Cells seeded within 70-μm grooves were able to demonstrate some reorientation to cyclic strain, in which the width of the groove may have been wide enough for cells to maneuver.

13.3.3 Cell Sources: Cell Phenotype and Secreted Extracellular Matrix

To truly restore normal function, tissue engineering in regenerative medicine must incorporate the cell—whether *in vivo*, *ex vivo*, or *in vitro*—into its constructs such that the cell can allow for short- and long-term remodeling. Ideally, cell sources for cardiovascular tissue engineering should be autologous, easy to obtain, and proliferative in culture. Autologous cell sources are desired to avoid immunogenicity, especially blood-contacting ECs; however, most cardiovascular cells (i.e., cardiomyocytes, ECs, SMCs) are terminally differentiated in adults and have limited proliferative/replicative capacity [7,205]. Thus, large numbers of cells cannot be biopsied, but to be viable, constructs must be seeded at a high density. Because of source availability, good survival, and some beneficial effects following implantation [17,206], autologous myoblasts and other multipotent adult stem cells have been implanted clinically (as reviewed elsewhere) [207], in hopes of inducing transdifferention into cardiomyocytes; however, the cells remained of a skeletal lineage [208] and did not electromechanically couple with the host myocardium [209]. In addition, producing collagenous matrix *in vitro* from adult vascular cells is too slow for developing robust engineered grafts, requiring culture times of 8–24 weeks or more [117,205]; however, adult vascular cells produce less collagen and elastin [210] than fetal and neonatal cells, and usually undergo senescence within 10–30 population doublings (whereas 45–60 population doublings is preferable for mechanically robust engineered tissues) [117,199,205]. As a result, cardiovascular tissue engineers have concentrated on using cells with greater cell potency: progenitor cells, multipotent stem cells (adult stem cells), and pluripotent stem cells (embryonic [ESCs] and induced pluripotent stem cells [iPSCs]) (Figure 13.2).

Progenitor cells (e.g., EPCs, bone marrow stromal cells) are capable of proliferating, migrating, and differentiating into the specific cell types (e.g., ECs), but are already pushed into a more specific lineage than multipotent stem cells and have limited self-renewal potential. In contrast, *stem cells* (e.g., ESCs) have unlimited self-renewal capability before terminally differentiating. Although pluripotent stem cells (ESCs and iPSCs) can differentiate into cell types from any of the three germ layers, teratomas (i.e., tumors with derivatives of all three germ layers) may form if undifferentiated cells are implanted [8,19]; thus, methods that can be scaled up to produce highly pure populations are desired. Adult stem cells are multipotent and reside undifferentiated in differentiated tissues and yield organ-specific cell types (e.g., hematopoietic cells can differentiate into multiple blood cell types [lymphocytes, monocytes, neutrophils]) [211]. However, longer culture times and increasing donor age can decrease the proliferative and differentiating capacities of multipotent adult stem cells. Studies that evaluate methods of differentiation for all of these cell types are numerous. We instead focus on (a) coculturing and (b) genetic manipulations, two methods that can influence hierarchical structure and increase ECM synthesis and maturation.

Most tissues are comprised of more than one cell type, and *coculturing* with supporting cells (e.g., fibroblasts) increases secretion of soluble factors and matrix. Although early cardiovascular tissue engineering efforts tended to remove noncontractile cells (e.g., fibroblasts) [78], many studies now show their inclusion to have a supportive role that enhances function of engineered tissues [56,212–214]. For example, Iyer and colleagues showed that, in comparison with enriched cardiomyocyte culture or simultaneous triculture, sequential culturing of fibroblasts and ECs for 2 days, followed by cardiomyocyte seeding, led to significantly increased cell viability and contractile function measured by field stimulation of synchronous contractions [56,212]. Contractile function for constructs made in sequential culture was comparable to constructs made from enriched cardiomyocytes alone, but had three times less cardiomyocytes [212]. Additionally, factors secreted from cells in close proximity may influence cell morphology or phenotype, such as ESCs cultured on mouse embryonic fibroblast feeder layers [215] or ECs that can influence SMCs present in the second layer of the blood vessel [63]. Multiple groups have tried to mimic the spatial arrangement of ECs and SMCs to research their interactions and implications to vascular tissue engineering [216–219]. Supporting cells can also significantly influence ECM remodeling and thus cell alignment and architecture. For example, Nichol and associates determined that cardiomyocytes cocultured with fibroblasts exhibit significantly increased cell alignment in comparison with cultures of enriched cardiomyocytes; when matrix metalloproteases were inhibited, the increase in alignment between the two cultures was eliminated [220].

Cell cocultures can also allow for vascular formation to increase tissue viability. Narmoneva and associates demonstrated increased survival of cardiomyocytes from cell-to-cell interaction with ECs [221]. Cardiomyocytes preferentially organized spatially along EC networks, which also promoted synchronous cardiomyocyte contraction. Sekiya and colleagues demonstrated that cardiac cell monolayers cocultured with ECs at a ratio of 9 : 1 sprouted and formed EC networks that were similar in appearance to networks from ECs cultured in Matrigel (BD Biosciences, San Jose, CA), in comparison with cardiac cell sheets alone that formed significantly fewer endothelial networks [222]. Contrary to most literature where host cells gradually replace graft-derived cells [223–225], graft transplantation of these constructs into the subcutaneous tissues of normal rats showed that blood vessel growth and integration with the host vascular network was mostly graft-derived [222]. A later study by Sekine and colleagues from the same group using the same techniques demonstrated angiogenic growth factor secretion from cocultured cardiomyocytes and ECs increased significantly in comparison with cardiomyocytes alone (with no significant secretion when fibroblasts were cocultured with ECs) [226]. Additionally, they showed that increasing EC density in a graft could significantly increase cardiac function recovery in ischemic hearts 4 weeks after transplantation, in comparison with sham controls, and decrease fibrotic connective tissue formation as well.

Genetic manipulations focus on (a) increasing proliferative capacity to expand the initial pool of cells, (b) increasing efficiency and yield of differentiated cells, and (c) modifying genes such that implanted cells can be tracked for further study. *Telomerase reverse transcriptase* can be inserted into adult cells to extend cellular life span and thus increase proliferative life span and consequently, ECM secretion [227,228]. However, the retroviral vectors used to infect cells may also randomly insert into other cells and activate oncogenes [229,230]. As a result, many groups have attempted to increase efficiency and selection of highly pure populations of differentiating cells [183,184,186,188,189,231]. *Fusion genes* consisting of a promoter (e.g., α-myosin heavy chain, Flk-1) and a drug selection gene (e.g., G418 or puromycin resistance) transfected into stem cells allow for preferentially differentiated cells to be isolated from a heterogeneous population with high selectivity and purity. In addition, fusion genes have also been used to insert *reporter genes*, such as enhanced green fluorescent protein (EGFP), (a) to enhance selection of pure populations of cells from heterogeneous populations (e.g., via fluorescent activated cell sorting [FACS]) [188,232–234] or (b) to allow for cell imaging tracking in implanted cells [235–237].

13.4 SCAFFOLD-FREE TISSUE ENGINEERING: 3-D TISSUES WITHOUT EXOGENOUS MATERIAL COMPLICATIONS

Despite the numerous advances over the past several decades, biomaterials-based strategies present many problems, such as scaffold material choice, host inflammatory response, degradation product toxicity, mechanical compliance mismatch, reduced force-generating ability, incorrect ECM deposition and alignment, and reduced cell–cell connections. Residual polymer fragments have been demonstrated to disrupt structural organization [238] and influence cell phenotype [239,240]. Consequently, some research groups have begun to explore *scaffold-free tissue engineering solutions*—constructs that are composed only of cells and the matrix they secrete. As with scaffold-based tissue engineering, scaffold-free tissue engineering also faces some of the same challenges: reliable, scalable production of thick tissues with precise replication and control of hierarchical structure.

Stevens and colleagues constructed scaffold-free, beating human cardiac tissue patches that were 300- to 600-μm thick with synchronous calcium transients [196]; however, the patches also suffered the same necrosis diffusion limits at patch centers. Further iterations included cocultures of human umbilical vein endothelial cells (HUVECs) and mouse embryonic fibroblasts (MEFs) to accelerate vascularization [195]; when passive mechanical properties were measured, patches with cocultured cells were four times stiffer than patches with only cardiomyocytes (7.9 ± 3.1 mN/mm^2 vs. 2.0 ± 0.7 mN/mm^2), but still an order of magnitude lower than the stiffness of neonatal pig myocardium (30.2 ± 3.5 mN/mm^2). The cocultured patches were also found to have secreted fivefold greater collagen per patch area, with more organized collagen fibrils. Additionally, cocultured patches exhibited higher sarcomeric organization and had greater survival in rodent heart graft studies than cardiomyocyte-only patches; however, sarcomeric organization in comparison with the host myocardium was immature.

Norotte and associates have introduced a scaffold-free bioprinting technology that allows rapid prototyping of 3-D structures using multicellular spheroids as building blocks that has been applied to tissue-engineered blood vessels [240]. Early iterations allowed layer-by-layer construction of fused spheroids onto a collagen gel, but were distorted with each successive layer and lacked precision [241]. Using agarose rods as template molds, multicellular spheroids were deposited in precisely molded patterns that controlled for tube diameter, wall thickness, and branching patterns in a microvascular structure, where the smallest tube assembled was 900 μm in diameter with a wall thickness of 300 μm. Following deposition, the structures could be cultured for 5–7 days to allow the spheroids to fuse, and the final tubular construct could have the agarose rods removed and be perfused for maturation. However, spheroid fusion was nonuniform; additionally, the technique is time-consuming, with preparing large quantities of spheroids and assembling in a sterile environment. Further iterations allowed for simple multicellular cylindrical extrusions from micropipettes, which sped up the printing process; however, the large diameter (300–500 μm) micropipettes led to apoptotic cells after 3 days of fusion. Although smaller micropipettes could be used to create smaller diameter vessels, the agarose rods are extracted from the lumen by manually pulling from the open end, a problem which also becomes more complex in branching geometries.

Similarly, Gwyther and coworkers created scaffold-free tubular structures 2–6 mm in diameter using SMC aggregates formed into rings that were fused together [242,243]. The rings were strong enough to withstand mechanical testing within 8 days of cell seeding into an agarose ring mold, with an average ultimate tensile strength ranging from 100 to 500 kPa [243] (compared with 16 kPa for SMCs cultured statically within collagen) [244], but still low in comparison with native arteries (e.g., 6.6 MPa for porcine carotid artery) [238]. As with the traditional paradigm for tissue engineering using scaffolds, methods to improve ECM synthesis, such as soluble factors or mechanical conditioning, may improve tissue strength.

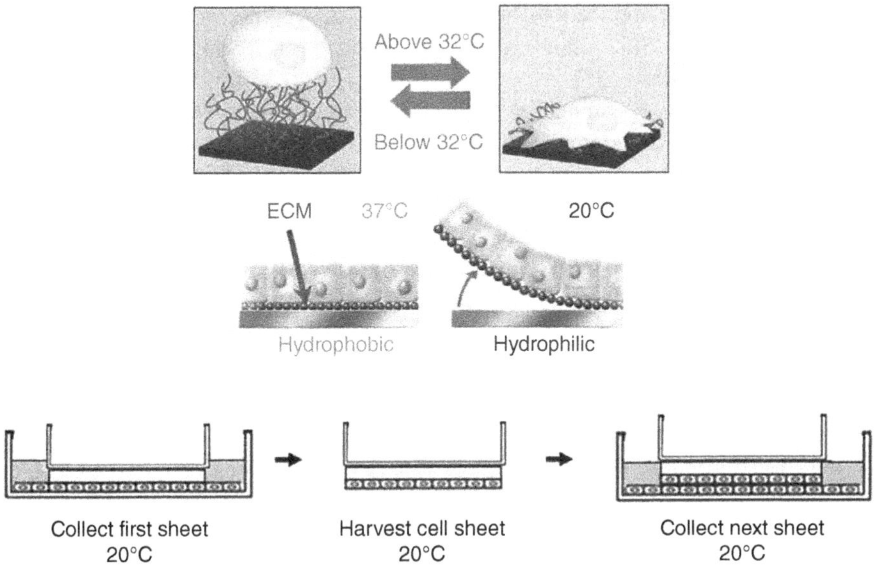

FIGURE 13.5 Cell sheet engineering technology using thermally responsive polymer poly(N-isopropylacrylamide) (P(NIPAAm)) allows cell attachment at 37°C and nonenzymatic cell detachment below lower critical solution temperature 32°C. Cell sheets can be harvested and layered in this manner. (Adapted from Elloumi-Hannachi *et al.* [259], Nakayama *et al.* [260], and Williams *et al.* [143].)

Likewise, this methodology also produced nonhomogeneous fusion between ring boundaries that may be improved with extended culture.

Other groups have explored scaffold-free tissue engineering using *cell sheet engineering technologies*, which employs a thermally responsive polymer poly(N-isopropylacrylamide) (P(NIPAAm)) that undergoes a reversible transition from hydrophobic (to allow cell culture at incubation temperatures) to hydrophilic (to allow spontaneous cell sheet detachment without the need for proteolytic enzymes when the polymer is lowered to room temperature). As a result, cell sheets can be harvested fully intact, with cell–cell and cell–matrix interactions fully maintained following detachment and transfer to another substrate or cell layer (Figure 13.5). Cell sheet monolayers have been cultured on and detached from multiple substrates modified with P(NIPAAm), including tissue-culture polystyrene (TCPS) [143,222,245–247], hydrogels [248], and PDMS [127,249]. Cell sheet technology also enables tissues to be fabricated with prevascularized networks, by way of sandwiching EC layers between cardiomyocyte layers [250] or cocultures of ECs with cardiomyocytes [222,226]. Electrical couplings have been found to be preserved within cell sheets, as well as between cell layers [251–253]. Likewise, cell sheets could be coupled with supporting cell types (e.g., fibroblasts) to increase robustness [253]. Cultured sheets can be stacked into various configurations for cardiovascular tissue applications and have been successfully implanted in several studies [80,222,254,255], with 10 times greater cell survival and retention than dissociated cell injections [255].

Current cell sheet engineering technologies aim to incorporate local microenvironment hierarchical structural cues into the tissue. Isenberg and coworkers were able to generate a microtextured TCPS substrate and modify the surface with P(NIPAAm) using electron beam grafting to allow for orientation of cell sheets that could subsequently be detached; orientation was well maintained post-transfer [246]. Using μCP fibronectin patterns on P(NIPAAm)-grafted TCPS dishes, Williams and colleagues from the same group demonstrated that cell orientation could be precisely controlled for individual cell sheets, which were then used as functional units and stacked to recapitulate complex tissue structure [143]. Both patterned and nonpatterned cell sheets could grow to 13- to 15-μm thickness, and preliminary mechanical tests showed that the patterned cell sheets had trends of increased stiffness and strength.

Although these technologies incorporate structural cues, culture time is still extensive, and methods to increase ECM synthesis to increase cell sheet robustness are desired. Lee and von Recum were the first to incorporate P(NIPAAm) copolymers into an elastic, PDMS-based substrate such that cells can be mechanically conditioned to induce cell alignment and increase ECM growth before cell sheet detachment, with cell–cell and cell–matrix junctions fully intact [127]. Similarly, Lin and associates were able to graft P(NIPAAm) onto PDMS microtextured with patterned lanes that mimicked native arterial organization, which allowed SMCs to elongate and align with high fidelity [249]. Cell sheets showed anisotropic shrinkage when detaching from patterned substrates in comparison with isotropic shrinkage detaching from nonpatterned substrates. The elasticity of the patterned substrates also can allow for future studies of maturing aligned cell sheets with mechanical conditioning.

13.5 CONCLUSION

In this chapter, we have discussed tissue engineering tools for imparting hierarchical structures into cardiovascular tissue constructs using scaffolds, bioreactors and biomolecules, and manipulations to cell sources, as well as methods of scaffold-less tissue engineering. These tools present opportunities for recreating tissue microenvironments so that we can begin to understand differences between native and diseased tissue, and the progression to the diseased state. Our understanding of the diseased state will help us build models that have the potential for screening therapeutics and reconstructing functional regenerative tissue replacements, the ultimate goal of tissue engineering.

Acknowledgments

The authors acknowledge the support of NIH/NIAID T32AI089673-01A1 Training Grant, NIH/NHLBI 1F32HL115999-01 Postdoctoral Fellowship, and The Hartwell Foundation.

REFERENCES

[1] Fung, Y.C., quoted in Preface to *Tissue engineering* (Eds: R. Skalak, C.F. Fox), R. Liss, New York: 1988.

[2] Go, A.S., Mozaffarian, D., Roger, V.L., Benjamin, E.J., Berry, J.D., Borden, W.B., Bravata, D.M., Dai, S., Ford, E.S., Fox, C.S., Franco, S., Fullerton, H.J., Gillespie, C., Hailpern, S.M., Heit, J.A., Howard, V.J., Huffman, M.D., Kissela, B.M., Kittner, S.J., Lackland, D.T., Lichtman, J.H., Lisabeth, L.D., Magid, D., Marcus, G.M., Marelli, A., Matchar, D.B., McGuire, D.K., Mohler, E.R., Moy, C.S., Mussolino, M.E., Nichol, G., Paynter, N.P., Schreiner, P.J., Sorlie, P.D., Stein, J., Turan, T.N., Virani, S.S., Wong, N.D., Woo, D., Turner, M.B. *Circulation* **2013**, *127*, e6.

[3] Reller, M.D., Strickland, M.J., Riehle-Colarusso, T., Mahle, W.T., Correa, A. *The Journal of Pediatrics* **2008**, *153*, 807.

[4] Roguin, N., Du, Z.D., Barak, M., Nasser, N., Hershkowitz, S., Milgram, E. *Journal of the American College of Cardiology* **1995**, *26*, 1545.

[5] Warnes, C.A., Williams, R.G., Bashore, T.M., Child, J.S., Connolly, H.M., Dearani, J.A., del Nido, P., Fasules, J.W., Graham, T.P., Jr., Hijazi, Z.M., Hunt, S.A., King, M.E., Landzberg, M.J., Miner, P.D., Radford, M.J., Walsh, E.P., Webb, G.D. *Circulation* **2008**, *118*, e714.

[6] Thygesen, K., Alpert, J.S., Jaffe, A.S., Simoons, M.L., Chaitman, B.R., White, H.D., Thygesen, K., Alpert, J.S., White, H.D., Jaffe, A.S., Katus, H.A., Apple, F.S., Lindahl, B., Morrow, D.A., Chaitman, B.R., Clemmensen, P.M., Johanson, P., Hod, H., Underwood, R., Bax, J.J., Bonow, J.J., Pinto, F., Gibbons, R.J., Fox, K.A., Atar, D., Newby, L.K., Galvani, M., Hamm, C.W., Uretsky, B.F., Steg, P.G., Wijns, W., Bassand, J.P., Menasche, P., Ravkilde, J., Ohman, E.M., Antman, E.M., Wallentin, L.C., Armstrong, P.W., Simoons, M.L., Januzzi, J.L., Nieminen, M.S., Gheorghiade, M., Filippatos, G., Luepker, R.V., Fortmann, S.P., Rosamond, W.D., Levy, D., Wood, D., Smith, S.C., Hu, D., Lopez-Sendon, J.L., Robertson, R.M., Weaver, D., Tendera, M., Bove, A.A., Parkhomenko, A.N., Vasilieva, E.J., Mendis, S., Bax, J.J., Baumgartner, H., Ceconi, C., Dean, V., Deaton, C., Fagard, R., Funck-Brentano, C., Hasdai, D., Hoes, A., Kirchhof, P., Knuuti, J., Kolh, P., McDonagh, T., Moulin, C., Popescu, B.A., Reiner, Z., Sechtem, U., Sirnes, P.A., Tendera, M., Torbicki, A., Vahanian, A., Windecker, S., Morais, J., Aguiar, C., Almahmeed, W., Arnar, D.O., Barili, F., Bloch, K.D., Bolger, A.F., Botker, H.E., Bozkurt, B., Bugiardini, R., Cannon, C., de Lemos, J., Eberli, F.R., Escobar, E., Hlatky, M., James, S., Kern, K.B., Moliterno, D.J., Mueller, C., Neskovic, A.N., Pieske, B.M., Schulman, S.P., Storey, R.F., Taubert, K.A., Vranckx, P., Wagner, D.R. *Journal of the American College of Cardiology* **2012**, *60*, 1581.

[7] Soonpaa, M.H., Field, L.J. *Circulation Research* **1998**, *83*, 15.

[8] Laflamme, M.A., Murry, C.E. *Nature Biotechnology* **2005**, *23*, 845.

[9] Brower, G.L., Gardner, J.D., Forman, M.F., Murray, D.B., Voloshenyuk, T., Levick, S.P., Janicki, J.S. *European Journal of Cardio-Thoracic Surgery* **2006**, *30*, 604.

[10] Janicki, J.S., Brower, G.L. *Journal of Cardiac Failure* **2002**, *8*, S319.

[11] Seo, K., Inagaki, M., Nishimura, S., Hidaka, I., Sugimachi, M., Hisada, T., Sugiura, S. *Circulation Research* **2010**, *106*, 176.

[12] Lim, Z.Y., Maskara, B., Aguel, F., Emokpae, R., Jr., Tung, L. *Circulation* **2006**, *114*, 2113.

[13] Organ Procurement and Transplantation Network (OPTN) and Scientific Registry of Transplant Recipients (SRTR). OPTN/SRTR 2011 Annual Data Report (Ed: Department of Health and Human Services, Health Resources and Services Administration, Healthcare Systems Bureau, Division of Transplantation), 2012.

[14] Wilhelmi, M., Giere, B., Harder, M. *Advances in Biochemical Engineering/ Biotechnology* **2012**, *126*, 105.

[15] McAllister, T.N., Maruszewski, M., Garrido, S.A., Wystrychowski, W., Dusserre, N., Marini, A., Zagalski, K., Fiorillo, A., Avila, H., Manglano, X., Antonelli, J., Kocher, A., Zembala, M., Cierpka, L., de la Fuente, L.M., L'Heureux, N. *Lancet* **2009**, *373*, 1440.

[16] Vorp, D.A., Maul, T., Nieponice, A. *Frontiers in Bioscience* **2005**, *10*, 768.

[17] Menasche, P. *Circulation* **2009**, *119*, 2735.

[18] Muller-Ehmsen, J., Whittaker, P., Kloner, R.A., Dow, J.S., Sakoda, T., Long, T.I., Laird, P.W., Kedes, L. *Journal of Molecular and Cellular Cardiology* **2002**, *34*, 107.

[19] Nussbaum, J., Minami, E., Laflamme, M.A., Virag, J.A., Ware, C.B., Masino, A., Muskheli, V., Pabon, L., Reinecke, H., Murry, C.E. *The FASEB Journal* **2007**, *21*, 1345.

[20] Lantz, G.C., Badylak, S.F., Coffey, A.C., Geddes, L.A., Blevins, W.E. *Journal of Investigative Surgery* **1990**, *3*, 217.

[21] Lantz, G.C., Badylak, S.F., Hiles, M.C., Coffey, A.C., Geddes, L.A., Kokini, K., Sandusky, G.E., Morff, R.J. *Journal of Investigative Surgery* **1993**, *6*, 297.

[22] Harris, L., O'Brien-Irr, M., Ricotta, J.J. *Journal of Vascular Surgery* **2001**, *33*, 528.

[23] Dardik, H., Wengerter, K., Qin, F., Pangilinan, A., Silvestri, F., Wolodiger, F., Kahn, M., Sussman, B., Ibrahim, I.M. *Journal of Vascular Surgery* **2002**, *35*, 64.

[24] Dardik, H., Ibrahim, I.M., Sussman, B., Kahn, M., Sanchez, M., Klausner, S., Baier, R.E., Meyer, A.E., Dardik, I.I. *Annals of Surgery* **1984**, *199*, 61.

[25] Calligaro, K.D., Syrek, J.R., Dougherty, M.J., Rua, I., Raviola, C.A., DeLaurentis, D.A. *Journal of Vascular Surgery* **1997**, *26*, 919.

[26] Lehalle, B., Geschier, C., Fieve, G., Stoltz, J.F. *Journal of Vascular Surgery* **1997**, *25*, 751.

[27] Ochiai, Y., Imoto, Y., Sakamoto, M., Kajiwara, T., Sese, A., Watanabe, M., Ohno, T., Joo, K. *European Journal of Cardio-Thoracic Surgery* **2009**, *36*, 63.

[28] Gauvin, R., Guillemette, M., Galbraith, T., Bourget, J.M., Larouche, D., Marcoux, H., Aube, D., Hayward, C., Auger, F.A., Germain, L. *Tissue Engineering. Part A* **2011**, *17*, 2049.

[29] Veith, F.J., Gupta, S.K., Ascer, E., White-Flores, S., Samson, R.H., Scher, L.A., Towne, J.B., Bernhard, V.M., Bonier, P., Flinn, W.R., *et al. Journal of Vascular Surgery* **1986**, *3*, 104.

[30] Baguneid, M.S., Goldner, S., Fulford, P.E., Hamilton, G., Walker, M.G., Seifalian, A.M. *Journal of Vascular Surgery* **2001**, *33*, 812.

[31] Berger, K., Sauvage, L.R., Rao, A.M., Wood, S.J. *Annals of Surgery* **1972**, *175*, 118.

[32] Sarkar, S., Salacinski, H.J., Hamilton, G., Seifalian, A.M. *European Journal of Vascular and Endovascular Surgery* **2006**, *31*, 627.

[33] Haruguchi, H., Teraoka, S. *Journal of Artificial Organs* **2003**, *6*, 227.

[34] Abbott, W.M., Megerman, J., Hasson, J.E., L'Italien, G., Warnock, D.F. *Journal of Vascular Surgery* **1987**, *5*, 376.

[35] Nerem, R.M. *Vox Sanguinis* **2004**, *87*(Suppl. 2), 158.

[36] Zaucha, M.T., Gauvin, R., Auger, F.A., Germain, L., Gleason, R.L. *Journal of the Royal Society, Interface* **2011**, *8*, 244.

[37] Nerem, R.M. *Yonsei Medical Journal* **2000**, *41*, 735.

[38] Badylak, S.F., Grompe, M., Caplan, A.I., Greisler, H.P., Guldberg, R.E., Taylor, D.A. *Annals of the New York Academy of Sciences* **2002**, *961*, 319.

[39] Freed, L.E., Guilak, F., Guo, X.E., Gray, M.L., Tranquillo, R., Holmes, J.W., Radisic, M., Sefton, M.V., Kaplan, D., Vunjak-Novakovic, G. *Tissue Engineering* **2006**, *12*, 3285.

[40] MacLellan, W.R., Wang, Y., Lusis, A.J. *Nature Reviews* **2012**, *9*, 172.
[41] Martorell, J., Santoma, P., Molins, J.J., Garcia-Granada, A.A., Bea, J.A., Edelman, E.R., Balcells, M. *Annals of the New York Academy of Sciences* **2012**, *1254*, 51.
[42] Taylor, C.A., Humphrey, J.D. *Computer Methods in Applied Mechanics and Engineering* **2009**, *198*, 3514.
[43] Ingber, D.E., Mow, V.C., Butler, D., Niklason, L.E., Huard, J., Mao, J., Yannas, I., Kaplan, D., Vunjak-Novakovic, G. *Tissue Engineering* **2006**, *12*, 3265.
[44] Humphrey, J.D. *Proceedings of the Royal Society of London. Series A* **2003**, *459*, 3.
[45] Abarbanell, A.M., Herrmann, J.L., Weil, B.R., Wang, Y., Tan, J., Moberly, S.P., Fiege, J.W., Meldrum, D.R. *The Journal of Surgical Research* **2010**, *162*, 239.
[46] Vilahur, G., Padro, T., Badimon, L. *Journal of Biomedicine and Biotechnology* **2011**, *2011*, 907575.
[47] Hearse, D.J., Sutherland, F.J. *Pharmacological Research* **2000**, *41*, 597.
[48] Doggrell, S.A., Chan, V. *Methods and Findings in Experimental and Clinical Pharmacology* **2001**, *23*, 457.
[49] Byrom, M.J., Bannon, P.G., White, G.H., Ng, M.K. *Journal of Vascular Surgery* **2010**, *52*, 176.
[50] Sefton, M.V. in *Advances in Chemical Engineering: Molecular and Cellular Foundations of Biomaterials*, Vol. 29 (Eds: M.V. Sefton, N.A. Peppas), Academic Press, San Diego, CA: 2004.
[51] Tranquillo, R.T. *Biochemical Society Symposium* **1999**, *65*, 27.
[52] Isenberg, B.C., Williams, C., Tranquillo, R.T. *Circulation Research* **2006**, *98*, 25.
[53] Feinberg, A.W., Alford, P.W., Jin, H., Ripplinger, C.M., Werdich, A.A., Sheehy, S.P., Grosberg, A., Parker, K.K. *Biomaterials* **2012**, *33*, 5732.
[54] LeGrice, I.J., Smaill, B.H., Chai, L.Z., Edgar, S.G., Gavin, J.B., Hunter, P.J. *The American Journal of Physiology* **1995**, *269*, H571.
[55] Banerjee, I., Yekkala, K., Borg, T.K., Baudino, T.A. *Annals of the New York Academy of Sciences* **2006**, *1080*, 76.
[56] Iyer, R.K., Chiu, L.L., Vunjak-Novakovic, G., Radisic, M. *Biofabrication* **2012**, *4*, 035002.
[57] Azzawi, M., Kan, S.W., Hillier, V., Yonan, N., Hutchinson, I.V., Hasleton, P.S. *Histopathology* **2005**, *46*, 314.
[58] Parker, K.K., Ingber, D.E. *Philosophical Transactions of the Royal Society of London* **2007**, *362*, 1267.
[59] Pope, A.J., Sands, G.B., Smaill, B.H., LeGrice, I.J. *American Journal of Physiology* **2008**, *295*, H1243.
[60] Young, A.A., Legrice, I.J., Young, M.A., Smaill, B.H. *Journal of Microscopy* **1998**, *192*, 139.
[61] Black, L.D., 3rd, Meyers, J.D., Weinbaum, J.S., Shvelidze, Y.A., Tranquillo, R.T. *Tissue Engineering. Part A* **2009**, *15*, 3099.
[62] Lekakou, C., Lamprou, D., Vidyarthi, U., Karopoulou, E., Zhdan, P. *Journal of Biomedical Materials Research* **2008**, *85*, 461.
[63] Song, Y., Feijen, J., Grijpma, D.W., Poot, A.A. *Clinical Hemorheology and Microcirculation* **2011**, *49*, 357.
[64] Chamley-Campbell, J., Campbell, G.R., Ross, R. *Physiological Reviews* **1979**, *59*, 1.
[65] Chamley-Campbell, J.H., Campbell, G.R. *Atherosclerosis* **1981**, *40*, 347.
[66] Hastings, G.V. *Cardiovascular Biomaterials*; Springer-Verlag, London: 1992.
[67] Wagenseil, J.E., Mecham, R.P. *Journal of Cardiovascular Translational Research* **2012**, *5*, 264.
[68] Brodsky, B., Ramshaw, J.A. *Matrix Biology* **1997**, *15*, 545.

[69] Buttafoco, L., Kolkman, N.G., Engbers-Buijtenhuijs, P., Poot, A.A., Dijkstra, P.J., Vermes, I., Feijen, J. *Biomaterials* **2006**, *27*, 724.

[70] Kelleher, C.M., McLean, S.E., Mecham, R.P. *Current Topics in Developmental Biology* **2004**, *62*, 153.

[71] Daamen, W.F., Hafmans, T., Veerkamp, J.H., Van Kuppevelt, T.H. *Biomaterials* **2001**, *22*, 1997.

[72] Patel, A., Fine, B., Sandig, M., Mequanint, K. *Cardiovascular Research* **2006**, *71*, 40.

[73] Palsson, B.O., Bhatia, S.N. *Tissue Engineering*; Prentice Hall, Upper Saddle River, NJ: 2004.

[74] Takai, E., Costa, K.D., Shaheen, A., Hung, C.T., Guo, X.E. *Annals of Biomedical Engineering* **2005**, *33*, 963.

[75] Takai, E., Landesberg, R., Katz, R.W., Hung, C.T., Guo, X.E. *Molecular and Cellular Biomechanics* **2006**, *3*, 1.

[76] Freed, L.E., Engelmayr, G.C., Jr., Borenstein, J.T., Moutos, F.T., Guilak, F. *Advanced Materials (Deerfield Beach, Fla.)* **2009**, *21*, 3410.

[77] Papadaki, M., Bursac, N., Langer, R., Merok, J., Vunjak-Novakovic, G., Freed, L.E. *American Journal of Physiology* **2001**, *280*, H168.

[78] Bursac, N., Papadaki, M., Cohen, R.J., Schoen, F.J., Eisenberg, S.R., Carrier, R., Vunjak-Novakovic, G., Freed, L.E. *The American Journal of Physiology* **1999**, *277*, H433.

[79] Carrier, R.L., Papadaki, M., Rupnick, M., Schoen, F.J., Bursac, N., Langer, R., Freed, L.E., Vunjak-Novakovic, G. *Biotechnology and Bioengineering* **1999**, *64*, 580.

[80] Shimizu, T., Sekine, H., Isoi, Y., Yamato, M., Kikuchi, A., Okano, T. *Tissue Engineering* **2006**, *12*, 499.

[81] Zimmermann, W.H., Fink, C., Kralisch, D., Remmers, U., Weil, J., Eschenhagen, T. *Biotechnology and Bioengineering* **2000**, *68*, 106.

[82] Pankajakshan, D., Agrawal, D.K. *Canadian Journal of Physiology and Pharmacology* **2010**, *88*, 855.

[83] Shin, H. *Biomaterials* **2007**, *28*, 126.

[84] Falconnet, D., Csucs, G., Grandin, H.M., Textor, M. *Biomaterials* **2006**, *27*, 3044.

[85] Kane, R.S., Takayama, S., Ostuni, E., Ingber, D.E., Whitesides, G.M. *Biomaterials* **1999**, *20*, 2363.

[86] Brauker, J.H., Carr-Brendel, V.E., Martinson, L.A., Crudele, J., Johnston, W.D., Johnson, R.C. *Journal of Biomedical Materials Research* **1995**, *29*, 1517.

[87] Boswell, C.A., Williams, S.K. *Journal of Biomaterials Science* **1999**, *10*, 319.

[88] Sharkawy, A.A., Klitzman, B., Truskey, G.A., Reichert, W.M. *Journal of Biomedical Materials Research* **1998**, *40*, 586.

[89] Engelmayr, G.C., Jr., Cheng, M., Bettinger, C.J., Borenstein, J.T., Langer, R., Freed, L.E. *Nature Materials* **2008**, *7*, 1003.

[90] Guillemette, M.D., Park, H., Hsiao, J.C., Jain, S.R., Larson, B.L., Langer, R., Freed, L.E. *Macromolecular Bioscience* **2010**, *10*, 1330.

[91] Wang, Y., Ameer, G.A., Sheppard, B.J., Langer, R. *Nature Biotechnology* **2002**, *20*, 602.

[92] Kharkar, P.M., Kiick, K.L., Kloxin, A.M. *Chemical Society Reviews* **2013**, *42*, 7335.

[93] Orlova, Y., Magome, N., Liu, L., Chen, Y., Agladze, K. *Biomaterials* **2011**, *32*, 5615.

[94] Xu, C., Inai, R., Kotaki, M., Ramakrishna, S. *Tissue Engineering* **2004**, *10*, 1160.

[95] Chen, S., Wang, P.P., Wang, J.P., Chen, G.Q., Wu, Q. *Journal of Biomedical Materials Research. Part A* **2008**, *86*, 849.

[96] Patil, P.B., Chougule, P.B., Kumar, V.K., Almstrom, S., Backdahl, H., Banerjee, D., Herlenius, G., Olausson, M., Sumitran-Holgersson, S. *Stem Cells Translational Medicine* **2013**, *2*, 307.

[97] Lehr, E.J., Rayat, G.R., Chiu, B., Churchill, T., McGann, L.E., Coe, J.Y., Ross, D.B. *The Journal of Thoracic and Cardiovascular Surgery* **2011**, *141*, 1056.

[98] Quint, C., Arief, M., Muto, A., Dardik, A., Niklason, L.E. *Journal of Vascular Surgery* **2012**, *55*, 790.

[99] Wainwright, J.M., Czajka, C.A., Patel, U.B., Freytes, D.O., Tobita, K., Gilbert, T.W., Badylak, S.F. *Tissue Engineering. Part C: Methods* **2010**, *16*, 525.

[100] Teebken, O.E., Bader, A., Steinhoff, G., Haverich, A. *European Journal of Vascular and Endovascular Surgery* **2000**, *19*, 381.

[101] Bader, A., Schilling, T., Teebken, O.E., Brandes, G., Herden, T., Steinhoff, G., Haverich, A. *European Journal of Cardio-Thoracic Surgery* **1998**, *14*, 279.

[102] Turner, W.S., Wang, X., Johnson, S., Medberry, C., Mendez, J., Badylak, S.F., McCord, M.G., McCloskey, K.E. *Journal of Biomedical Materials Research* **2012**, *100*, 2060.

[103] Badylak, S.F., Freytes, D.O., Gilbert, T.W. *Acta Biomaterialia* **2009**, *5*, 1.

[104] Williams, C., Liao, J., Joyce, E.M., Wang, B., Leach, J.B., Sacks, M.S., Wong, J.Y. *Acta Biomaterialia* **2009**, *5*, 993.

[105] Meyer, S.R., Nagendran, J., Desai, L.S., Rayat, G.R., Churchill, T.A., Anderson, C.C., Rajotte, R.V., Lakey, J.R., Ross, D.B. *The Journal of Thoracic and Cardiovascular Surgery* **2005**, *130*, 469.

[106] Gilbert, T.W., Sellaro, T.L., Badylak, S.F. *Biomaterials* **2006**, *27*, 3675.

[107] Barnes, C.A., Brison, J., Michel, R., Brown, B.N., Castner, D.G., Badylak, S.F., Ratner, B.D. *Biomaterials* **2011**, *32*, 137.

[108] Wang, B., Tedder, M.E., Perez, C.E., Wang, G., de Jongh Curry, A.L., To, F., Elder, S.H., Williams, L.N., Simionescu, D.T., Liao, J. *Journal of Materials Science* **2012**, *23*, 1835.

[109] Wang, B., Borazjani, A., Tahai, M., Curry, A.L., Simionescu, D.T., Guan, J., To, F., Elder, S.H., Liao, J. *Journal of Biomedical Materials Research. Part A* **2010**, *94*, 1100.

[110] Ott, H.C., Matthiesen, T.S., Goh, S.K., Black, L.D., Kren, S.M., Netoff, T.I., Taylor, D.A. *Nature Medicine* **2008**, *14*, 213.

[111] Gui, L., Muto, A., Chan, S.A., Breuer, C.K., Niklason, L.E. *Tissue Engineering. Part A* **2009**, *15*, 2665.

[112] He, M., Callanan, A. *Tissue Engineering. Part B, Reviews* **2013**, *19*, 194.

[113] Campbell, E.M., Cahill, P.A., Lally, C. *Journal of the Mechanical Behavior of Biomedical Materials* **2012**, *14*, 130.

[114] Piterina, A.V., Callanan, A., Davis, L., Meaney, C., Walsh, M., McGloughlin, T.M. *Bio-Medical Materials and Engineering* **2009**, *19*, 333.

[115] Nemcova, S., Noel, A.A., Jost, C.J., Gloviczki, P., Miller, V.M., Brockbank, K.G. *Journal of Investigative Surgery* **2001**, *14*, 321.

[116] Niklason, L.E., Abbott, W., Gao, J., Klagges, B., Hirschi, K.K., Ulubayram, K., Conroy, N., Jones, R., Vasanawala, A., Sanzgiri, S., Langer, R. *Journal of Vascular Surgery* **2001**, *33*, 628.

[117] L'Heureux, N., Paquet, S., Labbe, R., Germain, L., Auger, F.A. *The FASEB Journal* **1998**, *12*, 47.

[118] Weinberg, C.B., Bell, E. *Science (New York, N.Y.)* **1986**, *231*, 397.

[119] Hoerstrup, S.P., Zund, G., Sodian, R., Schnell, A.M., Grunenfelder, J., Turina, M.I. *European Journal of Cardio-Thoracic Surgery* **2001**, *20*, 164.

[120] Worth, N.F., Rolfe, B.E., Song, J., Campbell, G.R. *Cell Motility and the Cytoskeleton* **2001**, *49*, 130.

[121] LaFramboise, W.A., Scalise, D., Stoodley, P., Graner, S.R., Guthrie, R.D., Magovern, J.A., Becich, M.J. *American Journal of Physiology* **2007**, *292*, C1799.

[122] Tang, Z., Wang, A., Wang, D., Li, S. *Circulation Research* **2013**, *112*, 23.

[123] Dinardo, C.L., Venturini, G., Omae, S.V., Zhou, E.H., da Motta-Leal-Filho, J.M., Dariolli, R., Krieger, J.E., Alencar, A.M., Costa Pereira, A. *Biorheology* **2012**, *49*, 365.

[124] Nguyen, A.T., Gomez, D., Bell, R.D., Campbell, J.H., Clowes, A.W., Gabbiani, G., Giachelli, C.M., Parmacek, M.S., Raines, E.W., Rusch, N.J., Speer, M.Y., Sturek, M., Thyberg, J., Towler, D.A., Weiser-Evans, M.C., Yan, C., Miano, J.M., Owens, G.K. *Circulation Research* **2013**, *112*, 17.

[125] Driesen, R.B., Verheyen, F.K., Debie, W., Blaauw, E., Babiker, F.A., Cornelussen, R.N., Ausma, J., Lenders, M.H., Borgers, M., Chaponnier, C., Ramaekers, F.C. *Journal of Cellular and Molecular Medicine* **2009**, *13*, 896.

[126] Dispersyn, G.D., Geuens, E., Ver Donck, L., Ramaekers, F.C., Borgers, M. *Cardiovascular Research* **2001**, *51*, 230.

[127] Lee, E.L., von Recum, H.A. *Journal of Biomedical Materials Research. Part A* **2010**, *93*, 411.

[128] Zaucha, M.T., Raykin, J., Wan, W., Gauvin, R., Auger, F.A., Germain, L., Michaels, T.E., Gleason, R.L., Jr. *Tissue Engineering. Part A* **2009**, *15*, 3331.

[129] Engelmayr, G.C., Jr., Rabkin, E., Sutherland, F.W., Schoen, F.J., Mayer, J.E., Jr., Sacks, M.S. *Biomaterials* **2005**, *26*, 175.

[130] Syedain, Z.H., Weinberg, J.S., Tranquillo, R.T. *Proceedings of the National Academy of Sciences of the United States of America* **2008**, *105*, 6537.

[131] Zhao, Z., Qin, L., Reid, B., Pu, J., Hara, T., Zhao, M. *Stem Cell Research* **2012**, *8*, 38.

[132] Sadoshima, J., Jahn, L., Takahashi, T., Kulik, T.J., Izumo, S. *The Journal of Biological Chemistry* **1992**, *267*, 10551.

[133] van Wamel, A.J., Ruwhof, C., van der Valk-Kokshoorn, L.J., Schrier, P.I., van der Laarse, A. *Archives of Biochemistry and Biophysics* **2000**, *381*, 67.

[134] Ruwhof, C., van Wamel, A.E., Egas, J.M., van der Laarse, A. *Molecular and Cellular Biochemistry* **2000**, *208*, 89.

[135] Ruwhof, C., van der Laarse, A. *Cardiovascular Research* **2000**, *47*, 23.

[136] Park, T.H., Shuler, M.L. *Biotechnology Progress* **2003**, *19*, 243.

[137] Chen, C.S., Mrksich, M., Huang, S., Whitesides, G.M., Ingber, D.E. *Biotechnology Progress* **1998**, *14*, 356.

[138] Akbulut, O., Yu, A.A., Stellacci, F. *Chemical Society Reviews* **2010**, *39*, 30.

[139] Quist, A.P., Pavlovic, E., Oscarsson, S. *Analytical and Bioanalytical Chemistry* **2005**, *381*, 591.

[140] Bernard, A., Renault, J.P., Michel, B., Bosshard, H.R., Delamarche, E. *Advanced Materials* **2000**, *12*, 1067.

[141] Renault, J.P., Bernard, A., Juncker, D., Michel, B., Bosshard, H.R., Delamarche, E. *Angewandte Chemie (International ed. in English)* **2002**, *41*, 2320.

[142] Atmanli, A., Domian, I.J. *Journal of Visualized Experiments* **2013**, (73), e50288.

[143] Williams, C., Xie, A.W., Yamato, M., Okano, T., Wong, J.Y. *Biomaterials* **2011**, *32*, 5625.

[144] Radisic, M., Fast, V.G., Sharifov, O.F., Iyer, R.K., Park, H., Vunjak-Novakovic, G. *Tissue Engineering. Part A* **2009**, *15*, 851.

[145] Bursac, N., Parker, K.K., Iravanian, S., Tung, L. *Circulation Research* **2002**, *91*, e45.

[146] Radisic, M., Park, H., Shing, H., Consi, T., Schoen, F.J., Langer, R., Freed, L.E., Vunjak-Novakovic, G. *Proceedings of the National Academy of Sciences of the United States of America* **2004**, *101*, 18129.

[147] Tandon, N., Cannizzaro, C., Chao, P.H., Maidhof, R., Marsano, A., Au, H.T., Radisic, M., Vunjak-Novakovic, G. *Nature Protocols* **2009**, *4*, 155.

[148] Dupont, E., Matsushita, T., Kaba, R.A., Vozzi, C., Coppen, S.R., Khan, N., Kaprielian, R., Yacoub, M.H., Severs, N.J. *Journal of Molecular and Cellular Cardiology* **2001**, *33*, 359.

[149] Kostin, S., Rieger, M., Dammer, S., Hein, S., Richter, M., Klovekorn, W.P., Bauer, E.P., Schaper, J. *Molecular and Cellular Biochemistry* **2003**, *242*, 135.

[150] Eloff, B.C., Lerner, D.L., Yamada, K.A., Schuessler, R.B., Saffitz, J.E., Rosenbaum, D.S. *Cardiovascular Research* **2001**, *51*, 681.

[151] Rahimi, N., Molin, D.G., Cleij, T.J., van Zandvoort, M.A., Post, M.J. *Biomacromolecules* **2012**, *13*, 1448.

[152] Tranquillo, R.T., Girton, T.S., Bromberek, B.A., Triebes, T.G., Mooradian, D.L. *Biomaterials* **1996**, *17*, 349.

[153] Kiang, J.D., Wen, J.H., Del Alamo, J.C., Engler, A.J. *Journal of Biomedical Materials Research. Part A* **2013**, *101*, 2313.

[154] Bursac, N., Papadaki, M., White, J.A., Eisenberg, S.R., Vunjak-Novakovic, G., Freed, L.E. *Tissue Engineering* **2003**, *9*, 1243.

[155] Vunjak-Novakovic, G., Freed, L.E., Biron, R.J., Langer, R. *AIChE Journal* **1996**, *42*, 850.

[156] Pei, M., Solchaga, L.A., Seidel, J., Zeng, L., Vunjak-Novakovic, G., Caplan, A.I., Freed, L.E. *The FASEB Journal* **2002**, *16*, 1691.

[157] Akhyari, P., Fedak, P.W., Weisel, R.D., Lee, T.Y., Verma, S., Mickle, D.A., Li, R.K. *Circulation* **2002**, *106*, I137.

[158] Fink, C., Ergun, S., Kralisch, D., Remmers, U., Weil, J., Eschenhagen, T. *The FASEB Journal* **2000**, *14*, 669.

[159] Kim, B.S., Nikolovski, J., Bonadio, J., Mooney, D.J. *Nature Biotechnology* **1999**, *17*, 979.

[160] Zimmermann, W.H., Schneiderbanger, K., Schubert, P., Didie, M., Munzel, F., Heubach, J.F., Kostin, S., Neuhuber, W.L., Eschenhagen, T. *Circulation Research* **2002**, *90*, 223.

[161] Brown, T.D. *Journal of Biomechanics* **2000**, *33*, 3.

[162] Gooch, K.J., Tennant, C.J. *Mechanical Forces: Their Effects on Cells and Tissues*; Springer, New York: 1997.

[163] Lee, E.L., von Recum, H.A. in *Tissue Engineering: Principles and Practices* (Eds: J.P. Fisher, A.G. Mikos, J.D. Bronzino, D.R. Peterson), CRC Press, Boca Raton, FL: 2012.

[164] Dartsch, P.C., Betz, E. *Basic Research in Cardiology* **1989**, *84*, 268.

[165] Haghighipour, N., Tafazzoli-Shadpour, M., Shokrgozar, M.A., Amini, S., Amanzadeh, A., Khorasani, M.T. *Molecular and Cellular Biomechanics* **2007**, *4*, 189.

[166] Salameh, A., Wustmann, A., Karl, S., Blanke, K., Apel, D., Rojas-Gomez, D., Franke, H., Mohr, F.W., Janousek, J., Dhein, S. *Circulation Research* **2010**, *106*, 1592.

[167] Hayakawa, K., Hosokawa, A., Yabusaki, K., Obinata, T. *Zoological Science* **2000**, *17*, 617.

[168] Zhao, L., Sang, C., Yang, C., Zhuang, F. *Journal of Biomechanics* **2011**, *44*, 2388.

[169] Joung, I.S., Iwamoto, M.N., Shiu, Y.T., Quam, C.T. *Microvascular Research* **2006**, *71*, 1.

[170] Standley, P.R., Cammarata, A., Nolan, B.P., Purgason, C.T., Stanley, M.A. *American Journal of Physiology, Heart and Circulatory Physiology* **2002**, *283*, H1907.

[171] Kada, K., Yasui, K., Naruse, K., Kamiya, K., Kodama, I., Toyama, J. *Journal of Molecular and Cellular Cardiology* **1999**, *31*, 247.

[172] Salameh, A., Karl, S., Djilali, H., Dhein, S., Janousek, J., Daehnert, I. *Pharmacological Research* **2010**, *62*, 506.

[173] Wang, J.H., Grood, E.S. *Connective Tissue Research* **2000**, *41*, 29.
[174] Nguyen, T.D., Liang, R., Woo, S.L., Burton, S.D., Wu, C., Almarza, A., Sacks, M.S., Abramowitch, S. *Tissue Engineering. Part A* **2009**, *15*, 957.
[175] Gould, R.A., Chin, K., Santisakultarm, T.P., Dropkin, A., Richards, J.M., Schaffer, C.B., Butcher, J.T. *Acta Biomaterialia* **2012**, *8*, 1710.
[176] Gonen-Wadmany, M., Gepstein, L., Seliktar, D. *Annals of the New York Academy of Sciences* **2004**, *1015*, 299.
[177] Adamo, L., Garcia-Cardena, G. *Antioxidants and Redox Signaling* **2011**, *15*, 1463.
[178] Zweigerdt, R., Burg, M., Willbold, E., Abts, H., Ruediger, M. *Cytotherapy* **2003**, *5*, 399.
[179] Carpenedo, R.L., Sargent, C.Y., McDevitt, T.C. *Stem Cells (Dayton, Ohio)* **2007**, *25*, 2224.
[180] Sargent, C.Y., Berguig, G.Y., Kinney, M.A., Hiatt, L.A., Carpenedo, R.L., Berson, R.E., McDevitt, T.C. *Biotechnology and Bioengineering* **2010**, *105*, 611.
[181] Sargent, C.Y., Berguig, G.Y., McDevitt, T.C. *Tissue Engineering. Part A* **2009**, *15*, 331.
[182] He, W., Ye, L., Li, S., Liu, H., Wang, Q., Fu, X., Han, W., Chen, Z. *Biological and Pharmaceutical Bulletin* **2012**, *35*, 308.
[183] Niebruegge, S., Bauwens, C.L., Peerani, R., Thavandiran, N., Masse, S., Sevaptisidis, E., Nanthakumar, K., Woodhouse, K., Husain, M., Kumacheva, E., Zandstra, P.W. *Biotechnology and Bioengineering* **2009**, *102*, 493.
[184] Zandstra, P.W., Bauwens, C., Yin, T., Liu, Q., Schiller, H., Zweigerdt, R., Pasumarthi, K.B., Field, L.J. *Tissue Engineering* **2003**, *9*, 767.
[185] Fluri, D.A., Tonge, P.D., Song, H., Baptista, R.P., Shakiba, N., Shukla, S., Clarke, G., Nagy, A., Zandstra, P.W. *Nature Methods* **2012**, *9*, 509.
[186] Niebruegge, S., Nehring, A., Bar, H., Schroeder, M., Zweigerdt, R., Lehmann, J. *Tissue Engineering. Part A* **2008**, *14*, 1591.
[187] Yirme, G., Amit, M., Ląevsky, I., Osenberg, S., Itskovitz-Eldor, J. *Stem Cells and Development* **2008**, *17*, 1227.
[188] Kim, S., von Recum, H.A. *Tissue Engineering. Part A* **2009**, *15*, 3709.
[189] Kim, S., von Recum, H.A. *Tissue Engineering. Part A* **2010**, *16*, 1065.
[190] Chiu, B., Wan, J.Z., Abley, D., Akabutu, J. *Acta Astronautica* **2005**, *56*, 918.
[191] Festag, M., Sehner, C., Steinberg, P., Viertel, B. *Toxicology in Vitro* **2007**, *21*, 1619.
[192] Dang, S.M., Gerecht-Nir, S., Chen, J., Itskovitz-Eldor, J., Zandstra, P.W. *Stem Cells (Dayton, Ohio)* **2004**, *22*, 275.
[193] Gerecht-Nir, S., Cohen, S., Ziskind, A., Itskovitz-Eldor, J. *Biotechnology and Bioengineering* **2004**, *88*, 313.
[194] Gerecht-Nir, S., Cohen, S., Itskovitz-Eldor, J. *Biotechnology and Bioengineering* **2004**, *86*, 493.
[195] Stevens, K.R., Kreutziger, K.L., Dupras, S.K., Korte, F.S., Regnier, M., Muskheli, V., Nourse, M.B., Bendixen, K., Reinecke, H., Murry, C.E. *Proceedings of the National Academy of Sciences of the United States of America* **2009**, *106*, 16568.
[196] Stevens, K.R., Pabon, L., Muskheli, V., Murry, C.E. *Tissue Engineering. Part A* **2009**, *15*, 1211.
[197] Hahn, M.S., McHale, M.K., Wang, E., Schmedlen, R.H., West, J.L. *Annals of Biomedical Engineering* **2007**, *35*, 190.
[198] Huang, A.H., Niklason, L.E. *Journal of Visualized Experiments* **2011**, (52), pii: 2646; doi: 10.3791/2646.
[199] Niklason, L.E., Gao, J., Abbott, W.M., Hirschi, K.K., Houser, S., Marini, R., Langer, R. *Science (New York, N.Y.)* **1999**, *284*, 489.

[200] Tandon, V., Zhang, B., Radisic, M., Murthy, S.K. *Biotechnology Advances* **2013** , *31*, 722.

[201] Heidi Au, H.T., Cui, B., Chu, Z.E., Veres, T., Radisic, M. *Lab on a Chip* **2009**, *9*, 564.

[202] Au, H.T., Cheng, I., Chowdhury, M.F., Radisic, M. *Biomaterials* **2007**, *28*, 4277.

[203] Chiu, L.L., Janic, K., Radisic, M. *The International Journal of Artificial Organs* **2012**, *35*, 237.

[204] Houtchens, G.R., Foster, M.D., Desai, T.A., Morgan, E.F., Wong, J.Y. *Journal of Biomechanics* **2008**, *41*, 762.

[205] Sundaram, S., Niklason, L.E. *Cells, Tissues, Organs* **2012**, *195*, 15.

[206] Menasche, P. *Expert Review of Cardiovascular Therapy* **2004**, *2*, 21.

[207] Murry, C.E., Field, L.J., Menasche, P. *Circulation* **2005**, *112*, 3174.

[208] Reinecke, H., Poppa, V., Murry, C.E. *Journal of Molecular and Cellular Cardiology* **2002**, *34*, 241.

[209] Reinecke, H., MacDonald, G.H., Hauschka, S.D., Murry, C.E. *The Journal of Cell Biology* **2000**, *149*, 731.

[210] McMahon, M.P., Faris, B., Wolfe, B.L., Brown, K.E., Pratt, C.A., Toselli, P., Franzblau, C. *In Vitro Cellular and Developmental Biology. Animal* **1985**, *21*, 674.

[211] Penn, M.S., Mal, N. *Methods in Molecular Medicine* **2006**, *129*, 329.

[212] Iyer, R.K., Chiu, L.L., Radisic, M. *Journal of Biomedical Materials Research. Part A* **2009**, *89*, 616.

[213] Caspi, O., Lesman, A., Basevitch, Y., Gepstein, A., Arbel, G., Habib, I.H., Gepstein, L., Levenberg, S. *Circulation Research* **2007**, *100*, 263.

[214] Levenberg, S., Rouwkema, J., Macdonald, M., Garfein, E.S., Kohane, D.S., Darland, D.C., Marini, R., van Blitterswijk, C.A., Mulligan, R.C., D'Amore, P.A., Langer, R. *Nature Biotechnology* **2005**, *23*, 879.

[215] Ilic, D. *Regenerative Medicine* **2006**, *1*, 95.

[216] Fillinger, M.F., Sampson, L.N., Cronenwett, J.L., Powell, R.J., Wagner, R.J. *The Journal of Surgical Research* **1997**, *67*, 169.

[217] Powell, R.J., Cronenwett, J.L., Fillinger, M.F., Wagner, R.J., Sampson, L.N. *Annals of Vascular Surgery* **1996**, *10*, 4.

[218] Lavender, M.D., Pang, Z., Wallace, C.S., Niklason, L.E., Truskey, G.A. *Biomaterials* **2005**, *26*, 4642.

[219] Ziegler, T., Alexander, R.W., Nerem, R.M. *Annals of Biomedical Engineering* **1995**, *23*, 216.

[220] Nichol, J.W., Engelmayr, G.C., Jr., Cheng, M., Freed, L.E. *Biochemical and Biophysical Research Communications* **2008**, *373*, 360.

[221] Narmoneva, D.A., Vukmirovic, R., Davis, M.E., Kamm, R.D., Lee, R.T. *Circulation* **2004**, *110*, 962.

[222] Sekiya, S., Shimizu, T., Yamato, M., Kikuchi, A., Okano, T. *Biochemical and Biophysical Research Communications* **2006**, *341*, 573.

[223] Vajkoczy, P., Olofsson, A.M., Lehr, H.A., Leiderer, R., Hammersen, F., Arfors, K.E., Menger, M.D. *American Journal of Pathology* **1995**, *146*, 1397.

[224] Demarchez, M., Hartmann, D.J., Prunieras, M. *Transplantation* **1987**, *43*, 896.

[225] Linn, T., Schneider, K., Hammes, H.P., Preissner, K.T., Brandhorst, H., Morgenstern, E., Kiefer, F., Bretzel, R.G. *The FASEB Journal* **2003**, *17*, 881.

[226] Sekine, H., Shimizu, T., Hobo, K., Sekiya, S., Yang, J., Yamato, M., Kurosawa, H., Kobayashi, E., Okano, T. *Circulation* **2008**, *118*, S145.

[227] McKee, J.A., Banik, S.S., Boyer, M.J., Hamad, N.M., Lawson, J.H., Niklason, L.E., Counter, C.M. *EMBO Reports* **2003**, *4*, 633.

[228] Shao, R., Guo, X. *Biochemical and Biophysical Research Communications* **2004**, *321*, 788.
[229] Klinger, R.Y., Blum, J.L., Hearn, B., Lebow, B., Niklason, L.E. *Proceedings of the National Academy of Sciences of the United States of America* **2006**, *103*, 2500.
[230] European Society of Gene Therapy. *The Journal of Gene Medicine* **2003**, *5*, 82.
[231] Klug, M.G., Soonpaa, M.H., Koh, G.Y., Field, L.J. *The Journal of Clinical Investigation* **1996**, *98*, 216.
[232] Moore, J.C., Spijker, R., Martens, A.C., de Boer, T., Rook, M.B., van der Heyden, M.A., Tertoolen, L.G., Mummery, C.L. *The International Journal of Developmental Biology* **2004**, *48*, 47.
[233] Anderson, D., Self, T., Mellor, I.R., Goh, G., Hill, S.J., Denning, C. *Molecular Therapy* **2007**, *15*, 2027.
[234] van Laake, L.W., Qian, L., Cheng, P., Huang, Y., Hsiao, E.C., Conklin, B.R., Srivastava, D. *Circulation Research* **2010**, *107*, 340.
[235] Au, P., Tam, J., Fukumura, D., Jain, R.K. *Blood* **2008**, *111*, 4551.
[236] Xie, C.Q., Zhang, J., Xiao, Y., Zhang, L., Mou, Y., Liu, X., Akinbami, M., Cui, T., Chen, Y.E. *Stem Cells and Development* **2007**, *16*, 25.
[237] van Laake, L.W., Passier, R., Monshouwer-Kloots, J., Nederhoff, M.G., Ward-van Oostwaard, D., Field, L.J., van Echteld, C.J., Doevendans, P.A., Mummery, C.L. *Nature Protocols* **2007**, *2*, 2551.
[238] Dahl, S.L., Rhim, C., Song, Y.C., Niklason, L.E. *Annals of Biomedical Engineering* **2007**, *35*, 348.
[239] Higgins, S.P., Solan, A.K., Niklason, L.E. *Journal of Biomedical Materials Research. Part A* **2003**, *67*, 295.
[240] Norotte, C., Marga, F.S., Niklason, L.E., Forgacs, G. *Biomaterials* **2009**, *30*, 5910.
[241] Jakab, K., Norotte, C., Damon, B., Marga, F., Neagu, A., Besch-Williford, C.L., Kachurin, A., Church, K.H., Park, H., Mironov, V., Markwald, R., Vunjak-Novakovic, G., Forgacs, G. *Tissue Engineering* **2008**, *14*, 413.
[242] Gwyther, T.A., Hu, J.Z., Billiar, K.L., Rolle, M.W. *Journal of Visualized Experiments* **2011**, (57), e3366.
[243] Gwyther, T.A., Hu, J.Z., Christakis, A.G., Skorinko, J.K., Shaw, S.M., Billiar, K.L., Rolle, M.W. *Cells, Tissues, Organs* **2011**, *194*, 13.
[244] Seliktar, D., Black, R.A., Vito, R.P., Nerem, R.M. *Annals of Biomedical Engineering* **2000**, *28*, 351.
[245] von Recum, H.A., Okano, T., Kim, S.W., Bernstein, P.S. *Experimental Eye Research* **1999**, *69*, 97.
[246] Isenberg, B.C., Tsuda, Y., Williams, C., Shimizu, T., Yamato, M., Okano, T., Wong, J.Y. *Biomaterials* **2008**, *29*, 2565.
[247] Okano, T., Yamada, N., Okuhara, M., Sakai, H., Sakurai, Y. *Biomaterials* **1995**, *16*, 297.
[248] von Recum, H.A., Kim, S.W., Kikuchi, A., Okuhara, M., Sakurai, Y., Okano, T. *Journal of Biomedical Materials Research* **1998**, *40*, 631.
[249] Lin, J.B., Isenberg, B.C., Shen, Y., Schorsch, K., Sazonova, O.V., Wong, J.Y. *Colloids and Surfaces* **2012**, *99*, 108.
[250] Tsuda, Y., Shimizu, T., Yamato, M., Kikuchi, A., Sasagawa, T., Sekiya, S., Kobayashi, J., Chen, G., Okano, T. *Biomaterials* **2007**, *28*, 4939.
[251] Haraguchi, Y., Shimizu, T., Sasagawa, T., Sekine, H., Sakaguchi, K., Kikuchi, T., Sekine, W., Sekiya, S., Yamato, M., Umezu, M., Okano, T. *Nature Protocols* **2012**, *7*, 850.
[252] Haraguchi, Y., Shimizu, T., Yamato, M., Kikuchi, A., Okano, T. *Biomaterials* **2006**, *27*, 4765.

[253] Haraguchi, Y., Shimizu, T., Yamato, M., Okano, T. *Journal of Tissue Engineering and Regenerative Medicine* **2010**, *4*, 291.
[254] Shimizu, T., Yamato, M., Isoi, Y., Akutsu, T., Setomaru, T., Abe, K., Kikuchi, A., Umezu, M., Okano, T. *Circulation Research* **2002**, *90*, e40.
[255] Sekine, H., Shimizu, T., Dobashi, I., Matsuura, K., Hagiwara, N., Takahashi, M., Kobayashi, E., Yamato, M., Okano, T. *Tissue Engineering. Part A* **2011**, *17*, 2973.
[256] Bajpai, V.K., Andreadis, S.T. *Tissue Engineering. Part B: Reviews* **2012**, *18*, 405.
[257] Kikuchi, K., Poss, K.D. *Annual Review of Cell and Developmental Biology* **2012**, *28*, 719.
[258] Williams, C., Tsuda, Y., Isenberg, B.C., Yamato, M., Shimizu, T., Okano, T., Wong, J.Y. *Advanced Materials* **2009**, *21*, 2161.
[259] Elloumi-Hannachi, I., Yamato, M., Okano, T. *Journal of Internal Medicine* **2010**, *267*, 54.
[260] Nakayama, M., Yamada, N., Kumashiro, Y., Kanazawa, H., Yamato, M., Okano, T. *Macromolecular Bioscience* **2012**, *12*, 751.
[261] Kim, B.S., Nikolovski, J., Bonadio, J., Smiley, E., Mooney, D.J. *Experimental Cell Research* **1999**, *251*, 318.
[262] Iwai, S., Sawa, Y., Ichikawa, H., Taketani, S., Uchimura, E., Chen, G., Hara, M., Miyake, J., Matsuda, H. *The Journal of Thoracic and Cardiovascular Surgery* **2004**, *128*, 472.
[263] Sarasam, A., Madihally, S.V. *Biomaterials* **2005**, *26*, 5500.
[264] Venugopal, J., Ma, L.L., Yong, T., Ramakrishna, S. *Cell Biology International* **2005**, *29*, 861.
[265] Shum-Tim, D., Stock, U., Hrkach, J., Shinoka, T., Lien, J., Moses, M.A., Stamp, A., Taylor, G., Moran, A.M., Landis, W., Langer, R., Vacanti, J.P., Mayer, J.E., Jr. *The Annals of Thoracic Surgery* **1999**, *68*, 2298.
[266] Buttafoco, L., Boks, N.P., Engbers-Buijtenhuijs, P., Grijpma, D.W., Poot, A.A., Dijkstra, P.J., Vermes, I., Feijen, J. *Journal of Biomedical Materials Research* **2006**, *79*, 425.
[267] Kurane, A., Simionescu, D.T., Vyavahare, N.R. *Biomaterials* **2007**, *28*, 2830.
[268] Leach, J.B., Wolinsky, J.B., Stone, P.J., Wong, J.Y. *Acta Biomaterialia* **2005**, *1*, 155.

CHAPTER 14

EVALUATION OF BIO-INSPIRED MATERIALS FOR MINERALIZED TISSUE REGENERATION USING TYPE I COLLAGEN REPORTER CELLS

LIISA T. KUHN, EMILY JACOBS, and A. JON GOLDBERG

Reconstructive Sciences, University of Connecticut Health Center, Farmington, CT, USA

14.1 INTRODUCTION

Materials that enhance bone regeneration have a wealth of potential clinical applications. These applications are causing the market for biomaterials-based treatments in orthopedics to grow at a rapid rate. In the past, materials intended for implantation were designed to be bioinert; however, materials scientists have shifted toward the design of deliberately bioactive materials that integrate with cells or biological molecules and regenerate tissues. It is desirable that these new bioactive materials promote osteogenesis, that is, they should be able to influence and support the cells around them or the cells seeded directly on them. As bio-inspired biomaterials get more sophisticated and are being designed to interact with the biological environment and influence cell behavior, methods are needed to monitor the effects of biomaterials, particularly in a way that gives spatial information that can be related to the structure of the biomaterial. Stem cells are of increasing importance, and bio-inspired materials can influence stem cell behavior, particularly proliferation and differentiation. In the field of mineralized tissue regeneration, there is an emerging technology that greatly simplifies the analysis of osteogenic differentiation known as type I collagen/green fluorescent protein (GFP) reporter technology. In this chapter, we explain the basics of how the technology works, the methodology, and

Bio-inspired Materials for Biomedical Engineering, First Edition. Edited by Anthony B. Brennan and Chelsea M. Kirschner.

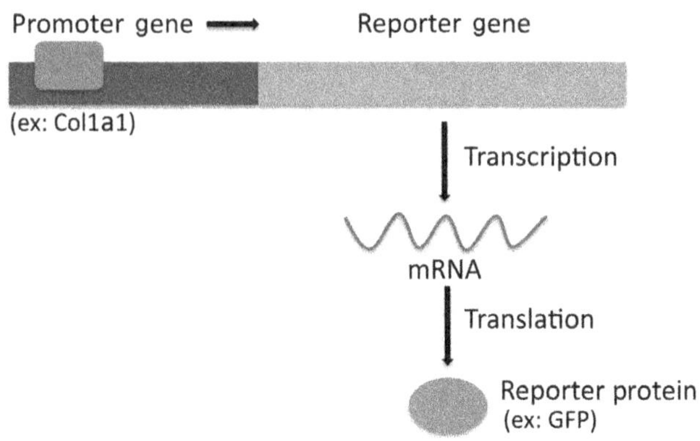

FIGURE 14.1 A reporter gene added to the DNA gene sequence produces a readily detectable fluorescent reporter protein, indicating that the functional promoter gene of interest has been expressed.

provide examples of use and advantages of this technology when applied to several types of bio-inspired biomaterials and cell therapy.

14.2 COLLAGEN 1 PROMOTER/GFP REPORTER TECHNOLOGY

14.2.1 Reporter Gene Systems

The use of a reporter gene system allows for the ability to track, monitor, and visualize cellular response to a given material or environment. A reporter gene is used to study the activity of a particular gene of interest called the promoter gene (Figure 14.1). A reporter gene system relies on the fusion of a promoter gene to a specific reporter gene and can be used to determine the promoter's activity under specific conditions or within a particular cell. In more simple terms, the reporter gene is attached to the coding region of the promoter gene within the DNA; when the promoter gene is biologically activated, it synthesizes not only its own regulatory proteins defined by that gene, but it activates the reporter protein expression. Ideally, the reporter gene should cause the production of a protein that is easily identified and monitored such as a protein that does not occur within the normal system of that animal. A protein that is widely used for this purpose is GFP that is normal in jellyfish, but absent in mammals. The expression of the GFP allows scientists to detect when and where the promoter gene is active in the genome. The reporter system allows for the study of the promoter gene at various developmental stages, or allows visualization of when the promoter gene is regulated by an external stimulation.

Shimomura *et al.* discovered GFP in the jellyfish *Aequorea victoria* in 1962 [1]. GFP is a protein that exhibits bright green fluorescence when exposed to light

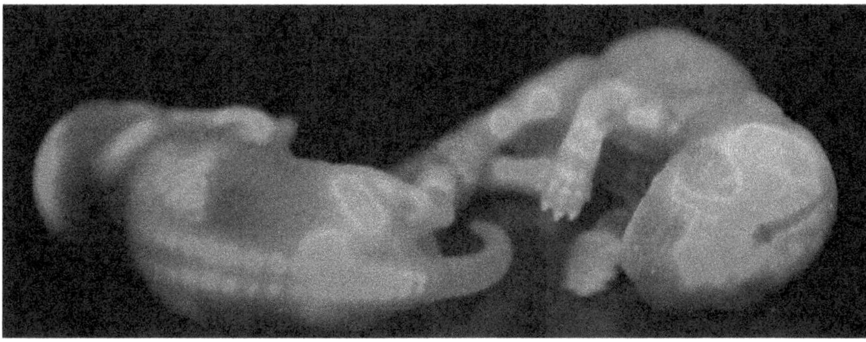

FIGURE 14.2 Under ultraviolet light, the bones of these transgenic mouse pups fluoresce green, indicating the successful incorporation of both the GFP reporter gene and the linked functional gene of interest. (Courtesy of Professor David Rowe of the University of Connecticut Health Center.) *(See insert for color representation of the figure.)*

in the blue to ultraviolet range. Since then, the GFP gene has been widely used as a reporter to study transcriptional activities within a broad range of hosts. If the cells are expressing GFP, the researcher knows that his or her promoter gene of interest also successfully made it into the reporter system and is being expressed. This is because GFP will be expressed only if the promoter is functionally active. GFP as a reporter can be used *in vitro* through the gene transfection or transduction process, or it can be used *in vivo* in transgenic animals. There are numerous benefits associated with the use of GFP as a reporter; however, the main benefits are the ability to visually monitor the promoter gene's activity, the broad host applicability, and the absence of cell lysis or necessary substrate addition that are common with other reporter genes [2].

Bioactive materials that enhance bone regeneration interact with and positively effect bone progenitor cells, also known as bone marrow stromal cells or mesenchymal stem cells (MSCs). In the field of orthopedic research, it is of particular interest to know if implanted biomaterials influence the osteogenic differentiation of progenitor cells. To track osteogenic differentiation of bone progenitor cells, there are several marker genes associated with osteogenesis, and each marker's promoter gene can be used to drive a reporter for event-specific imaging. Osteocalcin, type I collagen (Col1a1), and alkaline phosphatase (ALP) are important markers associated with matrix maturation at different stages of progenitor differentiation into the bone tissue [3]. Transgenic mice with GFP expression attached to a variety of promoter genes associated with musculoskeletal regeneration have been produced by Dr. David Rowe's lab of the University of Connecticut Health Center [4]. As seen in Figure 14.2, the bone progenitor cells within the mouse bones fluoresce depending on their level of differentiation, thereby allowing researchers to monitor the progression of bone progenitor cells as they differentiate into mature bone cells. This chapter will focus on the use of the Col1a1 promoter to drive GFP expression to identify various stages of osteogenic cell differentiation on biomimetic biomaterials.

14.2.2 How to Make a Reporter Gene System

14.2.2.1 Retrovirus There are two general categories of gene delivery systems: viral and nonviral. The majority of gene delivery systems use virus vectors. The reason why viral delivery is most commonly used is because viruses have evolved to be very efficient at replication and survival. One of the most crucial aspects of a virus is its ability to transport its genomic DNA to the nucleus of the host cells without degradation by lysosomes. Nonviral gene delivery systems include, for example, direct DNA delivery, liposomes, and DNA–protein complexes. The advantage of these nonviral systems is that they can be easily prepared in a uniform fashion and can be made in larger quantities; however, when compared to viral vector delivery systems, there is a reduced efficacy of expression [5]. Here we will focus on viral vectors, specifically retroviral vectors.

Retroviruses are enveloped RNA viruses that are members of the family Retroviridae, and they are widely distributed in various vertebrate species. These types of viruses are utilized because they contain the enzyme reverse transcriptase. Reverse transcriptase is essential for the conversion of single-strand RNA to double-strand DNA in the process of forming a provirus that is integrated into cellular DNA. Retrovirus vectors also integrate into the chromosomes of the host cells; therefore, they provide an opportunity for prolonged gene expression [5]. An essential step in the development of a virus vector is the creation of a recombinant virus that is capable of infecting a cell but is replication-defective. For retroviruses, this has been accomplished by deleting all the viral structural protein genes and inserting either a therapeutic gene or a marker gene (often called the transgene) [5]. Many retrovirus vectors are introduced *in vitro* into autologous stem cells, and then the cells are transplanted directly into the host for analysis *in situ*. It is expected that some of the transplanted cells will remain in an undifferentiated state and will act as a permanent reservoir of gene-corrected cells [6]. When the appropriate signals are received, the promoter gene will be activated and will cause expression of the reporter gene.

14.2.2.2 The Process of Creating Transgenic Mice Genetically manipulated mouse models make it possible to study gene functions in whole animals. Traditional gene knockout mice that represent a loss-of-function strategy have been widely used in research for many years. The transgenic mice discussed here represent a newer concept with a gain-of-function approach to define molecular and cellular functions of a gene of interest. This approach can be used to analyze tissue-specific or developmental stage-specific gene expression by introducing reporter genes, (such as GFP as previously discussed), under the control of a specific gene promoter. If the functional domain of a gene of interest is well characterized, transgenic mouse strains can be created [7].

Transgenic mice are most commonly generated by microinjecting the transgenic construct into the male pronuclei of a fertilized mouse egg. This method has five basic steps: (a) purifying the transgenic construct, (b) harvesting the donor eggs, (c) microinjecting the transgenic construct into the male pronuclei of the egg, (d) implanting the microinjected eggs into a surrogate female mouse, and (e) genotyping

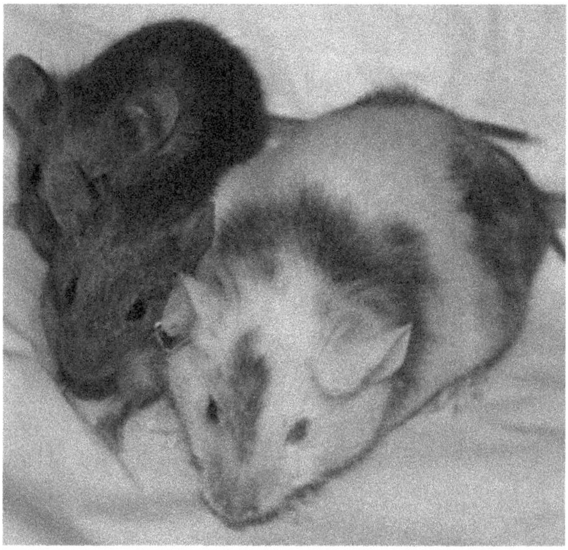

FIGURE 14.3 Transgenic mice are often developed from dark- and light-coated pairs to readily determine which of the offsprings contains the transgene. Shown here is the spotted chimeric mother and completely dark pups that are screened to be homozygous for the transgene. (Reprinted with permission from Reference 9.)

and analyzing the transgene expression in the offspring [8]. Usually, dark coated males and light coated females are selected to produce the initial fertilized egg because the researcher will be able to visualize which offspring contains the transgene; remember, the transgene is microinjected into the male pronuclei, therefore, mice with dark spots on their coats will contain the construct. The offspring of the surrogate are referred to as progeny or chimeric mice and will contain genetic material with and without the transgene and their coats will appear spotted (Figure 14.3 [9], mother). The progeny that screen positive for containing the transgene will be selected to mate. The screening is done by removing a small piece of tissue from the tail and examining its DNA for the desired gene, or by simply looking for GFP expression if GFP is the reporter and expressed in tissues near the skin surface. Mice that have the gene will be heterozygous (spotted) and are mated with wild-type mice. One out of four pups of the new litter will be homozygous for the transgene (will have a completely dark coat) (Figure 14.3, pups), and subsequent crosses (matings) of this mouse will produce the transgenic strain. Other ways to introduce the transgene into an egg is the use of a retrovirus vector as previously discussed, or to transfect a transgenic construct into mouse embryonic stem cells (ESCs). These ESCs can then be injected into mouse blastocysts. This method is useful to obtain a low copy number of the transgene in the mice, and it is also useful when embryonic death is expected in the resultant transgenic mice [7]. With both the retroviral and embryonic stem cell method, once the embryo has been implanted into a surrogate, the steps for producing the transgenic line are the same with the microinjection method.

Kalajzic *et al.* have developed transgenic mice by the use of a transgenic construct, in which GFP expression is under the control of the 3.6- and 2.3-kb Col1a1 promoter fragments [4]. GFP-driven 2.3-kb Col1a1 promoter fragment of the type I collagen gene (referred to as 2.3Col/GFP) is expressed in mature osteoblasts, harvested from the bony tissue of transgenic mice pups as described below, at the onset of mineralization. 3.6-kb Col1a1 promoter fragment driving a cyan variant of GFP (3.6Col/GFP) is expressed in preosteoblast cultures 5–7 days before the colonies develop into mineralizing multilayered nodules. The use of multiple promoters and fluorescent isomers of GFP makes it possible to monitor developmental stage-specific differentiation of the osteoblast lineage. Distinctly, different populations of cells within the osteoblastic lineage can be recognized—3.6Col/GFP indicating early osteoblast progenitors and 2.3Col/GFP indicating mature differentiated osteoblasts [4]. As a result of these transgenic mice, various stages of osteoblastic differentiation in cell culture and in intact bone can be assessed when evaluating novel, bioactive materials. This chapter is focused on describing the application of this technology to evaluate osteogenic differentiation.

14.3 PRIMARY CELL HARVEST AND IMAGE ANALYSIS OF THE COLLAGEN REPORTER CELLS FROM TRANSGENIC MICE

14.3.1 Harvesting Primary Osteoprogenitor Cells From Transgenic Mice Calvarium

In order to conduct *in vitro* studies with primary type I collagen/GFP reporter primary osteoblast cells, the cells must first be harvested from the bony tissues of transgenic mouse pups. Osteoprogenitor cells can be harvested from the bone marrow or the bones. The use of young mice greatly increases the yield of the progenitor cells, which are typically harvested from the calvarial bones. To obtain the cells from the calvaria, pups that are 3–6 days old are sacrificed with carbon dioxide. In a sterile environment, the euthanized pups are cleaned with 70% ethanol. The calvarial bones are gently cut out from the skull and removed as one piece, including both parietal bones and the sutures. Nonbony tissue is carefully scraped away and the sutures are then cut away and discarded. To release the cells from the bony matrix, the matrix and other tissues associated with the calvaria must be digested enzymatically using an enzyme solution prepared from a combination of collagenase I, phosphate buffered solution (PBS), and trypsin. Multiple digestions are performed to extract the cells. The first extract is discarded but the second through fifth solutions are kept. The digestion solution is added to equal parts medium with 10% fetal bovine serum to stop the enzymatic reaction and is stored on ice. Once the digestion steps are completed, the resulting cell solution is centrifuged and the cell pellet is resuspended in proliferative medium. Cells are counted and plated typically at 15,000 cells/cm^2. That is a suitable density to achieve confluency by day 7 and robust mineralization over the 21-day mineralization assay [10]. If the biomaterial degrades rapidly in

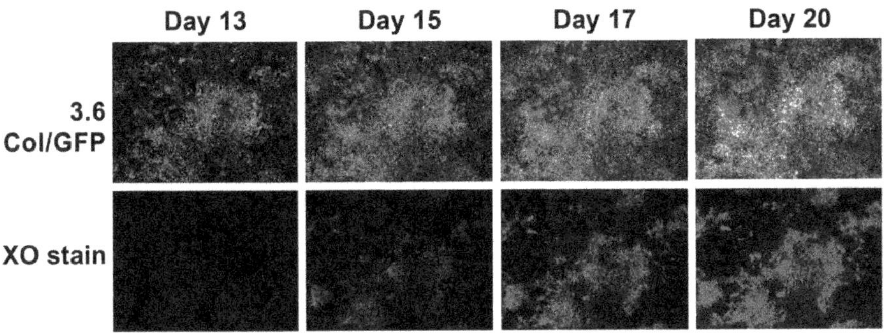

FIGURE 14.4 Fluorescence microscopy images of cultures expressing 3.6Col/GFP associated with preosteoblasts just prior to mineralization, and nontoxic xylenol orange (XO) staining of mineral taken from the same area in the cell culture plate at multiple time points. The cell reporter technology allows continuous monitoring of cell differentiation without requiring the use of dyes or antibody staining that require cell culture termination. (Courtesy of Yu-Hsiung Wang of the University of Connecticut Health Center.) *(See insert for color representation of the figure.)*

culture and is intended for short-term use only, the 3.6Col/GFP reporter system is preferred since the readout occurs earlier in the culture.

14.3.2 *In Vitro* Imaging and Analysis

Nondestructive fluorescence imaging can be utilized to view type I collagen/GFP-expressing cells during experiments without having to terminate plates. Not only can osteogenic cells be viewed, but the use of a fluorescent calcium chelating dye that binds to CaP mineral crystals, allows one to visualize the mineral associated with the GFP positive cells and estimate the calcium content in living cell cultures. One of the most commonly used dyes for assessing mineralization nondestructively is xylenol orange (XO). Calcium deposition or mineralization is one of several features that characterize bone formation or osteoblastic function/differentiation. In some cases, calcium deposition may be unrelated to osteoblast activity if extensive cell death occurs in the cell cultures, or if high amounts of osteogenic medium components are used that lead to artifactual precipitation. Distinguishing between calcium deposition associated with osteoblast-produced mineral and that from pathological or artifactual deposition requires additional structural and chemical characterization of the mineralized matrix and biological characterization of the cell. The use of the type I collagen/GFP reporter cells facilitates distinguishing mineral produced by osteoblasts. Using fluorescence imaging, one can determine if the stained mineralized nodules correspond, or overlap with, GFP expressing mature osteoblast cells. An example of colocalization of the GFP and XO is demonstrated in Figure 14.4. The stained mineralized matrix (red) overlaps with the green fluorescence representing osteoblast cells indicating that the GFP+ cells have formed the mineralized matrix. Using this reporter technology, one can also observe the increase in cell GFP expression with time without needing to terminate the cultures or apply any dyes or antibodies to the cultures. This is one key benefit of the technology.

In addition to visualizing cells and mineral deposition, using image analysis software allows the researcher to determine quantitative results such as percent GFP/XO-positive area, GFP/XO intensity, and calcium content. These values can be normalized to cell number by measuring total DNA content; however, DNA content is a destructive assay, therefore, it will require additional control plates. The imaging and analysis practice involves (a) collecting fluorescent microscopy images of the cells and (b) conducting image analysis of thresholded images of the standards and the samples to determine the area percentage and the mean intensity of the fluorescent areas. The use of a standard allows for the comparison between different samples and/or different time points.

To perform this type of imaging and analysis, the following equipment must be available:

- Fluorescent microscope with 10× objective and digital camera
- Appropriate filters for specific fluorescence of interest
- Camera and image collection software having a minimum resolution of 1000 × 1000 pixels, 14 bit
- Computer with image analysis software
- Image analysis software: National Institutes of Health has a publically available program called ImageJ that does not require a license to use

For each sample to be analyzed, a representative field of view must be acquired by collecting multiple images of the sample. An array of 1 × 10 adjacent images at 100× magnification (10× from the eyepiece and 10× from the objective) is sufficient for a six-well plate or smaller. The 10 adjacent images can then be stitched together into one image for the image analysis; stitching the images together saves analysis time, but is not required. In order to quantify the GFP or XO-positive areas within the images, the images are thresholded using an algorithm such as that in ImageJ to isolate the areas to be analyzed. Further explanation of the imaging process can be found in Kuhn *et al.* [10]

14.4 TYPE I COLLAGEN/GFP REPORTER SYSTEM WITH HUMAN CELLS

So far, this chapter has only discussed sourcing the cells for *in vitro* studies from transgenic mice harboring the type I collagen/GFP reporter systems; however, it is important to note that the type I collagen/GFP reporter system can also be used with human cells. Yin *et al.* have applied the type I collagen/GFP reporter system to two different *ex vivo* expanded human progenitor cell populations to assess their contribution to bone regeneration *in vivo* [11]. Regenerative medicine and tissue engineering approaches often involve the implantation of not fully differentiated progenitor cells which can give rise to a number of cell types, including osteoblasts (bone), chondrocytes (cartilage), adipocytes (fat). These cells may or may not be implanted with endothelial cells needed to reconstruct vascular beds. Knowledge of the exact mechanisms by which the implanted cells contribute to bone repair and regeneration is limited, and the fate of transplanted stem cells and the extent of their direct contribution to tissue regeneration remain controversial. Specifically, for *in vivo* bone formation, the exact contribution of implanted human cells remains uncertain, thus, Yin *et al.* applied the type I collagen/GFP reporter system to assess the contribution of human cells implanted in immunocompromised mice [11]. In these studies, the researchers produced a 2.3Col/GFP lentivirus and used that to transduce human cells to express GFP on type I collagen gene activation. They verified that the bone-specific promoter 2.3Col/GFP was able to define the fate of cells differentiating into the osteoblast lineage *in vitro* prior to the *in vivo* studies. Utilizing these transfected cells, Yin *et al.* were able to observe *in vivo* that, indeed, the implanted human cells had entered the osteoblast lineage [11]. The donor origin and cell-specific contribution to bone formation was specifically determined and currently, there is no other technology that offers this capability. This same procedure may be used in future studies to assess effects of biomaterials on the biological response of human progenitor stem cells.

14.5 EVALUATION OF BIOMIMETIC CHA THIN FILMS BY COLLAGEN/GFP REPORTER CELLS

14.5.1 Biomedical and Clinical Applications of Carbonated Hydroxyapatite

The mineral components of bones and teeth generally follow the hydroxyapatite (HA) composition ($Ca_{10}[PO]_6[OH]_2$), but with substitutions of carbonate for most of the OH– groups [12]. Accordingly, this has inspired the use of synthetic calcium phosphates (CaPs) and particularly synthetic HA in biomedical applications. Early investigations of HA implants demonstrated its biocompatibility and enhanced adherence of bone [13], which led to the use of HA coatings on implants [14], which is still a common practice today. In addition to coatings, CaPs are widely used in particulate form for dental restorative applications [15], for various bone filling and augmentation procedures, and more recently, for bone tissue engineering [16]. The results from clinical applications of HA are consistent with *in vitro* investigations, which have shown that HA enhances cell attachment [17], proliferation [18], and differentiation of the mineral-producing osteoblasts [19]. The cellular response to various bioactive ceramics has been reviewed elsewhere [20].

Understanding cell–CaP biomaterial interactions is complex, in part because there are many compositional and structural variations of CaP even within the subset of apatitic CaP [21]. Evaluation requires critical preparation and characterization of the CaP before the cell/tissue response can be interpreted. The variations among the forms of apatitic CaP are often subtle, but may strongly influence cell and tissue behavior. Accordingly, there is a continuing interest in synthesis methods, in

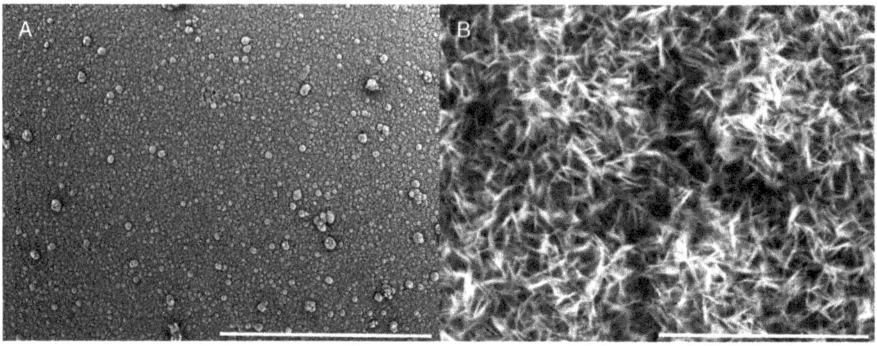

FIGURE 14.5 Scanning electron microscopy images of a carbonated hydroxyapatite coating at low and high magnification showing nanoscale, plate-like morphology. (A) Low mag. (100×) scale bar = 1 mm, (B) high mag. (5000×) scale bar = 20 μm.

understanding how CaP influences cell behavior, and in optimizing those effects for particular biomedical applications.

14.5.2 Synthesis and Characterization of Thin Films of Carbonated Hydroxyapatite

To demonstrate the use of the type I collagen reporter cells, here we describe studies conducted on thin films of carbonated hydroxyapatite (cHA). In biomedical applications, HA is used as particulates, coatings, in bulk form, and as a component of composites, but analysis of thin films allows more comprehensive characterization of the HA itself and facilitates the microscopic methods used to analyze the transgenic reporter cells. A modified method of direct deposition from a CaP solution [22] was used to obtain uniform, smooth, adherent coatings on tissue-culture plastic (TCPS) substrates [23]. In brief, disks of TCPS were sandblasted with Al_2O_3 powder to roughen the surface. The disks were then immersed in an ionic solution that included supersaturated amounts of calcium and phosphate ions. The solution is similar to blood plasma and is referred to as simulated body fluid (SBF) [24]. By controlling the pH, temperature, and concentration of the other ions, the Ca^{2+} and HPO_4^{2-} can be precipitated to form synthetic cHA. Conditions were controlled in a first solution to rapidly precipitate an amorphous cHA coating [25]. After drying, the disks were immersed in a second SBF solution with lower concentrations of apatite precipitation inhibitors to form a layer of crystalline cHA. The synthesis requires strict attention to all variables in the procedure. The high-quality surface is necessary for proper cell attachment and also because slight variations in the CaP stoichiometry can influence cell behavior [26].

Scanning electron micrographs (SEM) in Figure 14.5 show the expected morphology of nanoscale plate-like structures of the cHA coating [23]. Composition and crystallography were determined with energy dispersive x-ray spectroscopy (EDS), x-ray diffraction (XRD), and Fourier transform infrared spectroscopy (FTIR). All results were consistent with the formation of cHA. Specifically, there was no

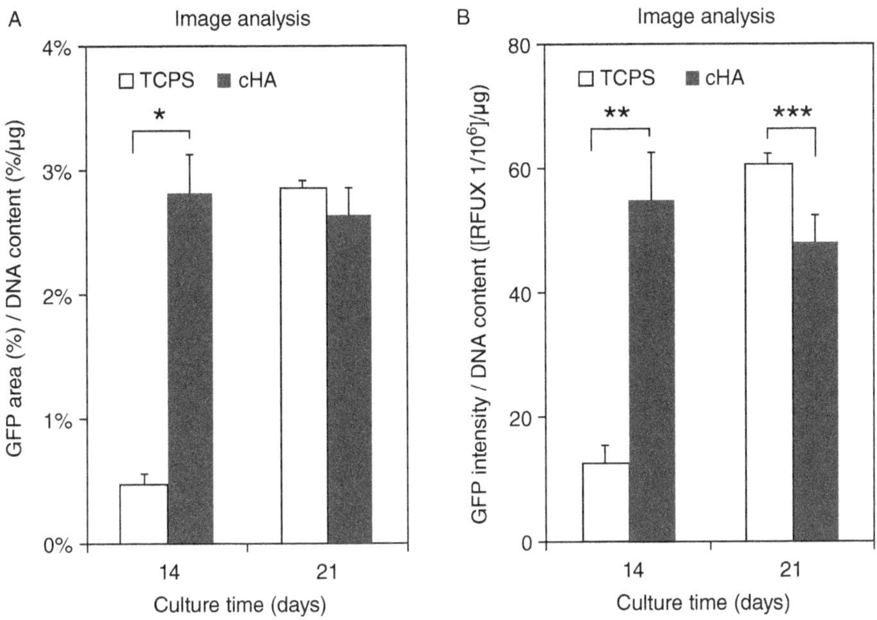

FIGURE 14.6 To assess differences between test groups, the GFP-positive area (A) or the intensity (B) can be quantified and normalized to DNA content. In this study, the carbonated hydroxyapatite surface accelerated differentiation and mineralization of the osteoprogenitor cells compared to the TCPS control as evidenced by more GFP expression at an earlier time point. (Reprinted with permission from Reference 10.)

evidence of potential alternative forms of CaP, including octacalcium phosphate, dibasic calcium phosphate dihydrate, or beta-tricalcium phosphate.

14.5.3 Assessment of Osteogenic Properties of cHA Thin Films Using Primary Type I Collagen/GFP Reporter Cells

Use of GFP reporter cells allows visualization and quantification of the extent of differentiation on a biomaterial surface. The images from fluorescence microscopy can be analyzed for area or intensity of GFP expression. Additionally, results may be normalized to the number of cells, depending on the intent/purpose of the study. Harvesting of the transgenic cells, seeding, culturing on the cHA biomaterial surface, imaging, and quantification methods were conducted as described earlier in the chapter. Figure 14.6 shows both the area and intensity of GFP expression on the cHA and TCPS control surfaces at 14 and 21 days determined by image analysis [10]. The results were normalized to DNA content to compare the effect of the surface on a per cell basis. The cHA accelerated differentiation relative to TCPS, as evidenced by earlier GFP expression in the cHA samples measured at 14 days. The cells on TCPS had reached the same level of differentiation by 21 days as evidenced by the comparable GFP expression at 21 days.

Figure 14.7 demonstrates some major advantages of the use of lineage-specific reporter cells in analyzing the effects of biomaterials on cell behavior. Bio-inspired

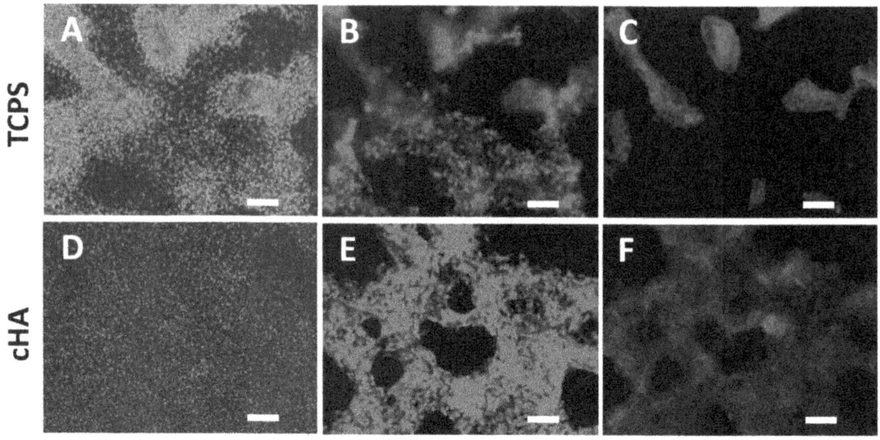

FIGURE 14.7 Fluorescence microscopy images of the calvarial cells from the transgenic mice on TCPS and cHA at 21 days with DAPI staining to show all cells (A,D), osteoblasts revealed by GFP expression (B,E), and XO staining for deposited mineral (C,F). Scale bar = 100 μm. (Reprinted with permission from Reference 10.) *(See insert for color representation of the figure.)*

materials, such as the cHA surface shown here, would be expected to influence more than simply attachment and proliferation. At 21 days, the blue DAPI stain which is taken up by all cell nuclei on TCPS shows the typical multicellular mineralized nodules, bright blue, (Figure 14.7A) that are confirmed to contain osteoblasts by GFP expression (Figure 14.7B) and mineral (Figure 14.7C). Mineralized nodules are recognizable in a TCPS cell culture dish to the trained eye even without any staining, but are difficult to assess on a cHA surface since both are a mineralized substance. Furthermore, in this experiment, the DAPI staining of cHA (Figure 14.7D) shows a uniform distribution of cells and no multicellular nodules typically associated with mineralization and hence, no indication of which cells have differentiated to osteoblasts and which have not. GFP expression distinguishes the differentiated cells (Figure 14.7E), and XO staining reveals that the mineralization pattern on cHA is more widespread and diffuse (Figure 14.7F) than on TCPS (Figure 14.7C). The XO staining is colocalized with the GFP expression, confirming that the mineral was produced by the osteoblasts. This finding could be determined through microscopic examination without complicated and destructive immunostaining. A major advantage of these reporter cells is that the cell cultures are not terminated, so images and quantitative analyses can be conducted over multiple time points. In this example, the spatial distribution information revealed important differences in the pattern of mineral formation. The cells were given an osteoinductive boost when grown on cHA, while cells on TCPS required cell–cell signaling possible in a nodule. This type of information would be even more important for evaluation of patterned or gradient surfaces, where the influence of composition or topography on differentiation and mineralization could be spatially and temporally continuously observed.

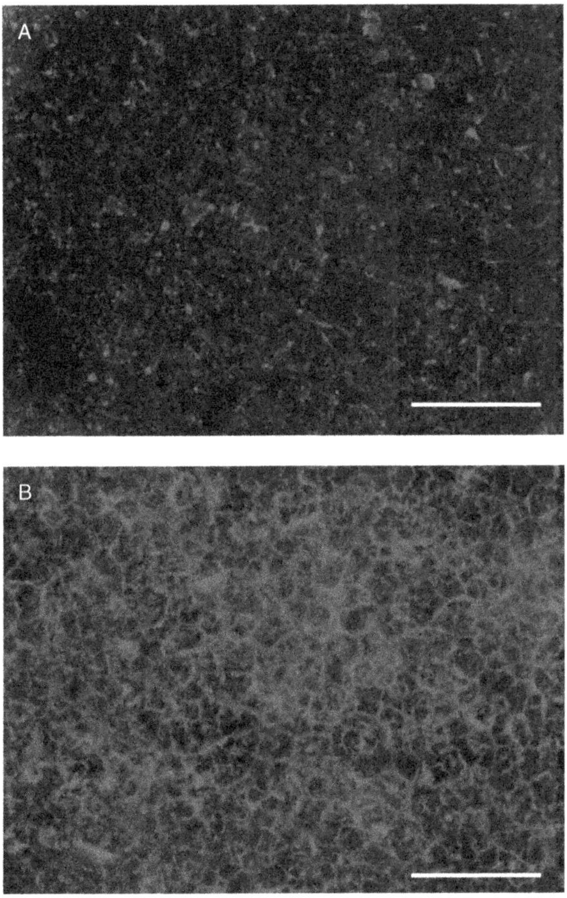

FIGURE 14.8 Differences in morphology revealed by the fluorescence microscopy images allow distinction between the cHA substrate (A) without cells and (B) with cell-deposited mineral. Scale bar = 200 μm. (Reprinted with permission from Reference 10.)

In evaluating mineral formation on HA biomaterial surfaces, some care is necessary to distinguish the cell-mediated mineral and the substrate. Additionally, nonphysiological mineral can be formed during cell culture of multipassaged cells [27]. The colocalization of XO staining with GFP expression, as shown in Figure 14.7, confirms that the mineral was cell-mediated. Traditional methods for detecting mineralization, such as von Kossa staining, lack this specific information and also require termination of the cultures. Furthermore, von Kossa staining cannot be used to evaluate cell response on CaP because of the high absorption of the von Kossa to the test substrate itself. While the cHA substrate also absorbs XO stain, as seen in Figure 14.8, the lower intensity, isolated islands on the XO-stained cHA substrate without cells (Figure 14.8A) could be distinguished from the more intense, continuous, web-like pattern of cell-deposited mineral (Figure 14.8B) [10]. Fluorimetry can also be used to quantify the extent of GFP expression and XO staining; however, image analysis of GFP-expressing cells and XO stain can reveal the spatial patterns. As an additional quantification method, cells from the reporter cultures could be removed and sorted by flow cytometry for more detailed molecular studies of pathways, receptors, and gene expression.

14.6 EVALUATION OF FIBRILLAR COLLAGEN THIN FILMS BY PRIMARY TYPE I COLLAGEN/GFP REPORTER CELLS

14.6.1 Biomedical and Clinical Applications of Fibrillar Collagen

Because of its prevalence throughout the extracellular matrix (ECM) and its ability to be processed into various forms, collagen is probably one of the original *bio-inspired* biomaterials. It is used in a wide range of biomedical applications. The mechanical characteristics, biocompatibility, and controllable degradation of collagen facilitates its use as a suture material, in drug delivery, and in fabrication of blood vessels and heart valves [28]. In addition, its ability to bind cells and influence cell behavior has made it well suited for a range of tissue engineering scaffold applications [29].

Type I collagen has a classic hierarchical structure where the polypeptide chains self-assemble into triple helices, which further assemble into fibrils. The high tensile strength of the fibers contributes to the mechanical properties of the ECM, but the structure, as well as the composition of the collagen, also influences cell behavior. Investigators have always appreciated the importance of the primary, secondary, and tertiary structures of collagen on cell behavior, but as the use of stem/progenitor cells has grown, there has been an increasing scrutiny of the effects of collagen's structure and particularly its fibrillar characteristics on cell behavior. Altering collagen's fibrillar structure and density can modify the morphology and proliferation of smooth muscle cells [30]. Others have studied the effect of collagen scaffolds on the progenitors of the bone-producing osteoblast cells. Altering the thickness of collagen fibers also affects osteoblasts [31]. Alignment of collagen fibers has been shown to influence the growth and differentiation of MSCs [32]. While not always carefully controlled, the method of preparing the collagen can significantly influence its structure, including the fibrillar morphology [33]. A group at the National Institute of Standards and Technology (NIST) has developed a standardized method for preparing thin collagen films to obtain a homogeneous, reproducible, fibrillar structure due to the demonstrated importance of its structure and the wide range of preparation methods in use [34].

The use of reporter cells to study this bio-inspired material is particularly appropriate since the progenitor cells of interest are likely to produce a heterogeneous population, not readily distinguished by traditional assays. Additionally, others studying osteoblast behaviors on collagen have reported formation of noncell mediated mineral [27,35], which may not be identified with standard methods. Accordingly, in this section, we describe the use of GFP reporter cells from transgenic mice to study the influence of NIST-standardized fibrillar collagen (FC) thin films on the differentiation behavior of osteoblast progenitor cells.

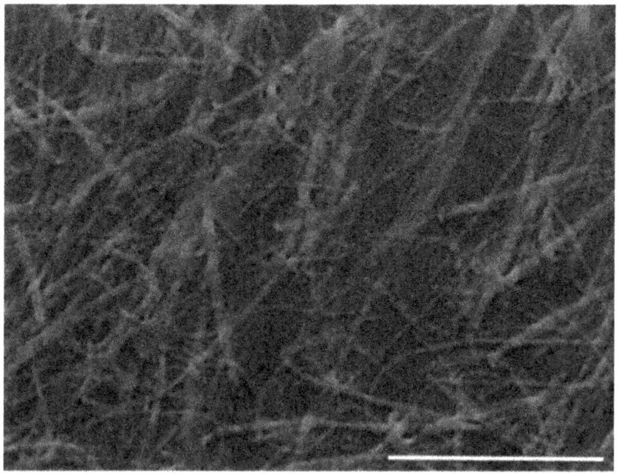

FIGURE 14.9 Scanning electron micrograph (SEM) of fibrillar collagen surface. Scale bar = 30 μm.

14.6.2 Synthesis and Characterization of Fibrillar Thin Films of Collagen

The standardized FC thin films were prepared following the method developed at NIST [30], except that here, the films were formed in 12-well nontreated tissue culture plates (BD Biosciences, San Jose, CA) instead of pretreated glass coverslips. In brief, a neutralized solution of bovine, type I collagen (PureCol, Advanced BioMatrix, San Diego, CA) was prepared in PBS. Next, 1.2 mL of the solution was placed in each well and maintained for 12–21 hours at 37°C. Excess solution was then removed and the adherent film was washed twice with PBS and nanopure water. Finally, the films were dried with a stream of nitrogen. Nonfibrillar films were formed in a similar fashion using a much lower concentration of the starting solution that prevents self-assembly.

For examination with SEM, samples of the thin films were fixed with 4% glutaraldehyde, dehydrated with sequentially increasing concentrations of ethanol, and then stored in a desiccator. The samples were sputter-coated with a 15-nm thick gold coating, mounted on stubs, and examined with SEM (TM-1000, Hitachi, Tokyo, Japan). The microstructure of the FC thin films (Figure 14.9) was consistent with earlier reports from NIST [30]. The diameter of the collagen fibrils ranged from 150 to 350 nm. The surface of the nonfibrillar collagen (NFC) films was featureless and devoid of fibers, except for occasional spherical clusters of collagen known to occur.

14.6.3 Results of Type I Collagen/GFP Reporter Cells Technology Applied to Fibrillar Collagen Films

Harvesting of the primary collagen type I reporter cell from the transgenic mice, seeding, culturing on the biomaterial surface, imaging, and quantification methods

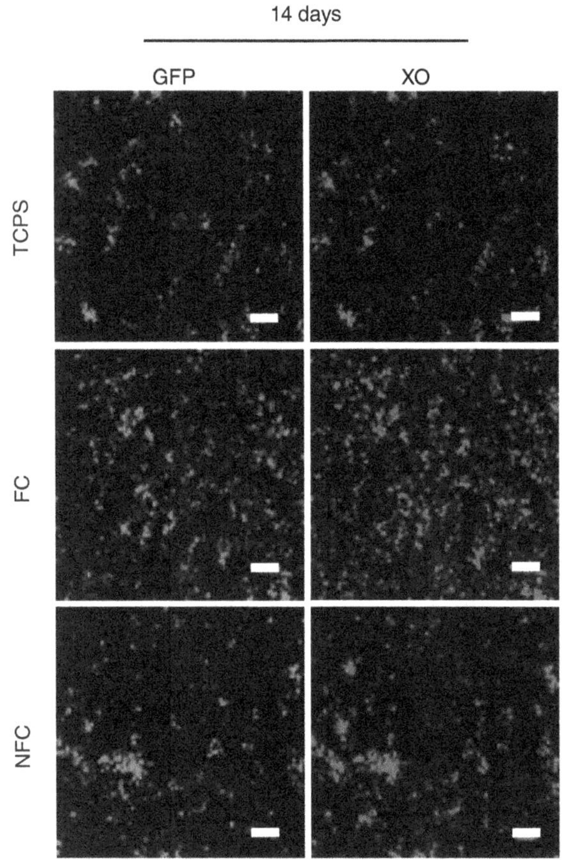

FIGURE 14.10 Fluorescence microscopy images of GFP positive (green) and xylenol orange staining (red) of mineralized matrix after 14 and 21 days in culture. Scale bars = 2 mm. *(See insert for color representation of the figure.)*

were conducted as described earlier in the chapter. Representative fluorescence images of GFP expression and XO staining on FC and NFC at 14 days in culture are shown in Figure 14.10. The various methods for analyzing the images were described in Section 14.3.2. Here, we chose to demonstrate relative intensities of GFP expression and XO staining on the FC, NFC, and TCPS control (Figure 14.11). Measurement of intensity requires selection of threshold values, which would normally be based on the fluorescence values of cultures at earlier time points.

The images and quantified values of GFP expression demonstrate the expected increase in cell differentiation to osteoblasts between 14 and 21 days on all surfaces. The increase in the intensity of the XO stain shows the expected accompanying increase in mineral formation. While the data are not shown here, expression of genes associated with osteoblasts were evaluated and found to be upregulated and consistent with the increases in GFP expression. At 14 days, GFP expression and

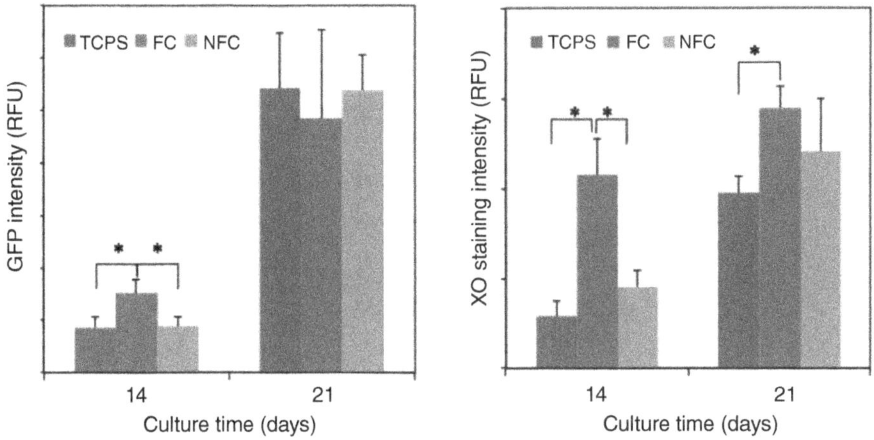

FIGURE 14.11 Relative intensity of GFP and XO staining fluorescence images of the 14- and 21-day cultures measured by NIH ImageJ. The fibrillar collagen (FC) surface accelerated differentiation and mineralization at the earlier time point.

XO staining were statistically greater on the FC film compared to the nonfibrillar film and TCPS control ($P < 0.05$). The earlier differentiation of the cells on the FC relative to the TCPS or NFC led to more mineralization matrix deposition. At 21 days in this analysis, differentiation was comparable on all three surfaces, although mineralization on the FC surface was statistically greater than on the TCPS control because the cells had an earlier start at forming the matrix ($P < 0.05$). Image analysis of the percent areas covered by GFP positive cells and positive XO staining was generally consistent with the results shown here. Normalization to DNA content demonstrated enhanced differentiation and mineralization of the FC surface at 21 days in culture. The data shown were obtained using freshly harvested cells from the transgenic mouse calvaria; however, experiments have been completed with reporter cells that were harvested, frozen, thawed, and plated on the FC and TCPS control. The frozen cells needed to be plated at a higher density ($1.5\times$) to reach confluency by day 7; however, they were also able to demonstrate differences in cellular response to the two different biomaterials. Thus, it may not be necessary to always use cells on the day of harvest. Passaging of the primary cells prior to plating for an experiment greatly reduces their GFP expression and is not recommended.

The GFP and XO images were merged to demonstrate colocalization, confirming that the mineral formation was cell-mediated in the FC and TCPS cultures. However, the high magnification inserts of the XO staining show an interesting exception on the NFC surface (Figure 14.12). Several isolated islands of red XO staining (center and lower right of insert) are not associated with the mineral-producing osteoblasts, indicating nonphysiological mineralization. Additionally, the GFP and XO fluorescence images could be compared with other physical characterization mapping methods, such as SEM topographical micrographs or EDS maps. This would allow a direct evaluation of the effects of chemical or topographical surface gradients on cell differentiation.

FIGURE 14.12 Merged fluorescence images showing colocalization of GFP expression and XO staining in cultures grown on nonfibrillar collagen (NFC). The inset image of XO staining shows isolated islands of mineral not associated with GFP positive differentiated cells. *(See insert for color representation of the figure.)*

14.7 *IN VIVO* USE OF TYPE I COLLAGEN/GFP REPORTER MICE TO SCREEN BIOMIMETIC COLLAGEN/ HYDROXYAPATITE SCAFFOLDS

14.7.1 Biomedical and Clinical Applications of Collagen/Apatite Scaffolds

Collagen–Apatite (Col–Ap) composites have been extensively studied for bone tissue engineering applications due to their biocompatibility and their resemblance to natural bone. Collagen has high affinity to cells and good resorbability but poor mechanical properties. Apatite, short for HA, exhibits good biocompatibility, osteoconductivity, and bone-bonding ability, but when used alone is brittle and slowly resorbing, thereby inhibiting bone in-growth. By adding apatite to collagen, the resulting composite's mechanical and resorption properties can be substantially increased [36]. Currently, there are commercially available collagen/apatite scaffolds used in orthopedic clinics, particularly for spine repair (e.g., Heolos®, DePuy Spine Inc., Raynham, MA). These scaffolds can be used in place of bone autograft but are less osteoinductive.

14.7.2 Synthesis and Characterization of the Collagen/Apatite Scaffolds

To make biomimetic Col–Ap composite, many researchers have used the coprecipitation approach; they allow type I collagen to mineralize in a CaP solution. This self-assembly closely mimics the biomineralization process found in nature. Coprecipitation allows for a high degree of control over the apatite content and crystal growth. In tissue engineering, it is important that the scaffold provide a three-dimensional (3-D) spatial and temporal structure to direct cell attachment, proliferation and differentiation, and guide tissue formation. Freeze casting, a method based on the physical properties of ice formation to form pores, is a popular method used to generate scaffolds with equiaxed or anisotropic lamellar structures. Xia *et al.* have recently investigated the freezing behavior of Col–Ap composites [36]. They have developed a fabrication process combining the novel biomimetic strategy with self-assembled collagen fibers with controllable freeze casting. The physical properties of the Col–Ap scaffolds were characterized via field emission electron microscopy (FESEM), FTIR, x-ray diffraction, and thermogravimetric analysis (TGA). Pore size and lamellar spacing were determined using a histological technique.

In vivo evaluation of the Col–Ap scaffolds' ability to promote osteogenesis was performed in 3.6Col/GFP transgenic mice [36]. One million calvarial osteoblast reporter cells were harvested from 3.6Col/GFP cyan transgenic mice (blue) and loaded onto the Col–Ap-35 wt% apatite with an equiaxed structure. The loaded scaffolds were implanted into calvarial defects made in 3.6Col/GFP topaz host mice (green). The mice were given an intraperitoneal injection of XO 24 hours prior to being euthanized to stain areas of active mineralization (red). Defects were allowed to heal for 4 weeks, then all mice were euthanized, and the defect area, together with the scaffold, was dissected from the surrounding tissue and prepared for histological analysis. Histology of frozen, nondecalcified sections is superior for visualizing both the GFP expression and the XO staining. GFP expression in the calvarial sections was determined to identify implanted cells and host cells as well as mineralized tissue. Cell nuclei were stained by Hoechst.

14.7.3 Results of Type I Collagen/GFP Reporter Cell Technology Applied to Assess *In Vivo* Osteogenesis of Collagen/Hydroxyapatite Scaffolds

After 4 weeks of implantation, more than 90% of the defect was filled with new bone. The 3.6 Col/GFP cyan-positive osteoblast cells (blue) were found distributed throughout the entire defect area; however, a few host cells (green) had also infiltrated and formed some of the new tissue. The defect area was highly cellular. An overlap of the XO stain and 3.6 Col/GFP cyan-positive osteoblast cells was clearly observed, and together, they marked the outer surface of the new bone. This suggests that new bone formation was mainly attributed to the implanted 3.6 Col/GFP cyan-positive donor cells. In comparison, very few host cells were observed in the defect area and they did not overlay with the mineral label (XO), suggesting the host cells did not strongly contribute to the new bone formation in the defect [36]. The use of multiple GFP reporter cells and transgenic mice expressing different colors allows the researcher to not only observe new bone formation, but it allows the researchers to be able to determine what specific cells are responsible for the new growth. Without the use of the two reporters, the specific contributions of the host cells and the implanted cells on the new bone formation could not be determined.

14.8 CONCLUSIONS AND FUTURE DIRECTIONS

Type I collagen/GFP reporter technology is an effective tool to evaluate osteogenesis of biologically active biomaterials and osteoprogenitor cells. The technology can be applied in several different ways as demonstrated in this chapter—either by using primary cells harvested from the tissues of transgenic mice that harbor the type I collagen/GFP reporter or by transducing normal cells that originally did not have the reporter. The cells can be used for *in vitro* studies and cultured on biomimetic biomaterials or seeded onto scaffolds and then implanted into mice for *in vivo* osteogenesis studies. The primary cells can be frozen for later use or plated immediately. *In vitro* it is not necessary to terminate the cell cultures before conducting an analysis since the GFP fluorescence is brightly visible and can be detected with a standard fluorescent microscope. The fluorescence does not photobleach and test samples can be reanalyzed at later time points without damage to the fluorescence. In combination with nontoxic XO, the colocalization of the collagen-associated GFP within the cells with the mineralized matrix allows for a conclusive determination of the cell-mediated mineralization that often confounds these studies. This is a particularly attractive technique for biomaterials scientists that may not have the expertise in biological assays such as RT-PCR or immunostaining, since it replaces those analyses. When applied to study the fate and contribution of implanted cells, it simplifies the analysis leading to conclusive spatial information about which cells were directly involved in bone formation. To conclude, type I collagen/GFP reporter technology is a sophisticated and useful tool that is helping to accelerate our understanding of the influence of biomimetic biomaterials on osteoprogenitor differentiation.

Acknowledgments

The authors would like to acknowledge the financial support from the International Team for Implantology (ITI; Basel, Switzerland) grant # 348/2004, the National Institutes of Health R01DE021103-01, and the State of Connecticut Stem Cell Initiative.

REFERENCES

[1] Shimomura, O., Johnson, F.H., Saiga, Y. *Journal of Cellular and Comparative Physiology* **1962**, *59*, 223.

[2] Ghim, C.M., Lee, S.K., Takayama, S., Mitchell, R.J. *BMB Reports* **2010**, *43*, 451.

[3] Corn, D.J., Kim, Y., Krebs, M.D., Mounts, T., Molter, J., Gerson, S., Alsberg, E., Dennis, J.E., Lee, Z. *Journal of Orthopaedic Research* **2013**, *31*, 871.

[4] Kalajzic, I., Kalajzic, Z., Kaliterna, M., Gronowicz, G., Clark, S.H., Lichtler, A.C., Rowe, D. *Journal of Bone and Mineral Research: The Official Journal of the American Society for Bone and Mineral Research* **2002**, *17*, 15.

[5] Wivel, N.A., Wilson, J.M. *Hematology/Oncology Clinics of North America* **1998**, *12*, 483.

[6] Stover, M.L. *Molecular Therapy* **2001**, *3*, 8.

[7] Haruyama, N., Cho, A., Kulkarni, A.B. *Current Protocols in Cell Biology* **2009**, *Chapter 19*, Unit 19 10.

[8] Cho, A., Haruyama, N., Kulkarni, A.B. *Current Protocols in Cell Biology*; **2009**, *Chapter 19*, Unit 19 11.

[9] Transgenic Core Facility. Photo gallery. 2013, National Institute of Health, National Institute of Mental Health.

[10] Kuhn, L.T., Liu, Y., Advincula, M., Wang, Y.H., Maye, P., Goldberg, A.J. *Tissue Engineering. Part C, Methods* **2010**, *16*, 1357.

[11] Yin, D., Wang, Z., Gao, Q., Sundaresan, R., Parrish, C., Yang, Q., Krebsbach, P.H., Lichtler, A.C., Rowe, D.W., Hock, J., Liu, P. *Molecular Therapy* **2009**, *17*, 1967.

[12] Loong, C.K., Rey, C., Kuhn, L.T., Combes, C., Wu, Y., Chen, S., Glimcher, M.J. *Bone* **2000**, *26*, 599.

[13] Jarcho, M., Kay, J.F., Gumaer, K.I., Doremus, R.H., Drobeck, H.P. *Journal of Bioengineering* **1977**, *1*, 79.

[14] Lacefield, W.R. *Annals of the New York Academy of Sciences* **1988**, *523*, 72.

[15] LeGeros, R.Z. *Advances in Dental Research* **1988**, *2*, 164.

[16] Zhou, H., Lee, J. *Acta Biomaterialia* **2011**, *7*, 2769.

[17] Sawyer, A.A., Hennessy, K.M., Bellis, S.L. *Biomaterials* **2007**, *28*, 383.

[18] Melville, A.J., Harrison, J., Gross, K.A., Forsythe, J.S., Trounson, A.O., Mollard, R. *Biomaterials* **2006**, *27*, 615.

[19] Kim, S.S., Park, M.S., Gwak, S.J., Choi, C.Y., Kim, B.S. *Tissue Engineering* **2006**, *12*, 2997.

[20] Knabe, C., Ducheyne, P. in *Bioceramics and Their Clinical Applications* (Ed.: T. Kokubo), Woodhead, Cambridge, UK: 2008.

[21] Elliott, J.C. *Structure and Chemistry of the Apatites and Other Calcium Orthophosphates*, Vol. 18; Elsevier, Amsterdam: 1994.

[22] Barrere, F., van Blitterswijk, C.A., de Groot, K., Layrolle, P. *Biomaterials* **2002**, *23*, 2211.

[23] Goldberg, A.J., Liu, Y.X., Advincula, M.C., Gronowicz, G., Habibovic, P., Kuhn, L.T. *Journal of Biomaterials Science—Polymer Edition* **2010**, *21*, 1371.

[24] Kokubo, T., Kushitani, H., Sakka, S., Kitsugi, T., Yamamuro, T. *Journal of Biomedical Materials Research* **1990**, *24*, 721.

[25] Habibovic, P., Barrere, F., van Blitterswijk, C.A., de Groot, K., Layrolle, P. *Journal of the American Ceramic Society* **2002**, *85*, 517.

[26] Chou, Y.-F., Huang, W., Dunn, C.Y., Miller, T.A., Wu, B.M. *Biomaterials* **2005**, *26*, 285.

[27] Boskey, A.L., Roy, R. *Chemical Reviews* **2008**, *108*, 4716.

[28] Yannas, I.V. *Biomaterials Science*; Elsevier, Amsterdam: 2004.

[29] Gomes, M., Azevedo, H., Malafaya, P., Silva, S., Oliveira, J., Silva, G., Sousa, R., Mano, J., Reis, R. *Tissue Engineering*; Elsevier, Amsterdam: 2008.

[30] Elliott, J.T., Tona, A., Woodward, J.T., Jones, P.L., Plant, A.L. *Langmuir* **2003**, *19*, 1506.

[31] Vigier, S., Helary, C., Fromigue, O., Marie, P., Giraud-Guille, M.M. *Journal of Biomedical Materials Research. Part A* **2010**, *94*, 556.

[32] Lanfer, B., Seib, F.P., Freudenberg, U., Stamov, D., Bley, T., Bornhauser, M., Werner, C. *Biomaterials* **2009**, *30*, 5950.

[33] Dupont-Gillain, C.C., Pamula, E., Denis, F.A., De Cupere, V.M., Dufrene, Y.F., Rouxhet, P.G. *Journal of Materials Science: Materials in Medicine* **2004**, *15*, 347.

[34] Elliott, J.T., Woodward, J.T., Langenbach, K.J., Tona, A., Jones, P.L., Plant, A.L. *Matrix Biology* **2005**, *24*, 489.

[35] Khouja, H.I., Bevington, A., Kemp, G.J., Russell, R.G. *Bone* **1990**, *11*, 385.

[36] Xia, Z., Yu, X., Jiang, X., Brody, H.D., Rowe, D.W., Wei, M. *Acta Biomaterialia* **2013**, *9*, 7308.

CHAPTER 15

LEARNING FROM TISSUE EQUIVALENTS: BIOMECHANICS AND MECHANOBIOLOGY

DAVID D. SIMON and JAY D. HUMPHREY

Department of Biomedical Engineering, Yale University, New Haven, CT, USA

15.1 INTRODUCTION

Advances in tissue engineering and regenerative medicine have been tremendous over the past decade. Clinical successes include a number of implantable tissues, with the majority of available products being in orthopedic and wound healing arenas [1]. Still, a major challenge in research and development is optimization, both in the performance of the tissue-engineered construct and in the scale-up of processes required for commercialization.

Tissue engineering approaches also benefit basic research in biomechanics and mechanobiology. Tissue equivalents represent excellent model systems for studying cellular responses to mechanical and chemical stimuli, both of which can be well controlled *in vitro*. Whereas biomechanics seeks to understand bulk material behavior in terms of microstructure, which can be manipulated in a tissue-engineered construct, mechanobiology seeks to correlate cellular responses with mechanical stimuli, which is often difficult to infer *in vivo* and even in native tissues *in vitro*. Mechanically stimulated cells can alter their local environment, often by working on and remodeling the local extracellular matrix (ECM) and hence, bulk material behavior. Such changes, in turn, can change the mechanical stimuli sensed by the cell and thereby lead to additional cell-mediated changes of the construct.

A combined approach using tissue equivalent-based experiments and computational models of cell–matrix mechanics promises to provide increased insight into these complex couplings. Quantitative measurement of cellular gene expression and protein synthesis can be correlated with mechanical stimuli via mechanical testing

Bio-inspired Materials for Biomedical Engineering, First Edition. Edited by Anthony B. Brennan and Chelsea M. Kirschner.

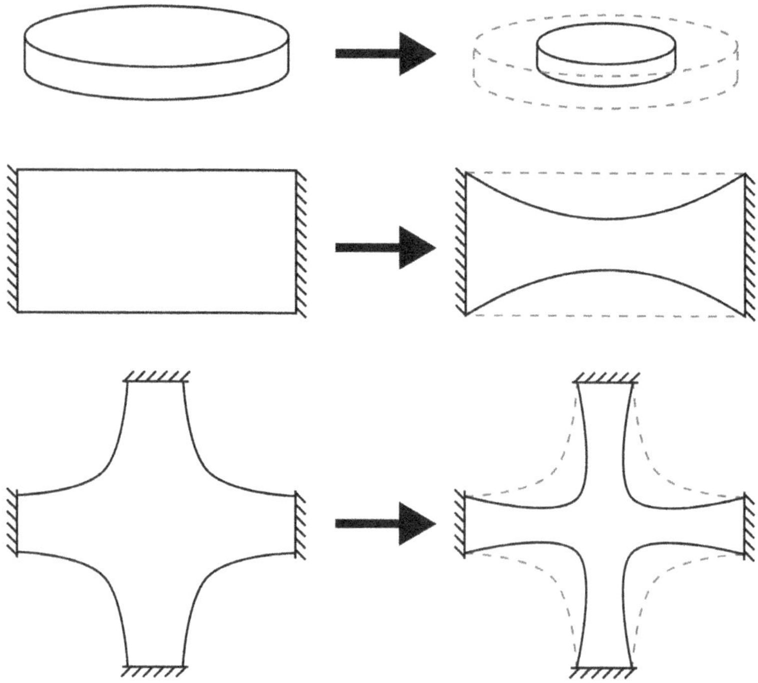

FIGURE 15.1 Schematic drawings of cell-mediated matrix compaction for free-floating (top), uniaxially constrained (middle), and biaxially constrained (bottom) tissue equivalents. From an initial geometry (left), cells begin to spread and retract, resulting in matrix compaction that changes the overall geometry of the tissue equivalents (right).

and characterization of tissue equivalents. This information can then inform computational simulations to correlate how the mechanical stimuli and subsequent cellular responses lead to matrix remodeling.

In this chapter, we review prior observations of three types of tissue equivalents (free-floating, uniaxially constrained, and biaxially constrained; see Figure 15.1), prior work in the mechanical characterization of the tissue equivalents, and prior theoretical work on mathematically modeling the growth and remodeling (G&R) of tissue equivalents. Finally, we propose a possible methodology for combining experiments and computational simulations to examine cell-mediated remodeling of tissue equivalents.

15.2 BACKGROUND

15.2.1 Matrix Composition and Integrins

The composition of the ECM varies depending on tissue type, *in vivo* location, and tissue health. Most components of the ECM are proteins, with collagen often being the most prevalent. Collagens consist of large domains of the repeating peptide sequence Gly-X-Y that are folded into a triple-helix structure, with X and Y representing different amino acids, often proline or hydroxyproline. The collagens are divided into subfamilies, mainly fibrillar (e.g., type I, II, and III) and network-like (e.g., type IV), with most of the other collagen types playing accessory roles for the two dominant subfamilies [2]. Of the fibrillar collagens, type I is found throughout the body and is a substantial structural component in tissues such as skin, arteries, tendons, and ligaments. Type II collagen is found primarily in cartilage, whereas type III collagen is often associated with type I, except during early wound healing when it precedes type I. During the formation of collagen fibers, lysyl oxidases convert some of the hydroxylysine and lysine residues into aldehydes, which allows the formation of covalent cross-links between fibrils and aids in the aggregation of collagen fibers [3]. Collagen is continually turned over by cells (i.e., degraded, removed, and replaced), which allows tissue to adapt to changes in both the mechanical and chemical environment [4].

Other common ECM components include elastic fibers, adhesion molecules, and proteoglycans (PGs), which consist of a protein core and associated glycosaminoglycans (GAGs). Elastic fibers consist primarily of elastin, but also associated glycoproteins such as the fibrillins and fibulins; elastic fibers endow tissues with resilience, recoil, and greater degrees of extensibility than does collagen alone. Elastin tends to be produced primarily during the perinatal period; hence, it has a minimal role in tissue adaptation and wound repair [5]. Among the many adhesion molecules, fibronectin is a glycoprotein that is important in the migration and localization of many cell types due to its repeated arginine–glycine–aspartic acid (RGD) domains, which serve as ligands for multiple cell–matrix adhesion receptors. Fibronectin can also bind a variety of growth factors that can contribute to cellular differentiation [5]. PGs and GAGs play diverse roles in tissue maintenance and adaptation. They are important in the assembly or organization of other matrix components, including collagen, through the creation of a swelling pressure due to abundant fixed negative charges. By sequestering water, they help resist compressive loads, which is particularly important in tissues such as cartilage. PGs and GAGs can also sequester growth factors and thereby modulate cell behavior [6].

Cell–matrix interactions are important in mechanical sensing by cells and the regulation of cellular gene expression and protein synthesis, which together enable physical remodeling of the matrix. The primary receptors responsible for cell attachment to the ECM are the integrins. These heterodimeric structures consist of one α and one β subunit, each of which spans the plasma membrane of cells. The extracellular domain binds to ECM molecules and the intracellular domain is anchored to the cytoskeleton. This association with the cytoskeleton allows the transduction of mechanical stimuli (outside-in signaling) as well as the transmission of actomyosin-based forces to the matrix (inside-out signaling), and can thus affect multiple signaling pathways [7]. The α and β subunit pairing determines the ligands with which the integrin will associate. For example, α_v-based integrins associate with the RGD peptide found in fibronectin and vitronectin, while $\alpha_1\beta_1$ and $\alpha_2\beta_1$ associate with collagen through the GFOGER peptide sequence.

15.2.2 Fibroblasts

The cell type often associated with connective tissue development, maintenance, and remodeling is the fibroblast. Fibroblasts are found in many tissues of the body and can display different cellular behaviors depending on embryonic origin, surrounding matrix, and biochemical cues [8]. Fibroblasts can produce a wide range of ECM proteins, including collagens, fibronectin, and a wide array of PGs and GAGs [9]. Fibroblasts can also differentiate into the myofibroblast phenotype, which is often indicated by the cellular expression of alpha smooth muscle actin, denoted α-SMA [10]. This phenotypic change to a myofibroblast appears to require a tensile mechanical environment and cell exposure to transforming growth factor beta (TGF-β1, one of three primary isoforms); it increases the magnitude of force the cell can apply to the surrounding matrix [11].

15.2.3 Growth Factors and Culture Media

The terms *cytokine* and *growth factor* are often used interchangeably. These biomolecules are soluble proteins or glycoproteins that are secreted by cells and act nonenzymatically to regulate cellular functions via both paracrine and autocrine stimulation [12]. Functions regulated by growth factors can be diverse, and a single growth factor can elicit opposite effects (e.g., proliferation or apoptosis) in the same cell type due to differences in other cellular cues like the matrix environment or the presence of other growth factors. Two growth factors of importance for fibroblasts and collagen maintenance are platelet-derived growth factor (PDGF) and TGF-β1. PDGF can act as a chemotactic agent, which is important for wound healing to induce fibroblasts and other cell types to infiltrate the wound site. Furthermore, PDGF can induce proliferation, and it can stimulate the production of matrix components and matrix metalloproteinases (MMPs) that contribute to the degradation of many ECM proteins. Hence, PDGF can play an important role in tissue maintenance and remodeling [13]. TGF-β1 has diverse effects on cells that depend strongly on other environmental factors. It has been shown, for example, to stimulate both proliferation and apoptosis, to be required for cell phenotypic modulation, and to increase matrix deposition while downregulating MMPs. In terms of cellular activity, TGF-β1 "seems capable of doing just about everything" [14]. Other cytokines of importance include the interleukins (e.g., IL-2 or IL-6). They are produced by leukocytes and often play important roles *in vivo* in tissue remodeling, wound healing, and disease progression.

Cell culture media for *in vitro* cell and tissue experiments is necessary to ensure the viability of resident cells, as well as to serve as a source of molecules needed for matrix production and other cell-mediated processes. Some commonly used formulations include Ham's nutrient mixture F12, minimal essential medium, and Dulbecco's modified Eagle's medium (DMEM; Sigma-Aldrich, St. Louis, MO). Basic constituents of culture media include organic salts that maintain osmolality and act as pH buffers, essential amino acids needed for protein synthesis, vitamins for metabolic activities, and glucose for energy. To provide other factors needed for cell maintenance, serum is often added to culture media. Serum provides a broad spectrum of growth factors and other proteins including albumin. In lieu of serum, individual growth factors and proteins can be added to allow delineation of their individual effects in a given experiment [15].

15.2.4 Mechanics and Mechanobiology

It has been known at least since the time of Borelli (1608–1678) that mechanics plays important roles in biology. Nevertheless, it was not until the mid-1970s that it was shown experimentally that cells respond directly to changes in their mechanical environment, often via changes in gene expression. Indeed, more recently, it was even shown that matrix stiffness can contribute to stem cell differentiation. Whereas continuum biomechanics focuses on mechanical responses of cells, tissues, and organs to applied loads (under conditions of interest), mechanobiology focuses on biological responses of cells to mechanical stimuli. Mechanics and mechanobiology are thus allied fields—biological responses by cells can change tissue geometry, properties, and even loads, which in turn can alter the mechanical properties and hence, mechanical responses by that tissue, which in turn can change the mechanical stimulus sensed by the cell. Mechanobiology often involves transduction (i.e., conversion of a mechanical stimulus to a chemical signal), transcription (i.e., an associated change in gene expression), and translation (i.e., the resulting production of a protein). Clearly, therefore, systems biology is also a natural ally of biomechanics and mechanobiology, and there is a pressing need for multiscale models that address the multiple levels of response [16]. One of the key questions in mechanobiology is actually how the mechanical stimuli are transduced via corresponding signaling pathways [17]. Toward this end, there is a need for continued quantification of both the mechanical loads that act on cells and the subsequent biological responses, and to correlate how the former affects the latter. Results from studies on tissue equivalents promise to improve tissue engineering and regenerative medicine through proper mechanical and biological characterization [18].

15.3 PRIOR EXPERIMENTS

15.3.1 Free-Floating Cell-Populated Lattice

Although observing cells plated on different ECM proteins can lead to the collection of considerable information, adherent cells behave differently within 3-D matrices than when attached to 2-D surfaces [19–22]. More than 30 years ago, Bell *et al.* introduced the circular, free-floating fibroblast-populated collagen lattice (FF-FPCL) [23]. Collagen gels containing embedded fibroblasts were suspended in culture media, and over a period of days, the cells compacted the lattices and thus reduced the gel diameter. The resulting tissue was noted as resembling the skin and having a rubber-like consistency. The rate and extent of compaction could be varied depending on a number of factors, including cell concentration, collagen density, cell passage number, and the presence of inhibitors of cell contractility such as cytochalasin B. Many investigators point to this work as the beginning of tissue engineering,

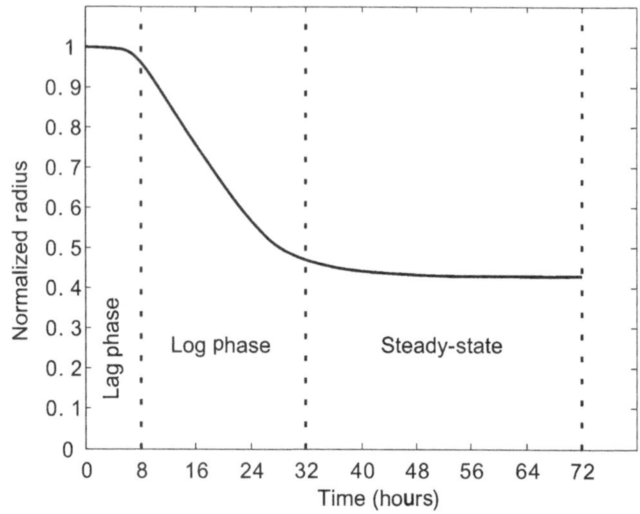

FIGURE 15.2 General trend of radius reduction during cell-mediated compaction of a free-floating cell-populated lattice. During the *lag* phase, cells adhere to and begin to spread within the matrix. The *log* phase is characterized by a rapid reduction in the radius. Eventually, the lattice reaches a steady state when compaction ceases. The duration of each phase is dependent on many factors and can vary considerably depending on experimental protocol.

but it also provided a simple experimental framework for studying cell–matrix interactions in a 3-D matrix.

The FPCL has been used predominantly as an *in vitro* model in wound healing research to examine cell–matrix interactions with an emphasis on understanding factors associated with wound contraction and closure. Detailed methods for the casting and culturing of FPCLs are described by Ehrlich [24]. Briefly, a solution containing known densities of collagen (usually type I) and fibroblasts is cast into a circular mold, often a bacteriological or tissue culture dish, allowed to undergo gelation, and then suspended in culture media. Over time (on the order of days to weeks), the embedded cells compact the collagen, which expels fluid from the matrix and reduces the overall volume of the lattice. The standard protocol described [24] results in an initial density of collagen $\sim$1.25 mg/mL and an initial concentration of fibroblasts of 50,000 cells/mL. The compaction of this standard configuration is characterized as occurring in two phases: an initial *lag* phase and a subsequent *log* phase (Figure 15.2) [25]. During the lag phase, there is little to no lattice compaction. This observation is likely due to the time required for the cells to begin adhering to and spreading within their surrounding matrix. After approximately 6–8 hours, lattices enter the log phase when the rate of compaction can increase greatly, then decrease gradually over multiple days until yielding an apparent steady state where compaction becomes minimal or ceases altogether. The compaction in this phase is believed to arise from cellular contractions (cycling of protrusion and retraction) associated with locomotion plus associated matrix reorganization. The degree of compaction is usually reported as a percent decrease in the circular surface area or diameter, which in standard experiments is on the order of 50–70%. These studies can be modified to examine changes in the rate of compaction, cell morphology and phenotype, and lattice organization due to a number of factors, including different cell types and concentrations, matrix compositions and densities, and the presence of exogenous growth factors.

In the initial experiments performed in Reference 23, differences in the rate and extent of compaction were noted when modifying either the collagen density or number of fibroblasts. Increasing the collagen density while using the same number of cells resulted in slower compaction, with the final extent of compaction reduced. The cessation of compaction is considered by some as a state of *equilibrium* at which the size and organization of collagen fibrils or fibers prevent further compaction [24]. For lattices with higher collagen densities, cells are surrounded locally by more collagen and can achieve such an equilibrium without reducing the overall volume of the lattice to the same extent as a lower density matrix.

Conversely, if the collagen density is held fixed and fibroblast concentration is increased, both the rate and the extent of lattice compaction increase [23]. When the cell concentration is increased in the standard protocol by an order of magnitude (0.5×10^5 to 5.0×10^5), the lag phase is decreased by 4–6 hours, and the log phase is completed after 6 hours rather than the 24 hours needed for the lower densities [25]. This change in the compaction profile is thought to be caused by forces generated by recently passaged cells that are spreading and elongating. Though spreading-related forces are weak when compared with cellular forces linked with migration and especially contractility, increased cell densities can increase greatly the total force applied to the lattice during cell spreading [26]. Of course, these forces may only affect the matrix in the immediate vicinity of a spreading cell. Yet, at high cell concentrations, individual cells may reside within close proximity to each other such that many collectively applied forces are exerted on the same volume of matrix during initial spreading.

Different cell types produce different compaction rates and final degrees of overall compaction. Fibroblasts from different species as well as those from different tissue sources compact collagen at varying rates [27]. Increased passage number for a given cell line has also been shown to reduce the rates at which collagen is compacted [23,27]. For example, when comparing normal rat skin fibroblasts with a line of transformed rat sarcoma cells, it was found that normal cells compacted the matrix much more than the transformed line did [28]. By 7 days, however, the total extent of compaction was approximately equal for both cell lines and remained equal for the remaining period of culture (14 days). This pattern of compaction is likely due to differences in cell proliferation, as the number of transformed cells at 7 days was almost 100 times greater than the normal cells. When comparing bovine vascular smooth muscle cells with human dermal fibroblasts, the latter compacted gels more quickly and to a greater extent [29]. This difference is possibly due to the fibroblasts having a more elongated morphology compared with the smooth muscle cells, which would allow cellular locomotion and contractility to influence more surrounding matrix. Aortic adventitial fibroblasts and medial smooth muscle cells from the cynomolgus monkey were found to compact collagen lattices similarly over a 24-hour period [30]. Cells from different aged donors as well as those from pathological conditions can also exhibit differing degrees and rates of compaction [31]. Smooth muscle cells isolated from balloon-induced intimal thickening in rat aortas compacted collagen to a lesser degree than smooth muscle cells from the underlying media or from normal aortas. If cultured in plasma-derived serum instead of fetal bovine serum, compaction was decreased for all cell types; plasma-derived serum similarly resulted in less compaction by newborn (4-day-old) rat smooth muscle cells compared with those from young (8–10 weeks) and old (16–18 months) adult rats.

Matrix composition similarly plays an important role in the rate and extent of the cell-driven compaction of lattices. Multiple fibroblast lines have been shown to compact lattices comprised of different collagen types at different rates and to varying degrees [32,33]. Dermal fibroblasts compact type III collagen more quickly and to a greater extent than type I, while type II collagen is compacted more slowly and to a lesser degree than either type I or type III [32]. These cells had similar morphologies across all three collagen types, however. Yet others have found the opposite to be true with MRC5 fibroblasts, compacting type I collagen more than type III [33]. These differences could be attributed to a number of factors, including duration of pepsin digestion during collagen isolation, differences in initial collagen and cell densities, and potential differences between dermal and MRC5 fibroblasts [32–34].

Dermal fibroblasts also tend to compact collagen lattices to a greater extent than fibrin lattices, whereas gingival fibroblasts compact collagen to the same degree as dermal fibroblasts while compacting fibrin more and completely degrading it after 7 days in culture [35]. The addition of ε-amino-caproic acid (an inhibitor of fibrinolysis) to gingival fibroblast–fibrin cultures resulted in compactions similar to the untreated dermal fibroblast–fibrin lattices. This finding suggests that gingival fibroblasts may have increased fibrinolytic capability compared with dermal fibroblasts.

The addition of GAGs and PGs to collagen lattices can affect the compaction of the lattices. The addition of hyaluronan to collagen lattices has been shown to increase the degree to which smooth muscle cells can compact the matrix compared with collagen alone [30]. When CD44 binding of hyaluronan is blocked, compaction is comparable to collagen alone, which indicates that the cells may be able to use hyaluronan as an indirect linker to collagen to expedite matrix compaction. In contrast, the addition of decorin can reduce lattice compaction by hypertrophic scar fibroblasts, likely through the sequestration of TGF-β1, which is highly expressed by these cells [36].

The ability of cells to cross-link collagen is also important in the compaction of collagen lattices. Collagen is enzymatically cross-linked primarily through the action of lysyl oxidase, but can also be cross-linked by transglutaminases [37]. Pretreatment of dermal fibroblasts with a lysyl oxidase inhibitor, β-aminopropionitrile (BAPN) delayed the onset of compaction, and subsequent exposure during culture reduced the final degree of compaction [38]. If BAPN is added only during lattice compaction, lung fibroblasts maintain normal levels of compaction early in culture ($\sim$2 days) but exhibit reduced compaction at later times compared with untreated lattices [39]. Early compaction of the lattices may proceed normally assuming that

BAPN does not inhibit cell spreading or attachment to surrounding collagen. That is, the cells may continue to actively pull in collagen and compact the matrix. With the inhibition of lysyl oxidase, however, prior data suggest the cells are unable to covalently cross-link the collagen following the initial compaction, and thereby, they cannot entrench any of applied deformations within the matrix. In the normal culture environment, cells could cross-link their surrounding collagen to entrench the local compaction, then detach and adhere to other collagen to continue the process of compaction and cross-linking. Cross-linking may thus allow resident cells to develop a *residual matrix tension* in an attempt to stiffen their local matrix environment to achieve a preferred mechanical environment; this process or mechanism has been referred to as *tensional homeostasis*, which appears to be an *in vitro* example of a general process of mechanical homeostasis that promotes tissue formation, maintenance, remodeling, and adaptation [40–42].

Compaction of collagen lattices by fibroblasts also depends on specific integrin–matrix interactions [43]. Multiple reports have shown that fibroblast compaction of lattices formed from reconstituted type I collagen is achieved primarily through cell–matrix interactions via the $\alpha_2\beta_1$ integrin [44–46]. Altered matrix composition can allow compaction even in the presence of $\alpha_2\beta_1$-directed antibodies through cell attachment to other matrix components that are entangled with the collagen [30]. Moreover, cells that are deficient of the $\alpha_2\beta_1$ integrin are still capable of compacting collagen lattices via the $\alpha_v\beta_3$ integrin [47,48]. This difference in adhesion can also lead to changes in mechanobiological and biochemical responses as, for example, PDGF increases compaction in $\alpha_2\beta_1$-mediated gels but not in those mediated via $\alpha_v\beta_3$ [47].

The culture environment of collagen lattices can greatly affect their compaction. It has been shown that serum is required for the compaction of FPCLs, likely due to the presence of myriad growth factors [28,49]. Lattices cultured in the absence of serum can also compact if the culture media is supplemented with growth factors such as TGF-β1 and PDGF [50–53]. Compaction of lattices exposed to PDGF suggests that part of the process may be mediated through cell migratory processes [52,54]. Exposure to TGF-β1 leads to lattice compaction, but the stimulated cellular mechanism may be augmented when cells express the αPDGF receptor [50,53]. This suggests that TGF-β1 may induce PDGF release and subsequently increase compaction [53,55]. TGF-β1 has also been shown to increase the expression of integrins, which would allow cells to adhere to more collagen and to transmit more force to the matrix [56,57]. TGF-β1 is also necessary for fibroblasts to differentiate into the more contractile myofibroblasts, thus increasing the contractile force they can apply to their surrounding matrix in the free-floating collagen lattice, possibly through increased actin expression [11,58].

Gene and protein expression have been examined in free-floating lattices and are usually compared with levels expressed by cells in lattices that are allowed to adhere to the surface of the culture dish. It has been assumed by many that adhered matrices develop a tensile mechanical environment while free-floating gels remain mechanically relaxed, though the mechanical environment has not been explicitly measured or derived [11]. Cells in floating lattices have a reduced response to PDGF compared with adhered gels as measured by level of receptor autophosphorylation

[59]. Type I collagen expression is also decreased in floating constructs compared with monolayer cell cultures while MMP-1 levels are increased [45]. Blocking $\alpha_1\beta_1$ and $\alpha_2\beta_1$ integrins further downregulates type I collagen expression and upregulates MMP-1 in floating gels [45]. Expression of MMPs by ocular fibroblasts has similarly been shown to change throughout culture time [60]. MMP-1 expression is elevated at 9 hours and increases to day 1 before dropping off by day 7. MMP-2 and MMP-3 showed similar trends; however, total expression of both at 9 hours is markedly lower compared with MMP-1, with MMP-2 levels being significantly higher than MMP-1 and MMP-3 at day 7. Total protein measures for MMP-2 were also much greater than levels of MMP-1 and MMP-3. The varying levels of MMP activity throughout lattice compaction could possibly arise from a continually evolving mechanical environment. In examining cellular contractile components, anchored and floating lattices show similar levels of β-actin expression, but anchored gels show higher levels of α-SMA. In both configurations, TGF-β1 increases the ratio of α-SMA to β-actin [51].

Measures of gene and protein expression show that there are global differences between adhered and floating lattice cultures, but such measurements do not account for the potential of local variations within the matrix. Free-floating lattice compaction has been shown to lead to regional variations in cell and matrix alignment as well as cell phenotype [25,61]. After 24 hours in culture, two distinct fibroblast phenotypes and states of matrix organization can arise. The center of the lattice is populated by randomly oriented fibroblasts with densely compacted, randomly oriented collagen, whereas the periphery of the gel contains fibroblasts that have differentiated into myofibroblasts with both cells and collagen aligned parallel to the outer edge of the lattice. Similar results were shown by Costa *et al.* [62] in various constrained geometries—cell compaction of collagen lattices leads to fiber alignment parallel to any free or unconstrained edge (Figure 15.3). Although there are no externally applied loads or constraints on the system, the peripheral alignment is potentially due to the development of different local mechanical environments that could lead to differences in local gene and protein expression.

15.3.2 Uniaxial Collagen Gels

Externally applied mechanical loads can greatly alter cellular activity, and tissue equivalents represent controllable systems suitable for examining the mechanobiological responses of cells to such applied loads. Uniaxially constrained collagen gels are simple systems wherein the axial, or in-line, force can be measured and the axial strain can be controlled accurately and precisely. One of the first experiments carried out on uniaxial tissue equivalents sought to characterize cellular forces associated with locomotion [63]. A collagen solution containing either normal or diseased skin fibroblasts from human or calf explants was cast in a rectangular mold and allowed to polymerize around Velcro constraints; one end was connected to a strain gauge to infer the force generation and the other to a mobile vernier to adjust the length of the gel. It was reported that tension developed within 30–60 minutes of polymerization and increased rapidly over 6–12 hours with increases in tension slowing or ceasing by 24–48 hours. During the period of observation, the geometry of the

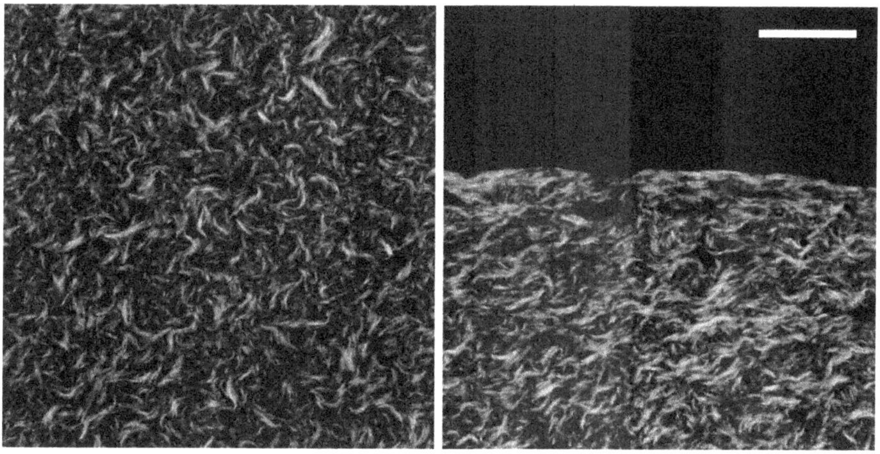

FIGURE 15.3 Picrosirius red stained free-floating fibroblast populated collagen lattice under circularly polarized light to show birefringent collagen. Central region (left) shows dense, randomly oriented collagen fibers while the outer edge of the lattice (right) shows aligned fibers for a lattice that has reached steady state (see Figure 15.2). Scale bar = 50 μm. *(See insert for color representation of the figure.)*

gel changed due to cellular compaction, including development of a parabolic shape (maximal gel width at the end restraints, minimal width at midaxial location) and reduction in thickness. After tension reached a steady state, the gels were lengthened or shortened to cause a step increase or decrease of 2 g of load. The cells restored the in-line force to the previous steady level in both cases within 1 hour.

This investigational setup has been modified, characterized, and used by multiple groups [41,64,65]. It can be used to perform experiments similar to those for the free-floating lattice to examine the effect of different initial or continual culture conditions on the mechanobiological responses to uniaxial constraints. Easy to implement variations include different cell types, cell concentrations, matrix type, matrix composition and density, static or cyclic loading conditions, and the effects of different growth factors and pharmacological agents. For example, Delvoye *et al.* noted that calf skin fibroblasts generated larger forces compared with human skin fibroblasts [63]. They also observed that calf dermatosparactic skin fibroblasts were less effective at generating tension within the uniaxial lattice when compared with normal fibroblasts, which likely results from reduced cell–matrix interactions observed previously in animals [66].

A similar experimental setup has been employed to measure forces generated by endothelial cells [64]. Instead of casting the cells within the collagen, an acellular collagen solution was cast, and following gelation, endothelial cells were seeded on the apical surface to form a monolayer. Whereas fibroblasts were found to reach a steady state force within 2 days, endothelial cells required 4–5 days in culture and only reached force levels an order of magnitude lower than the fibroblasts. This finding could have been due to the different casting methods, though no corresponding experiments were performed with fibroblast monolayers. Uniaxial equivalents also allow direct comparisons between cell types. For example, they have been employed to compare and characterize differences between human dermal and Tenon's capsule (ocular) fibroblasts [67]. The ocular fibroblasts showed a gradual increase in force generation that achieved only one-third of the force generated by dermal fibroblasts after 24 hours.

Tissue origin can also play a role in cellular responses as smooth muscle cells from different layers of piglet pulmonary arteries developed tensile forces in uniaxial tissue equivalents at different rates and to different extents in 24 hours [68]. Cells from the outer medial layers generated force more quickly than those from inner layers and they also generated more force. Cells sourced from hypoxic animals generated smaller forces overall, with cells from the inner media producing more tension than those from the outer layers.

Free-floating lattices have increased rates of compaction when cell concentrations are increased; the same holds for uniaxial tissue equivalents. Along the same line, increased collagen density increases the rate of force generation in tethered gels while reducing the rate and extent of compaction in free-floating constructs [63]. Increasing collagen density would increase initial substrate stiffness and apparently require less cellular compaction to achieve homeostatic or preferred forces [41]. The role of initial cell and matrix densities is likely important when trying to engineer potential regenerative therapies as well [69]. Too high of a cell seeding can lead to overly compacted and damaged constructs.

Uniaxial tissue equivalents also allow measurement of contractility by cells [64]. Similar to Reference 63, force generation began within 3 hours of casting fibroblast populated uniaxial tissue equivalents, increased rapidly over 24 hours, and reached a steady level in 48–72 hours that was maintained in cultures taken out to 7 days. To assess contractility, constructs were exposed to thrombin after reaching steady state, which induced an increase in force within 5–10 minutes that was sustained for several days. Once exposed to cytochalasin D to disrupt actin filaments, all tension in the system was lost within 10 minutes.

Whereas it was expected that disruption of actin would diminish tension development, it appears that microtubules can also influence cellular forces applied to matrices [64,70]. Disrupting microtubules within fibroblasts leads to an initial increase in force in uniaxial constructs that is gradually returned to prior steady-state levels. Microtubules help maintain cell shape and buffer the tension of actin filaments by acting as compressive supports. When disrupted, the tension in the actin cytoskeleton can no longer compress the microtubules, hence, leading to a net increase in the tensile force on the tissue equivalent. Over time, the cells seem to want to maintain a homeostatic level of tension and potentially relax intracellular tension to restore preferred levels [41,70].

As stated previously, cells appear to establish and then maintain a preferred level of tension, which is referred to as tensional homeostasis [41]. By cycling the length of uniaxial tissue equivalents, it became clear that the cells sought to restore at a preferred force level. On reaching steady state, constructs were then shortened abruptly, which resulted in a sharp decrease in force. Almost immediately thereafter, however, the force began to increase quickly before plateauing at a steady-state value.

Lattices were also cycled between lengthening and shortening. The cycling regimen started with an applied 0.6-mN increase above endogenous force, then followed a pattern of a 15-minute rest, a 1.2 mN unloading over 15 minutes, 15-minute rest, and a 1.2 mN loading over 15 minutes. This protocol kept the cycled force measurements centered about the endogenous force. Other variations of this cycling regimen have been used to cycle from the endogenous tension level to either 1.2 mN above or below it. When constructs were lengthened, force increased throughout loading and then immediately started to decrease during the resting cycle; when they were shortened, forces decreased and then gradually started to increase once at rest. These results again suggest that cells attempt to maintain a certain mechanical environment and they respond quickly to perturbations to actively restore a preferred state.

The idea of tensional homeostasis is supported further by the work of Marenzana *et al.* [40]. After adding cytochalasin D to release fibroblast-induced forces at various time points in culture, it was found that not all tension is released from the matrix and the amount of this residual matrix tension seems to increase linearly between 4 and 60 hours in culture. During this time, the total axial force in the uniaxial tissue equivalents reaches a steady level in 24 hours. It appears that the cells actively deform the matrix in an attempt to restore tensional homeostasis. To aid in this process, they will also remodel their resident matrix to build in tension (the aforementioned residual matrix tension) and potentially reduce the amount of active force they need to apply to maintain the preferred mechanical environment.

Growth factors and cytokines can also alter cell contractility and matrix compaction in uniaxial constructs. In particular, TGF-β1 has been investigated extensively [40,71,72]. TGF-β1 tends to increase the rate and extent of force generation, although it has been shown that high concentrations of this cytokine could potentially inhibit increases in force [71]. Exposure to TGF-β1 also appears to speed the development of the residual matrix tension and to increase fibroblast contractile forces greatly, possibly by inducing the myofibroblast phenotype [40]. TGF-β1 also alters cell responses to lengthening of the tissue equivalent as treated fibroblasts initially increase then reduce the tension relative to controls, which begin reducing matrix tension almost immediately following the perturbation [72].

The uniaxial mechanical environment can also lead to the up- or downregulation of multiple genes and proteins. If constructs are prestrained to increase matrix stiffness before the onset of force generation, the expression of multiple genes can be altered [73]. Prestrain significantly increases the expression of MMP-2 and tissue inhibitor of metalloproteinase (TIMP)-2 in human dermal fibroblasts while collagen type III expression is increased only at a high prestrain (10%), though not significantly due to a broad range of expression in control expression. Cyclic loading (one cycle per hour) of dermal fibroblast-populated constructs also leads to changes in protease expression [74,75]. MMP-2, MMP-9, and tissue plasminogen activator (tPA) levels are increased by cyclic loading while MMP-3 and urokinase-type plasminogen activator (uPA) levels are reduced. These levels can also be altered by construct geometry [75]. The upregulation of both MMPs and TIMPs due to increased mechanical loads suggests that the resident cells attempt to remodel the constructs to reduce the amount of tension in the system and restore tensional homeostasis.

15.3.3 Biaxial Collagen Gels

While an extensive number of studies have been carried out employing free-floating and uniaxially constrained collagen lattices, biaxial tissue equivalent-based investigations have been relatively few in number. Most of this work has focused on the design and characterization of biaxial systems, measurement of cell and mechanically induced tissue anisotropy, and mechanical characterization of equivalents following prescribed culture periods [76–78].

Similar to uniaxial tissue equivalents, a collagen solution containing cells is cast in a mold containing restraints, usually a porous plastic, to facilitate mechanical manipulation of the constructs. Two mold shapes are usually employed, either a square with restraints on each side or a cruciform shape with restraints at the end of each arm [76,77]. On gelation, the specimens can be mechanically loaded in culture either through the attachment of weights to the restraining bars to prescribe an isotonic load or by holding the specimen dimensions static and allowing the resident cells to compact the matrix isometrically and thereby develop endogenous loads [76,77]. Biaxial systems also allow multiple loading protocols, including equibiaxial stretching or loading (i.e., same stretch or force along each axis), non-equibiaxial stretching or loading (often proportionally), or strip biaxial stretching (one axis held fixed at original length while other axis is loaded).

Measuring biaxial strains in the central region (or arms of the cruciform specimens) can be achieved by tracking the positions of embedded markers within the gel, often microspheres [76]. The tracking of marker positions during both cell-mediated compaction and mechanical testing allows calculation of the deformation gradient tensor, $\mathbf{F}$, in 2-D and calculation of associated strain measures. Other groups have also used histological stains or ink on the specimen surface to track strains [78]. It has been shown that strain is often inhomogeneous across biaxial specimens during cell compactions, though they may be nearly homogeneous within the central region of the specimens [76].

Boundary conditions in culture can be used to change the organization of collagen fibers. In comparing the structure and mechanical behavior of biaxially and uniaxially constrained gels, it has been shown that uniaxial constraints result in a marked increase in tissue anisotropy following 3 days of cell-driven remodeling [79]. In biaxial gels, collagen fiber orientations remain randomly oriented in the central region during equibiaxial stretching or mechanical testing, indicative of an isotropic response. In contrast, uniaxial gels develop a preferred collagen alignment parallel to the constrained axis, which thus exhibits a markedly stiffer response in mechanical testing compared with the unconstrained axis.

The degree of tissue anisotropy may also be altered by imposing unequal initial stretches to constructs [77]. Applying two different magnitudes of stretch to each axis will increase collagen alignment toward the direction of larger initial stretch. With a cruciform-shaped sample, it is noted that the mechanical environment of the arms is uniaxial and thus, results in distinct alignment parallel to the corresponding axis regardless of the *biaxial* loading protocol. Moreover, if the arms of one axis of a cruciform sample are wider than the other axis, the fibers tend to align in the center of the tissue equivalent toward the axis of the wider arms [78]. As the ratio of widths of the arms is increased, the degree of alignment also increases. This finding is most likely due to a greater force generated by the larger number of cells in the wider arms of the gel compared with the narrower arms.

The application of external loads also leads to some initial alignment of collagen fibers [77,80]. As stated before, uniaxial constraining leads to alignment along the constrained axis during cell-mediated compaction. If that axis is then unloaded and the gel constrained or loaded perpendicular to the original constraint, the collagen lattice can become isotropically distributed [80]. If cultured out further, the cells will begin to reorganize the matrix and align collagen predominately with the new constrained axis. Polarized light microscopy has also been employed to monitor fiber alignment even during biaxial mechanical testing [81–83]. Displacement along one axis shows a shift in collagen fiber alignment toward the direction of the applied motion [81].

Fibroblast-populated fibrin tissue equivalents have also been used to examine the role of fiber alignment in a biaxial constrained environment [84]. After 10 weeks in culture, fibrin-based tissue equivalents showed marked changes in tissue composition. Collagen accounted for approximately 3–4% of the total dry weight of cruciforms at 10 weeks, with low levels of elastin present (~0.10–0.15% dry weight). Regional differences were also found, with the most pronounced being a higher percentage of collagen content in the narrow arms of cruciforms with geometrically induced fiber alignment.

15.4 PRIOR MECHANICAL ANALYSES

Given that embedded cells are highly responsive to their mechanical environment and changes therein, there is clear motivation to quantify the states of stress and strain imposed on a tissue equivalent during culture, as well as to quantify changes that result due to the action of the cells on the matrix. It is surprising, therefore, that appropriate mechanical quantification remains wanting in many regards. Indeed, not only has there been little attention to the macroscopic stress or strain fields, there has been less attention to the cell–matrix interactions.

15.4.1 Free-Floating Lattices

The first mechanical model of the free-floating collagen lattice sought to quantify the *traction force* associated with the cell [85]. A spherical geometry was employed as spherical symmetry simplifies both the problem formulation and solution. The ECM was modeled as a linear viscoelastic material described by the following stress–strain–strain rate relation:

$$\sigma_{\mathrm{ECM}} = \mu_1 \frac{\partial \epsilon}{\partial t} + \mu_2 \frac{\partial \theta}{\partial t} \boldsymbol{I} + \frac{E}{1+\nu}\left[\epsilon + \frac{\nu}{1-2\nu}\theta \boldsymbol{I}\right] \tag{15.1}$$

with material parameters μ_1 and μ_2 related to the shear, μ, and bulk, K, viscosities ($\mu_1 = 2\mu$, $\mu_2 = K - 2/3\mu$), E is the Young's modulus, ν is the Poisson's ratio, $\boldsymbol{I}$ is the identity tensor, ε is the infinitesimal strain tensor, and θ is the dilatation. The active stress (applied by the cells on the ECM) modeled as a *negative pressure* having the form

$$\boldsymbol{\sigma}_{\text{Cell/ECM}} = \tau_0 \frac{\rho n}{1+\lambda n^2} \boldsymbol{I} \tag{15.2}$$

where τ_0 is a traction parameter (units of force per matrix density, ρ, and cell density, n) and λ is a contact inhibition parameter. A constant value was applied for the cell traction parameter, and the boundary value problem was solved for the displacement field by imposing a traction-free (zero stress) boundary condition at the outer surface. The model showed that increasing the value of the traction parameter resulted in an increased rate of compaction or diameter reduction. Similar to experimental data, increasing the initial cell density also resulted in faster compaction. This model has been modified using a different material model and a time varying traction parameter to incorporate the lag in the onset of compaction in one variation as well as the incorporation of a cell–cell interaction force [86,87].

This viscoelastic model shows that the application of an active cellular stress can lead to overall compaction with corresponding changes in matrix and cell densities [85]. One drawback, however, is the assumption of linear material behavior and infinitesimal strains despite the actual gels experiencing large deformations (10% reduction in diameter). Moreover, the model was not implemented to explicitly solve for the stress within the gel, which could modulate cellular responses and lead to changes in the cell traction parameter. Finally, use of a viscoelastic model may be most appropriate to collagen gels that are subjected to step increases in force or displacement, whereas the time scale of cell-induced traction and contractility may be long enough that any viscous effects are negligible.

A second model of the free-floating collagen lattice employed finite element analysis [88]. The collagen matrix was modeled as a 3-D isotropic solid disk with a Young's modulus, E, between 25 and 50 kPa and a Poisson's ratio, ν, of 0.33 to allow compressibility. Cell traction was then simulated as a change in temperature, noting that most solids contract in response to a decrease in temperature. The simulation for a free-floating or unconstrained gel resulted in an inward contraction of the solid with negligible stress development. Although the model simulates a compaction of an unconstrained solid material, as noted by the authors, the use of thermal contraction has no physical meaning. That is, it models cell traction as an imposed internal strain or displacement, not as an internal or active force that leads to the deformation of the solid.

As stated previously, some have observed that the organization of the free-floating lattice varies regionally, namely, the cells and collagen fibers in the center remain randomly oriented, while at the periphery of the construct, they become highly aligned, parallel to the outer edge [25,61]. It appears, therefore, that the FPCL could be modeled as a cylindrical annulus exhibiting a nonlinear compressible material behavior, with axisymmetric properties that can vary with radial position [89]. Whereas an initial solution focused on the initiation of the compaction, assuming incompressibility, a subsequent development of the model assumed an extended

Blatz-Ko constitutive behavior to allow compressibility, resulting in the following general form for the Cauchy stress–stretch response:

$$t = \varphi_{iso}(r)\left(\frac{\mu}{III_C^{1/2}}\left(III_C^{1/2}\boldsymbol{I} - \boldsymbol{B}^{-1}\right) + t_a\boldsymbol{I}\right) + \varphi_{ani}(r)\left(\frac{c}{III_C^{1/2}}\left(IV_C^2 - IV_C\right) + t_c\right)\boldsymbol{e}_\vartheta \otimes \boldsymbol{e}_\vartheta \tag{15.3}$$

where μ and c are material parameters for the isotropic and anisotropic materials, respectively, and φ_{iso} and φ_{ani} are mass fractions for the respective materials. III_C and IV_C are invariants of the right Cauchy–Green tensor, $\boldsymbol{C} = \boldsymbol{F}^T \cdot \boldsymbol{F}$, with $III_C = \det\boldsymbol{C}$ and $IV_C = \boldsymbol{M} \cdot \boldsymbol{CM}$ where $\boldsymbol{M}$ is a unit vector defining the preferred direction for the anisotropic material. $\boldsymbol{B}$ is the left Cauchy–Green tensor, $\boldsymbol{B} = \boldsymbol{F} \cdot \boldsymbol{F}^T$. At any point along the radius, r, $\varphi_{iso} + \varphi_{ani} = 1$, with $\varphi_{iso} = 1$ at the center and $\varphi_{ani} = 1$ at the outer edge. The terms t_a and t_c are the actively applied stresses by the cells, which were determined from the solution, not prescribed a priori [90].

In satisfying equilibrium given the traction-free boundary conditions at the outer edge and the apical and basal surfaces, nontrivial deformations were admitted only in the presence of a residual-type stress field. The radial stress was compressive in the center of the gel and transitioned to a zero at the outer edge, satisfying the outer boundary condition. The circumferential stress was also compressive in the center, but tensile near the outer edge. The possible existence of this type of residual stress field was examined qualitatively through the experiments on compacted gels. A radial cut from the center of the gel to the outer edge resulted in a pronounced opening angle that is consistent with the release of residual stresses. Furthermore, creating a circular hole near the center of a compacted gel resulted in a narrowing of the hole, which is consistent with the presence of local compressive stresses and lengthening of compressed collagen fibers [90].

Although this analysis did not explicitly determine the stress state of a particular free-floating lattice, it provided a general characterization of the mechanical environment based on experimental observations of cell and matrix alignment. Although the assumption of global isotropy within the gel also mathematically admits a compaction that results in a zero stress field, such a solution requires the thickness of the matrix to decrease by the same degree as the radius, which is not realized experimentally [90]. Hence, it appears that the residual stress-type field is most likely.

The free-floating lattice construct can continue to serve as a great tool for understanding mechanobiological responses of cells, but the evolving mechanical environment still needs to be characterized precisely. That is, current mechanical models need to be extended to capture the evolution of the gel from a dilute isotropic construct to a dense, nonhomogeneous tissue-like structure, with improved estimates of applied cell tractions throughout. Prior generalized inferences on gene and protein expression may be confounded by a nonzero, radially varying stress field rather than the prior assumed relaxed or stress-free environment.

15.4.2 Uniaxial and Biaxial Lattices

The mechanical characterization of uniaxially and biaxially constrained collagen lattices follows more traditional methods of analysis as the nature of their construction provides a simple shift from culture to mechanical testing. One major detriment of the uniaxial setup, however, is these tissue equivalents are often only characterized uniaxially, which only provides information on material behavior in one dimension. Finally, careful consideration needs to be given with regard to a number of factors, such as how the sample is gripped, which can alter measured responses from the tissue [91,92].

Mechanical analysis is either carried out on a region of interest within the tissue construct or on the entire sample. Focusing analysis on the central region usually invokes Saint-Venant's principle; namely, the assumption that any inhomogeneity in force or displacement caused by loading at the boundaries of a sample is negligible at a sufficient distance from the sample edge, hence distributions of stresses and strains can be assumed to be homogeneous away from the edges. Assessment of whole constructs is more complicated as constrained tissue equivalents develop complex geometries, and the local distribution of cells and matrix can lead to complex material properties. Clearly, finite element methods become essential in such studies [93].

The primary focus of mechanical modeling of uniaxially and biaxially constrained tissue equivalents has been on the relation of matrix orientation to the mechanical state of the tissue. Some investigators have incorporated imaging systems into mechanical testing rigs to monitor matrix alignment throughout the testing. Imaging modalities include confocal and polarized light microscopy [94–96]. Both approaches have advantages and disadvantages. Confocal systems can provide detailed images of individual matrix fibers and cells as well as their volumetric distributions, but the long acquisition times may allow stress relaxation to occur. Polarized light microscopy allows a quicker acquisition of images and a larger field of view, but it does not provide volumetric assessment of alignment; it assesses alignment as perturbations to polarized light as it passes through a sample. Confocal microscopy may be more readily implemented in studies focusing on only the central region of constructs while the nature of polarized light microscopy makes it better suited for studies assessing the mechanics of entire tissue equivalents.

Mechanical models of constrained tissue equivalents typically employ either structural analog models or continuum models. Structural network-based models attempt to capture effects of individual matrix fibers [97]. Noting that collagen exhibits a nonlinear stress–strain behavior due to fiber crimp, an *effective* fiber stress–strain relation of the following form is often used:

$$S_f = A\left[\exp\left(BE_f\right) - 1\right] \tag{15.4}$$

Here, S_f is the second Piola–Kirchhoff stress of a fiber, E_f is the Green's strain of a fiber, and A and B are material parameters with A having units of stress. The next step is to prescribe an initial orientation to individual fibers, which is usually accomplished via a distribution function that describes fiber distributions acquired via imaging.

Fiber-based models have also been implemented in analytical continuum formulations when focusing on the central region of a tissue equivalent or through the use of representative volume elements (RVEs) in finite element models [81,82,93,98].

In a continuum, the total stress is equal to the sum of the stresses in all the modeled collagen fibers for a given deformation. However, such models do not account for interactions between individual fibers [98]. In some finite element-based approaches, RVEs are constructed to prescribe the microscopic fiber network at Gauss points of each element of the model. For more information on the implementation of this model, see Reference 93. This method allows modeling of entire tissue constructs having complex geometries and seeks to describe the macroscopic mechanical behavior via representative microscale fiber networks. A current drawback of these models is that they only account for the fibrous collagen network. During compaction and remodeling of collagen lattices, other matrix components may be deposited and may influence mechanical responses of the tissue.

Another method of modeling the mechanics of tissue equivalents is via traditional continuum approaches. This methodology typically involves specifying a strain energy function to describe the constitutive response. One model that could potentially apply to tissue-engineered constructs is an n-fiber family model that has been employed to characterize arterial mechanics; such models are meant to capture overall responses, not to describe fiber-level mechanics per se [99]. The strain energy takes the following form:

$$W = \frac{c_1}{2}(I_C - 3) + \sum_{\alpha}^{n} \frac{c_2^\alpha}{4c_3^\alpha}\left[\exp\left(c_3^\alpha\left(\lambda^{\alpha^2} - 1\right)^2\right) - 1\right] \tag{15.5}$$

where c_1, c_2, and c_3 are material parameters with c_1 and c_2 having units of Pa, α is the index of each fiber family, n is the total number of fiber families, λ^α is the stretch experienced by the α fiber family, and I_C is the first invariant $\boldsymbol{C}$, $I_C = \mathrm{tr}\boldsymbol{C}$. For the case of a constrained biaxial tissue equivalent, $n = 2$ fiber families could describe a matrix aligned with each of the principal axes. Again, however, there is a need in tissue equivalents to model the cell-mediated tractions.

15.5 GROWTH AND REMODELING (G&R) MODELS

A major need in the study of tissue equivalents is models capable of predicting the evolving nature of the constructs in terms of myriad factors, including externally applied loads (boundary conditions), internally applied loads (cell traction forces and contractility), cell-mediated matrix production and removal, and the effects of soluble factors on cellular activity. Toward this end, however, there is also a need for better experimental data. For example, the free-floating construct is mechanically simple and thus, mathematically tractable, yet there is currently no way to estimate the regionally varying force development within the matrix without applying a physical constraint to the system. Measuring the compaction of the lattice would provide some indirect information regarding cell-generated forces, yet the evolving residual matrix tension would complicate such inferences beyond the early stages of compaction.

There has been some initial work on modeling the G&R of uniaxially and biaxially constrained tissue equivalents, for which more data are available, including information on net forces generated during compaction. For example, matrix remodeling has been described using the theory of kinematic growth, which assumes that the total deformation gradient **F** can be decomposed into an elastic part (in response to externally applied loads) and a growth part (due to biological activity), which is assumed to occur in stress-free configurations [100]. Kroon adopted such an approach, with the contribution of the cells representing a rotation to realign matrix based on a mechanical cue (e.g., stress, strain, or stiffness) [101,102]. This approach indirectly accounts for cellular activity and models changes in the matrix as deformation instead of direct changes in the material constitution.

Other attempts to account for the evolution of tissue equivalents have sought to model changes in cell and matrix alignment as time varying entities without directly accounting for the mechanical cues that may drive or augment the remodeling process [80]. Such an approach is simple to implement as it only requires the assessment of cell and matrix orientation at multiple time points to fit a desired function, but it neglects contributions of other cell-mediated factors such as cross-linking and the development of residual matrix tension.

Another candidate approach to modeling the evolving properties of tissue equivalents, which could capture cell-driven alterations to the matrix in terms of changes to the strain energy function for the individual constituents, could be borrowed from modeling soft tissue G&R. A constrained mixture theory of G&R allows one to model separately the mechanical properties, rates and extents of turnover, and natural configurations of individual constituents [103]. Put simply, the stress in a body depends on the sum of the stresses in individual constituents, which need not be the same. Stress develops when a constituent deforms from its natural or stress-free configuration, which can evolve. For example, cross-linking of some fibers to build in a residual matrix tension could render their natural configuration different from the original natural configuration of similar fibers. The basic model requires that one construct a rule-of-mixtures strain energy function (i.e., $W = \sum W^{\alpha}$), where:

$$W^{\alpha}(s) = \frac{\rho^{\alpha}(0)}{\rho(s)} Q^{\alpha}(s) W^{\alpha}\left(\mathbf{C}^{\alpha}_{n(0)}(s)\right) + \int_0^s \frac{m^{\alpha}(\tau)}{\rho(s)} q^{\alpha}(s,\tau) W^{\alpha}\left(\mathbf{C}^{\alpha}_{n(\tau)}(s)\right) d\tau \quad (15.6)$$

The first part of the strain energy accounts for the contribution of constituents that were present at time zero and still are present at the current time s, where $\rho^{\alpha}(0)$ is the mass density of constituent α that was present at time zero, $\rho(s)$ is the mass density of the mixture at time s, $Q^{\alpha}(s)$ is the fraction of constituent α that was present at time zero and yet remains at time s, and $\mathbf{C}^{\alpha}_{n(0)}(s)$ is the right Cauchy–Green tensor that describes the deformation of constituent α present at time zero, relative to its individual natural configuration, to the current configuration at time s. The second part (integral term) of the strain energy accounts for contributions of constituents produced at any time $\tau \in [0, s]$ that remain at s, where $m^{\alpha}(\tau)$ is the mass density of constituent α produced at time τ, $q^{\alpha}(s, \tau)$ is the fraction of constituent α produced at time τ that remains at time s, and $\mathbf{C}^{\alpha}_{n(\tau)}(s)$ is the right Cauchy–Green tensor that describes the deformation of constituent α produced at time τ, relative to its individual natural configuration, to the current configuration at time s [104]. This approach has been used previously to model theoretical changes in the alignment of newly deposited matrix in a biaxial tissue equivalent under different loading conditions [105]. In these preliminary simulations, it was found that if new matrix is deposited in the direction of greatest principal stretch, the principal Cauchy stresses in the system can be restored to near homeostatic values. For the simulations, the production of a new matrix was a function of stress deviation from a preferred level of stress.

States of tissue reorganization with no deposition of new materials can also be accounted for in the G&R framework. This is particularly important in short-term (~5 days) culture of tissue equivalents where experimental evidence shows that there is little to no new matrix deposition [35,106]. Removal and deposition relations would still be needed, but for these early times, any removed constituent would need to be replaced by another one. Conceptually, this would amount to removing a collagen fiber at one orientation and replacing it with another fiber in a new orientation.

Current models for the evolution of tissue equivalents take into account some mechanical cues, but they do not directly account for chemical factors that modulate cell behavior. Recall that mechanical perturbations to tissue equivalents can lead to the cellular production of soluble factors that can have a paracrine or autocrine effect on cell-driven remodeling. Within the context of G&R, it would be possible to account naturally for the contribution of such factors in the mass production ($m^{\alpha}(\tau)$) and mass removal ($Q^{\alpha}(s)$, $q^{\alpha}(s, \tau)$) terms of the stored energy functions. A simple implementation of this approach would be to allow the concentrations of growth factors to modulate the rates at which a constituent is produced or removed. Similarly, the effectiveness of a soluble factor to modulate the tissue equivalent can also depend on the mechanical environment. For example, there is experimental evidence and subsequent network modeling simulations that have shown a reduced rate of enzymatic matrix degradation when a tissue equivalent is strained [107,108].

As with the free-floating lattice, constitutive models for actively applied forces or stresses are needed for constrained tissue equivalents. Similar to the mass production and removal terms, many factors can affect the ability of a cell to apply stress to its surrounding matrix. These can include, but are not limited to, matrix density, composition, and orientation, as well as soluble factors, cell phenotype, and expression levels of integrins and other cell surface proteins and binding domains. Of course, expressions for active cellular stresses would need to account for how the local mechanical environment in which the cells reside may modulate the level of stress a cell can apply.

The coupling of sequential mechanical testing data and G&R simulations may be possibly used to determine parameters related to matrix remodeling. This could be accomplished using an approach proposed to describe G&R of the lens capsule following cataract surgery [109]. Briefly, the mechanical behavior of the tissue equivalent could be described phenomenologically via a standard strain energy function (e.g., Fung-type) at any time during the culture period. This descriptor could then be used in a standard stress analysis to determine stresses and stretches within the construct to determine differences from homeostatic values that would be expected to modulate the compaction as well as deposition and removal of constituents, which could be captured with a G&R model. The results of the G&R could

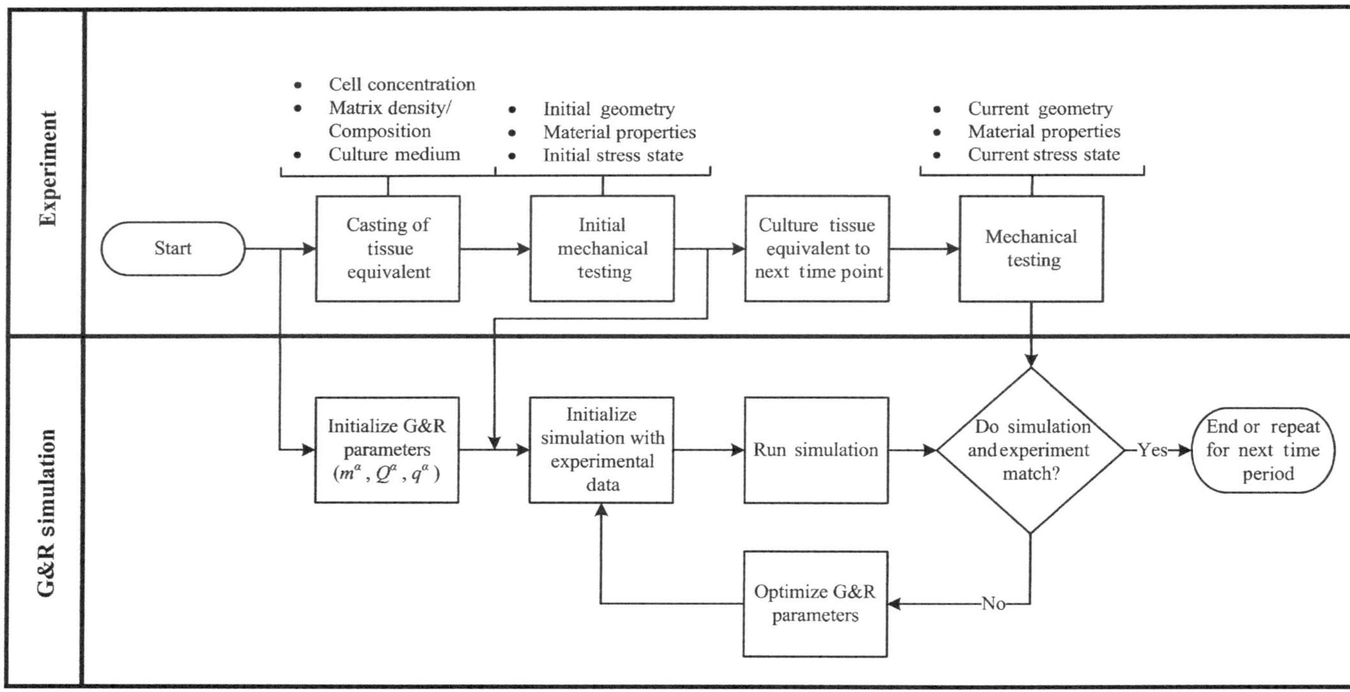

FIGURE 15.4 Example flow chart to implement a combined experimental–computational approach to study the mechanobiology of tissue equivalents.

TABLE 15.1 Summary of Prior Observations from Tissue Equivalent Experiments

Experiment type	Parameter	Observation	Reference
Free-floating	Cell concentration	If increased, increases rate/extent of compaction	[23,25]
	Cell type/origin	Varies rate/extent of compaction	[27–29,31]
	Matrix density	If increased, reduces rate/extent of compaction	[23]
	Matrix type	Can alter rate/extent of compaction	[32,33,35]
	Collagen I with hyaluronan	Increases rate/extent of compaction	[30]
	Collagen I with decorin	Decreases rate/extent of compaction (inhibits TGF-β)	[36]
	Matrix organization	Alignment parallel to edge, random orientation in center	[25]
	Covalent cross-linking	If blocked, decreases extent of compaction	[38,39]
	Integrins	Blocking $\alpha_2\beta_1$ and $\alpha_v\beta_3$ reduces/eliminates compaction	[44–47]
	Serum	Required for compaction	[28,49]
	Growth factors	PDGF and TGF-β1 increase rate/extent of compaction, induce compaction in lieu of serum	[50–53]
	Gene expression	MMPs vary throughout culture time	[60]
	Collagen I expression	Decreased (compared with adhered lattice)	[45]
	α-SMA expression	Decreased (compared with adhered lattice), upregulated by TGF-β1, can regionally vary	[51,61]
	Mechanical assessment	Residual-type stress field, compressive in center, tensile at outer edge	[89,90]
Uniaxial	Cell concentration	If increased, increases rate/extent of force generation	[63]
	Cell type/origin	Varies rate/extent of force generation	[63,64,67,68]
	Matrix density	If increased, increases rate/extent of force generation	[63]
	Contractility	Thrombin increases tensile force in the system	[64]
	Actin	Disruption eliminates or reduces tensile force in the system	[40,64]
	Microtubules	Disruption increases tensile force in the system	[64,70]
	Tensional homeostasis	Cells seek to maintain a preferred level of tension	[41]

(Continued)

TABLE 15.1 (*Continued*)

Experiment type	Parameter	Observation	Reference
	Residual matrix tension	Increases even after total force in the system plateaus	[40]
	Growth factors	TGF-β1 increases rate/extent of force generation, overexposure can be inhibitory	[40,71,72]
	Gene expression	MMP, TIMP levels affected by system prestrain and cyclic loading	[73–75]
Biaxial	Matrix realignment	Cell-driven, affected by initial geometry and loading during culture	[77–79]
	Load-induced matrix alignment	Mechanical loading results in alignment toward increased force	[81–83]
	Matrix production	In fibrin-based, collagen deposition can vary regionally based on initial conditions	[84]

then be assessed, or compared with experimental data, and used to recalculate the stress and strain state. Iteration of this process could allow one to describe or predict the mechanical evolution of the construct. Minimization of errors between experimentally and computationally determined material parameters via fine-tuning of the deposition and removal parameters would provide general measures of tissue reorganization and potential neotissue development and how these parameters may change in time due to changes in material composition (Figure 15.4).

15.6 SUMMARY

Mechanobiological research aims to understand how mechanical loads influence biological responses by cells. This search is often pursued as a one-step process, that is, mechanically stimulate a cell or cells and measure some biological entity of interest, as, for example, gene expression or protein production. This, however, is not the case *in vivo*. Cells often respond to changes in mechanical load by altering their surrounding environment, which in turn alters the mechanical behavior of the tissue and thus the local mechanical loading. Through this process, the cells appear to attempt to establish, maintain, or restore a preferred mechanical environment, a process that is referred to as mechanical homeostasis.

Tissue equivalents represent controllable experimental systems to explore the evolution of biomechanical and mechanobiological properties of tissue-engineered constructs and resident cells under well-controlled mechanical and chemical cues. They also provide a platform to examine the biological activity of cells within a 3-D matrix environment, which is particularly important for most connective tissue cell types (see Table 15.1 for summary). Finally, modeling of the evolution of tissue equivalents can provide new insight into the interplay between the mechanics and the biology. Models based on tissue equivalent development could also provide a means to simulate the progression of tissue engineering materials and enable more rational design rather than the often trial-and-error experimental approach. Conceptualization of such an approach has been proposed for tissue-engineered vascular grafts [110]. Clearly, much has been learned, but much remains to be accomplished. Tissue equivalents promise to remain fundamental to our understanding of cell–matrix interactions and the development and remodeling of soft tissues.

Acknowledgment

This work was supported in part by a grant from the National Science Foundation (CMMI-1161423).

REFERENCES

[1] Jaklenec, A., Stamp, A., Deweerd, E., Sherwin, A., Langer, R. *Tissue Eng. B Rev.* **2012**, *18*, 155–166.
[2] Prockop, J.D. *Annu. Rev. Biochem.* **1995**, *64*, 403–434.
[3] Birk, D.E., Brückner, P. in *Extracell. Matrix Overv* (Ed.: R.P. Mecham), Springer Berlin Heidelberg, 2011, pp. 77–115.
[4] Laurent, G.J. *Am. J. Physiol. Cell Physiol.* **1987**, *252*, C1–C9.
[5] Klingberg, F., Hinz, B., White, E.S. *J. Pathol.* **2013**, *229*, 298–309.
[6] Wight, T.N., Toole, B.P., Hascall, V.C. in *Extracell. Matrix Overv* (Ed.: R.P. Mecham), Springer Berlin Heidelberg, 2011, pp. 147–195.
[7] Barczyk, M., Carracedo, S., Gullberg, D. *Cell Tissue Res.* **2010**, *339*, 269–280.
[8] Sorrell, J.M., Caplan, A.I. in *Int. Rev. Cell Mol. Biol.* (Ed.: K.W. Jeon), Academic Press, 2009, pp. 161–214.
[9] Vedrenne, N., Coulomb, B., Danigo, A., Bonté, F., Desmoulière, A. *Pathol. Biol.* **2012**, *60*, 20–27.
[10] Darby, I., Skalli, O., Gabbiani, G. *Lab. Investig. J. Technol. Methods Pathol.* **1990**, *63*, 21–29.
[11] Tomasek, J.J., Gabbiani, G., Hinz, B., Chaponnier, C., Brown, R.A. *Nat. Rev. Mol. Cell Biol.* **2002**, *3*, 349–363.
[12] Nathan, C., Sporn, M. *J. Cell Biol.* **1991**, *113*, 981–986.
[13] Pierce, G.F., Mustoe, T.A., Altrock, B.W., Deuel, T.F., Thomason, A. *J. Cell. Biochem.* **1991**, *45*, 319–326.
[14] Sigal, L.H. *J. Clin. Rheumatol.* **2012**, *18*, 268–272.
[15] Jayme, D.W. in *eLS*; John Wiley & Sons, Ltd, 2001.
[16] Hayenga, H.N., Thorne, B.C., Yen, P., Papin, J.A., Peirce, S.M., Humphrey, J.D. in *Multiscale Comput. Model. Biomech. Biomed. Eng.* (Ed.: A. Gefen), Springer Berlin Heidelberg, 2013, pp. 209–240.
[17] Wang, J.H.C., Thampatty, B.P. *Biomech. Model. Mechanobiol.* **2006**, *5*, 1–16.
[18] Nagatomi, J. *Mechanobiology Handbook*; Taylor & Francis US, 2011.
[19] Discher, D.E., Janmey, P., Wang, Y. *Science* **2005**, *310*, 1139–1143.
[20] Chen, C.S. *J. Cell Sci.* **2008**, *121*, 3285–3292.
[21] Cukierman, E., Pankov, R., Stevens, D.R., Yamada, K.M. *Science* **2001**, *294*, 1708–1712.
[22] Chiquet, M., Gelman, L., Lutz, R., Maier, S. *Biochim Biophys Acta* **2009**, *1793*, 911–920.

[23] Bell, E., Ivarsson, B., Merrill, C. *Proc. Natl. Acad. Sci. U.S.A.* **1979**, *76*, 1274–1278.
[24] Ehrlich, H.P. *Methods Mol. Med.* **2003**, *78*, 277–291.
[25] Ehrlich, H.P., Rittenberg, T. *J. Cell. Physiol.* **2000**, *185*, 432–439.
[26] Harris, A.K., Wild, P., Stopak, D. *Science* **1980**, *208*, 177–179.
[27] Steinberg, B.M., Smith, K., Colozzo, M., Pollack, R. *J. Cell Biol.* **1980**, *87*, 304–308.
[28] Buttle, D.J., Ehrlich, H.P. *J. Cell. Physiol.* **1983**, *116*, 159–166.
[29] Ehrlich, H.P., Griswold, T.R., Rajaratnam, J. *Exp. Cell Res.* **1986**, *164*, 154–162.
[30] Travis, J.A., Hughes, M.G., Wong, J.M., Wagner, W.D., Geary, R.L. *Circ. Res.* **2001**, *88*, 77–83.
[31] Orlandi, A., Ferlosio, A., Gabbiani, G., Spagnoli, L.G., Ehrlich, P.H. *Exp. Cell Res.* **2005**, *311*, 317–327.
[32] Ehrlich, H.P. *Tissue Cell* **1988**, *20*, 47–50.
[33] Tiollier, J., Dumas, H., Tardy, M., Tayot, J.-L. *Exp. Cell Res.* **1990**, *191*, 95–104.
[34] Frey, J., Chamson, A., Raby, N., Rattner, A. *Biomaterials* **1995**, *16*, 139–143.
[35] Lorimier, S., Gillery, P., Hornebeck, W., Chastang, F., Laurent-Maquin, D., Bouthors, S., Droulle, C., Potron, G., Maquart, F.-X. *J. Cell. Physiol.* **1996**, *168*, 188–198.
[36] Zhang, Z., Garron, T.M., Li, X.J., Liu, Y., Zhang, X., Li, Y.Y., Xu, W.S. *Burns* **2009**, *35*, 527–537.
[37] Kagan, H.M. *Acta Trop.* **2000**, *77*, 147–152.
[38] Woodley, D.T., Yamauchi, M., Wynn, K.C., Mechanic, G., Briggaman, R.A. *J. Invest. Dermatol.* **1991**, *97*, 580–585.
[39] Redden, R.A., Doolin, E.J. *Skin Res. Technol.* **2003**, *9*, 290–293.
[40] Marenzana, M., Wilson-Jones, N., Mudera, V., Brown, R.A. *Exp. Cell Res.* **2006**, *312*, 423–433.
[41] Brown, R.A., Prajapati, R., McGrouther, D.A., Yannas, I.V., Eastwood, M. *J. Cell. Physiol.* **1998**, *175*, 323–332.
[42] Humphrey, J.D. *Cell Biochem. Biophys.* **2008**, *50*, 53–78.
[43] Dallon, J.C., Ehrlich, H.P. *Wound Repair Regen.* **2008**, *16*, 472–479.
[44] Schiro, J.A., Chan, B.M.C., Roswit, W.T., Kassner, P.D., Pentland, A.P., Hemler, M.E., Eisen, A.Z., Kupper, T.S. *Cell* **1991**, *67*, 403–410.
[45] Langholz, O., Röckel, D., Mauch, C., Kozlowska, E., Bank, I., Krieg, T., Eckes, B. *J. Cell Biol.* **1995**, *131*, 1903–1915.
[46] Kelynack, K.J., Hewitson, T.D., Nicholls, K.M., Darby, I.A., Becker, G.J. *Nephrol. Dial. Transplant.* **2000**, *15*, 1766–1772.
[47] Cooke, M.E., Sakai, T., Mosher, D.F. *J. Cell Sci.* **2000**, *113*(Pt 1), 2375–2383.
[48] Grundström, G., Mosher, D.F., Sakai, T., Rubin, K. *Exp. Cell Res.* **2003**, *291*, 463–473.
[49] Ehrlich, H.P., Buttle, D.J., Bernanke, D.H. *Exp. Mol. Pathol.* **1989**, *50*, 220–229.
[50] Montesano, R., Orci, L. *Proc. Natl. Acad. Sci. U.S.A.* **1988**, *85*, 4894–4897.
[51] Arora, P.D., Narani, N., McCulloch, C.A.G. *Am. J. Pathol.* **1999**, *154*, 871–882.
[52] Grinnell, F., Ho, C.-H., Lin, Y.-C., Skuta, G. *J. Biol. Chem.* **1999**, *274*, 918–923.
[53] Ikuno, Y., Kazlauskas, A. *Invest. Ophthalmol. Vis. Sci.* **2002**, *43*, 41–46.
[54] Jiang, H., Rhee, S., Ho, C.H., Grinnell, F. *FASEB J.* **2008**, *22*, 2151–2160.
[55] Battegay, E.J., Raines, E.W., Seifert, R.A., Bowen-Pope, D.F., Ross, R. *Cell* **1990**, *63*, 515–524.
[56] Ignotz, R.A., Massagué, J. *Cell* **1987**, *51*, 189–197.
[57] Heino, J., Ignotz, R.A., Hemler, M.E., Crouse, C., Massague, J. *J. Biol. Chem.* **1989**, *264*, 380–388.
[58] Leof, E.B., Proper, J.A., Getz, M.J., Moses, H.L. *J. Cell. Physiol.* **1986**, *127*, 83–88.

[59] Lin, Y.C., Grinnell, F. *J. Cell Biol.* **1993**, *122*, 663–672.
[60] Daniels, J.T., Cambrey, A.D., Occleston, N.L., Garrett, Q., Tarnuzzer, R.W., Schultz, G.S., Khaw, P.T. *Invest. Ophthalmol. Vis. Sci.* **2003**, *44*, 1104–1110.
[61] Ehrlich, H.P., Rajaratnam, J.B.M. *Tissue Cell* **1990**, *22*, 407–417.
[62] Costa, K.D., Lee, E.J., Holmes, J.W. *Tissue Eng.* **2003**, *9*, 567–577.
[63] Delvoye, P., Wiliquet, P., Levêque, J.-L., Nusgens, B.V., Lapière, C.M. *J. Invest. Dermatol.* **1991**, *97*, 898–902.
[64] Kolodney, M.S., Wysolmerski, R.B. *J. Cell Biol.* **1992**, *117*, 73–82.
[65] Eastwood, M., McGrouther, D.A., Brown, R.A. *Biochim. Biophys. Acta* **1994**, *1201*, 186–192.
[66] Mauch, C., van der Mark, K., Helle, O., Mollenhauer, J., Pfäffle, M., Krieg, T. *J. Cell Biol.* **1988**, *106*, 205–211.
[67] Porter, R.A., Brown, R.A., Eastwood, M., Occleston, N.L., Khaw, P.T. *Wound Repair Regen.* **1998**, *6*, 157–166.
[68] Hall, S.M., Soueid, A., Smith, T., Brown, R.A., Haworth, S.G., Mudera, V. *J. Tissue Eng. Regen. Med.* **2007**, *1*, 287–295.
[69] Nirmalanandhan, V.S., Levy, M.S., Huth, A.J., Butler, D.L. *Tissue Eng.* **2006**, *12*, 1865–1872.
[70] Brown, R.A., Talas, G., Porter, R.A., McGrouther, D.A., Eastwood, M. *J. Cell. Physiol.* **1996**, *169*, 439–447.
[71] Brown, R.A., Sethi, K.K., Gwanmesia, I., Raemdonck, D., Eastwood, M., Mudera, V. *Exp. Cell Res.* **2002**, *274*, 310–322.
[72] Bisson, M.A., Beckett, K.S., McGrouther, D.A., Grobbelaar, A.O., Mudera, V. *J. Hand Surg. [Am]* **2009**, *34*, 1102–1110.
[73] Karamichos, D., Brown, R.A., Mudera, V. *J. Biomed. Mater. Res. A* **2007**, *83*, 887–894.
[74] Prajapati, R.T., Chavally-Mis, B., Herbage, D., Eastwood, M., Brown, R.A. *Wound Repair Regen.* **2000**, *8*, 226–237.
[75] Prajapati, R.T., Eastwood, M., Brown, R.A. *Wound Repair Regen.* **2000**, *8*, 238–246.
[76] Knezevic, V., Sim, A.J., Borg, T.K., Holmes, J.W. *Biomech. Model. Mechanobiol.* **2002**, *1*, 59–67.
[77] Hu, J.-J., Humphrey, J.D., Yeh, A.T. *Tissue Eng. A* **2009**, *15*, 1553–1564.
[78] Jhun, C.-S., Evans, M.C., Barocas, V.H., Tranquillo, R.T. *J. Biomech. Eng.* **2009**, *131*, 081006.
[79] Thomopoulos, S., Fomovsky, G.M., Holmes, J.W. *J. Biomech. Eng.* **2005**, *127*, 742–750.
[80] Lee, E.J., Holmes, J.W., Costa, K.D. *Ann. Biomed. Eng.* **2008**, *36*, 1322–1334.
[81] Sander, E.A., Stylianopoulos, T., Tranquillo, R.T., Barocas, V.H. *IEEE Eng. Med. Biol. Mag.* **2009**, *28*, 10–18.
[82] Sander, E.A., Stylianopoulos, T., Tranquillo, R.T., Barocas, V.H. *Proc. Natl. Acad. Sci. U.S.A.* **2009**, *106*, 17675–17680.
[83] Raghupathy, R., Witzenburg, C., Lake, S.P., Sander, E.A., Barocas, V.H. *J. Biomech. Eng.* **2011**, *133*, 091011.
[84] Sander, E.A., Barocas, V.H., Tranquillo, R.T. *Ann. Biomed. Eng.* **2011**, *39*, 714–729.
[85] Moon, A.G., Tranquillo, R.T. *AIChE J.* **1993**, *39*, 163–177.
[86] Barocas, V.H., Moon, A.G., Tranquillo, R.T. *J. Biomech. Eng.* **1995**, *117*, 161–170.
[87] Ramtani, S. *J. Biomech.* **2004**, *37*, 1709–1718.
[88] John, J., Throm Quinlan, A., Silvestri, C., Billiar, K. *Ann. Biomed. Eng.* **2010**, *38*, 658–673.
[89] Simon, D.D., Humphrey, J.D. *Int. J. Non-Linear Mech.* **2012**, *47*, 173–178.

[90] Simon, D.D., Horgan, C.O., Humphrey, J.D. *J. Mech. Behav. Biomed. Mater.* **2012**, *14*, 216–226.
[91] Waldman, S.D., Lee, J.M. *J. Mater. Sci. Mater. Med.* **2002**, *13*, 933–938.
[92] Waldman, S.D., Lee, J.M. *Biomaterials* **2005**, *26*, 7504–7513.
[93] Stylianopoulos, T., Barocas, V.H. *Comput. Methods Appl. Mech. Eng.* **2007**, *196*, 2981–2990.
[94] Voytik-Harbin, S.L., Roeder, B.A., Sturgis, J.E., Kokini, K., Robinson, J.P. *Microsc. Microanal.* **2003**, *9*, 74–85.
[95] Roeder, B.A., Kokini, K., Robinson, J.P., Voytik-Harbin, S.L. *J. Biomech. Eng.* **2004**, *126*, 699–708.
[96] Tower, T.T., Neidert, M.R., Tranquillo, R.T. *Ann. Biomed. Eng.* **2002**, *30*, 1221–1233.
[97] Billiar, K.L., Sacks, M.S. *J. Biomech. Eng.* **2000**, *122*, 327.
[98] Thomopoulos, S., Fomovsky, G.M., Chandran, P.L., Holmes, J.W. *J. Biomech. Eng.* **2007**, *129*, 642–650.
[99] Baek, S., Gleason, R.L., Rajagopal, K.R., Humphrey, J.D. *Comput. Methods Appl. Mech. Eng.* **2007**, *196*, 3070–3078.
[100] Rodriguez, E.K., Hoger, A., McCulloch, A.D. *J. Biomech.* **1994**, *27*, 455–467.
[101] Kroon, M. *J. Biomech. Eng.* **2010**, *132*, 111008–111008.
[102] Kroon, M. *J. Theor. Biol.* **2010**, *264*, 66–76.
[103] Humphrey, J.D., Rajagopal, K.R. *Math. Models Methods Appl. Sci.* **2002**, *12*, 407–430.
[104] Baek, S., Rajagopal, K.R., Humphrey, J.D. *J. Biomech. Eng.* **2006**, *128*, 142–149.
[105] Humphrey, J.D., Wells, P.B., Baek, S., Hu, J.-J., McLeroy, K., Yeh, A.T. *Biomech. Model. Mechanobiol.* **2008**, *7*, 323–334.
[106] Nusgens, B., Merrill, C., Lapiere, C., Bell, E. *Coll. Relat. Res.* **1984**, *4*, 351–363.
[107] Bhole, A.P., Flynn, B.P., Liles, M., Saeidi, N., Dimarzio, C.A., Ruberti, J.W. *Philos. Trans. R. Math. Phys. Eng. Sci.* **2009**, *367*, 3339–3362.
[108] Hadi, M.F., Sander, E.A., Ruberti, J.W., Barocas, V.H. *Mech. Mater.* **2012**, *44*, 72–82.
[109] Pedrigi, R.M., Humphrey, J.D. *Ann. Biomed. Eng.* **2011**, *39*, 537–548.
[110] Niklason, L.E., Yeh, A.T., Calle, E.A., Bai, Y., Valentín, A., Humphrey, J.D. *Proc. Natl. Acad. Sci. U.S.A.* **2010**, *107*, 3335–3339.

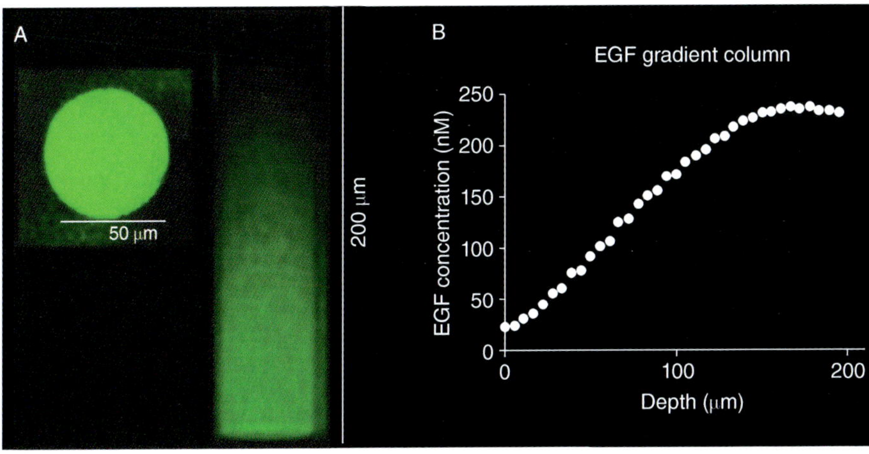

FIGURE 1.3 3-D photopatterning of EGF within a hyaluronic acid–PEG hydrogel. (A) Creation of a linear immobilized gradient of EGF. From the top of the hydrogel, the number of scans by the multiphoton laser are increased as it penetrates into the sample, corresponding to an increase in fluorescence intensity, and hence, an increase in protein immobilization. (B) The concentration of immobilized protein in the gradient was quantified by fluorescence intensity, showing a change in concentration from 25 nM at the top of the hydrogel to 250 nM at a depth of 150 μm in the hydrogel. EGF, epidermal growth factor. (Reproduced with permission from Owen, S.C., Fisher, S.A., Tam, R.Y., Nimmo, C.M., Shoichet, M.S. Langmuir 2013. Copyright 2013 American Chemical Society.)

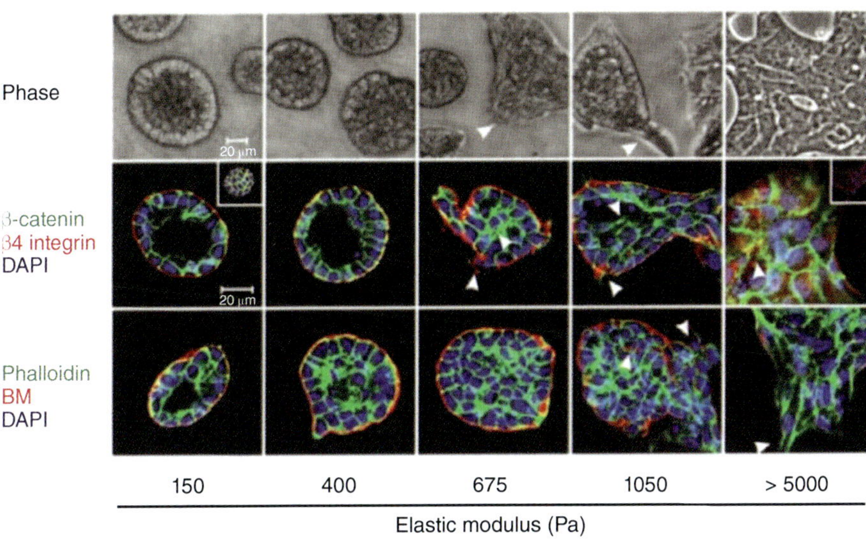

FIGURE 2.4 MCF10A acini polarization responds to extracellular matrix stiffness and is disturbed on being cultured on increased hydrogel stiffness. (Reproduced with permission from Reference 35.)

Bio-inspired Materials for Biomedical Engineering, First Edition. Edited by Anthony B. Brennan and Chelsea M. Kirschner.

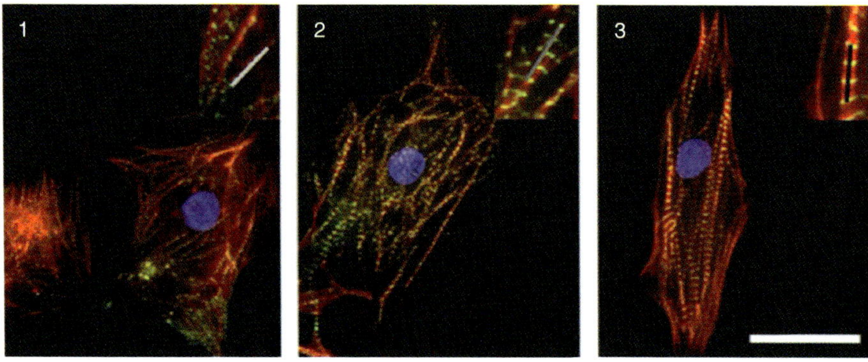

FIGURE 2.5 Immunofluorescent images of embryonic cardiomyocytes cultured on dynamic thiolated HA hydrogels at different developmental stages: premyofibril stage (1), maturing myofibrils (2), and mature cardiomyocytes (3). (Reproduced with permission from Reference 42.)

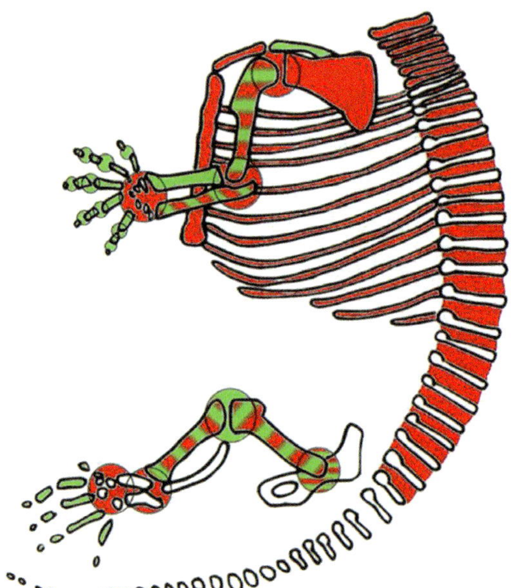

FIGURE 3.2 Effects of abnormal muscle force on skeletogenesis in mouse models. Red indicates effect on rudiment or joint due to abnormal muscle, green indicates no effect, striped red and green indicates findings of affected and unaffected aspects, and white indicates no data available. (Reproduced with permission from Reference 69.)

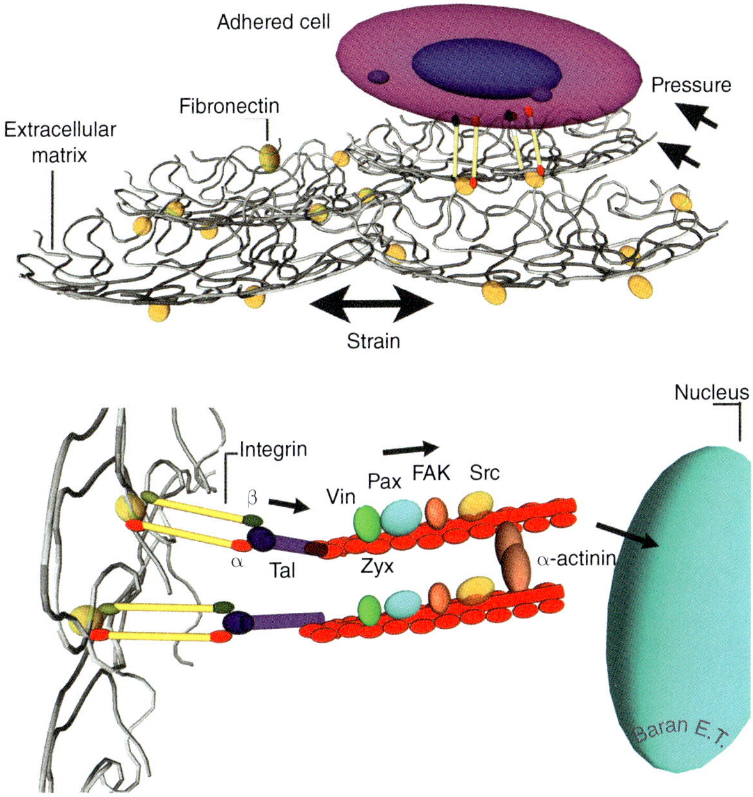

FIGURE 4.1 Molecular pathways mediating mechanotransduction signaling in a cell. In this pathway, mechanical forces such as, stretching, hydrostatic pressure, and shear stress stimulate the integrins on the cell membrane via extracellular matrix. In turn, the stimuli is transduced into the nucleus by engagement of anchorage proteins talin (tal), vinculin (vin), paxillin (pax), and α-actinin and signaling proteins FAK, Src, and zyxin (zyx).

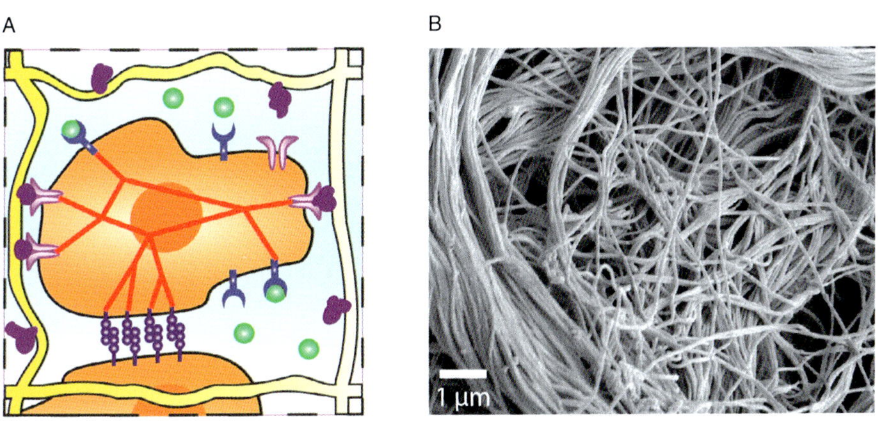

FIGURE 5.1 (A) In vivo cells receive biochemical and biophysical cues through interactions with the ECM, soluble factors, and neighboring cells. (See text for full caption.)

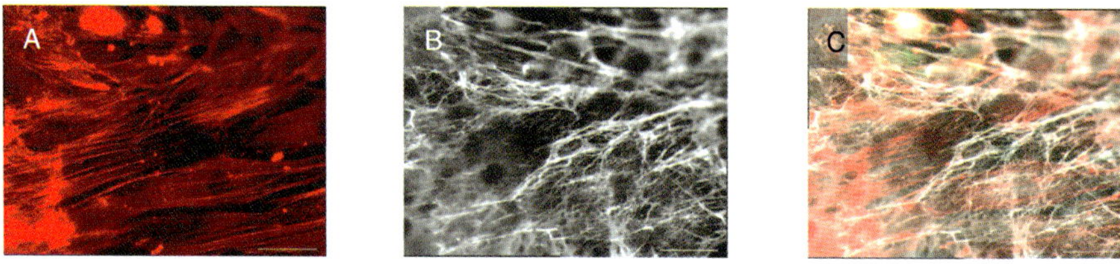

FIGURE 6.3 Assembly of FN matrix in hMSCs. Confluent layers of hMSCs assemble extensive FN matrices. (A) F-actin immunofluorescence; (B) FN labeled with anticellular FN antibody indicates that assembled FN was expressed in hMSCs; (C) Composite image of cell–matrix interactions.

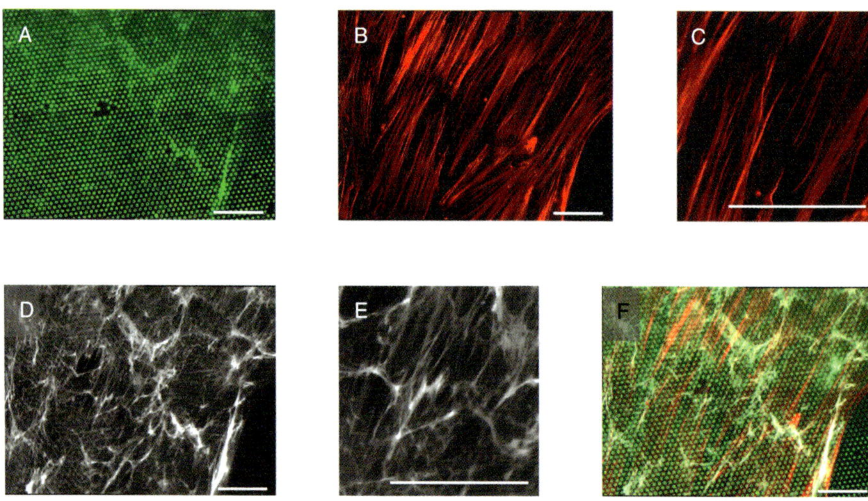

FIGURE 6.4 Assembly of FN matrix on a micropillar scaffold. Immunofluorescence images of a layer of human mesenchymal stem cells grown on a surface of micropillars for 10 days. (A) Fluorescently labeled pillars; (B) Actin cytoskeleton (red) (higher magnification shown in C); (D) Assembled fibronectin fibrils (higher magnification shown in E); (F) Composite image. Note that while there are visible spaces between cells in the actin image, they have formed a complete layer of ECM across the top surface of the pillars. Scale bar is 50 μm.

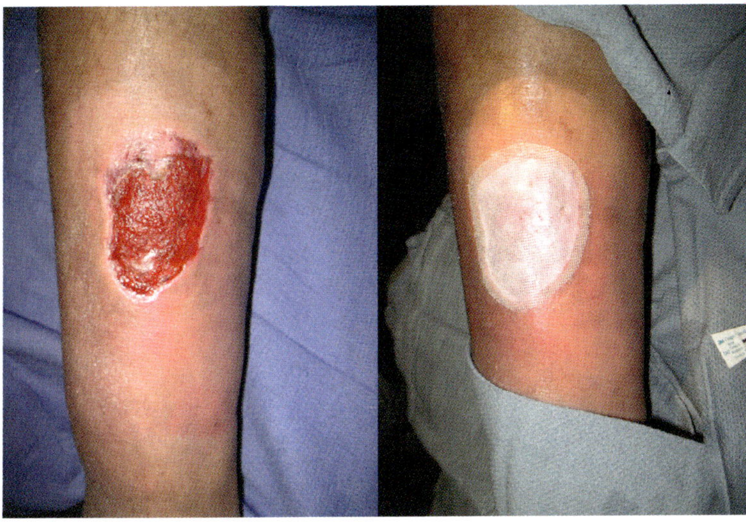

FIGURE 9.1 Application of allogenic product to leg wound. (Photos courtesy of Lauren R. Bayer, PA-C.)

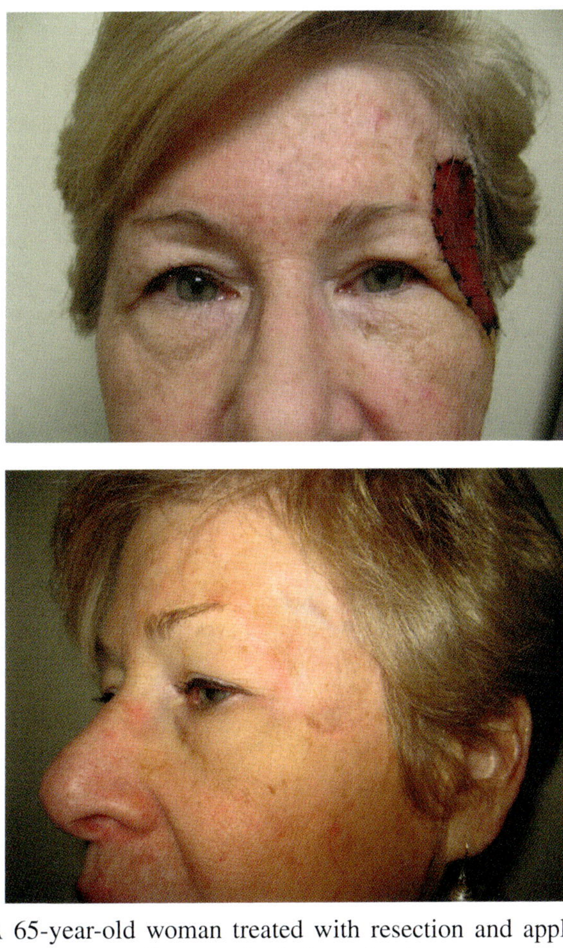

FIGURE 9.3 A 65-year-old woman treated with resection and application of a dermal regeneration template. (See text for full caption.)

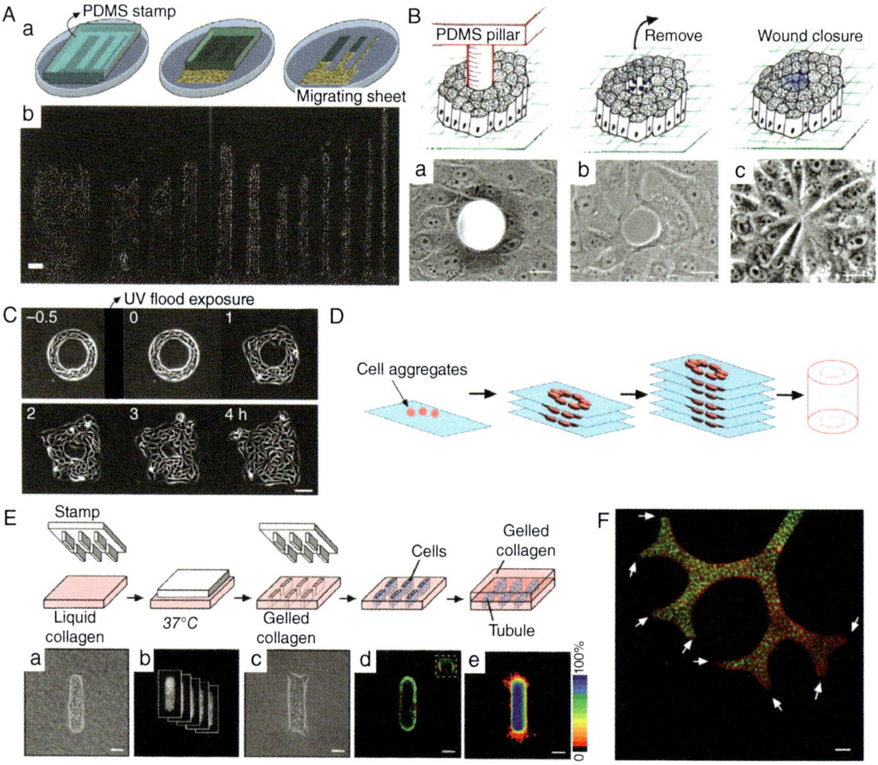

FIGURE 10.3 Engineering approaches that mimic epithelial morphogenesis. (A) Micropatterned adhesive substratum for investigating epithelial sheet migration. (B) Mimicking wounded epithelium using PDMS pillars. (C) Switchable substratum for expansion of epithelial sheets. (D) 3-D printing for constructing a biological tube. (E,F) 3-D micropatterned tubes for investigating branching morphogenesis. (Adapted from References 55, 57, 59, 71, and 72.)

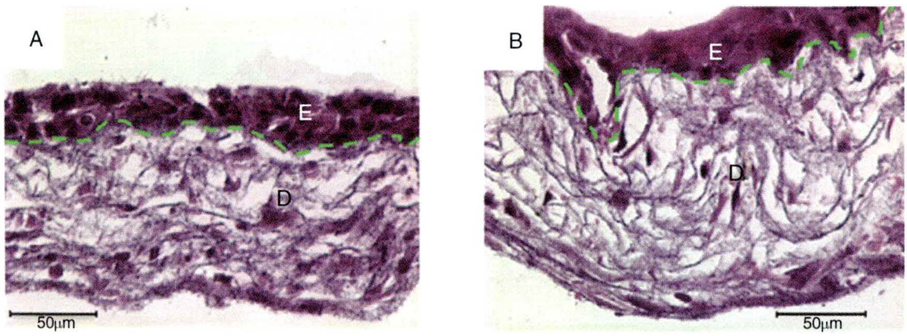

FIGURE 11.4 H&E-stained cross-sections of bilayer skin tissues composed of epidermal (E) and dermal (D) layers and formed by culturing L-b-L assembled cell/fiber constructs for 3 days (A) and 7 days (B). Green broken line outlines the border between E and D. (Reprinted from Reference 98, with permission from Mary Ann Liebert, Inc. Publishers.)

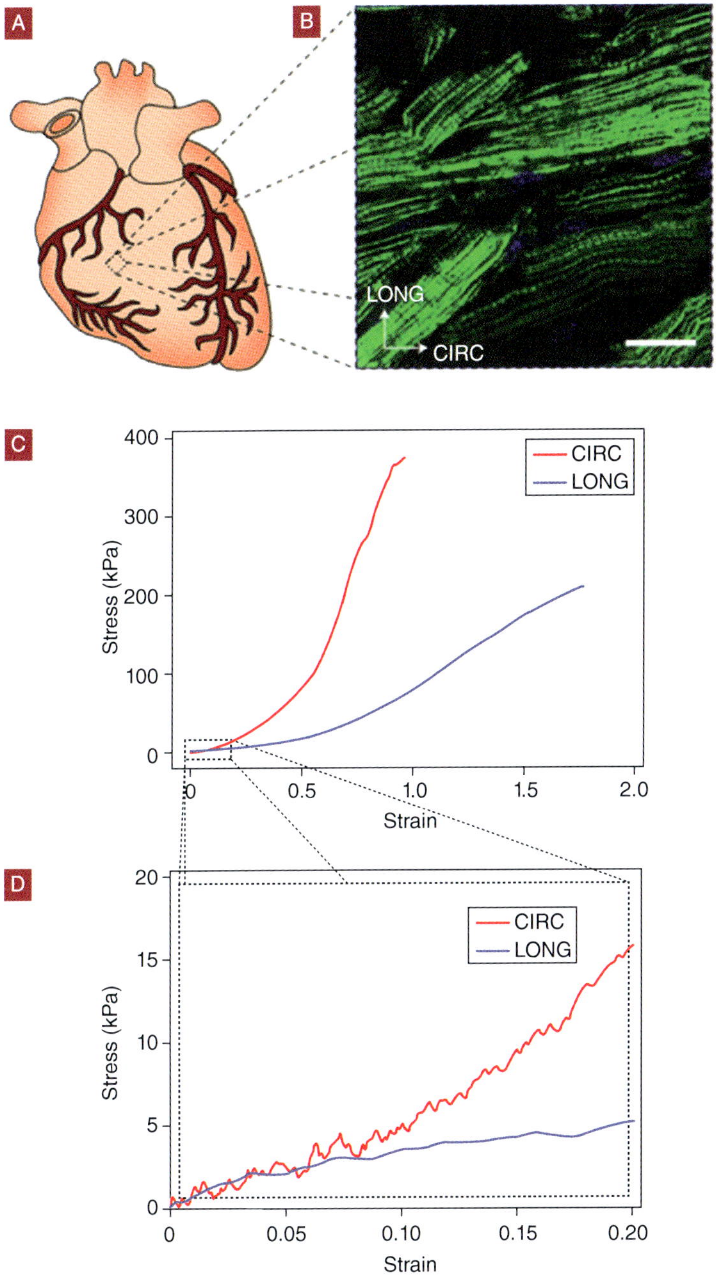

FIGURE 12.3 Mechanical and structural characteristics of the myocardium. (See text for full caption.)

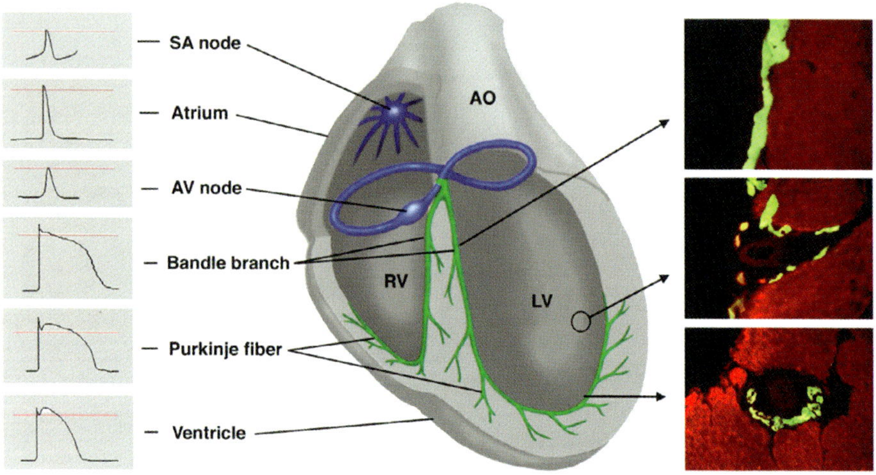

FIGURE 12.4 The cardiac conduction system. (See text for full caption.)

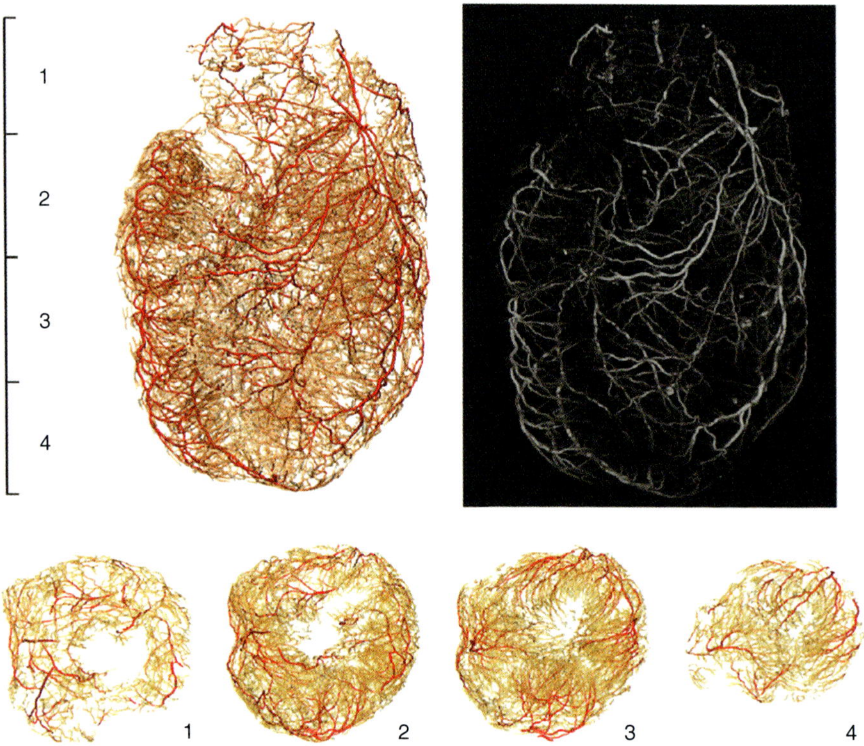

FIGURE 12.5 The whole vasculature of an adult rat heart was reconstructed (top left) from micro-CT data (top right). Transverse sections obtained in four planes (below) show the penetrating network of capillaries. The color of the rendered vessels corresponds to the intensity of the voxels in the original dataset. (Reprinted with permission from Reference 142.)

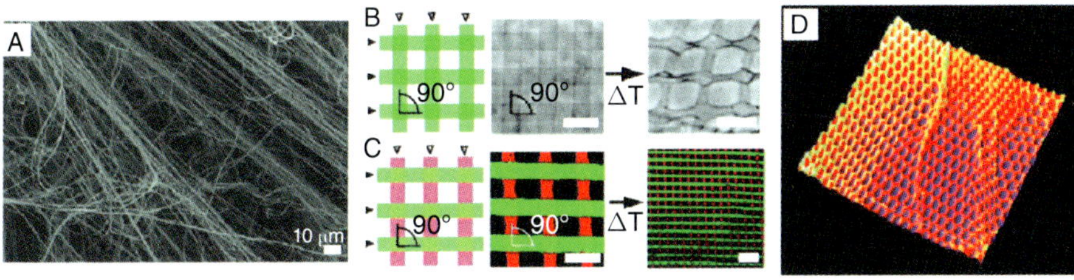

FIGURE 12.9 Alternatives to electrospinning to create micro- and nanofiber scaffolds. Rotary jet spinning uses a high-speed rotating spindle to draw fibers from synthetic and natural materials. (A) SEM images of gelatin fibers show a high degree of alignment. Surface-initiated assembly is a technique that mimics cell-mediated assembly and provides control of the scaffold nano- to macroscale structure and composition. (B) Schematic (left) and optical phase image (right) of two patterns of 20-μm width by 20-μm spacing fibronectin lines microcontact-printed orthogonally onto PIPAAm. After some time (ΔT) in cooling water, the mesh termed nanofabric is released and maintains its shape. (C) The same pattern was created with fibronectin (green) and laminin (red) and was released as a bicomponent nanofabric (right) showing that SIA can be used to control the architecture and composition of biomimetic ECM nanofabrics. (D) Three-dimensional, false-colored rendering of a fibronectin mesh with 20-μm wide elliptical holes observed by scanning electron microscopy. The nanofabric shows fishnet-like ripples. Scale bars are 40 μm in B and C and 100 μm for the released bicomponent nanofabrics. X, Y axes are 360 μm in D. (Adapted with permission from References 77 and 78.)

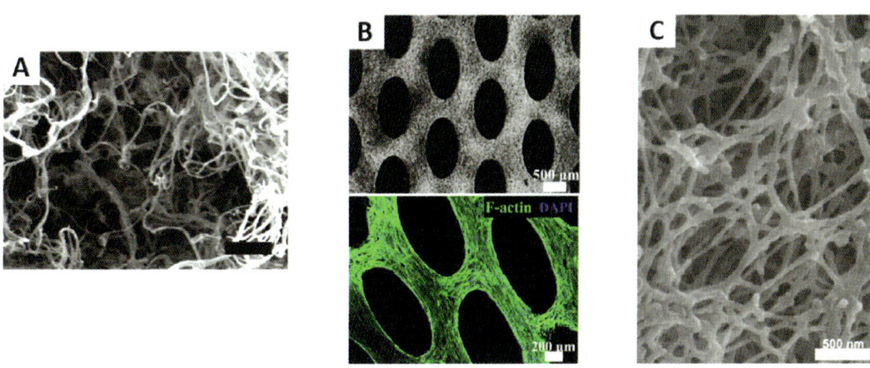

FIGURE 12.10 Examples of hydrogels for cardiac tissue engineering. Myocardial ECM gels can be obtained by decellularization, lyophilization, and enzymatic digestion. (A) The solubilized ECM components gel under physiological conditions into fibrous multicomponent hydrogels with ECM-like structure revealed by SEM. Scale bar is 1 μm. (B) Fibrin–matrigel hydrogels cast around an array of micropillars are remodeled by myocytes to form a contractile cardiac construct with local anisotropy. Scale bars are 500 μm (top) and 200 μm (bottom). (C) Synthetic polypeptides are designed to self-assemble into nanofibrous hydrogels and to mimic VEGF to induce angiogenesis. (Adapted with permission from References 89, 94, and 100.)

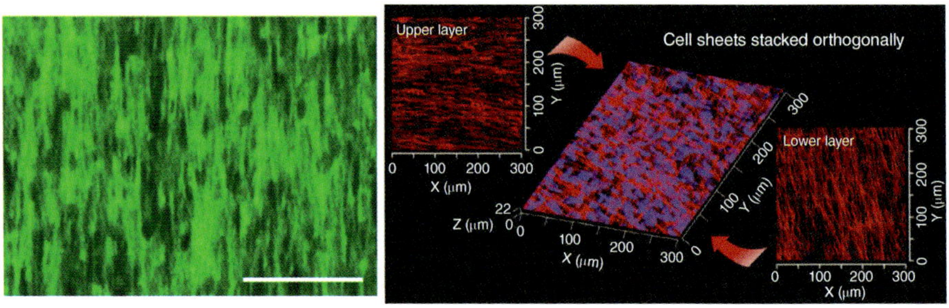

FIGURE 12.12 Cell sheet engineering is a scaffold-free approach to cardiac tissue engineering. Human fibroblasts were seeded on an anisotropic PIPAAm layer. After release, the cell sheets produced their own anisotropic ECM, revealed by observation of highly aligned collagen type I fibers (green, left). Aligned cell sheets can be stacked at different angles to create multilayer constructs (right) with F-actin (red) and nuclei (blue). Scale bar is 100 μm. (Adapted with permission from Reference 110.)

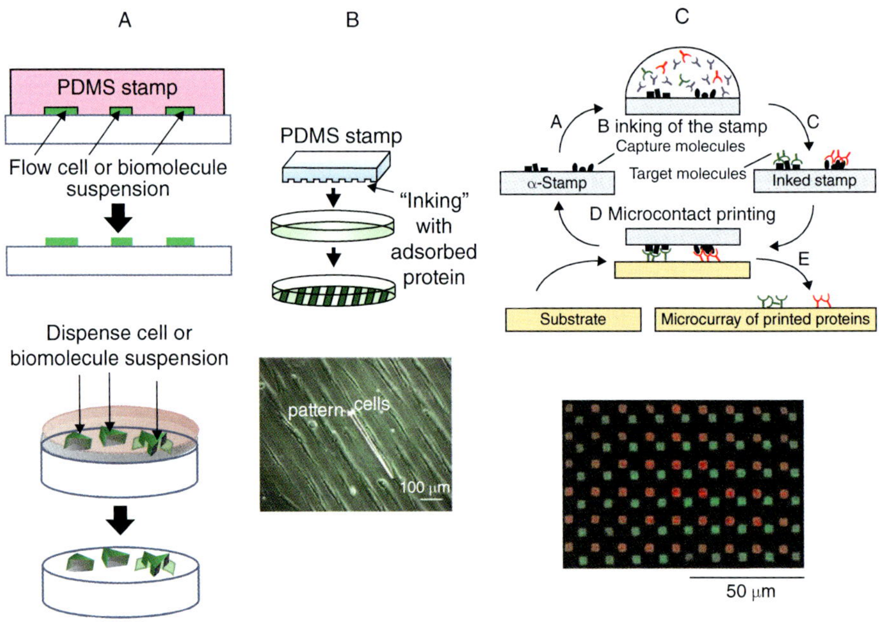

FIGURE 13.4 Schematic of soft lithography techniques used to create micropatterning on substrates using (A) blocking methods and solution dispensing with microchannels or stencils, (B) microcontact printing of adsorbed proteins using conformal contact, and (C) affinity contact printing using immobilized ligands for conformal contact printing of target biomolecules. (Panel A adapted from Park and Shuler [136]; panel B adapted from Williams *et al.* [143,258]; panel C from Renault *et al.* [141].)

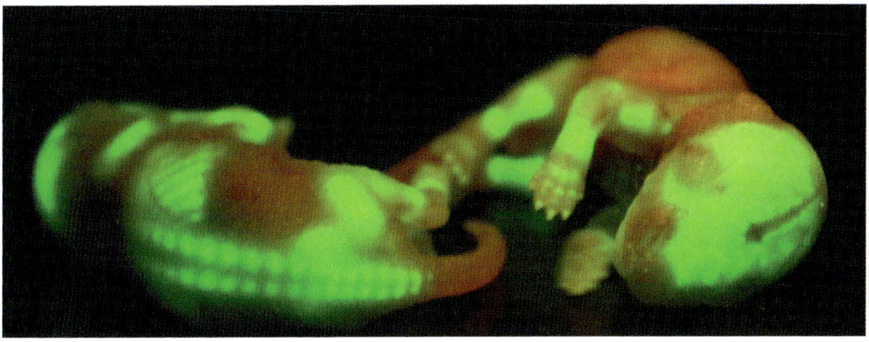

FIGURE 14.2 Under ultraviolet light, the bones of these transgenic mouse pups fluoresce green, indicating the successful incorporation of both the GFP reporter gene and the linked functional gene of interest. (Courtesy of Professor David Rowe of the University of Connecticut Health Center.)

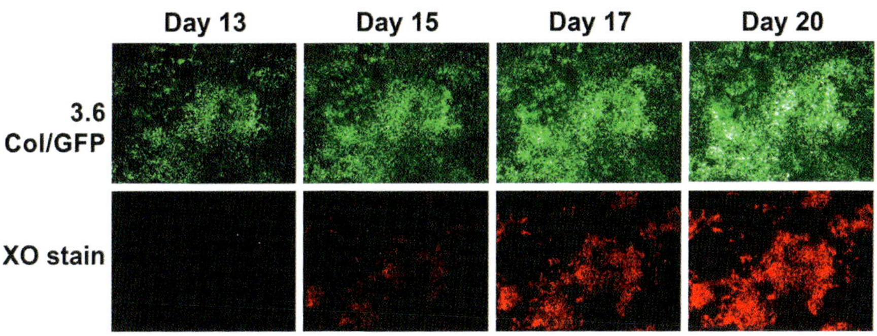

FIGURE 14.4 Fluorescence microscopy images of cultures expressing **3.6** Col/GFP associated with preosteoblasts just prior to mineralization, and nontoxic xylenol orange (XO) staining of mineral taken from the same area in the cell culture plate at multiple time points. The cell reporter technology allows continuous monitoring of cell differentiation without requiring the use of dyes or antibody staining that require cell culture termination. (Courtesy of Yu-Hsiung Wang of the University of Connecticut Health Center.)

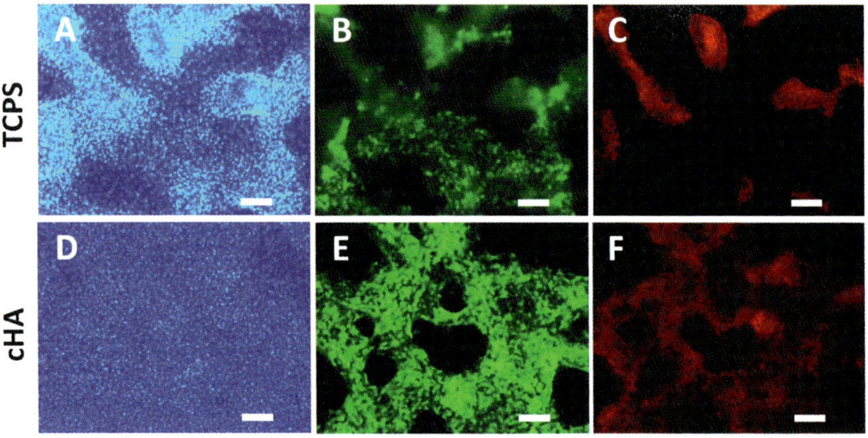

FIGURE 14.7 Fluorescence microscopy images of the calvarial cells from the transgenic mice on TCPS and cHA at 21 days with DAPI staining to show all cells (A,D), osteoblasts revealed by GFP expression (B,E), and XO staining for deposited mineral (C,F). Scale bar = 100 μm. (Reprinted with permission from Reference 10.)

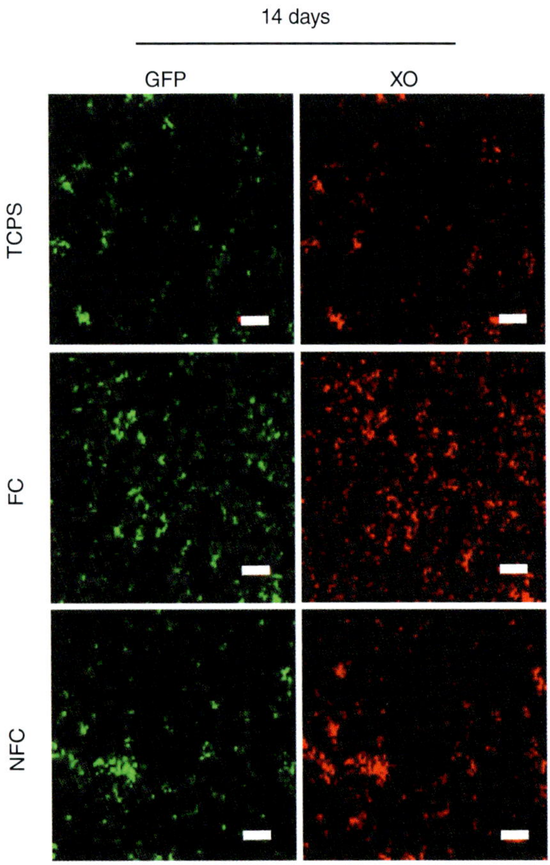

FIGURE 14.10 Fluorescence microscopy images of GFP positive (green) and xylenol orange staining (red) of mineralized matrix after 14 and 21 days in culture. Scale bars = 2 mm.

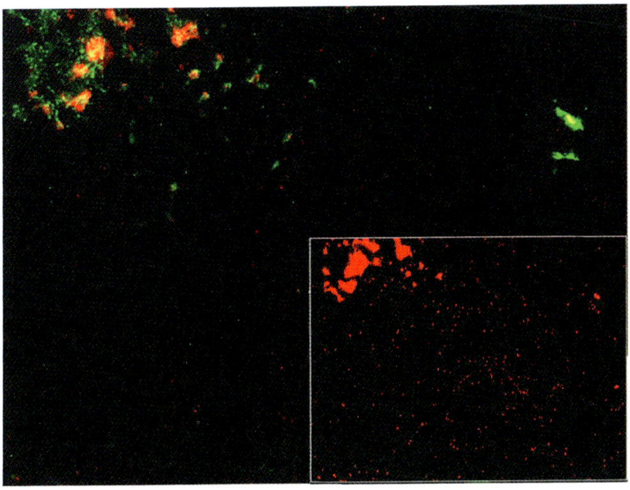

FIGURE 14.12 Merged fluorescence images showing colocalization of GFP expression and XO staining in cultures grown on nonfibrillar collagen (NFC). The inset image of XO staining shows isolated islands of mineral not associated with GFP positive differentiated cells.

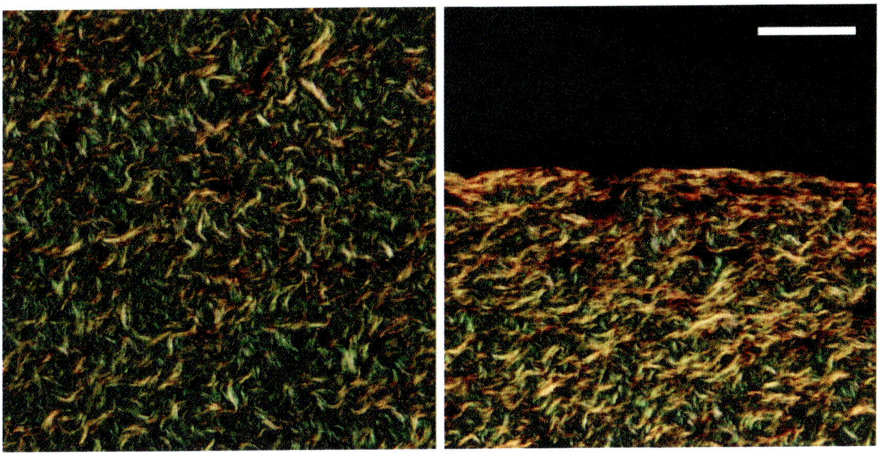

FIGURE 15.3 Picrosirius red stained free-floating fibroblast populated collagen lattice under circularly polarized light to show birefringent collagen. Central region (left) shows dense, randomly oriented collagen fibers while the outer edge of the lattice (right) shows aligned fibers for a lattice that has reached steady state (see Figure 15.2). Scale bar = 50 μm.

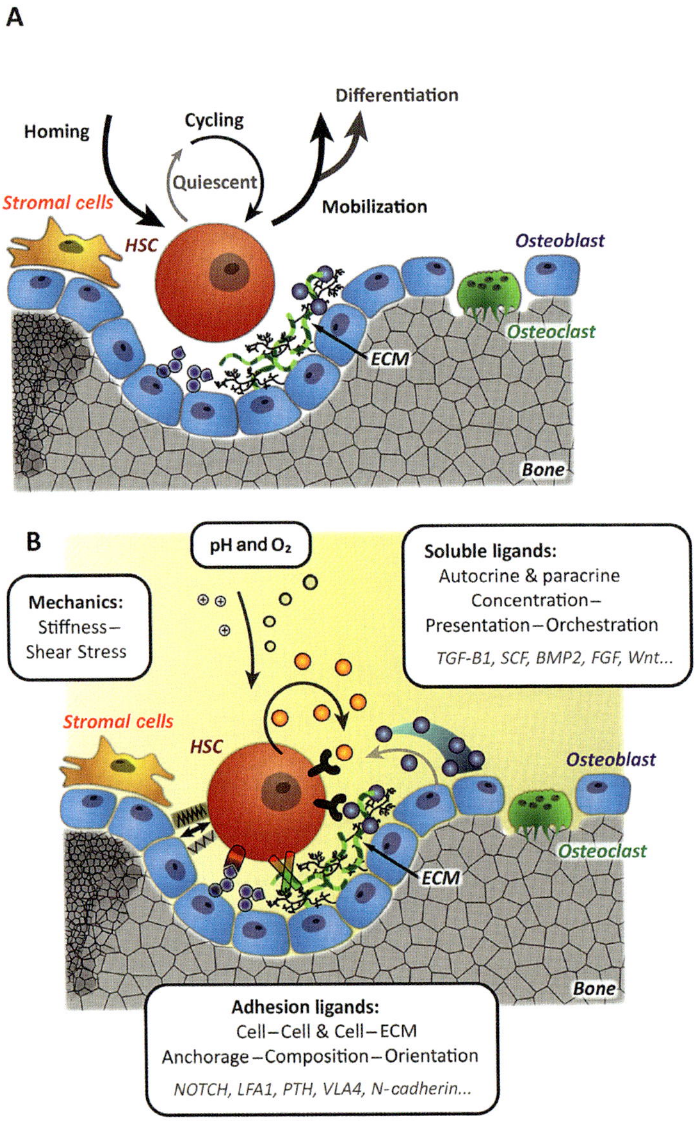

FIGURE 16.1 Concepts of the HSC niche. (A) Scheme of HSC regulation inside the niche microenvironment depicting the different HSC fate decisions which are orchestrated by the niche components. (B) Scheme of the different microenvironmental cues controlling HSC fate including biochemical, biophysical, and metabolic signals.

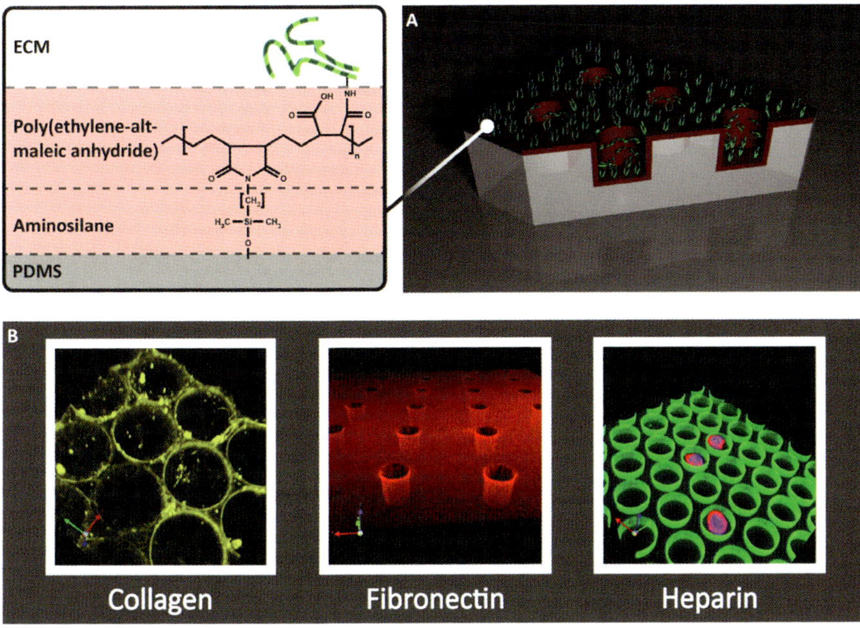

FIGURE 16.3 Protein immobilization on a microstructured surface for HSC culture. (A) Poly(dimethyl siloxane) (PDMS) microstructured with oxygen plasma activation are coated by aminosilane functionalization and maleic anhydride copolymer coating to immobilize components of the ECM. (B) Fluorescent images of ECM-modified PDMS microstructures.

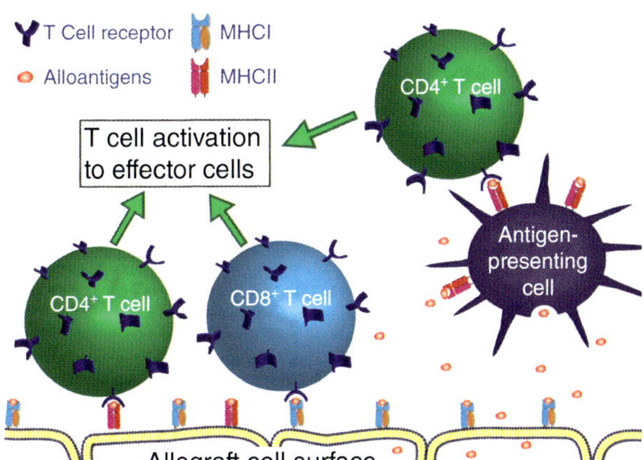

FIGURE 17.1 Summary of pathways for recognition of allograft by the adaptive immune system. (See text for full caption.)

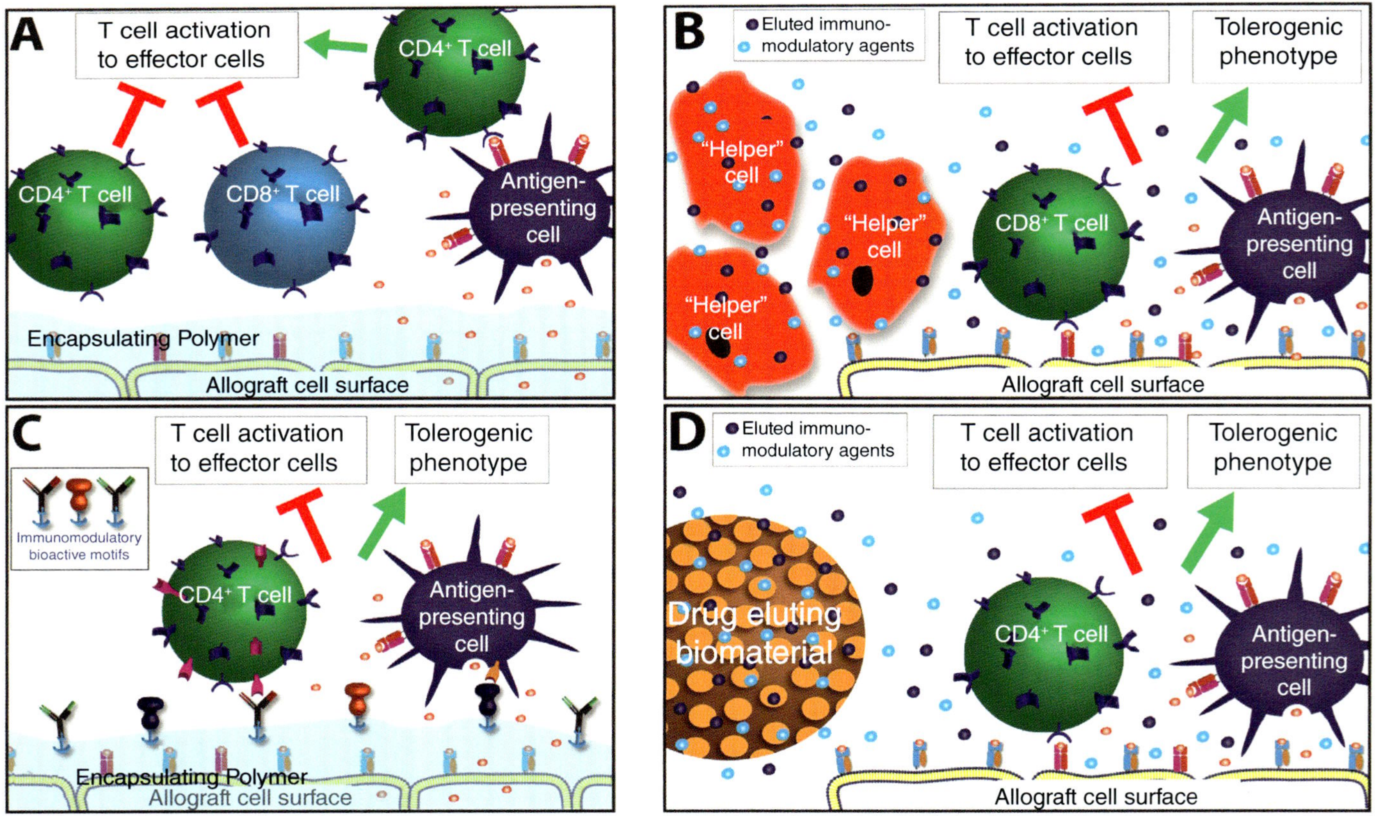

FIGURE 17.2 Engineering of immune response to allogeneic transplants can be conducted via (A) polymeric encapsulation, (B) cotransplantation with protective cells, (C) surface functionalization of transplanted biomaterials, and (D) codelivery of soluble factors. (See text for full caption.)

CHAPTER 16

MIMICKING THE HEMATOPOIETIC STEM CELL NICHE BY BIOMATERIALS

EIKE MÜLLER

Leibniz Institute of Polymer Research Dresden, Max Bergmann Center of Biomaterials, Dresden, Germany

MICHAEL ANSORGE

Universität Leipzig, Institute of Biochemistry, Leipzig, Germany

CARSTEN WERNER

Leibniz Institute of Polymer Research Dresden, Max Bergmann Center of Biomaterials, Technische Universität Dresden, Center for Regenerative Therapies Dresden, Dresden, Germany

TILO POMPE

Leibniz Institute of Polymer Research Dresden, Max Bergmann Center of Biomaterials, Dresden, Germany; Universität Leipzig, Institute of Biochemistry, Leipzig, Germany

16.1 INTRODUCTION

In the adult mammalian organism, hematopoietic stem cells (HSCs) reside primarily in the bone marrow (BM) and are the source for regeneration of all immune and blood cells. In order to fulfill this task, they have the ability to balance proliferation and quiescence as well as differentiation and self-renewal during the entire human life span, giving rise to a daily production of a trillion differentiated blood cells.

For that reason, the hematopoietic systems and especially the origin of their regeneration potential—the HSC—have attracted many efforts to explore regulating mechanisms *in vivo* and *in vitro*. Much evidence has been accumulated about the role of extrinsic signals in controlling HSC fate. The so-called *niche* concept refers

Bio-inspired Materials for Biomedical Engineering, First Edition. Edited by Anthony B. Brennan and Chelsea M. Kirschner.

to the physical and functional environment responsible for the convergence and integration of such signals [1]. Based on this concept, the HSC microenvironment is thought to encompass several stromal cell types, extracellular matrix (ECM) components, and a multitude of paracrine and endocrine signals. The organizational principles underlying the spatial arrangement of HSCs, osteoblasts, vasculature, and ECM, and in particular, their impact on the dynamics of HSC localization, function, and fate, are still not completely understood.

Recent progress in biotechnology and biomaterials science has fostered engineering approaches to explore the niche concept by means of bioartificial concepts *in vitro*. Here, specific cues of the *in vivo* BM are modeled in simple or complex setups with the aim of applying the full range of *in vitro* bioanalytical tools for the analysis of HSC behavior in such microenvironments. This chapter focuses on an introduction to existing HSC niche concepts and summarizes approaches to mimic them in an experimental manner *ex vivo*.

16.2 CONCEPTS OF HSC NICHES

Due to their unique regenerative potential, stem cell transplantations have become an attractive approach for the clinical therapy of numerous diseases such as cancer, heart failure, autoimmune disorders, or diabetes. As a consequence, the regulatory processes underlying stem cell growth and differentiation have gained a lot of interest. The local microenvironment—referred to as a *stem cell niche*—is widely accepted to play a major role in controlling stem cell fate [2]. The function of the niche is seen as a tight balance between proliferation and quiescence, and self-renewal and differentiation, regulated by cell intrinsic and extrinsic mechanisms, as sketched in Figure 16.1.

Similar to stem cell niches in other tissues, a stem cell niche is thought to control HSC fate decisions and is consequently involved in homeostasis of the blood system (Figure 16.1A). The clinical significance of the unique, local microenvironment surrounding HSC is evident from the involvement of alterations of this microenvironment in many diseases, for example, leukemia [3–5]. However, in contrast to other tissue types with an anatomically defined microenvironment for stem cells, the mammalian HSC niche is still only vaguely defined, as reflected by the ongoing debate regarding its localization and composition [6–10]. To date, there is evidence for the existence of at least two different characteristic regulatory regions within the BM [1]: the osteoblastic niche [11], located close to the endosteum, and a location known as the perivascular niche that is associated with the vasculature [12]. As key players of the niches, multiple stromal cells have been identified, in particular osteoblasts [13], (peri)vascular cells [14,15], and nestin-positive progenitors [16]. Apart from different cell types, the regulatory role of various ECM components has also been shown (Figure 16.1B) [17]. However, it is still unclear how many niches we have to consider when studying HSCs and how HSCs navigate between them. The most proposed model suggests that the osteoblastic niche provides long-term dormancy, and the perivascular niche is an intermediate home for activated HSCs that can generate differentiated progenitor cells or revert them to dormancy [18,19].

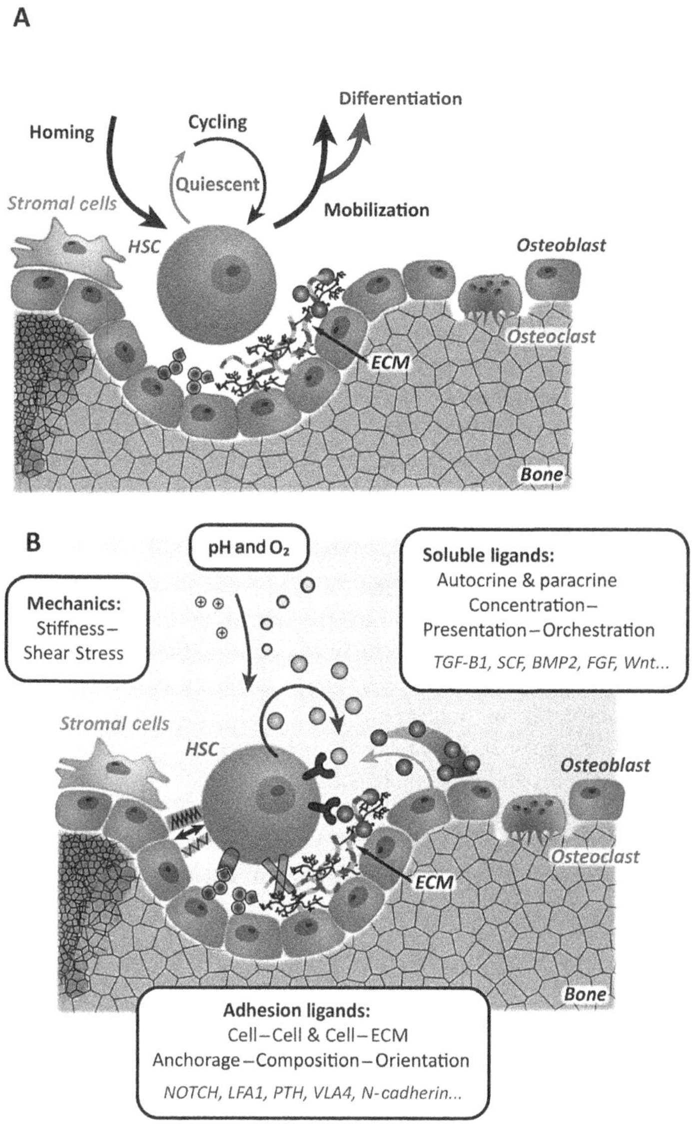

FIGURE 16.1 Concepts of the HSC niche. (A) Scheme of HSC regulation inside the niche microenvironment depicting the different HSC fate decisions which are orchestrated by the niche components. (B) Scheme of the different microenvironmental cues controlling HSC fate including biochemical, biophysical, and metabolic signals. *(See insert for color representation of the figure.)*

The complexity of the niche is reflected by the identification of numerous molecular pathways involved in the cross talk between HSC and its microenvironment (e.g., the notch pathway [13], angiopoietin-1 (ANG-1)-Tie2 [20], stromal cell-derived factor-1 (SDF1)-CXCR4 [21], thrombopoietin (TPO)-Mpl [22], stem cell factor (SCF)-c-Kit [23]). Despite initial reports claiming beta-catenin, the key component of canonical Wnt signaling, to be unnecessary for haematopoiesis [24], a growing number of studies indicate that Wnt signaling does influence HSC function [25–30].

While HSCs were initially assumed to be suspension cells, different adhesion receptors and ligands such as cadherins, integrins, fibronectin, osteopontin, heparan sulfate, and others have been shown to regulate HSC function as well as localization and homing [31–33]. Even though the occurrence of specific ECM components inside the BM was investigated by some groups [17,34], their detailed 3-D distribution is still unclear.

Despite this large amount of data on HSC–niche interaction, the mechanisms controlling HSC fate within different BM regions and cellular environments as well as the systemic interplay of the different niche types are far from being understood. In particular, this refers to the dynamic changes in the niche such as variations in direct cell–cell contacts, the interrelation of ECM components with regulatory molecules secreted by HSC or niche cells, and the spatial organization of niche components. Although little direct evidence has been collected about the effect of niche dynamics on HSC fate [28], niche alterations have been described following BM conditioning, transplantation, and aging [30,35,36]. Moreover, variations in HSC density and HSC progeny should not be underestimated in playing a role in HSC fate regulation. As stem and progenitor cells have distinct metabolic states, and the transition from stem to progenitor cell corresponds to a critical metabolic change, the dynamic balance of energy, oxygen, and redox status, essential for HSC maintenance, must also be considered as an important regulating parameter [37].

16.3 BIOMATERIAL APPROACHES TO CREATE BIOMIMETIC HSC NICHES

The development of artificial microenvironments mimicking important stem cell–niche interactions *in vitro* has received increased attention during the last decade to gain new insights into the orchestrating and regulatory roles of the niche components. Bioengineered platforms are envisioned to verify hypotheses generated *in vivo* explicitly under strictly defined conditions. They should help to provide mechanistic insights into HSC regulation, and to explore pronounced effects on HSC fate with minimalistic setups compared with an *in vivo* situation. Because of the wide range of proven and potential HSC regulators and the complexity of their interactions, high-throughput screening platforms are indispensable in identifying and screening the most relevant components and their multifaceted interrelations in a combinatorial manner.

Engineering approaches for biomimetic HSC microenvironments are based on biomaterial strategies using the full range of available options including

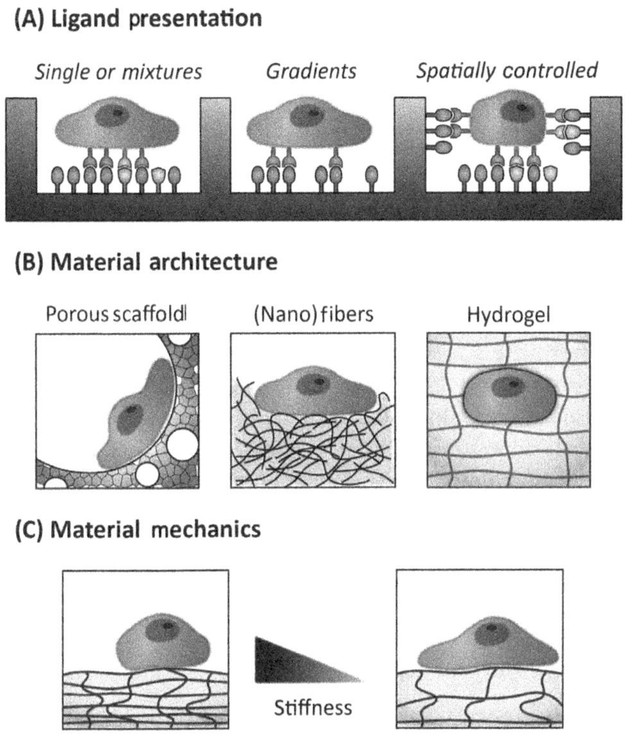

FIGURE 16.2 Biomaterial approaches to mimic signals of niche microenvironment for control of stem cell fate. (A) Distinct options to present adhesion ligands or growth factors in microstructured biomaterials scaffolds for *in vitro* experiments allowing for one type of ligand and mixtures, gradients of ligands, as well as spatial control of presentation mode, for example, 2-D versus 2.5-D. (B) Varying the architecture of the biomaterials allows to mimic different morphological architectures of the niche microenvironment including open pores, fibrous substrates, and hydrogel entrapment. (C) Biomaterials mechanics can be modified to alter mechanotransduction pathways of stem cells including stiffness-dependent cell differentiation. (Inspired by Reference 39.)

decellularized *ex vivo* matrices, biopolymer-based structures, biohybrid, and fully synthetic materials. For an excellent review on the general topic of biomaterial strategies for engineering extracellular microenvironments, the reader is referred to other recent publications [38]. In this section, we want to highlight topics relevant to the HSC niche concept as given earlier.

Biomaterials have been used to create a suitable microenvironment by providing spatial proximity of cells to each other as well as scaffolding structures, presenting adhesion receptor ligands and growth factors in defined spatiotemporal pattern, producing mechanical signals, and enabling cell-driven matrix reorganization [39]. Mimics of stem cell microenvironments have been developed, which include a wide range of important exogenous cues (see also Figure 16.2):

- spatial and temporal control of the presentation of growth and differentiation factors
- soluble and immobilized gradients of signaling molecules inside scaffolds
- regulation of adhesion ligands by specificity, composition, density, and spacing
- viscoelasticity of the ECM
- scaffold topography ranging from nanometer to micrometer scale
- geometrical constraints and guidance in 2-D and 3-D.

Based on the success in other areas of engineering biomaterials scaffolds, most of these parameters were considered for HSC microenvironments. It has to be mentioned that adhesion-related features are underrepresented in these studies as HSCs were thought to be suspension cells. However, the last 15 years have proven cell–ECM and cell–cell adhesion to be an essential feature of HSC regulation. This change in the understanding of HSC regulation is mirrored by a recent broadening in the range of bioengineering strategies of HSC microenvironments. To date, however, the cues most frequently addressed primarily cover the following:

- growth factor delivery
- adhesion ligand presentation
- growth factor presentation.

Emphasis has been given to the delivery of growth factors, which are often seen as key regulators of the niche. In addition to the common adjustment of levels and combinations of growth factors in conventional cell culture experiments, biomaterials have been used to deliver growth factors from matrices in a controlled manner and to confine the molecules to the surface of cell culture carriers by covalent or noncovalent conjugation. Biomaterial scaffolds have been used to present growth factors in an active orientation to trigger receptor activation in a specific, persistent, and more physiological manner [40,41]. Soft lithography techniques and microfluidics have been applied to present growth factors in spatially and temporally controlled patterns [42]. The gradients of released signaling molecules, like SDF-1, were adjusted by sustainable delivery from biohybrid matrices as an interesting vehicle to control chemotaxis of HSCs to SDF-1 gradients [43,44]. SDF-1 is known as a common signaling molecule binding to the angiopoietin-1 (ANG-1) thrombopoietin (TPO) stem cell factor (SCF) cell receptor. It activates different processes including a guided migration of cells toward the SDF-1 gradient [45]. This signaling pathway HSCs use *in vivo* to be attracted toward the BM is a process called stem cell homing [14,46].

Although HSCs are considered only weakly adhesive, their adhesion-dependent behavior has been investigated by providing ligand structures of the ECM or cell–cell contacts at biomaterial surfaces. Adsorptive deposition of multicomponent layers of adhesive proteins from culture media (such as fibronectin and vitronectin) or more dedicated modes of immobilization of ECM proteins (e.g., collagen, laminin, or peptide sequences) and receptors of cell–cell contacts were used not only to trigger simple adhesion but also to engage specific receptors of signaling cascades [47–49].

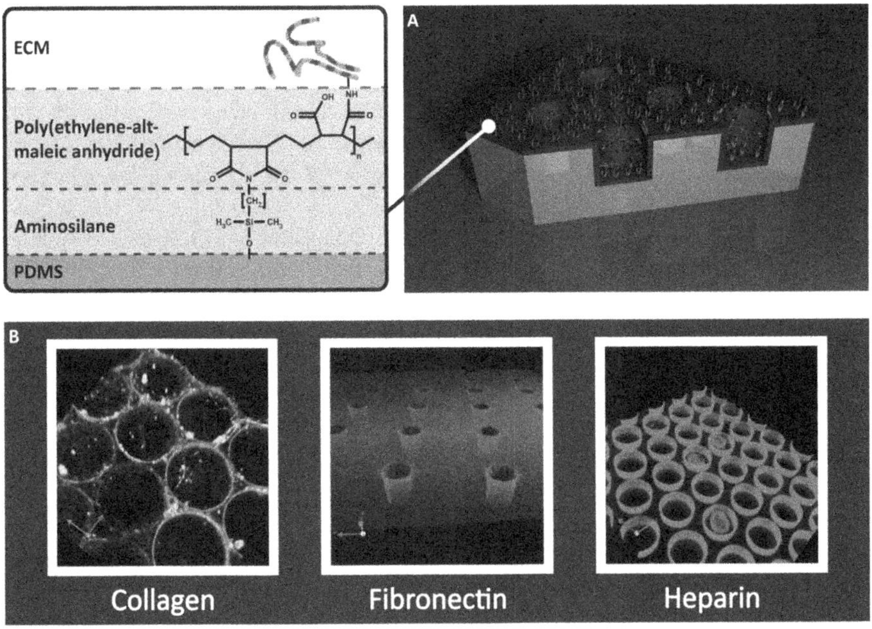

FIGURE 16.3 Protein immobilization on a microstructured surface for HSC culture. (A) Poly(dimethyl siloxane) (PDMS) microstructured with oxygen plasma activation are coated by aminosilane functionalization and maleic anhydride copolymer coating to immobilize components of the ECM. (B) Fluorescent images of ECM-modified PDMS microstructures. *(See insert for color representation of the figure.)*

Scaffolds with nanometer substrate features, such as nanofibers or nanoscale-spaced and clustered adhesion ligands, were developed to study the nanotopographical cues in HSC adhesion and regulation [50,51]. These studies suggested an improved adhesion and expansion of HSCs.

Topographical features of the HSC microenvironment in the micrometer range were addressed by *in vitro* strategies for the presentation of growth factors, ECM components, as well as cell receptors. Microstructures of rigid silicone or soft hydrogel were used to mimic single or multicell microenvironments with an example shown in Figure 16.3 [40,48,49].

Inspired by distinct mechanical properties of the niche, cell culture scaffolds of varying stiffness have been developed to investigate the impact of viscoelastic materials properties on HSC self-renewal [52]. These studies showed that a higher substrate stiffness promotes HSC expansion.

In the context of the design of microstructures for HSC microenvironments, array techniques are currently being developed to implement high-throughput approaches for screening larger sets and combinations of factors in their influence on stem cell dynamics and progeny [53]. These new generations of cell culture platforms build on the rapid progress in robotically controlled microfabrication technologies such as microfluidics, photo, and soft lithography. In this manner, protein microarrays and (a)symmetric immobilized or soluble gradients have been utilized to mimic the spatially controlled display of niche ligands. Recently, a microfluidic platform containing thousands of nanoliter-scale chambers combined with an automated medium exchange enabled the configuration of well-defined culture conditions with respect to cell culture parameters like growth factor concentration [54].

A microfluidic device consisting of an array of hydrodynamic traps hosting single cells was developed to study fate decisions at the single-cell level in combination with screening techniques such as cytometry or on-chip polymerase chain reaction [55,56]. Bio-inspired scaffolds for the expansion of HSCs in combination with novel (semi)automatic time-lapse imaging and single-cell tracking techniques have been applied to microstructured surfaces or in microfluidic devices providing new avenues to analyze mechanisms governing HSC cell growth and fate decisions at the single-cell level [54,56–58].

16.4 HSC CONTROL *EX VIVO*: FROM HSC EXPANSION TO BIOMIMETIC NICHES

As outlined above, biomaterial approaches have been developed to explore exogenous cues of the HSC niche *ex vivo*. In this section, we will illustrate a few examples, showing their ability to mimic relevant features of *in vivo* stem cell behavior. We begin with early approaches of HSC culture and HSC expansion. Additionally, a more complex setup of adhesion ligands and growth factors presentations as well as coculture systems is discussed.

16.4.1 *Ex Vivo* Expansion

HSCs were one of the first examples showing the regenerative potential of stem cells [59–61]. Within the introduction of BM transplantation by Donnall Thomson, HSCs were used to cure severe conditions in leukemia treatment [62]. Despite this long history of HSC research and therapeutic application, many questions remain in HSC biology and about their clinical use. In particular, it would be highly desirable to transplant a much larger number of HSCs when treating leukemia. At present, an expensive and resource-consuming pooling of several apheresis blood samples is used for collecting sufficient cells for a transplant. One way to overcome this HSC shortage is to use *ex vivo* expansion strategies or the controlled differentiation of embryonic stem cells toward blood lineages [63,64]. However, these new strategies are awaiting solid evidence for an improved clinical outcome in HSC transplantation [65,66].

Most approaches of *ex vivo* HSC expansion rely on the knowledge about the tight control of HSC behavior by the local microenvironment of the cells *in vivo* [1,2,67,68]. As HSCs were long considered to be suspension cells, the main focus has been on soluble cues in their microenvironment, for example, cytokines and oxygen pressure. Consequently, growth factor compositions of the cell culture media in static and perfusion cultures were examined in early expansion strategies [69–72]. To date, various promising growth factor *cocktails* have been developed and are still being developed on the basis of new screening tools. Besides the growth factors supplemented to the media, cell-released factors were also discovered to be substantial in the expansion of HSCs. A fed-batch approach was developed to control secreted factor concentrations demonstrating an improvement in HSC expansion [73]. Therein, a process of stepwise volume increase of cell culture media was introduced, so-called fed-batch, which dilutes cell-released factors and improves HSC expansion.

16.4.2 Growth Factor Presentation

Apart from identifying and adjusting concentration levels, the precise control over presentation and release of signaling molecules in a temporally controlled and oriented manner has more recently come into focus to improve HSC expansion. Biomaterial surfaces and hydrogel scaffolds were designed with growth factors (e.g., SCF, bone morphogenetic protein 4, and TPO) and cell surface ligands (e.g., Delta-1, Delta-4, and Jagged) being either (non)covalently tethered to the surface or incorporated into the biomaterials. These strategies were, in part, inspired by the function of glycosaminoglycans (e.g., heparan sulfate), which have been found to bind relevant growth factors specifically and present them to HSCs [74,75]. They show promising results in *ex vivo* HSC expansion and embryonic stem cell differentiation toward blood lineages [41,64,76,77]. Furthermore, it was shown that the release of SDF-1α from poly(lactic-co-glycolic acid) (PLGA) scaffolds can increase the invasion of stem cell populations (both mesenchymal stem cells and HSCs) to the site of porous scaffolds after implantation [44].

Recently, microfluidic devices, in some cases combined with micropatterned surfaces, were applied to temporally vary growth factor concentrations in the medium. Prolonged initial exposure (up to 16–24 hours) of HSCs to low concentrations of SCF (1 ng/mL) was found to irreversibly decrease cell viability as it could no longer be rescued by admitting high levels of SCF (300 ng/mL) [54]. By that specific local cues of HSC niche, like temporal and spatial variations and gradients, are envisioned to be mimicked in the *in vitro* settings.

Specific effects of immobilized signaling molecules on HSC maintenance were mimicked using a microstructured polyethylene glycol hydrogel platform and protein immobilization. This is considered to be more suitable in providing soft and highly hydrated cellular microenvironments, which are thought to mimic the BM more accurately than silicone materials [40]. HSCs could thereby be exposed to soluble and immobilized, well-defined niche signals to study how single signaling molecules or their combinations affect HSC self-renewal. In this manner, the interaction with niche cells was explored by testing immobilized cell–cell adhesion receptors without the need for complex cocultures. Pronounced effects on HSC maintenance were revealed for cell–cell contact receptors N-cadherin and the Wnt3a ligand by the demonstration of a successful BM engraftment in a mouse model after the *ex vivo* HSC culture. HSCs maintained their multipotency after *ex vivo* cell division or kept stem cell potential in a quiescent state with N-cadherin adherence. The rate and synchrony of stem cell division *in vitro* could be shown to correlate with the regeneration potential *in vivo* and was proposed as prediction criteria of stem cell function.

16.4.3 ECM Adhesion Ligands

In addition to the modulation of cytokine signals in regulating cell fate and cycle status, the direct adhesive interactions between ECM components (e.g., fibronectin, osteopontin, and heparan sulfate) or cell surface ligands in cell–cell interactions (e.g., cadherins and notch/Delta) also strongly influence HSC fate [78,79]. Investigations of the BM ultrastructure and composition, as well as *in vitro* and *in vivo* experiments, have revealed that many ECM components of the BM niche regulate HSCs [13,17,33,80].

Reactive polymer layers of maleic anhydride copolymers were used to covalently immobilize ECM proteins like collagen I and IV, fibronectin, laminin, osteopontin, as well as glycosaminoglycans like heparan sulfate, hyaluronic acid, and heparin. Adherent HSCs were assessed by reflection interference contrast microscopy (RICM) and intense cell–matrix interactions were found on surfaces coated with fibronectin, heparin, heparan sulfate, and fibrillar collagen I. The approach was further expanded to functionalize microstructures as discussed below [47–49].

Other studies showed collagen I to promote HSC proliferation *in vitro* [81], or the administration of hyaluronic acid to improve hematopoietic recovery and survival after HSC transplantation *in vivo* [82]. To gain more insights into the nanotopographical features of ECM components, HSC adhesion and regulation were investigated with respect to its dependence on the modulation of nanometer spacing and clustering of adhesion ligands [50]. A spacing of 32 nm was found to be the maximum tolerated distance between peptide adhesion ligands (i.e., arginine–glycine–aspartic acid (RGD); a cell-adhesive amino acid sequence found in several ECM proteins). Additionally, an increased redistribution of lipid rafts, CD34, and CD133 surface markers and the integrins $\alpha V\beta 5$ or $\alpha 5\beta 1$ was detected with a decreased spacing of the adhesion ligands [83].

16.4.4 3-D Topology and Topography

Early screening studies showed the impact of topographical features of biomaterials scaffolds on HSC expansion and differentiation [84]. As such, the need for bioengineering strategies in the spatial design of HSC culture systems was postulated for a targeted expansion of HSCs, supported by the 3-D ultrastructure of bone and BM. Initially, these systems were restricted to specific topological structures (porosity, foam structure, and fiber networks) of certain materials (polymers and titanium oxide) and simple coatings with ECM proteins (fibronectin and collagen I) on polystyrene cultureware [81,85–87]. To mimic the fibrous microenvironment in the niche, nanofiber meshes and films functionalized with amino groups were investigated and found to increase HSC expansion drastically [51]. 3-D honeycombed hydrogels, with adjustable internal dimensions and stiffness mimicking the bone spongosia were developed more recently to analyze the tissue migration of promyelocytic leukemia cells *in vitro* [88]. Another approach used naturally derived biopolymer matrices exhibiting an *in vivo*-like topology and composition, with promising results, as cell culture scaffolds concerning HSC expansion [89].

In vivo, HSCs are exposed to diverse matrix topologies with varying (fibrous) ECM composition, mesh size and degree of mineralization and, as a consequence, varying matrix stiffness. Inspired by exciting results on differentiation of mesenchymal stem cells using hydrogels of varying stiffness [90], it has been suggested that material properties might trigger HSC expansion as well. Although the weak adhesion characteristics of HSCs might impede such influences, studies have shown promising results from modulated mechanical characteristics of polyacrylamide hydrogels and thin layers of ECM proteins [52,91]. HSC expansion was enhanced on very soft and stretchable tropoelastin substrates. Furthermore, HSC morphology (spreading, shape) and viability might be influenced by biophysical properties of collagen I-coated polyacrylamide hydrogels because HSC spreading increased with increasing substrate stiffness.

A number of microstructured biomaterial approaches were carried out to unravel principles of the 3-D microenvironmental regulation of HSCs. Triggered by the pronounced differences in HSC adhesion strength and patterns on covalently tethered ECM components including fibronectin, collagen I, fibrillar collagen I, collagen IV, as well as heparin and hyaluronic acid on planar surfaces [47], the specific adhesion characteristics were combined with topographical micropatterns of silicone scaffolds to mimic a *quasi* 3-D microenvironment [49]. Additionally, these scaffolds were used to investigate the combined action of adhesion (i.e., fibronectin) and soluble cytokines on HSC fate. In line with previous reports, the studies showed that an increased engagement of cell adhesion receptors to fibronectin-coated single-cell HSCs compartments downregulated HSC proliferation and kept them in a more quiescent and undifferentiated state as exemplified in Figure 16.4 [48,92]. Examination of the cell cycle kinetics of HSCs cultivated on top of these artificial microstructures revealed an initial promotion of cell cycle entry, followed by a deceleration of cycling kinetics at later time points indicated by a decreased total DNA synthesis in microcavities hosting single HSC. Furthermore, these studies revealed a synergistic interplay of adhesion-related and cytokine signaling. Very high cytokine concentrations, far above physiological conditions, override the dependency of the adhesion signal of the microcavities due to the drastically increased cell cycling. The findings suggest that the 3-D arrangement of ECM adhesion ligands at moderate cytokine levels might conserve an immature state of HSC *in vitro*. This underpins the idea of a spatially induced quiescent state in the BM over long time periods.

These biomimetic polymer scaffolds were used to investigate dynamic HSC behavior further on a single-cell level *in vitro*. Such approaches are needed because available populations of human HSCs (CD133+ from peripheral blood after granulocyte colony-stimulating factor mobilization) are highly heterogeneous and contain only a small portion of real stem cells, despite large amounts of committed progenitor cells. It was hypothesized that single-cell tracking would enable improved quantification of the impact of 3-D topographical features on HSC behavior. In an automated tracking approach using optical time-lapse microscopy, HSCs were followed on a single-cell level over several days in terms of cell cycling, size, migration, and localization in the biomimetic niche microenvironments, depicted in Figure 16.5 [57]. The analysis could demonstrate a high symmetry in the characteristics of daughter cells after division over the full range of investigated parameters pointing

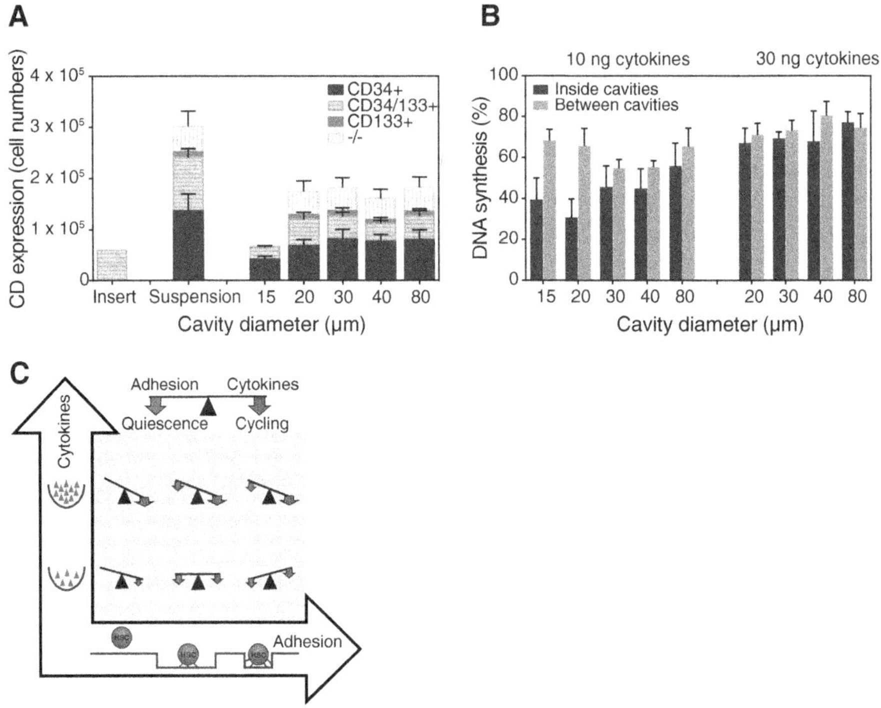

FIGURE 16.4 Synergistic action of adhesive micropatterns and soluble cytokines on HSC proliferation and differentiation. Single-cell microcavities at low cytokine levels maintain HSC in a quiescent and undifferentiated state. (A) Surface marker expression and cell number after 7 days of cell culture on fibronectin-coated microstructures. (B) Quantification of cell cycling by means of DNA synthesis (bromodeoxyuridine (BrdU) incorporation) directly on microstructures at low and high cytokine concentrations in the media. (C) Scheme of the balance of synergistic signals from soluble and adhesive cues. (Adapted from Reference 48, with permission from The Royal Society of Chemistry.)

to symmetric cell division events. A further analysis showed this symmetry in stem cell behavior to be controlled by the local microenvironment because topographical features disturb the symmetry characteristics of daughter cells. Hence, these investigations using single-cell tracking in biomimetic microenvironments permit the analysis of several features of HSC behavior including the type of cell division, cell cycling, migration characteristics, and their control by microenvironmental cues like ECM topography.

16.4.5 Coculture Systems

Coculture systems should also be mentioned as another approach to biomimetic HSC niche microenvironments. Early on, it was demonstrated that coculture of HSCs with stromal cells (e.g., mesenchymal stem cells) or other cell lines (feeder cell lines or

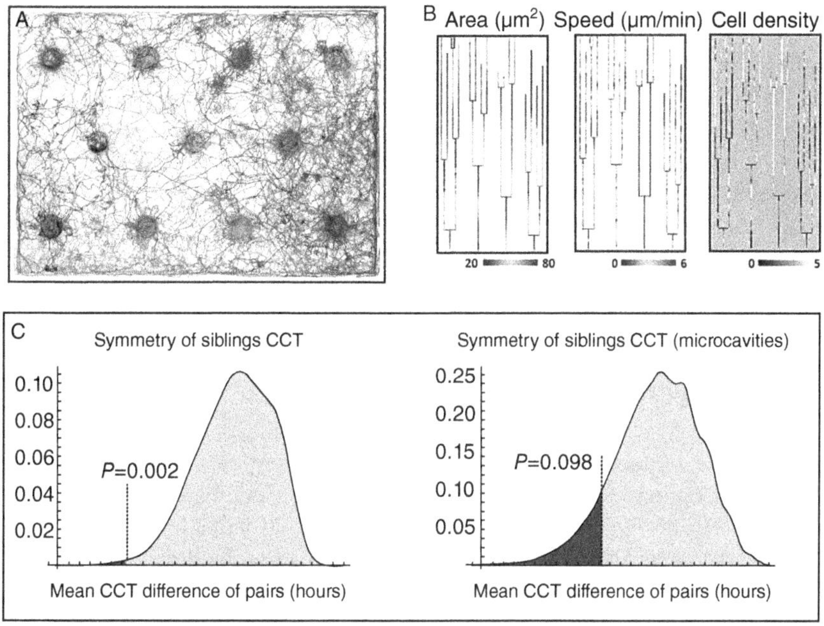

FIGURE 16.5 *In vitro* single-cell tracking of HSC in biomimetic microenvironments. (A) Time-evolved migration patterns of HSC in a microwell with inserted microcavities. Each color corresponds to a track of one single cell. (B) Cellular genealogies can be revealed on a single cell level. The time evolution (bottom to top) of four different cells is shown in respect to cell area, speed, and cell density. The color code of the lines indicates the respective cell properties in relation to the scale bar beneath the plot. (C) Statistical analysis proves highly symmetric characteristics of daughter cells by a permutation analysis, which are disturbed by micropatterns of the scaffolds. (Adapted from Reference 57. with permission from Elsevier.)

Dexter cultures), with and without direct contact, improved the expansion of HSCs [93,94]. (For details we refer to the review of coculture systems in Reference 95). Based on those observations, several bioengineering strategies tried to improve HSC coculture systems by using biomaterial scaffolds to support and enhance the function of feeder layers. For example, because of the critical role of mesenchymal stem cells in HSC function and self-renewal, matrices of collagen I and cancellous bone were used to stimulate the differentiation of mesenchymal stem cells toward osteoblast lineages as a supportive feeder layer for HSC expansion [96,97]. Furthermore, to mimic major components of the complex *in vivo,* BM niche 3-D scaffolds of either soft collagen hydrogels or hard, porous hydroxyapatite ceramics have been used to coculture HSCs and mesenchymal cells with variable differentiation characteristics [98,99]. Here, the coculture with BM mesenchymal stem cells supported the migration and engraftment of HSCs inside the collagen gels and increased the expansion of primitive HSCs. Accordingly, these scaffolds may aid the better understanding of the interaction of niche cells and HSCs in *in vitro* studies [99].

16.4.6 Toward Dynamic Control of HSC Microenvironments

As already mentioned earlier, the niche microenvironment has to be considered very complex with dynamic localization of HSC, as well as local and temporal gradients of growth factors and signaling molecules. To be effective, environmental constraints, summarized as the *metabolic niche*, should also be considered. This is why biomaterial scaffolds mimicking and modulating HSC microenvironments *in vitro* need to be combined with advanced bioreactors or perfusion systems for the dynamic control of growth factor levels and cell-produced signaling molecules, as well as other more general nutrition parameters [73,98,100].

16.5 OUTLOOK

This overview of biomaterial strategies for developing HSC microenvironments demonstrates options for mimicking *in vivo* stem cell microenvironments like the HSC niche. The design of advanced cell culture scaffolds will allow a deeper mechanistic understanding of stem cell function and interaction with their microenvironment, as well as the development of therapeutic strategies by maintenance and expansion of HSC *ex vivo*.

An interesting aspect of the expansion of these biomimetic options is their application in basic studies on the *in vivo* microenvironmental control of cancer stem cells. The concept of a cancer stem cell was first proposed by John Dick and Dominique Bonnet when studying acute myeloid leukemia [101], implying similar control mechanisms to those for HSC. The ability of cancer cells to evade chemotherapy and contribute to disease relapse might well be triggered by the state of the stem cell microenvironment—a possible cancer stem cell niche—with similar regulation mechanisms as discussed in this chapter.

We believe that the tremendous progress in bioengineering technologies will allow an increasingly complex design of stem cell microenvironments in the future, leading to a comprehensive understanding of the microenvironmental regulation of stem cell behavior. A synergistic application of new technologies in high-throughput arrangements with system approaches of modeling stem cell behavior in such environments will be a key in triggering progress in this field. Examples of these advanced technologies comprise microsystems and microfluidics [42], time-lapse/*in situ* analysis of cells [57,102], biohybrid hydrogel materials [103,104], and *in silico* modeling of HSC behavior [105–107].

Acknowledgments

E.M. and T.P. acknowledge support by a grant on Medical Systems Biology "HaematoSys" of the Bundesministerium für Forschung und Bildung (BMBF-FKZ 0315452). M.A. is supported by the Deutsche Forschungsgemeinschaft within the Graduate School "BuildMoNa." M.A. and T.P. are supported by the Human Frontier Science Program (RGP0051/2011). C.W. is supported by the Deutsche Forschungsgemeinschaft within the Priority Program "From Cells to Tissues" (SFB 655).

REFERENCES

[1] Schofield, R. *Blood Cells* **1978**, *4*, 7–25.
[2] Morrison, S.J., Spradling, A.C. *Cell* **2008**, *132*, 598–611.
[3] Raaijmakers, M.H.G.P., Mukherjee, S., Guo, S.Q., Zhang, S.Y., Kobayashi, T., Schoonmaker, J.A., Ebert, B.L., Al-Shahrour, F., Hasserjian, R.P., Scadden, E.O., Aung, Z., Matza, M., Merkenschlager, M., Lin, C., Rommens, J.M., Scadden, D.T. *Nature* **2010**, *464*, 852–U858.
[4] Colmone, A., Amorim, M., Pontier, A.L., Wang, S., Jablonski, E., Sipkins, D.A. *Science* **2008**, *322*, 1861–1865.
[5] Nwajei, F., Konopleva, M. *Adv. Hematol.* **2013**, *2013*, 953982.
[6] Askmyr, M., Sims, N.A., Martin, T.J., Purton, L.E. *Trends Endocrinol. Metab.* **2009**, *20*, 303–309.
[7] Kiel, M.J., Morrison, S.J. *Nat. Rev. Immunol.* **2008**, *8*, 290–301.
[8] Roeder, I., Lorenz, R. *Stem Cell Rev.* **2006**, *2*, 171–180.
[9] Loeffler, M., Roeder, I. *Cells Tissues Organs* **2002**, *171*, 8–26.
[10] Moore, K.A., Lemischka, I.R. *Science* **2006**, *311*, 1880–1885.
[11] Nilsson, S.K., Johnston, H.M., Coverdale, J.A. *Blood* **2001**, *97*, 2293–2299.
[12] Kiel, M.J., Yilmaz, O.H., Iwashita, T., Yilmaz, O.H., Terhorst, C., Morrison, S.J. *Cell* **2005**, *121*, 1109–1121.
[13] Calvi, L.M., Adams, G.B., Weibrecht, K.W., Weber, J.M., Olson, D.P., Knight, M.C., Martin, R.P., Schipani, E., Divieti, P., Bringhurst, F.R., Milner, L.A., Kronenberg, H.M., Scadden, D.T. *Nature* **2003**, *425*, 841–846.
[14] Sugiyama, T., Kohara, H., Noda, M., Nagasawa, T. *Immunity* **2006**, *25*, 977–988.
[15] Ding, L., Saunders, T.L., Enikolopov, G., Morrison, S.J. *Nature* **2012**, *481*, 457–U465.
[16] Mendez-Ferrer, S., Michurina, T.V., Ferraro, F., Mazloom, A.R., MacArthur, B.D., Lira, S.A., Scadden, D.T., Ma′ayan, A., Enikolopov, G.N., Frenette, P.S. *Nature* **2010**, *466*, 829–U859.
[17] Nilsson, S.K., Debatis, M.E., Dooner, M.S., Madri, J.A., Quesenberry, P.J., Becker, P.S. *J. Histochem. Cytochem.* **1998**, *46*, 371–377.
[18] Malhotra, S., Kincade, P.W. *Cell Stem Cell* **2009**, *4*, 27–36.
[19] Trumpp, A., Essers, M., Wilson, A. *Nat. Rev. Immunol.* **2010**, *10*, 201–209.
[20] Arai, F., Hirao, A., Ohmura, M., Sato, H., Matsuoka, S., Takubo, K., Ito, K., Koh, G.Y., Suda, T. *Cell* **2004**, *118*, 149–161.
[21] Sipkins, D.A., Wei, X., Wu, J.W., Runnels, J.M., Cote, D., Means, T.K., Luster, A.D., Scadden, D.T., Lin, C.P. *Nature* **2005**, *435*, 969–973.
[22] Yoshihara, H., Arai, F., Hosokawa, K., Hagiwara, T., Takubo, K., Nakamura, Y., Gomei, Y., Iwasaki, H., Matsuoka, S., Miyamoto, K., Miyazaki, H., Takahashi, T., Suda, T. *Cell Stem Cell* **2007**, *1*, 685–697.
[23] Driessen, R.L., Johnston, H.M., Nilsson, S.K. *Exp. Hematol.* **2003**, *31*, 1284–1291.
[24] Cobas, M., Wilson, A., Ernst, B., Mancini, J.C., MacDonald, H.R., Kemler, R., Radtke, F. *J. Exp. Med.* **2004**, *199*, 221–229.
[25] Qian, Z.J., Chen, L.N., Fernald, A.A., Williams, B.O., Le Beau, M.M. *J. Exp. Med.* **2008**, *205*, 2163–2175.
[26] Reya, T., Duncan, A.W., Ailles, L., Domen, J., Scherer, D.C., Willert, K., Hintz, L., Nusse, R., Weissman, I.L. *Nature* **2003**, *423*, 409–414.
[27] Trowbridge, J.J., Xenocostas, A., Moon, R.T., Bhatia, M. *Nat. Med.* **2006**, *12*, 89–98.
[28] Goessling, W., North, T.E., Loewer, S., Lord, A.M., Lee, S., Stoick-Cooper, C.L., Weidinger, G., Puder, M., Daley, G.Q., Moon, R.T., Zon, L.I. *Cell* **2009**, *136*, 1136–1147.

[29] Renstrom, J., Istvanffy, R., Gauthier, K., Shimono, A., Mages, J., Jardon-Alvarez, A., Kroger, M., Schiemann, M., Busch, D.H., Esposito, I., Lang, R., Peschel, C., Oostendorp, R.A.J. *Cell Stem Cell* **2009**, *5*, 157–167.
[30] Hooper, A.T., Butler, J.M., Nolan, D.J., Kranz, A., Iida, K., Kobayashi, M., Kopp, H.G., Shido, K., Petit, I., Yanger, K., James, D., Witte, L., Zhu, Z.P., Wu, Y., Pytowski, B., Rosenwaks, Z., Mittal, V., Sato, T.N., Rafii, S. *Cell Stem Cell* **2009**, *4*, 263–274.
[31] Zhang, J.W., Niu, C., Ye, L., Huang, H.Y., He, X., Tong, W.G., Ross, J., Haug, J., Johnson, T., Feng, J.Q., Harris, S., Wiedemann, L.M., Mishina, Y., Li, L.H. *Nature* **2003**, *425*, 836–841.
[32] Gottschling, S., Saffrich, R., Seckinger, A., Krause, U., Horsch, K., Miesala, K., Ho, A.D. *Stem Cells* **2007**, *25*, 798–806.
[33] Nilsson, S.K., Johnston, H.M., Whitty, G.A., Williams, B., Webb, R.J., Denhardt, D.T., Bertoncello, I., Bendall, L.J., Simmons, P.J., Haylock, D.N. *Blood* **2005**, *106*, 1232–1239.
[34] Lichtman, M.A. *Exp. Hematol.* **1981**, *9*, 391–410.
[35] Spiegel, A., Kalinkovich, A., Shivtiel, S., Kollet, O., Lapidot, T. *Cell Stem Cell* **2008**, *3*, 484–492.
[36] Kollet, O., Dar, A., Shivtiel, S., Kalinkovich, A., Lapid, K., Sztainberg, Y., Tesio, M., Samstein, R.M., Goichberg, P., Spiegel, A., Elson, A., Lapidot, T. *Nat. Med.* **2006**, *12*, 657–664.
[37] Suda, T., Takubo, K., Semenza, G.L. *Cell Stem Cell* **2011**, *9*, 298–310.
[38] Lutolf, M.P., Blau, H.M. *Adv. Mater.* **2009**, *21*, 3255–3268.
[39] Lutolf, M.P., Gilbert, P.M., Blau, H.M. *Nature* **2009**, *462*, 433–441.
[40] Lutolf, M.P., Doyonnas, R., Havenstrite, K., Koleckar, K., Blau, H.M. *Integr. Biol. (Camb.)* **2009**, *1*, 59–69.
[41] Toda, H., Yamamoto, M., Kohara, H., Tabata, Y. *Biomaterials* **2011**, *32*, 6920–6928.
[42] Kobel, S., Lutolf, M.P. *Curr. Opin. Biotechnol.* **2011**, *22*, 690–697.
[43] Prokoph, S., Chavakis, E., Levental, K.R., Zieris, A., Freudenberg, U., Dimmeler, S., Werner, C. *Biomaterials* **2012**, *33*, 4792–4800.
[44] Thevenot, P.T., Nair, A.M., Shen, J.H., Lotfi, P., Ko, C.Y., Tang, L.P. *Biomaterials* **2010**, *31*, 3997–4008.
[45] Lapidot, T., Dar, A., Kollet, O. *Blood* **2005**, *106*, 1901–1910.
[46] Jung, Y., Wang, J., Schneider, A., Sun, Y.X., Koh-Paige, A.J., Osman, N.I., McCauley, L.K., Taichman, R.S. *Bone* **2006**, *38*, 497–508.
[47] Franke, K., Pompe, T., Bornhauser, M., Werner, C. *Biomaterials* **2007**, *28*, 836–843.
[48] Kurth, I., Franke, K., Pompe, T., Bornhauser, M., Werner, C. *Integr. Biol. (Camb.)* **2009**, *1*, 427–434.
[49] Franke, K., Kurth, I., Bornhauser, M., Werner, C., Pompe, T. *Soft Matter* **2009**, *5*, 3505–3510.
[50] Lee-Thedieck, C., Spatz, J.P. *Macromol. Rapid Commun.* **2012**, *33*, 1432–1438.
[51] Chua, K.N., Chai, C., Lee, P.C., Tang, Y.N., Ramakrishna, S., Leong, K.W., Mao, H.Q. *Biomaterials* **2006**, *27*, 6043–6051.
[52] Holst, J., Watson, S., Lord, M.S., Eamegdool, S.S., Bax, D.V., Nivison-Smith, L.B., Kondyurin, A., Ma, L., Oberhauser, A.F., Weiss, A.S., Rasko, J.E. *Nat Biotechnol* **2010**, *28*, 1123–1128.
[53] Kobel, S., Lutolf, M.P. *Biotechniques* **2010**, *48*, Ix–Xxii.
[54] Lecault, V., VanInsberghe, M., Sekulovic, S., Knapp, D.J.H.F., Wohrer, S., Bowden, W., Viel, F., McLaughlin, T., Jarandehei, A., Miller, M., Falconnet, D., White, A.K., Kent, D.G., Copley, M.R., Taghipour, F., Eaves, C.J., Humphries, R.K., Piret, J.M., Hansen, C.L. *Nat. Methods* **2011**, *8*, 581–U593.

[55] Faley, S.L., Copland, M., Wlodkowic, D., Kolch, W., Seale, K.T., Wikswo, J.P., Cooper, J.M. *Lab Chip* **2009**, *9*, 2659–2664.
[56] Kobel, S.A., Burri, O., Griffa, A., Girotra, M., Seitz, A., Lutolf, M.P. *Lab Chip* **2012**, *12*, 2843–2849.
[57] Scherf, N., Franke, K., Glauche, I., Kurth, I., Bornhauser, M., Werner, C., Pompe, T., Roeder, I. *Exp. Hematol.* **2012**, *40*, 119–130.
[58] Dykstra, B., Ramunas, J., Kent, D., McCaffrey, L., Szumsky, E., Kelly, L., Farn, K., Blaylock, A., Eaves, C., Jervis, E. *Proc. Natl. Acad. Sci. U.S.A.* **2006**, *103*, 8185–8190.
[59] Till, J.E., Mc, C.E. *Radiat. Res.* **1961**, *14*, 213–222.
[60] Becker, A.J., McCulloch, E.A., Till, J.E. *Nature* **1963**, *197*, 452–454.
[61] Wu, A.M., Till, J.E., Siminovitch, L., McCulloch, E.A. *J. Exp. Med.* **1968**, *127*, 455–464.
[62] Sorrentino, B.P. *Nat. Rev. Immunol.* **2004**, *4*, 878–888.
[63] Dahlberg, A., Delaney, C., Bernstein, I.D. *Blood* **2011**, *117*, 6083–6090.
[64] Purpura, K.A., Bratt-Leal, A.M., Hammersmith, K.A., McDevitt, T.C., Zandstra, P.W. *Biomaterials* **2012**, *33*, 1271–1280.
[65] Williams, S.F., Lee, W.J., Bender, J.G., Zimmerman, T., Swinney, P., Blake, M., Carreon, J., Schilling, M., Smith, S., Williams, D.E., Oldham, F., VanEpps, D. *Blood* **1996**, *87*, 1687–1691.
[66] Haylock, D., Simmons, P. in *Handbook on Adult Stem Cell Biology*; Elsevier Academic Press, San Diego, CA: 2004.
[67] Shiozawa, Y., Havens, A.M., Pienta, K.J., Taichman, R.S. *Leukemia* **2008**, *22*, 941–950.
[68] Zon, L.I. *Nature* **2008**, *453*, 306–313.
[69] Zhang, C.C., Kaba, M., Ge, G.T., Xie, K., Tong, W., Hug, C., Lodish, H.F. *Nat. Med.* **2006**, *12*, 240–245.
[70] Nielsen, L.K. *Annu. Rev. Biomed. Eng.* **1999**, *1*, 129–152.
[71] Lemischka, I.R. *Stem Cells* **1997**, *15*, 63–68.
[72] Kohler, T., Plettig, R., Wetzstein, W., Schaffer, B., Ordemann, R., Nagels, H.O., Ehninger, G., Bornhauser, M. *Stem Cells* **1999**, *17*, 19–24.
[73] Csaszar, E., Kirouac, D.C., Yu, M., Wang, W.J., Qiao, W.L., Cooke, M.P., Boitano, A.E., Ito, C., Zandstra, P.W. *Cell Stem Cell* **2012**, *10*, 218–229.
[74] Roberts, R., Gallagher, J., Spooncer, E., Allen, T.D., Bloomfield, F., Dexter, T.M. *Nature* **1988**, *332*, 376–378.
[75] Gupta, P., McCarthy, J.B., Verfaillie, C.M. *Blood* **1996**, *87*, 3229–3236.
[76] Lahmar, M., Catelain, C., Poirault, S., Dorsch, M., Villeval, J.L., Vainchenker, W., Albagli, O., Lauret, E. *Stem Cells* **2008**, *26*, 621–629.
[77] Delaney, C., Heimfeld, S., Brashem-Stein, C., Voorhies, H., Manger, R.L., Bernstein, I.D. *Nat. Med.* **2010**, *16*, 232–U143.
[78] Ohishi, K., Varnum-Finney, B., Bernstein, I.D. *J. Clin. Invest.* **2002**, *110*, 1165–1174.
[79] Varnum-Finney, B., Halasz, L.M., Sun, M.Y., Gridley, T., Radtke, F., Bernstein, I.D. *J. Clin. Invest.* **2011**, *121*, 1207–1216.
[80] Gupta, P., Oegema, T.R., Brazil, J.J., Dudek, A.Z., Slungaard, A., Verfaillie, C.M. *Blood* **2000**, *95*, 147–155.
[81] Oswald, J., Steudel, C., Salchert, K., Joergensen, B., Thiede, C., Ehninger, G., Werner, C., Bornhauser, M. *Stem Cells* **2006**, *24*, 494–500.
[82] Matrosova, V.Y., Orlovskaya, I.A., Serobyan, N., Khaldoyanidi, S.K. *Stem Cells* **2004**, *22*, 544–555.

[83] Altrock, E., Muth, C.A., Klein, G., Spatz, J.P., Lee-Thedieck, C. *Biomaterials* **2012**, *33*, 3107–3118.
[84] Laluppa, J.A., McAdams, T.A., Papoutsakis, E.T., Miller, W.M. *J. Biomed. Mater. Res.* **1997**, *36*, 347–359.
[85] Ehring, B., Biber, K., Upton, T.M., Plosky, D., Pykett, M., Rosenzweig, M. *Cytotherapy* **2003**, *5*, 490–499.
[86] Feng, Q., Chai, C., Jiang, X.S., Leong, K.W., Mao, H.Q. *J. Biomed. Mater. Res. A* **2006**, *78*, 781–791.
[87] Chua, K.N., Chai, C., Lee, P.C., Ramakrishna, S., Leong, K.W., Mao, H.Q. *Exp. Hematol.* **2007**, *35*, 771–781.
[88] da Silva, J., Lautenschlager, F., Sivaniah, E., Guck, J.R. *Biomaterials* **2010**, *31*, 2201–2208.
[89] Badylak, S.F., Freytes, D.O., Gilbert, T.W. *Acta Biomater.* **2009**, *5*, 1–13.
[90] Engler, A.J., Sen, S., Sweeney, H.L., Discher, D.E. *Cell* **2006**, *126*, 677–689.
[91] Choi, J.S., Harley, B.A.C. *Biomaterials* **2012**, *33*, 4460–4468.
[92] Kurth, I., Franke, K., Pompe, T., Bornhauser, M., Werner, C. *Macromol. Biosci.* **2011**, *11*, 739–747.
[93] Dexter, T.M., Allen, T.D., Lajtha, L.G. *J. Cell. Physiol.* **1977**, *91*, 335–344.
[94] Wagner, W.G., Roderburg, C., Wein, F., Diehlmann, A., Frankhauser, M., Schubert, R., Eckstein, V., Ho, A.D. *Stem Cells* **2007**, *25*, 2638–2647.
[95] Dexter, T.M., Spooncer, E., Simmons, P., Allen, T.D. *Kroc Found. Ser.* **1984**, *18*, 57–96.
[96] Seib, F.P., Muller, K., Franke, M., Grimmer, M., Bornhauser, M., Werner, C. *Tissue Eng. A* **2009**, *15*, 3161–3171.
[97] Schneider, R.K., Puellen, A., Kramann, R., Raupach, K., Bornemann, J., Knuechel, R., Perez-Bouza, A., Neuss, S. *Biomaterials* **2010**, *31*, 467–480.
[98] Di Maggio, N., Piccinini, E., Jaworski, M., Trumpp, A., Wendt, D.J., Martin, I. *Biomaterials* **2011**, *32*, 321–329.
[99] Leisten, I., Kramann, R., Ferreiraa, M.S.V., Bovi, M., Neuss, S., Ziegler, P., Wagner, W., Knuchel, R., Schneider, R.K. *Biomaterials* **2012**, *33*, 1736–1747.
[100] Cross, M., Alt, R., Niederwieser, D. *Cells Tissues Organs* **2008**, *188*, 150–159.
[101] Bonnet, D., Dick, J.E. *Nat. Med.* **1997**, *3*, 730–737.
[102] Rieger, M.A., Hoppe, P.S., Smejkal, B.M., Eitelhuber, A.C., Schroeder, T. *Science* **2009**, *325*, 217–218.
[103] Sommer, J.U., Dockhorn, R., Welzel, P.B., Freudenberg, U., Werner, C. *Macromolecules* **2011**, *44*, 981–986.
[104] Freudenberg, U., Sommer, J.U., Levental, K.R., Welzel, P.B., Zieris, A., Chwalek, K., Schneider, K., Prokoph, S., Prewitz, M., Dockhorn, R., Werner, C. *Adv. Funct. Mater.* **2012**, *22*, 1391–1398.
[105] Wilson, A., Laurenti, E., Oser, G., van der Wath, R.C., Blanco-Bose, W., Jaworski, M., Offner, S., Dunant, C.F., Eshkind, L., Bockamp, E., Lio, P., MacDonald, H.R., Trumpp, A. *Cell* **2008**, *135*, 1118–1129.
[106] Kirouac, D.C., Ito, C., Csaszar, E., Roch, A., Yu, M., Sykes, E.A., Bader, G.D., Zandstra, P.W. *Mol. Syst. Biol.* **2010**, *6*, 1–16.
[107] Roeder, I., Loeffler, M., Glauche, I., Participants, O. *Blood Cell Mol. Dis.* **2011**, *46*, 308–317.

CHAPTER 17

ENGINEERING IMMUNE RESPONSES TO ALLOGRAFTS

ANTHONY W. FREI

Diabetes Research Institute, Leonard M. Miller School of Medicine, University of Miami, Miami, FL, USA; Department of Biomedical Engineering, College of Engineering, University of Miami, Coral Gables, FL, USA

CHERIE L. STABLER

Diabetes Research Institute, Department of Surgery, Leonard M. Miller School of Medicine, University of Miami, Miami, FL, USA; Department of Biomedical Engineering, College of Engineering, University of Miami, Coral Gables, FL, USA

17.1 INTRODUCTION

Tissue engineering, the combination of cells, biomaterials, and bioactive factors to repair, replace, or maintain a dysfunctional tissue, has tremendous potential in the treatment of numerous pathological conditions. The source of the cellular components within the tissue-engineered implant is a critical issue in clinical translation, whereby disparate sources, that is, autologous, allogeneic, or xenogeneic, are met with distinct challenges. Allogeneic third-party cells or tissues have the highest marketing and translational potential in that they may be banked or precultured for alacrity of use. Furthermore, they have the advantage over autologous, self-sourced cells, in that they avoid secondary trauma due to isolation and are of more consistent quality, as often the desired cells of the patient are damaged, aged, or limited. Allograft transplantation, however, is restricted by the need for systemic immunosuppressive drugs to prevent rejection of the foreign cells. While xenogeneic sources have large appeal due to their availability, viral transmission risks and limits on phenotypic compatibility restrict their widespread use. As such, when the cell source is either limited or dysfunctional in the patient, for example, organ failure, allogeneic transplantation is currently the only clinically relevant option.

Bio-inspired Materials for Biomedical Engineering, First Edition. Edited by Anthony B. Brennan and Chelsea M. Kirschner.

Whole organ allogeneic transplantation is the most common type of allograft, beginning with the first clinical kidney transplant in 1956 [1]. Currently, over 25,000 whole organ transplants are performed each year in the United States, with 15,152 kidneys and 5527 livers transplanted in 2012 alone [2]. The field of allotransplantation has expanded to encompass transplanted tissues, such as skin grafts, cells, bone marrow, or islets of Langerhans. In the future, it is envisioned that engineered tissues comprised of allogeneic cells will make up a large proportion of future transplants customized to fit the needs of the patient.

One of the biggest challenges to the clinical success of allogeneic transplants is the interaction between the host immune system and the foreign transplanted cells. Before delving into the unique response of the immune system to allografts, one should first understand the basic components of the immune system and its function under homeostasis (for a more extensive review, see Reference 3). The human immune system consists of the innate and adaptive immune systems, in which each branch plays a particular role in combating disease and infection.

17.1.1 Key Components and Pathways of the Immune System

The innate immune system provides intrinsic protection against microbes and consists of epithelia, phagocytes, natural killer (NK) cells, cytokines, and the complement system. Epithelia provide the physical barrier to pathogens entering tissues. Phagocytes are cells that ingest and break down microbes and include neutrophils, monocytes, and macrophages (MΦ). The primary role of NK cells is to eradicate microbe-infected host cells and release interferon-gamma (IFN-γ), a cytokine that directs MΦ to lyse their phaged substances. Last, the complement system consists of a family of proteins whose roles include: coating microbes to tag them for phagocytosis, and forming pore-inducing membrane complexes to induce microbe lysis. The recognition of pathogens occurs via molecular pattern recognition, in which pathogen-recognition receptors (PRRs), encoded in the germline, recognize generalized patterns expressed by infected cells. These patterns include damage-associated molecular patterns (DAMPs) and common molecules conservatively expressed on microbes, such as pathogen-associated molecular patterns (PAMPs). Although it is based on receptors to sequences found ubiquitously on microbes or stressed cells and is limited in adaptation, the innate immune system is still highly targeted, potent, and evolving. Perhaps the most well-known function of the innate immune system is instigating inflammation.

Acute inflammation is typically the primary response to any foreign invasion or trauma. Neutrophils are the dominant first responders, which initiate phagocytosis of microbes and release cytotoxic signals to degrade damaged cell components. The neutrophil response is followed by monocyte migration to the site. Without prompt resolution, the acute phase transitions to chronic inflammation, hallmarked by the prolonged presence of monocyte-derived MΦ. Chronic inflammation inevitably leads to damage of the native tissue. If the phagocytes are unable to phage the foreign substance, MΦ will fuse into clusters known as foreign body giant cells, large multinuclear cells characteristic of chronic inflammation. Ideally, chronic inflammation will be avoided in an allograft and the immune response can resolve as granulation tissue.

The adaptive immune system consists of cell-mediated immunity, primarily carried out by T lymphocytes, and humoral immunity, which involves antibodies generated by B lymphocytes. Contrary to the innate immune system, adaptive immunity is highly specific, as receptors expressed by a particular T or B cell clone only recognize a specific antigen. When the receptors of a particular clone bind antigen, specific clonal expansion occurs, permitting an aggressive and targeted attack of specific pathogens. Additionally, immunological memory facilitates enhanced protection following initial exposure to a particular antigen. Prior to antigen activation, T and B cells are considered naïve. Following activation, these lymphocytes are termed effector cells.

Cell-mediated immunity is conducted by T cells, which contain the cluster of differentiation (CD) marker CD3. These cells mature in the thymus to their distinct phenotype, the primary types being $CD8^+$ cytotoxic T (T_c) cells and $CD4^+$ helper T (T_h) cells. T_c cells bind to antigen presented by major histocompatibility complex (MHC) I receptors. T_c cells are subsequently activated when costimulated with cell-surface proteins CD80 or CD86, which complex with the CD28 receptor on the T cell. On activation, T_c cells are capable of directly killing the infected cell by releasing perforins, which create pores in the membrane of invading cells and by releasing apoptotic factors inducing granzymes. T_h cells are activated on encountering antigens presented by MHC II receptors on professional antigen-presenting cells (APCs). APCs include dendritic cells (DCs) and MΦ, which either reside in the peripheral tissue or are recruited to the site via chemotactic gradients. APCs present antigen to T_h cells, typically in the lymph node, whereby activated T_h cells secrete cytokines and assist other cells involved in the adaptive immune response, such as T_c, MΦ, and B cells. Each subtype of effector T_h cells, for example T_h1 or T_h2, releases a signature cocktail of cytokines that leads to a characteristic immune response. Another subtype of $CD4^+$ T_h cell is the $CD25^+FoxP3^+$ regulatory T cell (T_{reg}), which facilitates downregulation of the immune response following elimination of the antigen. T_{regs} instruct effector T cells via immunosuppressive cytokines, such as transforming growth factor beta 1 (TGF-β1), interleukin-10 (IL-10), and IL-35, and/or cell surface receptors such as galectin-1 [4]. Furthermore, T_{regs} interact with and suppress the function of APCs through a number of mechanisms, including cytotoxic T lymphocyte antigen 4 (CTLA-4), a T_{reg} cell surface receptor which results in decreased costimulation by APCs [5].

In humoral immunity, B cells and antibodies, that is, immunoglobulins (Igs), are the primary players. B cells are first generated in the bone marrow and then they undergo migration to the secondary lymphoid tissues. B cells are able to recognize antigens directly without MHC-restricted presentation by infected cells or APCs via their B cell receptor (BCR), which is an immobilized antibody molecule (IgM) specific to a particular B cell clone. Following the binding of the BCR by complementary antigen and with support from T_h cells, B cells differentiate into either effector plasma B cells or memory B cells. Effector plasma B cells secrete Ig, primarily IgG, for humoral targeting of microbes, tagging them for phagocytic destruction. Meanwhile, memory B cells remain in the body following antigen exposure to facilitate a more robust response if reinfection occurs.

Components of the immune system must undergo self-selection to learn which is a friend and which is a foe. Central tolerance for self-antigens is achieved via clonal selection, which occurs in the bone marrow for B cells and the thymus for T cells. Autoimmune diseases, such as rheumatoid arthritis, multiple sclerosis, and diabetes, occur when the immune system fails to tolerate itself and mounts an antigen-specific attack of host cells and/or tissues. Under normal conditions, lymphocyte clones with B and T cell receptors expressing strong affinity for self-antigens undergo negative selection and are targeted for apoptosis. In autoimmunity, this process of negative selection goes awry, and self-reactive lymphocytes survive and attack native tissue. Often, infections can play a role in instigating autoimmune disorders, as they can cause the presentation of self-antigen to self-reactive immune cells to occur with a costimulatory signal [6].

17.1.2 Immune Response to an Allograft

Once the components of the immune system and their standard response to antigen are understood, one can examine the immune response in the context of an allograft transplant. On transplantation of cells with or without a material, the initial response will be acute inflammation, whereby resident APCs respond to the transplant and monocytes are directed to the site. For allogeneic cells, the host recognizes some of the surface proteins on transplanted cells or released cellular components as nonnative, thus treating them as antigens. The principal antigens recognized are MHC I and II. Every human expresses six MHC I and six MHC II alleles, each of which contains hundreds of subtypes, resulting in transplanted cells expressing proteins recognized as foreign to the host. The host may recognize alloantigens via two pathways, as shown in Figure 17.1: direct antigen presentation or indirect antigen presentation. Direct antigen presentation is where a T cell, specific for the alloantigen, contacts the specific surface proteins of the transplanted cell. If the foreign cell is an APC, then $CD4^+$ T helper cells can be activated. If the foreign cell is not a professional APC, then $CD8^+$ T cytolytic cells are activated. While it was previously thought that $CD4^+$ T cells were the primary players in alloantigen recognition, it is now known that both $CD4^+$ and $CD8^+$ T cells play critical roles [7]. For indirect activation, graft rejection is initiated by alloantigens released by the allograft, the most potent being alloantigens released during cellular necrosis, which are coreleased with DAMPs [8]. DCs and other APCs in the periphery can phage these alloantigens, migrate to the lymphatics, and subsequently present antigen to generate clonally specific effector T cells. Alloantigens present in the interstitium can also undergo transport through the lymph, where they can bind to complementary B cells or be phaged by residing APCs to be presented to complementary T cells. Depending on the local cytokine environment which alters APC costimulatory receptors and signaling pathways, the resulting T cell response can vary from a characteristic T_h1, T_h2, T_h17, or regulatory response, as defined by the phenotype of the prevalent $CD4^+$ T cells [9]. T_h1 $CD4^+$ cells are the primary players in causing allograft rejection, through the production of the cytokine IL-2, which leads to the proliferation of

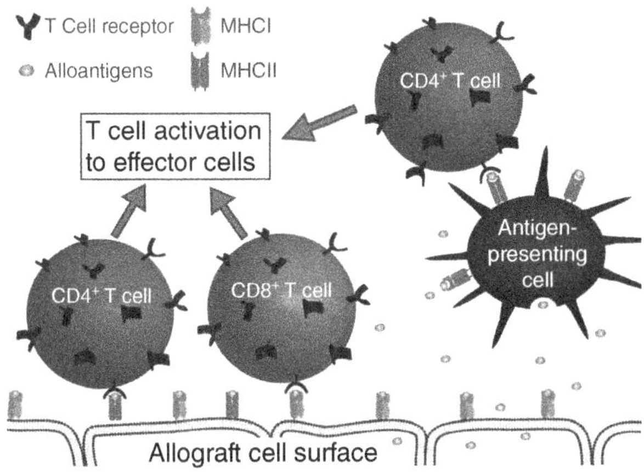

FIGURE 17.1 Summary of pathways for recognition of allograft by the adaptive immune system. In direct antigen presentation, $CD4^+$ T_h cells or $CD8^+$ T_c cells recognize foreign antigens directly on the allograft cell surface of the transplanted tissue. T_h cells recognize antigen presented by antigen-presenting cells via MHC II (typically presented by professional antigen-presenting cells residing in the donor tissue). T_c cells recognize antigen presented by nucleated cells via MHC I. Once the specific TCR/MHC binding occurs, pathways are initiated to result in generation of alloantigen-specific effector T cells. T_h assist other adaptive immune cells to enhance alloantigen clearance, while T_c can directly kill the foreign allogeneic cells. The indirect antigen-presenting pathway results from processing of foreign alloantigens by host antigen-presenting cells (APC), which present, via MHC II, to antigen-specific $CD4^+$ T_h to result in effector T cells. Alloantigens are shed by the transplanted cells through normal processes, but are elevated and more reactive during cellular stress or necrosis. TCR, T cell receptor. *(See insert for color representation of the figure.)*

effector $CD8^+$ cytolytic T cells, although T_h2 and T_h17 responses also play a role in mediating transplant rejection [10]. Occasionally, immune senescence occurs, where the host immune system is still primed to react against the allograft antigens, but constant exposure and immune challenges lead to the immune system reducing its attack, creating a symbiosis between the transplant and the host tissue [11]. Overall, however, the standard immune response results in allograft rejection, involving alloantigen uptake by APCs, presentation of antigen to generate effector T_h cells and alloantibody producing B cells, and culminating in cytolytic T cells inducing cell death of the transplanted cells via direct cell–cell contact.

Occasionally, specialized circumstances cause additional immune problems with allografts, such as cases of graft-versus-host disease (GVHD). In these transplants, the host rejects not only the graft, but donor immune cells present in the graft recognize the host as foreign and begin to attack the host, generating a dangerous and potentially life-threatening situation [12]. The transplantation of a cell type in which the original native cells were destroyed via autoimmune responses represents another unique immunological condition. For instance, allogeneic islets are transplanted in the treatment of type 1 diabetes [13], an autoimmune disorder that results in select immunological destruction of beta cells within the pancreatic islets of Langerhans. The resulting immunological response is particularly aggravated, as the recipient is primed with autoreactive T and B cells specifically targeting the type of cells being transplanted, in addition to the standard response targeting alloantigens.

17.1.3 Current Approaches to Mitigating Allograft Rejection

Once an allograft has been implanted, there are two main goals. The first is to achieve adequate function, and the second is to maintain function, which requires immune acceptance of the graft. Currently, allograft recipients rely on the administration of systemic immunosuppressant or antirejection drugs. The use of these agents is required to prevent, or at least dampen, T and B cell activation to the foreign antigens. Systemic immunosuppression typically involves two stages, an induction stage, followed by a maintenance regimen. For kidney transplants, the most common induction routine is a combination of a $CD3^+$ T cell depleting regimen (e.g., Thymoglobulin, Genzyme Corp., Cambridge, MA), followed by administration of anti-IL-2 receptor (e.g., basiliximab), which prevents IL-2 signaling and subsequent lymphocyte proliferation. For maintenance, a calcineurin inhibitor (e.g., tacrolimus) and a mechanistic target of rapamycin (mTOR) inhibitor (e.g., sirolimus), which block IL-2 secretion and responses, respectively, are used in combination with an anti-inflammatory glucocorticoid (e.g., prednisone) [14]. While protocols vary based on individual patient circumstances and the type of allograft performed, the goal is to dampen overall immune activation, resulting in systemic suppression of the immune system. As such, the general use of systemic immunosuppressants has a number of crucial disadvantages, such as increased susceptibility to disease, infection, and cancer [15]. Other side effects include increased risk of cardiovascular disease, hypertension, hyperlipidemia, and new-onset diabetes [16]. Transplant patients are resigned to a heavy drug regimen for the duration of their life, a difficult challenge requiring continual patient diligence. Furthermore, the financial burden of lifelong antirejection therapy is substantial, costing several thousands of dollars annually. For example, the antiproliferative agent mycophenolatemofetil, commonly prescribed for organ transplants, has an average wholesale price of $14,921 for an annual supply of the branded product and $11,578 for the generic [17]. Finally, even with long-term use of immunosuppressive therapy, a smoldering, chronic graft rejection typically occurs, leading to progressive decline of graft function [18].

Due to the weighty disadvantages associated with systemic immunosuppression, researchers are developing alternative treatment options, with a focus on generating a graft microenvironment that favors immunological tolerance. Immunological or operational tolerance is defined as a lack of reactivity to donor alloantigens in the absence of long-term immunosuppression, while retaining the capacity of the immune system to react to other foreign antigens or insults [19]. Immunological tolerance is a dynamic state, in which the plasticity of immune cells results in the ability to induce tolerance or activation. Thus, the success of protocols designed to induce tolerance has been unpredictable. However, a recent focus on engineering a tolerogenic microenvironment at the allograft transplant site may prove to be more successful, as peripheral tolerance is induced locally.

17.2 ENGINEERING STRATEGIES FOR IMMUNE ACCEPTANCE

Immune acceptance of an allograft can be engineered using a number of complementary approaches, with each approach seeking to generate a graft site conducive to long-term acceptance. One approach is changing the anatomical location of the graft, an adjustable parameter when transplanting cells and smaller tissues. Variations in resident inflammatory or tolerogenic cells, as well as the degree of immunological access, can greatly affect the severity of the initial immunological response to the transplant, with some areas deemed *immune privileged*. Alternatively, the use of methods to *camouflage* transplanted cells from the native immune system and prevent antigen recognition is common. These cloaking strategies include encapsulation and the grafting of cell surfaces with benign polymers. Another approach is the creation of a protective and immune-tolerant zone surrounding the allograft to provide instruction to nearby immune cells. These tolerogenic microenvironments can be achieved through the cotransplantation of cells with immunosuppressive properties, the delivery of immunomodulatory drugs or cytokines, and/or the generation of biomimetic surfaces capable of inducing immunoregulatory pathways, as summarized in Figure 17.2.

17.2.1 Selection of Allograft Site

Select areas of the body are deemed *immune privileged*, making them attractive transplant sites for allografts. These sites are identified by a decreased immunological response to antigen exposure. As such, these sites exhibit enhanced allograft survival when compared with the typical rejection profile observed elsewhere in the body, although the use of these sites does not prevent of graft rejection. In general, immune-privileged sites are natively endowed with unique characteristics, such as a blood barrier, the expression of molecules that actively inhibit immune cells, the increased presence of regulatory immune cells, and a generalized decrease in MHC I expression (for a cohesive review, see Reference 20). Tissues that have expressed this privilege include the eye, testis, brain, and pregnant uterus [21].

Immune privilege in the eye was first noted by the observation that corneal allografts fared better than skin allografts in similarly mismatched patients. The eye contains a blood–aqueous barrier and a blood–retinal barrier consisting of tight junctions between epithelial cells, which not only functions to help maintain intraocular pressure, but also separates the blood and lymphatic systems from the eye. Selected tissues within the eye contain cells that release mediating cytokines, such as the myeloid cells in the retina [22]. Retinal pigment epithelial (RPE) cells are known to express coinhibitory molecules, such as Fas ligand (FasL), programmed death 1 ligand 1(PD-L1), and CTLA-4, which suppress effector T cells [23]. Furthermore, parenchymal cells of the eye have been shown to be capable of inducing T_{regs} that suppress $CD4^+$ T_h1 effector cells [24]. With this dampened immunological activity, the eye has been explored as an allograft implantation site. In a study by Hatchell *et al.*, pancreatic islet allografts were delivered to the subretinal space of cat eyes, where the islets survived without rejection for up to 3 weeks [25]. The risk of

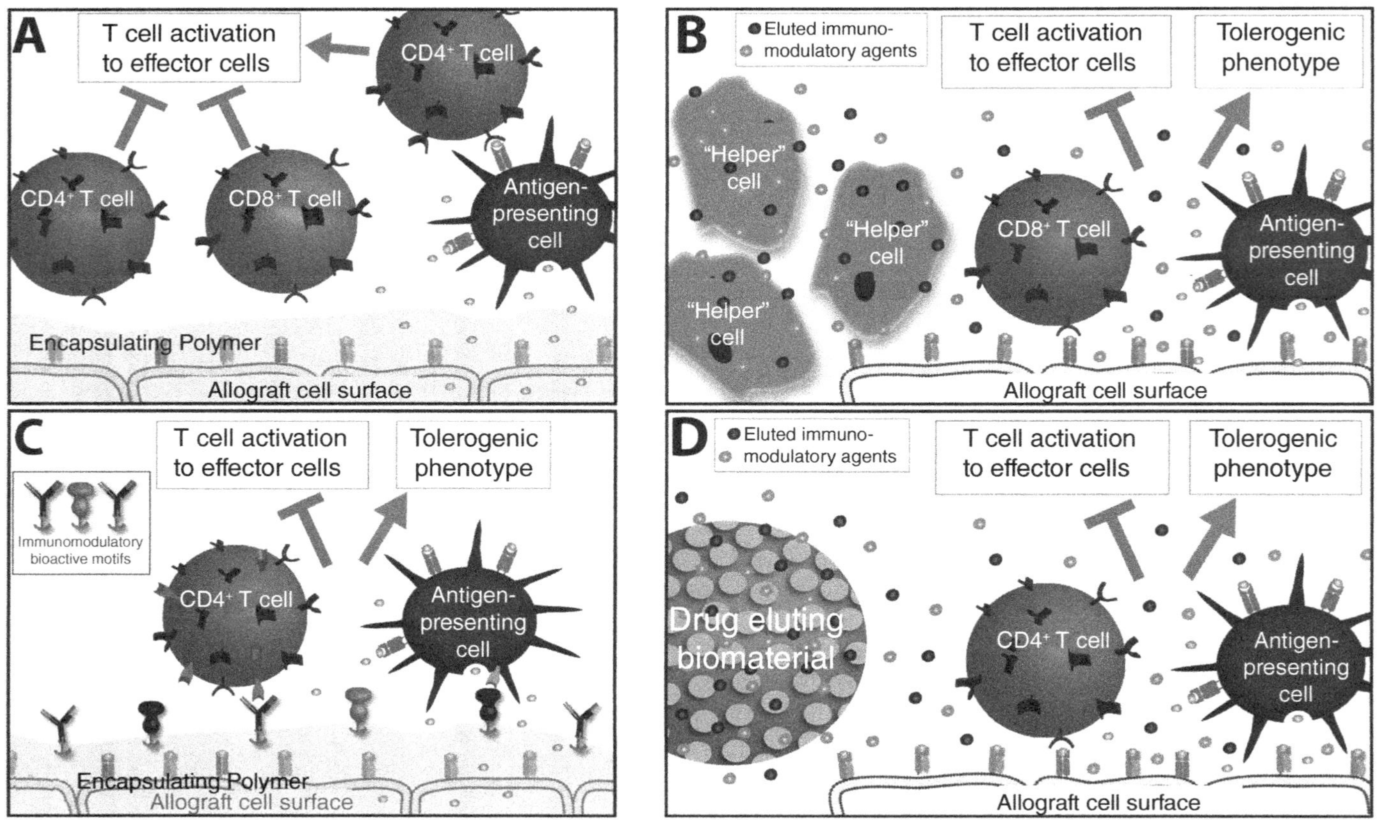
A
T cell activation to effector cells
CD4+ T cell
CD4+ T cell
CD8+ T cell
Antigen-presenting cell
Encapsulating Polymer
Allograft cell surface
B
Eluted immuno-modulatory agents
T cell activation to effector cells
Tolerogenic phenotype
"Helper" cell
"Helper" cell
"Helper" cell
CD8+ T cell
Antigen-presenting cell
Allograft cell surface
C
Immunomodulatory bioactive motifs
T cell activation to effector cells
Tolerogenic phenotype
CD4+ T cell
Antigen-presenting cell
Encapsulating Polymer
Allograft cell surface
D
Eluted immuno-modulatory agents
T cell activation to effector cells
Tolerogenic phenotype
Drug eluting biomaterial
CD4+ T cell
Antigen-presenting cell
Allograft cell surface

FIGURE 17.2 Engineering of immune response to allogeneic transplants can be conducted via (A) polymeric encapsulation, (B) cotransplantation with protective cells, (C) surface functionalization of transplanted biomaterials, and (D) codelivery of soluble factors. (A) Masking of surface antigens via polymeric encapsulation blocks direct antigen pathway recognition; however, shed alloantigens still permeate the capsule and can be phaged by host APCs to activate indirect antigen presentation. (B) Codelivery of immunomodulatory cells with allogeneic cells can generate a localized tolerogenic microenvironment through the delivery of multiple factors and/or the expression of surface motifs that impart immune cell deactivation and/or induction of tolerogenic phenotypes. (C) Tethering of immunomodulatory motifs to the polymeric capsule can locally direct immune cell deactivation or induction of tolerogenic phenotypes through surface mediated responses (e.g., blockage of costimulation receptor and/or activation of coinhibitory receptors). (D) Codelivery of materials capable of eluting soluble immunomodulatory agents can generate a tolerogenic microenvironment, resulting in decreased immune cell activation and/or induction of tolerogenic phenotypes. Individual strategies may, in turn, be combined for further immune modulation. *(See insert for color representation of the figure.)*

damaging the eyesight, however, restricts the use of this site for cell transplantation. To avoid this concern, selected immunoregulatory cells found within the eye have been exploited to induce tolerance, such as the combination of retinal progenitor cells and biodegradable polymers to engineer an immunoprivileged site [26]. Resulting grafts, transplanted in subrenal capsules of allograft recipients, demonstrated enhanced tolerance.

The testis, which is exposed to a number of unique proteins during sperm production, maintains an immune-privileged environment in order to protect the process of spermatogenesis from immune attack. This privilege has been observed through the resistance to rejection of allograft testes, ectopically transplanted under the skin of Balb/c mice without systemic immunosuppression [27]. Similar to the eye, the testis contain a number of compounding factors that contribute to immune privilege, such as a blood–testis barrier, a decrease in inflammatory MΦ, increased numbers of T_{regs}, and a lack of resident B cells [21]. The transplantation of cells and tissues to this site has shown increased survival in the absence of immunosuppression, such as the transplantation of islets [28], parathyroid, and skin [29]. The delay in allograft rejection at this site, however, has been modest and limited to small animal models [30]. Alternatively, the isolation and cotransplantation of selected cells found in the testis, notably Sertoli cells, have fared better. The cotransplantation of Sertoli cells and various other allogeneic cells with allografts have demonstrated beneficial effects, as highlighted in later sections.

The benefits of transplanting into an immune-privileged site must be weighed against the optimal site for graft efficacy and function. For example, pancreatic islets, used for the treatment of type 1 diabetes, require adequate vascularization to achieve optimal oxygenation and glucose responsiveness [31]. At these immune-privileged sites, the blood barrier that imparts immune protection may impede adequate cellular engraftment. As a result, the use of immunoprivileged sites for allograft transplants remains a promising avenue worthy of exploration, but this approach is not the silver bullet of engineering graft immune acceptance. Furthermore, the designation of *immune privilege* is a misnomer as an immune response is still typically observed, particularly when the blood barrier is breeched during transplantation. For example, the transplantation of fully mismatched allograft islets into the eye resulted in a significant T cell infiltrate surrounding the graft site at 14-days posttransplantation. While full graft rejection was slightly delayed compared with transplantation into the kidney capsule (average rejection time of 30 days vs. 15 days, respectively), full allograft rejection invariably occurred [32].

17.2.2 Cellular Encapsulation

The use of biomaterials to encapsulate cells accomplishes a key set of desired outcomes. In tissue engineering, the presence of a biomaterial gives cells a substrate to attach to and grow on, acting as a scaffold. Scaffolds provide mechanical support for implanted cells to proliferate, deposit extracellular matrix, and integrate with the native tissue. In the context of allografts, cells can be seeded within materials, not only for support, but also for masking immune recognition. Although the complexity and robustness of the immune system makes camouflaging foreign antigens through material encapsulation challenging, strategies are being developed to enhance the effectiveness of these engineered immunological barriers while still allowing for healthy allograft function.

While allogeneic cells can be encapsulated into various geometries, from sheets to rods to discs, the most common geometry employed are spheres (typically 600–1000 μm in diameter), given their high surface to volume ratio and ease of transplantation [33]. When selecting biomaterials for encapsulation, desired properties of materials include biocompatibility, cytocompatibility, mild scaffold production conditions, and biostability. Biocompatibility is essential for encapsulation, as even a mild foreign body response (FBR) to the coating can lead to significant decreases in nutrient delivery to the embedded cells. Cytocompatibility and mild encapsulation conditions are critical to ensure that the process of material encapsulation does not impart damage to the allogeneic cells; cell damage and necrosis leads to the release of danger signals and cellular lysis products into the local graft microenvironment, instigating inflammatory processes and increasing alloantigen dumping. The importance of biostability of the resulting coating is clear, as the duration of immunoprotection must be long-term. Commonly used materials for encapsulation include alginate, agarose, and polyethylene glycol (PEG) [34]. These materials are fabricated into hydrogels, which are water-containing hydrophilic polymers supported by physical or chemical cross-links. Biomaterials fashioned into hydrogels are highly functional in allowing encapsulated cells to perform their roles, for example, permitting appropriate levels of nutrient and the secretion of agents to and from the cells while simultaneously acting as a barrier to host cell to donor cell interactions [35].

Alginates are polysaccharide biopolymers of D-mannuronic acid and L-guluronic acid residues and are often used for the encapsulation of cells. Isolated from seaweed, their attractive properties include their biocompatibility, their nonimmunogenicity, and their ability to form hydrogels with instant gelation using multivalent cations. As cells do not adhere to alginate naturally, this material can be modified with cellular adhesion proteins to facilitate cell attachment [36]. While not inherently degradable, the ionic gelation process does result in gels that weaken over time due to ion exchange [37]. As such, efforts to stabilize the resulting capsule have been explored, such as the incorporation of covalent cross-links within the hydrogel [38]. For example, the functionalization of alginate with alkyl azide, followed by the addition of PEG cross-linkers functionalized with 1-methyl-2-diphenylphosphino-terephthalate (MDT), results in the formation of amide bonds via Staudinger ligation, a spontaneous, bioorthogonal reaction between an azide and phosphine known for its benignity. This approach demonstrated increased mechanical stabilization of the alginate, with stable hydrogels even following the removal of ionic cross-links [39]. Cytocompatibility of the cross-linked gels was high, with no adverse effects on viability or function observed for both beta cells and islets [40].

Evaluation of alginate-based capsules for immunoprotection of allografts in rodent models has been extensive, thus, complete review of all promising approaches would be lengthy [41]. In all, multiple groups have demonstrated long-term protection of alginate-based capsules within rodent allogeneic models, both for chemically induced and spontaneous diabetic models. Notably, barium cross-linked, islet loaded, alginate capsules, transplanted into syngeneic nonobese diabetic (NOD) mice, have shown extended protection, with graft function demonstrated over 350 days [42]. In larger animal models, published reports are limited, with most evaluating porcine islets and function observed when combined with immunosuppressive agents [43]. One study of particular interest, however, transplanted alginate encapsulated porcine islets into nonimmunosuppressed primates, with islet survival and function up to 6 months posttransplantation [44]. Of note, only chemical-induced diabetic models are available for nonhuman primate, thus autoimmune effects cannot be evaluated at this time in these animal models.

Agarose, a neutral polysaccharide, is biologically inert and does not adsorb proteins or cells, making it an attractive candidate for use in cellular encapsulation. The mechanical properties of agarose can be tailored by varying its concentration, and the material undergoes gelation on exposure to reduced temperatures [45]. Agarose has been used for immunoisolation in geometries ranging from slabs to microbeads, although controlling the size of the resulting microbeads is challenging. Agarose gels have demonstrated durability *in vivo*. For example, islet loaded 5% agarose microbeads remained intact 400 days posttransplantation, with functional islets and no immune cell infiltration in a mouse model of autoimmune diabetes [46].

PEG is a biocompatible molecule often used for encapsulation. PEG is highly inert, with flexibility in molecular weights and geometries from unbranched to multiarm configurations. PEG is easily tunable via functionalization of end groups, resulting in a wide variation of mechanical, chemical, and bioactive properties [47]. Commonly, photopolymerization has been utilized to achieve gelation; however, care must be taken in the duration and concentration of agents to minimize formation of free radicals that can damage encapsulated cells [48]. In an alternative approach, Teramura *et al.* utilized PEG hydrogels functionalized with maleimide or thiol to microencapsulate mouse islets. Ligation between the maleimide and thiol groups occurs efficiently, resulting in covalently cross-linked microsized coatings [49]. This microcapsule formulation was shown to retain islet function and reduce immune complement activation compared with controls; however, the coating uniformity and thickness was unpredictable. An additional facet of PEG is the use of the polymeric backbone to incorporate bioactive molecules and motifs. Lin *et al.* utilized this tunability to incorporate beta cell adhesion receptor EphA5 and ligand ephrinA5 within hydrogels, resulting in decreased anoikis of the cells following encapsulation [50].

The clinical implementation of encapsulation has the potential to treat numerous conditions, from diabetes to Parkinson's to myocardial infarction, whereby the allogenic cells, embedded within the polymer, can respond to host cues and secrete desired factors. Progress in the clinical area has primarily been in the field of islet transplantation to treat type 1 diabetes, although success has been limited [51]. In a clinical trial taking place in Italy, four patients implanted with islet intraperitoneal allografts microencapsulated in sodium alginate were examined for their immune response to the allograft up to 5 years posttransplantation [52]. These patients did not receive systemic immunosuppression and no directed immune response was found to the encapsulated islets, measured by the lack of antibodies to MHC and islets. Living Cell Technologies (LCT) has also performed clinical islet transplantation of encapsulated porcine islets using porcine islets sourced from specialized herds. In these limited trials, LCT has claimed a marked decrease in hypoglycemic episodes in these patients [53]. Clinical outcomes of encapsulated islets, from either allo- or xeno- sources, have yet to result in long-term independence from exogenous insulin.

One of the challenges in standard cellular microencapsulation is the substantial barrier generated between the cells and the surrounding transplant environment. Given that the average distance between a cell and its nearest capillary is on the order of 200 μm, encapsulation of cells within barriers that range from 600 to over 1000 μm in diameter impose too large of a barrier for optimal nutrient delivery. This presents a greater challenge when encapsulating highly metabolically active cells, such as islets, where slow nutrient delivery leads to starvation-induced apoptosis and significantly, impairs function of the encapsulated islets [54]. Any variability of tissue cluster size also leads to further challenges in applying standard encapsulation technologies, in which droplets of fixed diameter are produced. Attempts to reduce droplet size to minimize barriers typically lead to inadequate encapsulation. Furthermore, the transplantation of microbeads in larger animal models and humans is challenged by a significant increase in the transplant volume required to implant a therapeutic number of islets due to the added volume of encapsulation materials. This restricts transplant sites to those able to accommodate such large volumes, such as the peritoneal cavity, where nutrient levels are significantly lower than vascularized sites.

In an effort to retain the immunological masking advantages of barrier fabrication but avoid the transport limitations standard microcapsules impose, researchers are exploring methods for reducing the capsule size from hundreds of microns to tens of microns, or even to the nanoscale. These methods create coatings that are independent of tissue size and whose shape conforms to the irregular shape of the cell or cell cluster. Earlier approaches have shown promise [55]; however, most of these methods rely on emulsions or free radical-initiated gelation, which expose cells to detrimental agents or result in incomplete coatings. More recent strategies include: PEGylation, layer-by-layer electrostatic, or self-assembly via polyionic or lipid layers. PEGylation of the cell surface is commonly achieved through the use of a PEG (MW > 2000 g/mol) terminated with an activated ester, N-hydroxyl-succinimidyl ester (NHS), which reacts with free amines found on the extracellular matrix and cell membranes of the cell or cell cluster. Groups have found benefits of red blood cell or islet PEGylation on dampening inflammation [56] and enhancing allograft survival [57]. Additionally, PEG grafting has been used to present additional agents to modulate immune recognition and inflammation, such as glucagon-like peptide-1 (GLP-1), albumin, and thrombomodulin [58,59]. Alternatively, layer-by-layer assembly of polymers on the allograft surface may generate more robust capsules. While electrostatic assembly is the most common method used for coating surfaces, the use of polycations on cells is complicated by their well-documented cytotoxicity and lack of biocompatibility [60]. This toxicity has been mitigated through approaches such as PEG grafting [61]; however, balancing layer uniformity and cytotoxicity is challenging. Alternative approaches for layer-by-layer assembly include hydrogen bonding [62], molecular recognition [63,64], or combinations thereof [65]. While numerous coating approaches have been published [58,62,63,65,66], translation of coatings to animal models has been limited, with success of tested coatings negligible [63] or moderate (~30-day efficacy) [67]. Of great concern for layers formed using these approaches is the limited stability of the resultant layers and their ability to undergo dynamic structural changes. Due to the fact that they are held together by noncovalent interactions, disintegration of these layers is common in unfavorable conditions, such as mechanical stress or swift changes in the ionic environment [68]. Given that these coatings cover a dynamic viable cell surface and reside in the harsh *in vivo* environment, the long-term stability of such layers is limited [69]. Therefore, while ultrathin polymeric coatings have tremendous potential, more robust coatings are required to enhance longevity *in vivo*.

Despite the concentrated efforts of researchers, the implantation of encapsulated cells still results in the activation of immunological responses. On implementation of any biomaterial, an inflammatory response is inevitably initiated, simply due to trauma induced via implantation. The degree of the FBR depends greatly on the biocompatibility of the material. Biocompatibility is typically assessed via innate pathway activation, for example, monocyte and macrophage activation; however, DC responses must also be taken into consideration to fully evaluate the potential of a material for encapsulation [70]. If the material instigates a strong FBR, frustrated MΦ will lyse or fuse into foreign body giant cells, further generating an inflammatory milieu and releasing cytokines that promote collagen and fibroblast deposition. This chronic FBR commonly results in the development of a fibrous capsule, choking out the encapsulated cells [71]. Furthermore, if the foreign cells are stressed prior to or during transplant, they may release cytokines or cellular products that will further elevate immune responses. If great care is taken in the selection of the biomaterial, the implant site, and the health of the transplanted cells, the innate immune response can be greatly minimized.

The adaptive immune response to the foreign cells is greatly damped via material camouflage of the cells' surface antigens, which prevents activation via the direct antigen-presenting pathway. This strategy, however, does not lead to complete avoidance of adaptive immune activation (see Figure 17.2A). As the allogeneic cells proliferate, function, and die, alloantigens are released from the capsule into the surrounding microenvironment. As outlined earlier, these shed alloantigens are picked up by resident APCs, likely DCs, and presented to $CD4^+$ T helper cells to initiate a host response directed to the alloantigens. While the exact mechanisms for allograft destruction when they are embedded within these hydrogels have not been completely elucidated, it is known that activation of adaptive immune responses leads to a graft microenvironment that can be toxic to the underlying cells. While a robust response is not typically observed in rodent models, the lack of success of encapsulated cells in more advanced immune systems, such as nonhuman primates and humans, leads researchers to believe that indirect activation of adaptive responses will eventually lead to graft failure. This is particularly true for xenografts, in which the degree of reactive xenoantigens is substantially higher when compared with an allograft. In order to combat this response, the exploration of methods to dampen indirect activation, as well as to generate a tolerogenic microenvironment of the graft site, is critical. This could be achieved through numerous methods, such as codelivery of immunomodulatory cells, the controlled release of bioactive signals, such as immunomodulatory cytokines or immunosuppressive agents, and/or the incorporation of immunomodulatory agents onto the biomaterial surface (Figure 17.2). These approaches could be highly synergistic with cellular encapsulation, providing full immune camouflage in advanced immune systems, or possibly inducing alloantigen tolerance.

17.2.3 Codelivery of Immunomodulatory Cells

In engineering an immunosuppressive response to an allogeneic transplant, an alternative option is the cotransplantation of a disparate cell population capable of delivering an immunoprotective response. These suppressive cells direct the host's immunological response via direct cell-to-cell contact or through the release of a cocktail of cytokines and factors. Cells are injected intravenously or directly into the local transplant site. The intravenous method relies on *cell homing* to reach the target site and interact [72], while local injections would ease cellular delivery. Alternatively, for engineered tissues, cells could be cocultured within the allograft prior to transplant, which would provide more control over presentation of the cells within the graft microenviroment. The use of cells has advantages over the engineered controlled release of factors, particularly in the ability of cells to deliver multiple factors, many of which are yet to be fully characterized, as well as their ability to adapt and respond dynamically to the host immune response. With the implantation of an additional cell source, however, the results become less controlled and a range of other variables, risks, and uncertainties are introduced.

A common cell used for inducing localized immune suppression is the mesenchymal stem cell (MSC). MSCs are multipotent, adherent adult stem cells, popular due to their regenerative potential [73]. Additionally, MSCs express

TABLE 17.1 Immunomodulatory Paracrine Factors Released by MSCs

Soluble factor	Mechanism of action	Immune effect	Reference
IDO	Tryptophan depletion and kynurenine synthesis	Inhibition of T cell function and proliferation	[120]
TGF-β	Multiple mechanisms, regulates differentiation and cell growth	Induction of $CD4^{+}CD25^{hi}FoxP3^{+}T_{regs}$	[76,121]
IL-10	Multifaceted suppressor, inhibits IL-2, IFN-γ, IL-12, MHC, and costimulatory proteins	Reduction of inflammation and adaptive immune activation	[76,122]
PGE_2	Elevation of cAMP inhibiting IL-2 production	Suppression of T cell proliferation	[123]
NO	STAT5 inhibition, cGMP-mediated TNF-α inhibition	Reduced T cell proliferation, inhibition of DC maturation	[74,124]

cAPM, cyclic adenosine monophosphate; cGMP, cyclic guanosine monophosphate; IDO, indoleamine-pyrrole 2,3-dioxygenase; STAT5, signal transducer and activator of transcription 5.

immunosuppressive properties. MSCs have been shown to inhibit T cell proliferation through the soluble release of cytokines, including: indoleamine 2,3,-dioxygenase (IDO), TGF-β, IL-10, prostaglandin E_2 (PGE_2), and nitric oxide (NO) [74]. MSCs have also been shown to suppress NK cell proliferation and inhibit the differentiation of monocytes or other immune cells into antigen-presenting DCs [75]. Of note, MSCs dampen not only activation, but can also induce tolerance via promotion of T_{reg} differentiation [76]. Table 17.1 summarizes theorized roles of MSCs in directing immune responses through the release of paracrine factors. In addition to releasing soluble factors, MSCs can contribute to immune suppression through cell–cell contact interactions. For instance, MSCs lack expression of costimulatory markers, such as CD80, CD86 and CD40, thereby inducing an anergic response on interaction with T cells. Additionally, when exposed to inflammation, MSC elevate expression of program death ligand 1 (PD-L1) (alternatively known as B7-H1), which serve to induce apoptotic pathways in T cells following MHC binding [77]. The use of MSCs as a cotransplanted therapy for allografts is of growing interest, with a number of phase I, II, and III clinical trials exploring their immunosuppressive effects. Most current trials are evaluating their potential in combating GVHD, although clinical trials are also under way regarding autoimmune disorders and whole-organ kidney transplants [78]. In type 1 diabetes, the transplantation of MSCs along with islets has shown to improve graft outcomes in nonhuman primate models and early-phase clinical trials [79].

In the testes, sertoli cells release factors assisting in spermatogenesis and confer immune privilege, shielding testes antigens from immune attack, making them an attractive option for cotransplantation with allografts. Sertoli cells have shown to be effective in cotransplantation with allogeneic islets to confer protection of the graft from immune rejection, both in rodents and in nonhuman primates [80].

They appear to be able to change the composition of T cells and modulate the cytokine profile of the allograft recipient. The exact mechanisms by which sertoli cells exert their immunomodulatory functionality remain to be elucidated, but may involve FasL pathway and TGF-β1 release [81].

Hepatic stellate cells (HpSCs), vitamin A storing cells, which make up approximately 15% of the liver, are also being investigated for their immunotherapeutic potential. Due to the constant exposure of the liver to antigens and to escape an overbearing immune response, it is postulated that this organ exhibits immunomodulatory properties via HpSCs [81]. HpSCs have been shown to express an elevated amount of PD-L1 when exposed to inflammatory cytokine IFN-γ. In addition to promoting T cell apoptosis, HpSCs have been shown to induce regulatory responses, via recruitment of T_{regs} and myeloid-derived suppressor cells (MDSCs) [82]. In a study by Chen *et al.*, the transplantation of HpSCs, in combination with a bone marrow and splenic T cell allograft, led to prolonged survival in a murine model of GVHD [83].

MDSCs themselves have also been investigated as a cell source for cotransplantation, as they suppress effector cells through NO synthase and arginase 1 expression, and for their use of heme oxygenase 1 to inhibit the maturation of DCs [78]. The cotransplantation of MDSCs with islet allografts in mice led to allograft protection in the absence of immunosuppression, demonstrating the capacity of these cells to modulate allograft immunity [84].

While many immunomodulatory strategies seek to recruit or induce T regulatory cells, a more direct approach is the infusion of a population of T_{regs} on allograft transplantation. The $CD4^{+}CD25^{hi}FoxP3^{+}T_{regs}$ exert their suppressor functions through the release of soluble factors such as IL-10 and TGF-β1, as well as cell surface receptors such as CTLA-4 [5]. These cells have been used in clinical trials for bone marrow transplant patients and have shown encouraging results in reducing the occurrence of GVHD [78]. The main challenge facing the use of T_{regs} cell therapy is the difficulty in cell sourcing, as the cells are often contaminated with other effector T cells. Recent progress in efficient purification of these cells should enhance the potency of these cells in the clinical arena [85].

Overall, the use of cell therapy as an immunosuppressive modality for allografts is a promising treatment avenue. By harnessing the innate capabilities of cells responsible for maintaining tolerance in the periphery, the immune response to allografts can be effectively mitigated. The double-edged sword of using cells is their complexity; they come endowed with a wide array of beneficial mechanisms, but their plasticity can lead to unpredictable and undesirable responses. For example, the presence of plasmacytoid DCs has been correlated with transplant tolerance, but the use of a DC therapy is risky due to their primary role of activating the immune system [78]. Regulatory MΦ have exhibited promise, leading to a reduction in required immunosuppressive drugs when cotransplanted in kidney transplant patients [86]; however, in a graft environment exhibiting significant inflammation, there is a risk of these MΦ switching to an activated phenotype [87]. Furthermore, the use of cells increases the metabolic burden of the transplant site, which could lead to overall impairment of nutritional profiles within the graft, particularly during the initial engraftment period. With these drawbacks, cotransplantation might not be the optimal approach; however, commonalities among the mechanisms of immunoregulation (e.g., PD-L1 and TGF-β expression, and T_{reg} and MDSC recruitment) can provide a novel template for future endeavors seeking to engineer immunomodulatory surfaces.

17.2.4 Local Delivery of Soluble Immunomodulatory Agents

The release of soluble immunosuppressive factors holds great interest in the field of allografts and cell transplantation. The body is host to a plethora of cytokines and growth factors, each playing a specialized role. Several of these cytokines exert immunosuppressive effects, when distributed in the correct dosages over a period of time in the proper location. The local release of soluble agents at the graft site provides multiple advantages over systemic administration, including reduced cost, reduction of side effects, and increased potency of response. This strategy may further increase agent efficacy by eliminating common systemic sinks, such as metabolic degradation and excretion. Local delivery is particularly important for cytokines, as most cytokines direct a wide array of effector functions at variable concentrations in the body. For instance, TGF-β1, which can serve as an immunomodulatory cytokine, is also used for cartilage differentiation [88]. Additionally, the elevated expression or systemic delivery of TGF-β1 instigates fibrosis in various tissues, leading to pathological conditions [89]. Thus, the local delivery of TGF-β1 serves to mitigate the risk that this agent would produce an undesired effect. Furthermore, limiting the suppressive effects to the local graft microenvironment leaves the remaining immune system fully competent. To provide local delivery of these soluble agents, materials must be engineered to provide sustained release of the desired dosage. This requires careful engineering of parameters such as material type, doping amounts, and total volume delivered.

Perhaps the most popular and effective platform for localized controlled release of immunomodulatory factors is the use of biodegradable microparticles and nanoparticles. Polymers, such as poly(lactic-co-glycolic acid) (PLGA), are commonly used as materials to encapsulate factors for local delivery. PLGA and other polymers can be easily fabricated into particles of varying sizes, in which proteins or agents can be doped within the particle during fabrication. The particles are tailored to degrade over a time frame ranging from weeks to months, which facilitates the release of immunomodulatory factors in a desired concentration profile. Design criteria affecting the soluble factor release kinetics include particle size, polymer molecular weight, and chemical interactions between the encapsulated cytokines and the particles [90].

Particles designed for soluble release are created to achieve a characteristic cytokine release profile, but engineers must also be mindful of the effect of these particles on cells *in vivo*. For instance, the degradation by-products of PLGA are acidic and the local pH may decrease as the particles break down, a phenomenon also exhibited during an inflammatory response [91]. Of concern, the body may perceive this acidic microenvironment as a disruption of homeostasis and mount an immune response against the site to investigate. When producing particles for soluble release, the size of the particles is typically dictated by the desired release profile of the encapsulated factor; however, the influence of particle size on cell interactions should also be taken into consideration. If a particle is large enough, typically greater than 10 μm, it will be retained at the site of delivery. If particles are too small, they may be phagocytosed or may cross barriers to enter the vasculature [90]. By keeping in mind the interaction of transplanted materials with host cells and tissue, engineers can effectively design therapies, avoiding potential problems while maintaining adequate functionality.

Due to the complexity of paracrine signaling, the release of multiple cytokines is often required to achieve a desired immune response. To accomplish this, particles that release multiple factors or various particles doped with different factors are introduced [90]. These particle formulations are designed to elicit particular immune-directed responses, a popular choice being the recruitment or proliferation of T_{regs}. As discussed in earlier sections, the generation of T_{regs} can dampen alloreactive T cells and even facilitate allograft tolerance. Given that the differentiation of naïve $CD4^+$ T cells into induced T regulatory cells (iT_{regs}) is facilitated by TGF-β1 in the presence of IL-2 and retinoic acid (RA) [92], engineering particles that elute some or all of these cytokines into the graft microenvironment could facilitate generation of cells of this phenotype. Jhunjhunwala *et al.* used PLGA to generate IL-2, TGF-β, and rapamycin eluding microparticles, each with release profiles of up to four weeks. Coincubation of these three microparticle types with naïve $CD4^+$ T cells isolated from B6 or CD45.1 mice was found to induce T_{regs} with comparable efficacy to culturing the cells in media containing the factors in a soluble form. The resulting induced T_{regs} were able to suppress the proliferation of naïve T cells *in vitro* [93]. Alternatively, the capacity of leukemia inhibitory factor (LIF) eluting PLGA nanoparticles to expand induced T_{regs} *in vitro* was evaluated. In combination with low levels of soluble IL-2 and TGF-β1, LIF-PLGA nanoparticles were found to induce T_{regs} generation from naïve $CD4^+$ cells isolated from mice. Furthermore, these LIF-PLGA nanoparticles demonstrated functionality *in vivo* via delayed rejection of MHC-mismatched heart allografts in mice [94]. A similar approach was also explored in the protection of islet allograft transplants, which resulted in enhanced protection [95]. The Little Lab at the University of Pittsburgh investigated the controlled release of CCL22, a chemokine released by tumor cells to recruit T_{regs} and evade the immune system. Incorporation of CCL22 within porous PLGA microparticles resulted in a steady release of the chemokine over a period of 35 days. Injection of CCL22-PLGA microparticles into mouse triceps, followed by the intravenous infusion of allogeneic T_{regs}, led to the recruitment of the T_{regs} to the site of microparticle injection. Additionally, codelivery of CCL22 microparticles with lung carcinoma cell allografts resulted in delayed immune rejection of the graft [96]. Overall, these studies demonstrate the effectiveness of combining the release of multiple factors within degradable PLGA microspheres to create a local environment conducive to the induction, recruitment, and proliferation of T_{regs}.

In addition to targeting the adaptive immune response to the allograft, the therapeutic localized release of anti-inflammatory drugs combats the response of the innate immune system. While it is the cell-mediated response that leads to allograft failure and immune rejection, reducing inflammation can decrease tissue damage and inhibit cross talk between innate and adaptive components, lowering the level of overall immune activation. PGE_2 produced by APCs has a wide array of physiological functions, including suppression of T_h1-mediated immune responses and inflammation. Controlled release formulations of encapsulated PGE_2 in PLGA microparticles have been tested, resulting in the diminished release of the proinflammatory cytokine tumor necrosis factor-alpha (TNF-α) from activated MΦ [97]. Alternatively, the delivery of anti-inflammatory drugs using nanoparticle poly (amidoamine) dendrimers with conjugated folic acid was found to target MΦ. This biomaterial formulation was successful in reducing inflammation in an arthritis model [98]. Similar approaches may be effective in preventing inflammation postallograft.

While the soluble release of immunomodulatory agents has proven to be a versatile tool for dampening the immune attack on transplanted cells, questions remain as to its long-standing efficacy. The local activation of the immune response is reduced through the soluble release of factors, forming a protective bubble around the allograft site; however, as the amount of implanted agents is limited, their protective effect may wane following complete elution of the agent. This controlled localized release approach has the potential to be most effective when inducing a particular cellular approach, such as T_{reg} induction, as the resulting regulatory cells may persist and function to maintain the graft long after doped particles have expired. The combination of the localized release of factors with other approaches, such as encapsulation and modified suppressive material surfaces, could have a highly synergistic effect. When added together, each of these complementary approaches composes part of the puzzle and provides a solution to obtaining functional allografts without systemic immunosuppression.

17.2.5 Surface Modification with Immunomodulatory Motifs

The modification of surfaces with bioactive proteins, drugs, and receptors is a rapidly growing approach to combat the immune response to allografts. By tethering bioactive agents to stable biomaterial surfaces, these molecules may be able to retain their activity and function long-term *in vivo*. While most approaches seek to engineer materials to link bioactive motifs capable of directing cellular adhesion and migration (e.g., RGD) [99], differentiation (e.g., BMP-2) [100], or vascularization (e.g., VEGF, PDGF) [101], the concept of generating immunomodulatory biomaterials capable of instructing immune responses has gained legitimacy in recent years [9,102]. A significant advantage of this approach is that it serves as a means to bypass side effects associated with delivery of these agents. Instead, promising agents are tethered to the material surface, where they are either localized to the site of implantation or designed for delivery to targeted external locations, such as the lymph node or spleen. Moreover, this approach provides a means to express active proteins on the material surface without the need to genetically modify cells, bypassing the complexities associated with gene therapy, such as inefficient gene delivery, low expression, and safety concerns. Finally, by providing early and local instruction to the microenvironment, the potential for tolerance is elevated.

Advancements in biomaterial technology, protein engineering, and ligation schemes have generated biomaterial platforms that illustrate the potential of bioactive biomaterials to instruct immune cells, such as APCs and T cells [9,102–105]. While most of the field has focused on the development of superior vaccines or targeted cancer therapies, these approaches establish the proof that materials can be engineered to direct an immunological response. In the modulation of allograft immune responses, several approaches for bio-inspired surface modification are being explored. Approaches have varied from the incorporation of immunosuppressive motifs within hydrogels designed to diminish responses to the encapsulated allograft to the production of artificial APC (aAPC) particles that directly instruct immune cells to induce suppressive responses.

As outlined previously, the encapsulation of allografts with hydrogels is likely not sufficient to completely prevent alloantigen immune activation, as illustrated in Figure 17.2A. Resulting cytotoxic molecules produced by activated T cells and other small cytokines can subsequently infiltrate the encapsulated material and induce damage or death to the encapsulated allogeneic cells. Combining encapsulation with immunomodulatory agents, however, may serve to decrease this activation. These tunable coatings have been engineered to target an array of immune processes, including inflammation, immune cell chemotaxis, DC maturity, and the Fas pathway to induce T cell apoptosis.

The Fas/Fas ligand pathway is comprised of a regulatory receptor Fas, which is present on effector T cells, and FasL, which stimulates T cell apoptosis when bound to Fas. In one approach, anti-Fas IgG was covalently attached to the encapsulating PEG hydrogel. Using Jurkat immortalized T cells in an *in vitro* study, anti-Fas hydrogels were shown to induce significant T cell apoptosis. When combining this approach with the cellular adhesion molecule ICAM-1, linked via living radical polymerization, the degree of apoptosis was further enhanced, suggesting that cell-surface adhesion motifs led to increased interaction between Fas and surface anti-Fas [106]. Despite encouraging results, the use of the living radical polymerization method is difficult to use for cellular encapsulation. As an alternative fabrication method, a glucose oxidase coating with incorporated Fe^{2+} has been investigated [107]. Hydroxyl radicals formed in the presence of glucose mediate the polymerization of this coating. The incorporation of anti-Fas and ICAM-1 within glucose oxidase–coated PEG hydrogels led to high levels of T cell apoptosis, while the viability of the encapsulated beta cells was retained. In an alternative approach, Yolcu *et al.* tethered a specialized chimeric form of FasL protein, streptavidin (SA)-FasL, to the surface of an islet biotinylated via PEG grafting [108]. The FasL-coated islet, in combination with the systemic delivery of low-dose rapamycin, exhibited complete protection in a fully mismatched murine model, with long-term function of the islet grafts observed. The authors propose enhanced T_{reg} migration to the site as the mechanism of allograft protection. Thus, through deletion of effector T cells on the graft periphery, a more favorable microenvironment for the allogeneic cells can be generated.

Another avenue of immune modulation is the manipulation of DC responses. The maturity and/or activation status of the antigen-presenting DC influences the resulting responses to the particular presenting antigen. Specifically, when a mature DC encounters and presents antigen, MHC and costimulatory expression is increased, and cytokines such as IL-12 are released, leading to a T_h1 immune response; however, when immature DCs encounter antigen in the presence of immunosuppressive cytokines such as TGF-β or IL-10, an anergic or T_{reg} response occurs [92]. Thus, resulting T cell responses can be shifted from effector to tolerogenic through manipulation of DC maturity. In one approach, Hume *et al.* functionalized hydrogels with TGF-β1 and IL-10 via thiolation of bioactive agents and subsequent photopolymerization within PEG diacrylate hydrogels. *In vitro*, tethered cytokines retained their bioactivity and were shown to reduce maturation of DC and the ability of DCs to activate T cells [103]. Alternatively, microparticles engineered with DC receptors and doped with desired antigen can be used to direct antigen presentation within DCs without inducing activation. As illustrated by Lewis *et al.* [104], the presentation of benign PLGA-based microparticles tethered with DC receptors can direct the efficient phagocytosis of particles and subsequent presentation of desired antigen in an inactive DC. This approach, while geared toward autoimmune vaccinations, could be applied to enhancing allograft tolerance. Overall, these approaches highlight the potential of DC targeting to inhibit effector T cell responses.

The surfaces of hydrogels can also be manipulated with a variety of molecules to sequester factors that are secreted by effector immune cells. By functionalizing encapsulating hydrogels with bioactive motifs capable of binding inflammatory cytokines, the material can serve dual roles: blocking direct antigen presentation pathways and quenching cytokines prior to their interaction with underlying allogeneic cells. In this manner, Lin *et al.* developed PEG hydrogels functionalized with WP9QY, a peptide antagonist to the inflammatory cytokine TNF-α, which mimics the recognition loop of TNF receptor 1. Resulting functionalized gels were able to bind soluble TNF-α, with encapsulated mouse islets demonstrating improved viability and function on exposure to this cytokine [109]. In another study, interleukin-1 receptor inhibitory peptide (IL-1RIP) was tethered to PEG hydrogels via native chemical ligation. Resulting gels were used to encapsulate MIN6 cells, a beta cell line. When challenged with soluble inflammatory cytokines IL-1β, TNF-α, and IFN-γ, the IL-1RIP-functionalized gels provided superior protection to the encapsulated cells, exhibited by improved viability and insulin function [110]. Last, affinity peptides, selected for their ability to bind and sequester monocyte chemotactic protein-1 (MCP-1), a potent chemokine which promotes monocyte, DC, and memory T cell migration, were tethered to PEG hydrogels via thiol–acrylate photopolymerization. When encapsulated MIN6 cells were exposed to a cytokine environment designed to stimulate the production of MCP-1 *in vitro*, the gels containing MCP-1 affinity peptide resulted in the presence of significantly lower levels of MCP-1 in the media [111]. In general, the surface modification of encapsulating systems with agents capable of sequestering toxic immune agents represents a novel and promising strategy for the protection of allograft transplants.

An alternate approach that has shown great potential is the creation of biomimetic particles, which act as aAPCs. This biomimetic technology functions by attaching protein ligands to the surface of particles, which on contact with T cells, direct lymphocyte responses. While the aAPC approach has been primarily applied for *in vitro* studies of immune cell responses and mechanisms, researchers are beginning to investigate these platforms for engineering immune responses *in vivo*. Within the domain of promoting immunosuppression, two different strategies have been employed: that of targeting general T cell suppressor pathways, such as PD-1, and an alloantigen-specific approach.

The PD-1/PD-L has been implicated as an important avenue by which effector T cell responses are diminished [112]. The PD-1 receptor is expressed by both $CD4^+$ and $CD8^+$ T cells, as well as other leukocytes. When the PD-1 receptors on T cells are engaged by one of its ligands, PD-L1 or PD-L2, T cell proliferation and cytokine release is inhibited. The use of this pathway to induce a suppressive environment seems to work best with weakly activated T cells, suggesting that combining the PD-1–PD-L signal pathway with secondary suppressive signals may achieve superior results compared with the use of PD-L1 or PD-L2 alone [113]. In an *in vitro* study examining this phenomenon, aAPCs were created using epoxy beads covalently conjugated with PD-L1, as well as anti-CD3 and anti-CD28 to provide primary and costimulatory signals to T cells. On combining these functionalized beads with low levels of soluble TGF-β, a suppressive cytokine, naïve $CD4^+$ T cells in culture were highly converted into induced T_{reg} cells, whose functionality was demonstrated through their suppression of $CD4^+$ effector cells *in vitro* [114].

One concern with using cell-based APC therapy postallograft is biosafety, as the use of genetically engineered DCs may result in undesired responses, such as the deletion of all T cells or an increased immune response. The creation of aAPCs avoids this variability, and allows for the production of killer artificial APCs (KaAPCs) designed to eliminate antigen specific $CD8^+$ T cells. A study attempting to create allograft specific KaAPCs used latex beads conjugated with anti-Fas and MHC I antigen, with the MHC molecule being expressed by only the allograft tissue and not by the host. As such, the particles are designed so that the allograft-specific $CD8^+$ T cells will bind to the MHC and undergo apoptosis mediated by the binding of Fas with the bead-bound Fas antibody. In an *in vivo* mouse skin allograft model, the injection of these KaAPCs led to a 60% decrease in alloreactive $CD8^+$ cells after 2 days and prolonged graft rejection by an average of 6 days [115]. This study highlights the importance of multiple signals in directing T cell behavior, as either anti-Fas alone or MHC I alone results in a less optimal outcome than the combination of both signals.

The use of aAPC therapy to create an immunosuppressive environment postallograft is advantageous due to the consistency and control afforded by the use of engineered biomaterials. This approach opens the possibility for an off-the-shelf product, which eliminates uncertainties in culture and response when using live cells. Currently, a wide array of materials is used for fabricating aAPCs, including nondegradable and degradable polymers [90]. When using biodegradable particles for aAPCs, the particles should be fabricated in a way that prevents masking or degradation of the conjugated ligands as the particle degrades. This outcome has been explored through the use of avidin–palmitic acid constructs within PLGA microparticles, which are then bound to biotinylated ligands through avidin–biotin binding. As a dual approach, these particles have been formulated with encapsulated cytokines, providing signals both through conjugated receptors and released soluble factors [116]. Another aspect of aAPC engineering being explored is the spatial distribution of ligands on the surface of the aAPCs. While current aAPCs involve mostly random uniform surface coatings of ligands, new approaches, such as patterned Janus particles or lipid-coated particles, attempt to reflect the spatial distribution of proteins seen on cellular APCs with improved mimicry [117].

Of additional interest are more recent efforts in the delivery of antigen via apoptotic cells. In this avenue, the natural process of apoptotic cell processing, by which antigens can be presented in a tolerogenic manner, has been harnessed by either engineering antigens with binding sites for cells that are cyclically cleared via apoptotic pathways (e.g., erythrocytes) [118] or via conjugation of antigens to microparticles for splenocyte presentation [119]. In both cases, harnessing the inherent apoptotic cell processing pathway resulted in antigen-specific deletional responses in $CD4^+$ and $CD8^+$ T cells, which could provide a reproducible and cell-free means to hamper autoimmune or protein reactive responses. Translating this approach to allografts is more complex, given the variance of alloantigens. In this scenario, the induction of T regulatory cells is likely more favorable than the deletion of reactive T cells.

The modification of biomaterial surfaces with bioactive motifs provides a useful platform for directing interactions with the immune system. The presentation of ligands and cytokines by materials can drastically alter the immune response to a biomaterial and allograft transplant. This work is still in its development stage and will benefit from additional *in vivo* experimentation and improved fabrication techniques, such as improved bioactive coatings and biomimicry. Nevertheless, the use of surface modification of biomaterials is an attractive option due to the versatility of the approach, as bound factors and receptors can be tailored to target various responses, from specific effector cell apoptotic pathways to reducing inflammation.

17.3 CONCLUSION

The use of allografts to treat diseases is one of modern medicine's great achievements; however, the robust immune response to allogeneic cells necessitates the use of potent systemic immunosuppression regimen to prevent rejection. To promote operational tolerance and graft acceptance, engineers are developing an array of technologies that may result in the long-term functionality of allografts in the absence of long-term, systemic immunosuppressive drugs. These approaches are highly bio-inspired and include the selection of an immune-privileged graft site, immunocloaking through functional biomaterial encapsulation, the cotransplantation of suppressor cells, the controlled release of immunomodulatory factors, and the implementation of highly engineered biomaterials designed to elicit a suppressive response. By drawing on expertise in biology, materials science, and medicine, engineers are designing the next generation of novel therapeutics targeting disease and modulating host responses through cell therapy.

Acknowledgments

Funding support for this research currently comes from the National Institutes of Health through the Type 1 Diabetes Pathfinder Award Program (1DP2-DK083096-01) and SBIR Phase I/II (R43 DK093145-01), the Juvenile Diabetes Research Foundation (17-2012-361), and the Diabetes Research Institute Foundation.

REFERENCES

[1] Klein, A., Lewis, C., Madse, J. *Organ Transplantation: A Clinical Guide*; Cambridge University Press, New York: 2011.

[2] Health Resources and Services Administration, U.S. Department of Health and Human Services. U.S. Organ Procurement and Transplantation Network Data as of February 25th, 2013. http://optn.transplant.hrsa.gov (accessed February 25, 2013).

[3] Delves, P.J., Roitt, I.M. *Roitt's Essential Immunology*; Wiley-Blackwell, Chichester, West Sussex; Hoboken, NJ: 2011.

[4] Benoist, C., Mathis, D. *Cold Spring Harbor Perspectives in Biology* **2012**, *4*, a007021.

[5] Shevach, E.M. *Immunity* **2009**, *30*, 636.

[6] Abbas, A.K., Lichtman, A.H. *Basic Immunology: Functions and Disorders of the Immune System*; Saunders/Elsevier, Philadelphia: 2009.

[7] Kreisel, D., Krupnick, A.S., Gelman, A.E., Engels, F.H., Popma, S.H., Krasinskas, A.M., Balsara, K.R., Szeto, W.Y., Turka, L.A., Rosengard, B.R. *Nature Medicine* **2002**, *8*, 233.

[8] Ferguson, T.A., Choi, J., Green, D.R. *Immunological Reviews* **2011**, *241*, 77.

[9] Hubbell, J.A., Thomas, S.N., Swartz, M.A. *Nature* **2009**, *462*, 449.

[10] Liu, Z., Fan, H., Jiang, S. *Immunological Reviews* **2013**, *252*, 183.

[11] Wood, K.J., Bushell, A., Jones, N.D. *Immunological Reviews* **2011**, *241*, 119.

[12] Blazar, B.R., Murphy, W.J., Abedi, M. *Nature Reviews. Immunology* **2012**, *12*, 443.

[13] Barton, F.B., Rickels, M.R., Alejandro, R., Hering, B.J., Wease, S., Naziruddin, B., Oberholzer, J., Odorico, J.S., Garfinkel, M.R., Levy, M., Pattou, F., Berney, T., Secchi, A., Messinger, S., Senior, P.A., Maffi, P., Posselt, A., Stock, P.G., Kaufman, D.B., Luo, X., Kandeel, F., Cagliero, E., Turgeon, N.A., Witkowski, P., Naji, A., O'Connell, P.J., Greenbaum, C., Kudva, Y.C., Brayman, K.L., Aull, M.J., Larsen, C., Kay, T.W., Fernandez, L.A., Vantyghem, M.C., Bellin, M., Shapiro, A.M. *Diabetes Care* **2012**, *35*, 1436.

[14] Yabu, J.M., Vincenti, F. *Advances in Chronic Kidney Disease* **2009**, *16*, 226.

[15] Fan, H., Cao, P., Game, D.S., Dazzi, F., Liu, Z., Jiang, S. *Seminars in Immunology* **2011**, *23*, 453.

[16] Golshayan, D., Buhler, L., Lechler, R.I., Pascual, M. *Transplant International: Official Journal of the European Society for Organ Transplantation* **2007**, *20*, 12.

[17] Ensor, C.R., Trofe-Clark, J., Gabardi, S., McDevitt-Potter, L.M., Shullo, M.A. *Pharmacotherapy* **2011**, *31*, 1111.

[18] Sayegh, M.H., Carpenter, C.B. *The New England Journal of Medicine* **2004**, *351*, 2761.

[19] Bluestone, J.A. *Immunological Reviews* **2011**, *241*, 5.

[20] Niederkorn, J.Y. *Nature Immunology* **2006**, *7*, 354.

[21] Li, N., Wang, T., Han, D. *Frontiers in Immunology* **2012**, *3*, 152.

[22] Forrester, J.V., Xu, H. *Frontiers in Immunology* **2012**, *3*, 338.

[23] Stein-Streilein, J. *Trends in Immunology* **2008**, *29*, 548.

[24] Sugita, S., Horie, S., Yamada, Y., Keino, H., Usui, Y., Takeuchi, M., Mochizuki, M. *Investigative Ophthalmology and Visual Science* **2010**, *51*, 2529.

[25] Hatchell, D.L., Embabi, S.N., Maeno, T., Saloupis, P., Olson, G., Braun, R.D., Toth, C.A. *Transplantation Proceedings* **1998**, *30*, 593.

[26] Ng, T.F., Lavik, E., Keino, H., Taylor, A.W., Langer, R.S., Young, M.J. *Stem Cells* **2007**, *25*, 1552.
[27] Ma, P., Ge, Y., Wang, S., Ma, J., Xue, S., Han, D. *Reproduction (Cambridge, England)* **2004**, *128*, 163.
[28] Bobzien, B., Yasunami, Y., Majercik, M., Lacy, P.E., Davie, J.M. *Diabetes* **1983**, *32*, 213.
[29] Head, J.R., Neaves, W.B., Billingham, R.E. *Transplantation* **1983**, *36*, 423.
[30] Stein-Streilein, J. *Infection, Immune Homeostasis and Immune Privilege*; Springer, Basel; New York: 2012.
[31] Pedraza, E., Coronel, M.M., Fraker, C.A., Ricordi, C., Stabler, C.L. *Proceedings of the National Academy of Sciences of the United States of America* **2012**, *109*, 4245.
[32] Abdulreda, M.H., Faleo, G., Molano, R.D., Lopez-Cabezas, M., Molina, J., Tan, Y., Echeverria, O.A., Zahr-Akrawi, E., Rodriguez-Diaz, R., Edlund, P.K., Leibiger, I., Bayer, A.L., Perez, V., Ricordi, C., Caicedo, A., Pileggi, A., Berggren, P.O. *Proceedings of the National Academy of Sciences of the United States of America* **2011**, *108*, 12863.
[33] Giraldo, J.A., Weaver, J.D., Stabler, C.L. *Journal of Diabetes Science and Technology* **2010**, *4*, 1238.
[34] Pareta, R., Sanders, B., Babbar, P., Soker, T., Booth, C., McQuilling, J., Sivanandane, S., Stratta, R.J., Orlando, G., Opara, E.C. *Expert Review of Clinical Immunology* **2012**, *8*, 685.
[35] Kloxin, A.M., Kasko, A.M., Salinas, C.N., Anseth, K.S. *Science* **2009**, *324*, 59.
[36] Jeon, O., Alsberg, E. *Tissue Engineering. Part A* **2013**.
[37] Thu, B., Bruheim, P., Espevik, T., Smidsrod, O., Soon-Shiong, P., Skjak-Braek, G. *Biomaterials* **1996**, *17*, 1069; Benson J.P., Papas K.K., Constantinidis I., Sambanis A. *Cell Transplantation* **1997**, *6*, 395.
[38] Rogalsky, A.D., Kwon, H.J., Lee-Sullivan, P. *Journal of Biomedical Materials Research. Part A* **2011**, *99*, 367; Kuo C.K., Ma P.X. *Biomaterials* **2001**, *22*, 511; Chandy T., Mooradian D.L., Rao G.H. *Artificial Organs* **1999**, *23*, 894; Rokstad A.M., Donati I., Borgogna M., Oberholzer J., Strand B.L., Espevik T., Skjak-Braek G. *Biomaterials* **2006**, *27*, 4726; Jeon O., Bouhadir K.H., Mansour J.M., Alsberg E. *Biomaterials* **2009**, *30*, 2724; Dusseault J., Leblond F.A., Robitaille R., Jourdan G., Tessier J., Menard M., Henley N., Halle J.P. *Biomaterials* **2005**, *26*, 1515; Sakai S., Kawakami K. *Acta Biomaterialia* **2007**, *3*, 495.
[39] Gattas-Asfura, K.M., Fraker, C.A., Stabler, C. *Journal of Biomedical Materials Research. Part A* **2011**, *99*, 47; Gattas-Asfura K.M., Stabler C.L. *Biomacromolecules* **2009**, *10*, 3122.
[40] Hall, K.K., Gattas-Asfura, K.M., Stabler, C.L. *Acta Biomaterialia* **2011**, *7*, 614.
[41] Krishnamurthy, N., Gimi, B. *Critical Reviews in Biomedical Engineering* **2011**, *39*, 473.
[42] Duvivier-Kali, V.F., Omer, A., Parent, R.J., O'Neil, J.J., Weir, G.C. *Diabetes* **2001**, *50*, 1698.
[43] Cardona, K., Korbutt, G.S., Milas, Z., Lyon, J., Cano, J., Jiang, W., Bello-Laborn, H., Hacquoil, B., Strobert, E., Gangappa, S., Weber, C.J., Pearson, T.C., Rajotte, R.V., Larsen, C.P. *Nature Medicine* **2006**, *12*, 304; Cardona K., Milas Z., Strobert E., Cano J., Jiang W., Safley S.A., Gangappa S., Hering B.J., Weber C.J., Pearson T.C., Larsen C.P. *American Journal of Transplantation* **2007**, *7*, 2260.
[44] Dufrane, D., Goebbels, R.M., Saliez, A., Guiot, Y., Gianello, P. *Transplantation* **2006**, *81*, 1345.
[45] Kumachev, A., Greener, J., Tumarkin, E., Eiser, E., Zandstra, P.W., Kumacheva, E. *Biomaterials* **2011**, *32*, 1477.

[46] Kobayashi, T., Aomatsu, Y., Iwata, H., Kin, T., Kanehiro, H., Hisanga, M., Ko, S., Nagao, M., Harb, G., Nakajima, Y. *Cell Transplantation* **2006**, *15*, 359.
[47] Tessmar, J.K., Gopferich, A.M. *Macromolecular Bioscience* **2007**, *7*, 23.
[48] Lin, C.C., Raza, A., Shih, H. *Biomaterials* **2011**, *32*, 9685.
[49] Teramura, Y., Oommen, O.P., Olerud, J., Hilborn, J., Nilsson, B. *Biomaterials* **2013**, *34*, 2683.
[50] Lin, C.C., Anseth, K.S. *Proceedings of the National Academy of Sciences of the United States of America* **2011**, *108*, 6380.
[51] Tuch, B.E., Keogh, G.W., Williams, L.J., Wu, W., Foster, J.L., Vaithilingam, V., Philips, R. *Diabetes Care* **2009**, *32*, 1887.
[52] Basta, G., Montanucci, P., Luca, G., Boselli, C., Noya, G., Barbaro, B., Qi, M., Kinzer, K.P., Oberholzer, J., Calafiore, R. *Diabetes Care* **2011**, *34*, 2406.
[53] Garkavenko, O., Wynyard, S., Nathu, D., Quane, T., Durbin, K., Denner, J., Elliott, R. *Xenotransplantation* **2012**, *19*, 6; Elliott R.B., Living Cell T. *Current Opinion in Organ Transplantation* **2011**, *16*, 195.
[54] de Groot, M., Schuurs, T.A., van Schilfgaarde, R. *The Journal of Surgical Research* **2004**, *121*, 141; Schrezenmeir J., Kirchgessner J., Gero L., Kunz L.A., Beyer J., Mueller-Klieser W. *Transplantation* **1994**, *57*, 1308.
[55] Khademhosseini, A., May, M.H., Sefton, M.V. *Tissue Engineering* **2005**, *11*, 1797; May M.H., Sefton M.V. *Annals of the New York Academy of Sciences* **1999**, *875*, 126; Cruise G.M., Hegre O.D., Lamberti F.V., Hager S.R., Hill R., Scharp D.S., Hubbell J.A. *Cell Transplantation* **1999**, *8*, 293.
[56] Teramura, Y., Iwata, H. *Bioconjugate Chemistry* **2008**, *19*, 1389.
[57] Lee, D.Y., Park, S.J., Lee, S., Nam, J.H., Byun, Y. *Tissue Engineering* **2007**, *13*, 2133; Chen A., Scott M. *Biodrugs: Clinical Immunotherapeutics, Biopharmaceuticals and Gene Therapy* **2001**, *15*, 833; Scott M.D., Chen A.M. *Transfusion Clinique et Biologique: Journal de la Societe Francaise de Transfusion Sanguine* **2004**, *11*, 40.
[58] Kizilel, S., Scavone, A., Liu, X., Nothias, J.M., Ostrega, D., Witkowski, P., Millis, M. *Tissue Engineering. Part A* **2010**, *16*, 2217.
[59] Contreras, J.L., Xie, D., Mays, J., Smyth, C.A., Eckstein, C., Rahemtulla, F.G., Young, C.J., Anthony Thompson, J., Bilbao, G., Curiel, D.T., Eckhoff, D.E. *Surgery* **2004**, *136*, 537; Stabler C.L., Sun X.L., Cui W., Wilson J.T., Haller C.A., Chaikof E.L. *Bioconjugate Chemistry* **2007**, *18*, 1713.
[60] Fischer, D., Li, Y., Ahlemeyer, B., Krieglstein, J., Kissel, T. *Biomaterials* **2003**, *24*, 1121; Hong S., Leroueil P.R., Janus E.K., Peters J.L., Kober M.-M., Islam M.T., Orr B.G., Baker J.R., Banaszak Holl M.M. *Bioconjugate Chemistry* **2006**, *17*, 728.
[61] Wilson, J.T., Cui, W., Kozlovskaya, V., Kharlampieva, E., Pan, D., Qu, Z., Krishnamurthy, V.R., Mets, J., Kumar, V., Wen, J., Song, Y., Tsukruk, V.V., Chaikof, E.L. *Journal of the American Chemical Society* **2011**, *133*, 7054.
[62] Kozlovskaya, V., Baggett, J., Godin, B., Liu, X., Kharlampieva, E. *ACS Macro Letters* **2012**, *2012*, 384.
[63] Wilson, J.T., Cui, W., Chaikof, E.L. *Nano Letters* **2008**, *8*, 1940.
[64] Luan, N.M., Teramura, Y., Iwata, H. *Biomaterials* **2011**, *32*, 6487.
[65] Teramura, Y., Kaneda, Y., Iwata, H. *Biomaterials* **2007**, *28*, 4818.
[66] Wilson, J.T., Krishnamurthy, V.R., Cui, W., Qu, Z., Chaikof, E.L. *Journal of the American Chemical Society* **2009**, *131*, 18228; Zhi Z.L., Liu B., Jones P.M., Pickup J.C. *Biomacromolecules* **2010**, *11*, 610; Miura S., Teramura Y., Iwata H., *Biomaterials* **2006**, *27*, 5828; Krol S., del Guerra S., Grupillo M., Diaspro A., Gliozzi A., Marchetti P. *Nano Letters* **2006**, *6*, 1933.
[67] Zhi, Z.L., Kerby, A., King, A.J., Jones, P.M., Pickup, J.C. *Diabetologia* **2012**, *55*, 1081.

[68] Schmitt, J., Griinewald, T., Decher, G., Pershan, P., Kjaer, K., Losche, M. *Macromolecules* **1993**, *26*, 7058; Yuan W., Ji J., Fu J., Shen J., *Journal of Biomedical Materials Research. Part B, Applied Biomaterials* **2008**, *85*, 556.
[69] Teramura, Y., Kaneda, Y., Totani, T., Iwata, H. *Biomaterials* **2008**, *29*, 1345.
[70] Kou, P.M., Babensee, J.E. *Journal of Biomedical Materials Research. Part A* **2011**, *96*, 239.
[71] Thevenot, P.T., Baker, D.W., Weng, H., Sun, M.W., Tang, L. *Biomaterials* **2011**, *32*, 8394.
[72] Wu, G.D., Nolta, J.A., Jin, Y.S., Barr, M.L., Yu, H., Starnes, V.A., Cramer, D.V. *Transplantation* **2003**, *75*, 679.
[73] Jackson, W.M., Nesti, L.J., Tuan, R.S. *Stem Cells Translational Medicine* **2012**, *1*, 44.
[74] Shi, M., Liu, Z.W., Wang, F.S. *Clinical and Experimental Immunology* **2011**, *164*, 1.
[75] Jiang, X.X., Zhang, Y., Liu, B., Zhang, S.X., Wu, Y., Yu, X.D., Mao, N. *Blood* **2005**, *105*, 4120; Sotiropoulou P.A., Perez S.A., Gritzapis A.D., Baxevanis C.N., Papamichail M. *Stem Cells* **2006**, *24*, 74.
[76] Soleymaninejadian, E., Pramanik, K., Samadian, E. *American Journal of Reproductive Immunology* **2012**, *67*, 1.
[77] Tipnis, S., Viswanathan, C., Majumdar, A.S. *Immunology and Cell Biology* **2010**, *88*, 795.
[78] Wood, K.J., Bushell, A., Hester, J. *Nature Reviews. Immunology* **2012**, *12*, 417.
[79] Wang, J., Liao, L., Tan, J. *Expert Opinion on Biological Therapy* **2011**, *11*, 893; Berman D.M., Willman M.A., Han D., Kleiner G., Kenyon N.M., Cabrera O., Karl J.A., Wiseman R.W., O'Connor D.H., Bartholomew A.M., Kenyon N.S. *Diabetes* **2010**, *59*, 2558.
[80] Fallarino, F., Luca, G., Calvitti, M., Mancuso, F., Nastruzzi, C., Fioretti, M.C., Grohmann, U., Becchetti, E., Burgevin, A., Kratzer, R., van Endert, P., Boon, L., Puccetti, P., Calafiore, R. *The Journal of Experimental Medicine* **2009**, *206*, 2511; Dufour J.M., Rajotte R.V., Korbutt G.S., Emerich D.F. *Immunological Investigations* **2003**, *32*, 275.
[81] Charles, R., Lu, L., Qian, S., Fung, J.J. *Immunotherapy* **2011**, *3*, 1471.
[82] Chou, H.S., Hsieh, C.C., Yang, H.R., Wang, L., Arakawa, Y., Brown, K., Wu, Q., Lin, F., Peters, M., Fung, J.J., Lu, L., Qian, S. *Hepatology* **2011**, *53*, 1007; Jiang G., Yang H.R., Wang L., Wildey G.M., Fung J., Qian S., Lu L. *Transplantation* **2008**, *86*, 1492.
[83] Chen, C.H., Shu, K.H., Su, Y.H., Tang, K.Y., Cheng, C.H., Wu, M.J., Yu, T.M., Chuang, Y.W., Hu, C. *Transplantation Proceedings* **2010**, *42*, 971.
[84] Chou, H.S., Hsieh, C.C., Charles, R., Wang, L., Wagner, T., Fung, J.J., Qian, S., Lu, L.L. *Transplantation* **2012**, *93*, 272.
[85] Pahwa, R., Jaggaiahgari, S., Pahwa, S., Inverardi, L., Tzakis, A., Ricordi, C. *Journal of Immunological Methods* **2010**, *363*, 67.
[86] Hutchinson, J.A., Riquelme, P., Sawitzki, B., Tomiuk, S., Miqueu, P., Zuhayra, M., Oberg, H.H., Pascher, A., Lutzen, U., Janssen, U., Broichhausen, C., Renders, L., Thaiss, F., Scheuermann, E., Henze, E., Volk, H.D., Chatenoud, L., Lechler, R.I., Wood, K.J., Kabelitz, D., Schlitt, H.J., Geissler, E.K., Fandrich, F. *The Journal of Immunology* **2011**, *187*, 2072.
[87] Mosser, D.M., Edwards, J.P. *Nature Reviews. Immunology* **2008**, *8*, 958.
[88] DeFail, A.J., Chu, C.R., Izzo, N., Marra, K.G. *Biomaterials* **2006**, *27*, 1579.
[89] Pannu, J., Gardner, H., Shearstone, J.R., Smith, E., Trojanowska, M. *Arthritis and Rheumatism* **2006**, *54*, 3011; Kubo M., Ihn H., Yamane K., Tamaki K. *Arthritis and Rheumatism* **2001**, *44*, 731.
[90] Balmert, S.C., Little, S.R. *Advanced Materials* **2012**, *24*, 3757.

[91] Lu, L., Stamatas, G.N., Mikos, A.G. *Journal of Biomedical Materials Research* **2000**, *50*, 440.
[92] Zhou, L., Chong, M.M., Littman, D.R. *Immunity* **2009**, *30*, 646.
[93] Jhunjhunwala, S., Balmert, S.C., Raimondi, G., Dons, E., Nichols, E.E., Thomson, A.W., Little, S.R. *Journal of Controlled Release: Official Journal of the Controlled Release Society* **2012**, *159*, 78.
[94] Park, J., Gao, W., Whiston, R., Strom, T.B., Metcalfe, S., Fahmy, T.M. *Molecular Pharmaceutics* **2011**, *8*, 143.
[95] Dong, H., Fahmy, T.M., Metcalfe, S.M., Morton, S.L., Dong, X., Inverardi, L., Adams, D.B., Gao, W., Wang, H. *PLoS ONE* **2012**, *7*, e50265.
[96] Jhunjhunwala, S., Raimondi, G., Glowacki, A.J., Hall, S.J., Maskarinec, D., Thorne, S.H., Thomson, A.W., Little, S.R. *Advanced Materials* **2012**, *24*, 4735.
[97] Nicolete, R., Lima Kde, M., Junior, J.M., Jose, P.J., Sanz, M.J., Faccioli, L.H. *European Journal of Pharmaceutics and Biopharmaceutics: Official Journal of Arbeitsgemeinschaft fur Pharmazeutische Verfahrenstechnik e.V* **2008**, *70*, 784.
[98] Thomas, T.P., Goonewardena, S.N., Majoros, I.J., Kotlyar, A., Cao, Z., Leroueil, P.R., Baker, J.R., Jr. *Arthritis and Rheumatism* **2011**, *63*, 2671.
[99] Place, E.S., Evans, N.D., Stevens, M.M. *Nature Materials* **2009**, *8*, 457.
[100] Lutolf, M.P., Gilbert, P.M., Blau, H.M. *Nature* **2009**, *462*, 433; Trappmann B., Gautrot J.E., Connelly J.T., Strange D.G., Li Y., Oyen M.L., Cohen Stuart M.A., Boehm H., Li B., Vogel V., Spatz J.P., Watt F.M., Huck W.T. *Nature Materials* **2012**, *11*, 642.
[101] Phelps, E.A., Garcia, A.J. *Current Opinion in Biotechnology* **2010**, *21*, 704.
[102] Franz, S., Rammelt, S., Scharnweber, D., Simon, J.C. *Biomaterials* **2011**, *32*, 6692.
[103] Hume, P.S., He, J., Haskins, K., Anseth, K.S. *Biomaterials* **2012**, *33*, 3615.
[104] Lewis, J.S., Zaveri, T.D., Crooks, C.P., 2nd, Keselowsky, B.G. *Biomaterials* **2012**, *33*, 7221.
[105] Hori, Y., Winans, A.M., Irvine, D.J. *Acta Biomaterialia* **2009**, *5*, 969; Doh J., Irvine D.J. *Proceedings of the National Academy of Sciences of the United States of America* **2006**, *103*, 5700; Ali O.A., Emerich D., Dranoff G., Mooney D.J. *Science Translational Medicine* **2009**, *1*, 8ra19.
[106] Hume, P.S., Anseth, K.S. *Biomaterials* **2010**, *31*, 3166.
[107] Hume, P.S., Bowman, C.N., Anseth, K.S. *Biomaterials* **2011**, *32*, 6204.
[108] Yolcu, E.S., Zhao, H., Bandura-Morgan, L., Lacelle, C., Woodward, K.B., Askenasy, N., Shirwan, H. *The Journal of Immunology* **2011**, *187*, 5901.
[109] Lin, C.C., Metters, A.T., Anseth, K.S. *Biomaterials* **2009**, *30*, 4907.
[110] Su, J., Hu, B.H., Lowe, W.L., Jr., Kaufman, D.B., Messersmith, P.B. *Biomaterials* **2010**, *31*, 308.
[111] Lin, C.C., Boyer, P.D., Aimetti, A.A., Anseth, K.S. *Journal of Controlled Release: Official Journal of the Controlled Release Society* **2010**, *142*, 384.
[112] Riella, L.V., Paterson, A.M., Sharpe, A.H., Chandraker, A. *American Journal of Transplantation* **2012**, *12*, 2575.
[113] Sharpe, A.H., Wherry, E.J., Ahmed, R., Freeman, G.J. *Nature Immunology* **2007**, *8*, 239.
[114] Francisco, L.M., Salinas, V.H., Brown, K.E., Vanguri, V.K., Freeman, G.J., Kuchroo, V.K., Sharpe, A.H. *The Journal of Experimental Medicine* **2009**, *206*, 3015.
[115] Shen, C., He, Y., Cheng, K., Zhang, D., Miao, S., Zhang, A., Meng, F., Miao, F., Zhang, J. *Immunology Letters* **2011**, *138*, 144.
[116] Han, H., Peng, J.R., Chen, P.C., Gong, L., Qiao, S.S., Wang, W.Z., Cui, Z.Q., Yu, X., Wei, Y.H., Leng, X.S. *Biochemical and Biophysical Research Communications* **2011**, *411*, 530.

[117] Moon, J.J., Huang, B., Irvine, D.J. *Advanced Materials* **2012**, *24*, 3724.
[118] Kontos, S., Kourtis, I.C., Dane, K.Y., Hubbell, J.A. *Proceedings of the National Academy of Sciences of the United States of America* **2013**, *110*, E60.
[119] Getts, D.R., Martin, A.J., McCarthy, D.P., Terry, R.L., Hunter, Z.N., Yap, W.T., Getts, M.T., Pleiss, M., Luo, X., King, N.J., Shea, L.D., Miller, S.D. *Nature Biotechnology* **2012**, *30*, 1217.
[120] Meisel, R., Zibert, A., Laryea, M., Gobel, U., Daubener, W., Dilloo, D. *Blood* **2004**, *103*, 4619.
[121] English, K., Ryan, J.M., Tobin, L., Murphy, M.J., Barry, F.P., Mahon, B.P. *Clinical and Experimental Immunology* **2009**, *156*, 149.
[122] Matsuda, R., Kezuka, T., Nishiyama, C., Usui, Y., Matsunaga, Y., Okunuki, Y., Yamakawa, N., Ogawa, H., Okumura, K., Goto, H. *Investigative Ophthalmology and Visual Science* **2012**, *53*, 7235.
[123] Chen, L., Zhu, Y., Zhang, G., Gao, C., Zhong, W., Zhang, X. *Proceedings of the National Academy of Sciences of the United States of America* **2011**, *108*, 18778; Anastassiou E.D., Paliogianni F., Balow J.P., Yamada H., Boumpas D.T. *Journal of Immunology* **1992**, *148*, 2845.
[124] Paolucci, C., Rovere, P., De Nadai, C., Manfredi, A.A., Clementi, E. *The Journal of Biological Chemistry* **2000**, *275*, 19638.

CHAPTER 18

IMMUNOMIMETIC MATERIALS

JAMAL S. LEWIS and BENJAMIN G. KESELOWSKY

J. Crayton Pruitt Family Department of Biomedical Engineering, University of Florida, Gainesville, FL, USA

18.1 INTRODUCTION

The development of biomimetic/bio-inspired materials is rooted in the fields of tissue engineering and regenerative medicine. Synthetic scaffolds are often used to provide mechanical support and time-limited architecture for neotissue regeneration *in vivo*. Although synthetic scaffolds demonstrate favorable chemical and physical qualities, their lack of bioactivity is a limitation. Conversely, materials from natural sources with innate biomolecular recognition ability typically lack customizability, in addition to concerns regarding sterilization and immunogenicity. Biomimetic materials are a relatively new class where synthetic materials are adapted to possess biological motifs that direct specific responses at the molecular and cellular levels. These hybrid materials combine advantageous traits of both levels to provide well-controlled physical and mechanical properties, as well as biochemical and biological cues that influence material–cell communication.

Degradable biomimetic materials, as an example seen in numerous applications, aim to mirror many characteristics of the extracellular matrix (ECM) found during normal tissue formation and wound healing [1, 2]. The ECM plays a critical role in the morphogenesis, homeostasis, and regeneration of tissues throughout the human body. It is an ordered three-dimensional array of structural and functional biomolecules that differs in microstructure and biological activity for each tissue and organ. The ECM is secreted by the local population of cells and not only bestows structural integrity but also drives cell and tissue phenotype by sequestering and releasing biochemical signals in a spatially and temporally dependent manner. Furthermore, the ECM is a dynamic structure constantly undergoing remodeling due to environmental forces such as mechanical loads, oxygen debt, and pH [3]. One of the major biomimetic material strategies involves incorporation of ECM characteristics into synthetic structures to develop biointeractive constructs.

Bio-inspired Materials for Biomedical Engineering, First Edition. Edited by Anthony B. Brennan and Chelsea M. Kirschner.

In addition to mimicking characteristics of the ECM in which cells reside, recapitulation of certain features associated with cells themselves has also served as inspiration for biomaterial scientists. One of the most basic aspects of cells is that of compartmentalization. Selective isolation of biomolecules allows for exquisite control of metabolic reactions in a spatiotemporal context required for cell functionality. While accomplished primarily through phospholipid membranes by cells, a plethora of synthetic and biologic materials have been used to fashion artificial cell membranes with variable permeability, blood compatibility, surface properties, and moieties [4, 5]. For instance, McPhail *et al.* used a mixture of palmitol glycol chitosan and cholesterol to prepare an artificial outer membrane for their vesicle system [6], whereas, others have utilized biomaterials such as derivatized poly(lactide-co-glycolide) (PLGA) and derivatized poly(dimethysiloxane) to generate polymerosomes, with membrane-like compartmentalization capability [7–9]. A wide array of components can be incorporated into biomaterials-based compartments. These include live cells [10–12], enzyme systems [13–15], pharmacological drugs [16, 17], and antigens [18–20]. The number of biomedical applications is vast, ranging from red blood cell substitutes [21] to drug delivery systems enhancing targeted delivery to cells and organs [18].

These two broad biomimetic material strategies are also beginning to be explored in order to solve problems associated with immune cells, which is a wide-ranging field. For example, it has long been recognized that implantation of biomaterials triggers a profound reaction of host immune responses, collectively referred to as the foreign body reaction [22]. The physical injury due to implantation of biomaterials elicits an inflammatory response, considered to be part of the normal wound healing process. The presence of a biomaterial typically exacerbates this response, resulting in foreign body giant cell formation and antigen release (when a biological component is present) at the site of implantation [22]. Briefly, interaction with bodily fluid leads to protein adsorption on the surface of the biomaterial and can initiate the coagulation cascade, the complement system (which can polarize immune cells toward an inflammatory response), and the formation of a provisional matrix. These phenomenon have been extensively investigated on different biomaterial surfaces and it is thought that they are correlated to the physicochemical surface properties of the biomaterial, thereby linking biomaterial properties with host immune cell responses [23]. Following matrix formation, antigen-presenting cells, including macrophages and dendritic cells (DCs), can be recruited to the implant site by chemokines released by the matrix as well as surrounding cells. Macrophages, in particular, persist at the implantation site, adhering to the implant surface and coalescing with neighboring macrophages to form a giant cell body, which attempts to engulf the material. Within this encapsulation, macrophages secrete a number of inflammatory mediators, including reactive oxygen species and degradative enzymes that can be detrimental to the structure and functionality of the implanted biomaterial [23, 24].

More recently, biomedical engineers have taken aim at modulating host immune responses, including the foreign body reaction, to desired outcomes for improved diagnostic and therapeutic applications. This marriage of materials engineering and immunobiology has led to new immunomodulatory materials. This chapter highlights development and application of a number of immunomodulatory materials, categorized by the following general approaches: (a) surface motifs targeting cell surface receptors to direct immune cell responses; (b) morphogenic factor-related materials that release growth, differentiation, chemotactic, and immune-modulating factors; (c) stimuli-responsive materials that influence immune cell responses based on environmental conditions; and (d) self-assembly motifs, that when assembled, influence immune responses.

18.2 SURFACE MOTIFS

Surface modification with biomolecules is one of the most prevalent methods used to convey bioactivity to synthetic materials. There is a long history of literature establishing the influence of adsorbed adhesive proteins on cell adhesion, morphology, and migration at the material surface [25–27]. Adhesive proteins found naturally in the ECM, such as fibronectin, vitronectin, and laminin, provide multiple cues that direct diverse cellular processes via integrin binding [28–31]. Integrins, heterodimeric transmembrane proteins, are the key mediators between the ECM molecular signals and intracellular signaling pathways. On binding, integrins cluster and associate with various signal transduction molecules, thereby activating specific intracellular signaling cascades [25, 32–34].

The discovery of integrin-binding signaling domains, only several amino acids long, within ECM proteins prompted the development of material surfaces with ligated oligopeptide sequences. Short oligopeptide sequences have advantages over complete functional proteins, including their biospecificity, stability, lack of immunogenicity, and lower production/sterilization costs. Native ECM proteins tend to be immobilized on material surfaces in an uncontrolled fashion, such that presentation of specific domains is limited, as is selectivity for specific integrins and subsequent cellular interactions [2]. However, short adhesive peptides often have a relative decrease in activity due to loss of native conformation compared with the complete protein [35]. More recently, efforts have focused on the use of small protein fragments and conformationally constrained peptides, aiming to maintain biological activity while retaining the advantages of stability, lack of immunogenicity, and lower production cost [36–38].

To date, peptides containing the amino acid sequence arginine–glycine–aspartic acid (RGD), originally derived from fibronectin and found in numerous other adhesive proteins, have been the most extensively investigated as a bioadhesive motif [39, 40]. Other widely investigated ligands immobilized on material surfaces include tyr-ile-gly-ser-arg (YIGSR; derived from laminin), arg-glu-asp-val (REDV; derived from fibronectin), and val-pro-gly-ile-gly (VPGIG; derived from elastin) [41, 42]. Various materials (e.g., glass, quartz, metals, metal oxides, self-assembled monolayers, and polymers) have been used as model substrates for oligopeptide surface conjugation and the subsequent cellular responses on these surfaces were characterized [2, 39, 43, 44]. *In vitro* cell culture on biomimetic surfaces has resulted in enhanced cellular adhesion, spreading, focal contact, and cytoskeletal organization for several cell types including neurons [45], smooth muscle cells [46], endothelial cells [47], fibroblasts [38, 39], and osteoblasts [27, 48]. Although numerous investigations have clearly demonstrated (e.g., using RGD peptides) the difficulty in translating *in vitro* findings to *in vivo* scenarios [49, 50], application-specific successes have been demonstrated [51–53]. Nonfouling scaffolds, such as polyethylene glycol (PEG) hydrogels, thwart nonspecific cell adhesion and have been engineered to present various adhesive peptides [40]. The extent of integrin-mediated adhesion has been shown to be dependent on receptor–ligand affinity and on the density of the oligopeptide [42, 54–57]. For example, both cell migration and neurite extension have been correlated to adhesion strength, modulated in unimodal fashion by peptide density [58]. Through careful selection, it is possible to optimize for specific advantageous cellular response when designing scaffolds functionalized with adhesive ligands.

Biomaterial scaffolds engineered to present ECM proteins can also have a profound impact on host immune responses and have begun to be investigated, particularly for the attenuation of inflammatory responses. For instance, Acharya *et al.* demonstrated that culture of bone marrow-derived DCs on various ECM protein substrates (fibronectin, laminin, fibrinogen, serum, etc.) can result in differential levels of expression of proinflammatory surface molecules and pro- and anti-inflammatory cytokine secretion by DCs [28]. Historically, to modulate inflammatory responses to biomaterial implants, material scientists have manipulated material type, surface chemistry, and topography. These properties can influence protein deposition, provisional matrix formation, immune cell attachment, and ultimately, host immune responses [59]. Studies as those by Acharya *et al.* suggest that surface immobilization of naturally derived bioligands is an alternative approach with much promise for the mitigation of inflammatory responses to biomaterial implants [28]. Hume *et al.* adopted a similar approach, functionalizing PEG hydrogels with covalently immobilized immunosuppressive biological agents—transforming growth factor beta 1 (TGF-β1) and interleukin-10 (IL-10) to reduce inflammatory responses to cell-laden material carriers. Hume and coworkers immobilized TGF-β1 and IL-10 into PEG hydrogel networks via thiol–acrylate polymerization and showed that the presence of these two anti-inflammatory cytokines significantly downregulates DC maturation markers including IL-12 and MHC-II [60].

Functionalization of biomaterial surfaces with biologically derived molecules is also being applied to various immunotherapy approaches for cancer, infectious disease, autoimmune disease, and transplant rejection treatment. T cell adoptive immunotherapy, in particular, has emerged as a strategy with great potential for the treatment of a number of immune-related conditions. For instance, different T cell adoptive immunotherapies are now under clinical trial for treatment of human immunodeficiency virus (HIV), as well as different malignancies including breast carcinomas [61]. In this personalized medicine approach, patient-derived T cells specific for relevant antigens are isolated, engineered, expanded, and then reintroduced to the patient in an effort to amplify T cell responses to the malignant cell or infectious agent presenting the antigen [62, 63]. Conventionally, the *ex vivo* expansion of antigen-specific T cells is accomplished using autologous antigen-presenting cells, which have variability in phenotype and quantity. Additional drawbacks to the use of autologous antigen-presenting cells for T cell expansion include the time- and labor-intensive manufacturing process and its accompanying costs [64, 65]. As an alternative, material scientists have designed artificial antigen-presenting cells (aAPCs) as a solution to the use of autologous antigen-presenting cells. These acellular systems are typically based on polymeric microspheres [66], magnetic beads [67], or liposomes [68, 69]. For example, Tham *et al.* demonstrated that 5 μm polystyrene latex microparticles with surface immobilized B7-fusion proteins and a peptide–MHC complex can be an effective aAPC for *in vitro* antigen-specific T cell expansion [70].

Polymeric microspheres with surface-tethered immunoligands are also gaining widespread attention for their ability to manipulate DCs in DC-based immunotherapy, and more recently, for autoimmune applications. Lewis *et al.* reported that poly(d lactide-co-glycolide) microspheres with surface immobilized ligands (DEC205 and CD11c antibodies, and P-D2 peptide) are capable of enhanced DC targeting *in vitro* and *in vivo* without stimulating DC activation [71]. Similarly, Bandyopadhyay *et al.* demonstrated that DEC205-labeled PLGA nanoparticles not only efficiently target DCs for uptake but also increases production of IL-10 in DCs [72]. The findings from both of these studies could be instructive for the use of aAPCs to prevent and reverse autoimmune diseases. Along this line, researchers prevented and reversed diabetes in non-obese diabetic (NOD) mice using peptide–MHC complex coated iron oxide (FeO) nanoparticles. Santamaria and coworkers demonstrated that nanoparticles coated with disease-relevant peptide-major histocompatibility type I complexes (pMHC-I-nanoparticles) expanded cognate autoregulatory $CD8^+$ T cells in animals, suppressed the recruitment of noncognate specificities, prevented disease in prediabetic mice, and restored normoglycemia in diabetic animals [73]. This study demonstrates the enormous potential of engineering surface motifs in biomaterials applications.

18.3 MORPHOGENIC FACTOR-RELATED MATERIALS

The development of synthetic materials able to sequester and deliver morphogenic/growth factors at the right place and right time is another established paradigm in biomimetic scaffolds for tissue engineering. Natural ECM has the capacity to bind not only large quantities of diverse growth factors and cell-inducing agents, but also to release these morphogens in a controlled manner, which results in coordinated protein expression, cell migration, proliferation, and differentiation. Proteoglycans are a subset of non-collagenous glycoproteins that contain glycosaminoglycan side chains. Examples include decorin, heparin, and chondroitin. The large reservoir of growth factors bound by proteoglycans is another vital feature of the ECM and is implicated in tissue remodeling, growth, and differentiation. This function of the ECM is crucial for neotissue formation and repair [3, 74]. Synthetic analogs have replicated growth factor activity by regulating their display, their release kinetics, and their biostability. Sakiyama-Elbert *et al.* demonstrated that cell-mediated release of heparin-binding growth factors from fibrin scaffolds with immobilized heparin can enhance neurite extension [75]. This model capitalizes on the affinity of heparin for growth factors such as transforming growth factor beta (TGF-β) and fibroblast growth factor (FGF), controlling their presentation and delivery rate [76].

A more common delivery approach involves the sustained, controlled release of soluble morphogenic factor, previously dispersed into the matrix of the scaffold during preparation. This approach affords sustained, localized delivery of tissue inductive factors in a spatiotemporal manner to direct specific tissue responses as well as minimize off-target effects typically associated with systemic delivery. A wide array of matrices and scaffolds has been used to accommodate delivery of drugs that regulate multiple processes of cell chemotaxis, attachment, proliferation, differentiation, and morphogenesis [25, 77–79]. For instance, Lucas *et al.* showed that the incorporation of a water-soluble bone morphogenetic protein (BMP) extract from bone matrix into a synthetic polyanhydride matrix promoted chondrogenesis and osteogenesis when implanted ectopically *in vivo* [80]. Notably, this study demonstrated that only in combination are the polyanhydride matrix and osteogenic protein extract able to effect cartilage and bone formation. Additionally, their results highlighted the short half-life of proteins *in vivo*. Pharmaceutical scientists have developed ways to avoid this drawback such as PEGylation, which through the covalent linkage of PEG, stabilizes proteins, helps them resist protein adsorption and aggregation, and reduces their uptake by cells and therefore, immune system intervention [81, 82]. Alternatively, approaches utilizing the protein-making machinery of cells have been explored. Gene transfection is a potent and promising technique that involves the *in vitro* or *in vivo* incorporation of genetic material such as DNA, RNA, and RNAi [83–86] into cells for experimental and therapeutic purposes. Gene delivery provides for sustained therapeutic levels of protein and targeting of multiple cellular processes. Moreover, several investigators have reported successful tissue regeneration by gene transfection of tissue-building cells [87, 88]. However, there are immunological and safety concerns associated with viral vectors used in these studies, motivating development of nonviral vector systems [59].

For this reason, gene delivery via particulate systems fabricated from different biomaterials has received attention for the last 15 years. Particulate systems allow for delivery of genetic material in a defined spatial and temporal manner for therapeutic applications in regenerative medicine and degenerative conditions. For example, Phillips *et al.* demonstrated that PLGA microspheres loaded with antisense oligonucleotides for co-stimulatory molecules, passively targeted DCs and manipulated their immunoregulatory function. Phillips and associates successfully protected from type 1 diabetes in NOD mice through *in vivo* targeting with this oligonucleotide-loaded microparticle vaccine that genetically modify antigen-presenting cells upon interception [89]. Interestingly, Singh *et al.* used an injectable, polymer matrix–particulate composite system for tuned simultaneous delivery of chemoattractant (MIP3α), IL-10 siRNA, and antigen plasmid DNA to modulate infiltrating APC phenotype for *in vivo* cancer immunotherapy. More specifically, they demonstrated that the low cross-link density PEG hydrogels that made up the polymer matrix were degraded within 2–7 days *in vitro* and released chemokines in a sustained manner, which attracted significant numbers of DCs over a sustained period *in vitro*, compared with an equivalent bolus dose. Furthermore, DCs that migrated to the injection site were able to infiltrate the hydrogels and efficiently phagocytose the siRNA–DNA carrying microparticles which resulted in IL-10 gene knockdown in migrated primary DCs *in vitro* [90]. These studies demonstrate the versatility of this approach as well as its huge promise for *in vivo* specific immunotherapy for immune-related disease.

18.4 STIMULI-RESPONSIVE MATERIALS

The ECM is a dynamic three-dimensional structure constantly being remodeled due to environmental stimuli. This remodeling is critical for new tissue formation, homeostasis, and repair. ECM remodeling is primarily directed by a group of proteases collectively known as matrix metalloproteinases (MMPs). Additionally, serine proteases and hyaluronidases degrade proteoglycan elements of the matrix. Protease degradation has significant implications on cell migration, proliferation, differentiation, and survival due to ECM component breakdown and release of growth factors, cytokines, and other cell-modulating agents [91, 92].

Biomimetic material approaches have mimicked this ECM enzymatic responsiveness by incorporating proteolytic peptide sequences in synthetic scaffolds so as to accomplish directed cell invasion. Enzymatically sensitive peptide sequences derived from ECM proteins, such as pro-val-gly-leu-ile-gly (PVGLIG), are cleaved by secreted proteases and activated by cells locally, clearing a path in the scaffold through which the cells can migrate. This process occurs at the cell surface boundary and is tightly regulated by membrane-associated protease inhibitors (e.g., tissue inhibitor of metalloproteinases 2 [TIMP-2]) [25, 93]. Hydrogels engineered with proteolytic sites have shown enhanced cell migration both *in vivo* and *in vitro* [94]. The degree to which cells invade the construct is dependent on protease substrate activity, enzyme-sensitive ligand concentration, and network cross-linking density [95].

Enzyme-sensitive synthetic materials are one part of a large body of developmental work on stimuli-responsive polymers that respond to changes in the local environment with significant alterations of their physicochemical properties. In an effort to further synchronize drug delivery with therapeutic or diagnostic need and anatomical site, polymeric materials have been designed that respond to local physical, chemical, and biological cues including pH, temperature, redox potential, monosaccharides, magnetic field, ultrasound, and light [96–98]. Most of this work has focused on using polymeric particulate systems, which can be tailored to release therapeutics at the subcellular, cellular, tissue, or organ levels [99]. These programmable carriers have tremendous therapeutic potential for numerous diseases including cancerous tumors, which typically have a slightly acidic environment (pH 6.5–7.2) [100]. Further, the enhanced accumulation of nanosized materials in tumor sites due to the *leaky* nature of associated arterioles has been documented [101]. Building on the platform of these observations, polymeric nanoparticle delivery systems that release their contents at pH slightly below physiological level were designed and developed. For example, He *et al.* fabricated acid-labile nanoparticles from an α-cyclodextran polymer which showed low toxicity and good biocompatibility *in vitro* and *in vivo*. More importantly, these nanoparticles loaded with paclitaxel, an established anticancer drug, showed significant antitumor activity and lower side effects when injected in melanoma-bearing nude mice [102].

In context of the immunology and immunotherapy fields, the application of these stimuli-responsive particulate systems may also prove impactful. In efforts to develop more effective vaccines against infectious agents and cancerous tissues, experts have recognized the importance of delivery of protein antigens, where they can be assimilated into the MHC-I presentation pathway of antigen-presenting cells. Endosome-disrupting particulate systems are promising solutions to achieve cytosolic delivery of protein therapeutics, which act by escaping the acidic and degradative endosomal/lysosomal compartment following uptake [103, 104]. These systems can also be engineered to target phagocytic cells including DCs, which are critical to adaptive immunity initiation. Foster *et al.* illustrated that pH-responsive poly(propylacrylic acid) (PPAA) particulate systems with incorporated protein antigen could enhance $CD8^+$ cytotoxic T cell generation as well as humoral responses to antigen challenge in thymoma-bearing wild type mice [105].

18.5 SELF-ASSEMBLY MOTIFS

The fundamental function of the ECM is to provide a structural scaffold that is resistant and resilient to mechanical loading. Natural ECM is fortified with macromolecules such as collagens, many of which form relatively long, stiff fibrils, thereby conferring mechanical strength; and elastins, which form a network of fibers and sheets, imparting elasticity to the matrix [25]. These structural molecules often also play a large role in cell behavior and tissue physiology through domains that interact with cell surface receptors (e.g., integrins). Through these linkages, mechanical stimuli propagating through the matrix can translate into intracellular biochemical signals, which can direct cell shape, adhesion, and migration [29].

Structural motifs derived both chemically and by genetic engineering have been integrated into synthetic constructs, attempting to mimic mechanical capabilities of natural ECM. For instance, the pentapeptide val-pro-gly-val-gly (VPGVG), found repeatedly in mammalian elastin, has been bioconjugated with PEG to form self-assembling viscoelastic hydrogel biomaterials [106, 107]. Additionally, structures resembling natural ECM architecture have been developed using amphiphilic peptides that can self-assemble to form supramolecular nanofibers *in situ*. Self-assembly is based on noncovalent interactions between molecules and can be influenced by temperature, pH, ionic strength, or light [108]. Precise three-dimensional architectures can be easily fabricated by controlling peptide sequence. Holmes *et al.* designed self-assembling peptides, which in the presence of physiological media or salt solution, formed a hydrogel capable of supporting neurite attachment and outgrowth [109]. Biofunctionality may be inherent to these structures or can be incorporated by including adhesive, protease-degradable, or other bioactive domains. For example, Stupp and colleagues used heparin-binding amphiphilic peptides to develop a self-assembling gel that sequesters vascular endothelial growth factor and fibroblast growth factor-2. This nanofiber scaffold was shown to significantly improve revascularization and islet engraftment in diabetic mice with transplanted islet of Langerhans [110].

Self-assembling peptides have also been explored as immune adjuvants that physically present immunorelevant T cell and B cell epitopes. These self-assembling peptides are appealing as tissue engineering scaffolds because of their low immunogenicity, in addition to their ability to form spatially resolved structural networks [111]. Notably, Rudra *et al.* illustrated the broad utility of self-assembling peptides as chemically defined adjuvants. Their investigation demonstrated that a short fibrilizing peptide (Q11; Ac-QQKFQFQFEQQ-Am) with covalently bound OVA323-339 epitope (ISQAVHAAHAEINEAGR) antigen is capable of raising high antibody titers against the antigen in the absence of adjuvant in a mouse model [112, 113]. This research is promising for the development of adjuvant-free vaccine systems for vaccine applications.

18.6 CONCLUSIONS AND OUTLOOK

The limitation of conventional synthetic biomaterial scaffolds being unable to specifically interact with their biological environment has prompted exploration into a new field of biomaterials useful for tissue engineering and regenerative medicine—biomimetic/bio-inspired materials. Numerous approaches recapitulating structural and biological characteristics of natural ECM, cells, and tissue, using synthetic materials, have been investigated, demonstrating promise and broad utility. Development in the field has been extensive and has crossed over into disciplines other than tissue engineering/regenerative medicine. Immunology, because of its integral role in (patho)physiological functions, and its close association with the form and function of implanted materials, has prompted the development of a new class of immunomodulatory materials. Today, this material class is widely considered to have tremendous implication for numerous applications in medicine. The hard-won successes in terms of commercialization (e.g., manufacturing and storage considerations) of tissue engineering and regenerative medicine over the past decade should be able to parlay into establishing widespread use of immunomimetic materials. Limitations in current immuno-inspired material systems may be aided by combinatorial approaches [114, 115] and implementation of rigorous *in vivo* assessment for specific clinical applications. Critically, clinical use of these types engineered combination products will require extensive research in human safety, efficacy, and delivery to therapeutically relevant sites, with emphasis on safety.

REFERENCES

[1] Rosso, F., Marino, G., Giordano, A., Barbarisi, M., Parmeggiani, D., Barbarisi, A. *Journal of Cellular Physiology* **2005**, *203*(3), 465–470.

[2] Shin, H., Jo, S., Mikos, A.G. Biomimetic materials for tissue engineering. *Biomaterials* **2003**, *24*(24), 4353–4364.

[3] Badylak, S. *Clinical Techniques in Equine Practice* **2004**, *3*(2), 173–181.

[4] Chang, T.M.S. *Applied Biochemistry and Biotechnology* **1984**, *10*, 5–24.

[5] Marguet, M., Bonduelle, C., Lecommandoux, S. *Chemical Society Reviews* **2013**, *42*(2), 512–529.

[6] McPhail, D., Tetley, L., Dufes, C., Uchegbu, I.F. *International Journal of Pharmaceutics* **2000**, *200*(1), 73–86.
[7] Fu, Z., Ochsner, M.A., de Hoog, H.P., Tomczak, N., Nallani, M. *Chemical Communications* **2011**, *47*(10), 2862–2864.
[8] Yu, Y., Pang, Z., Lu, W., Yin, Q., Gao, H., Jiang, X. *Pharmaceutical Research* **2012**, *29*(1), 83–96.
[9] Zhang, Y., Yuan, W., Su, J., Wu, F., Du, Z. *Progress in Modern Biomedicine* **2010**, *10*(16), 3123–3125.
[10] Khattak, S.F., Bhatia, S.R., Roberts, S.C. *Tissue Engineering* **2005**, *11*(5–6), 974–983.
[11] Lukas, J., Fenclova, T., Mokry, J., Karbanova, J. *Chemicke Listy* **2004**, *98*(1), 14–21.
[12] Wang, N., Adams, G., Buttery, L., Falcone, F.H., Stolnik, S. *Journal of Biotechnology* **2009**, *144*(4), 304–312.
[13] Brady, D., Jordaan, J., Simpson, C., Chetty, A., Arumugam, C., Moolman, F.S. *BMC Biotechnology* **2008**, *8*, 1–11.
[14] Jones, M.N., Hill, K.J., Kaszuba, M., Creeth, J.E. *International Journal of Pharmaceutics* **1998**, *162*(1–2), 107–117.
[15] Chang, T.M.S., Prakash, S. *Molecular Biotechnology* **2001**, *17*(3), 249–260.
[16] Zeng, J., Yu, J., Huang, J., Chang, P.R. *Journal of Dispersion Science and Technology* **2012**, *33*(1–3), 293–306.
[17] Liang, X.F., Wang, H.J., Luo, H., Tian, H., Zhang, B.B., Hao, L.J., *et al. Langmuir* **2008**, *24*(14), 7147–7153.
[18] Gupta, R.K., Chang, A.C., Siber, G.R. *Developments in Biological Standardization* **1998**, *92*, 63–78.
[19] Koppolu, B., Zaharoff, D.A. *Biomaterials* **2013**, *34*(9), 2359–2369.
[20] Bal, S.M., Hortensius, S., Ding, Z., Jiskoot, W., Bouwstra, J.A. *Vaccine* **2011**, *29*(5), 1045–1052.
[21] Chang, T.M.S. *Haematology* **2000**, *13*(4), 651–667.
[22] Anderson, J.M. *Annual Review of Materials Research* **2001**, *31*, 81–110.
[23] Anderson, J.M., Rodriguez, A., Chang, D.T. *Seminars in Immunology* **2008**, *20*(2), 86–100.
[24] Anderson, J.M. *Current Opinion in Hematology* **2000**, *7*(1), 40–47.
[25] Lutolf, M.P., Hubbell, J.A. *Nature Biotechnology* **2005**, *23*(1), 47–55.
[26] Anderson, J.M., McNally, A.K. *Seminars in Immunopathology* **2011**, *33*(3), 221–233.
[27] Garcia, A.J., Keselowsky, B.G. *Critical Reviews in Eukaryotic Gene Expression* **2002**, *12*(2), 151–162.
[28] Acharya, A.P., Dolgova, N.V., Clare-Salzler, M.J., Keselowsky, B.G. *Biomaterials* **2008**, *29*(36), 4736–4750.
[29] Hynes, R.O. *Cell* **1992**, *69*(1), 11–25.
[30] Keselowsky, B.G., Wang, L., Schwartz, Z., Garcia, A.J., Boyan, B.D. *Journal of Biomedical Materials Research. Part A* **2007**, *80A*(3), 700–710.
[31] Keselowsky, B.G., Bridges, A.W., Burns, K.L., Tate, C.C., Babensee, J.E., LaPlaca, M.C., *et al. Biomaterials* **2007**, *28*(25), 3626–3631.
[32] Hynes, R.O. *Cell* **2002**, *110*(6), 673–687.
[33] Keselowsky, B.G., Collard, D.M., Garcia, A.J. *Biomaterials* **2004**, *25*(28), 5947–5954.
[34] Keselowsky, B.G., Garcia, A.J. *Biomaterials* **2005**, *26*(4), 413–418.
[35] Kao, W.J., Lee, D., Schense, J.C., Hubbell, T.A. *Journal of Biomedical Materials Research* **2001**, *55*(1), 79–88.

[36] Cutler, S.M., Garcia, A.J. *Biomaterials* **2003**, *24*(10), 1759–1770.
[37] Raynor, J.E., Petrie, T.A., Garcia, A.J., Collard, D.M. *Advanced Materials* **2007**, *19*(13), 1724–1729.
[38] Reyes, C.D., Petrie, T.A., Garcia, A.J. *Journal of Cellular Physiology* **2008**, *217*(2), 450–458.
[39] Massia, S.P., Hubbell, J.A. *Analytical Biochemistry* **1990**, *187*(2), 292–301.
[40] Yang, F., Williams, C.G., Wang, D.A., Lee, H., Manson, P.N., Elisseeff, J. *Biomaterials* **2005**, *26*(30), 5991–5998.
[41] Fittkau, M.H., Zilla, P., Bezuidenhout, D., Lutolf, M., Human, P., Hubbell, J.A., *et al.* *Biomaterials* **2005**, *26*(2), 167–174.
[42] Massia, S.P., Hubbell, J.A. *The Journal of Cell Biology* **1991**, *114*(5), 1089–1100.
[43] Causa, F., Netti, P.A., Ambrosio, L. *Biomaterials* **2007**, *28*, 5093–5099.
[44] Keselowsky, B.G., Collard, D.M., Garcia, A.J. *Proceedings of the National Academy of Sciences of the United States of America* **2005**, *102*(17), 5953–5957.
[45] Gunn, J.W., Turner, S.D., Mann, B.K. *Journal of Biomedical Materials Research. Part A* **2005**, *72A*(1), 91–97.
[46] Mann, B.K., Gobin, A.S., Tsai, A.T., Schmedlen, R.H., West, J.L. *Biomaterials* **2001**, *22*(22), 3045–3051.
[47] Sagnella, S., Anderson, E., Sanabria, N., Marchant, R.E., Kottke-Marchant, K. *Tissue Engineering* **2005**, *11*(1–2), 226–236.
[48] Rezania, A., Healy, K.E. *Journal of Biomedical Materials Research* **2000**, *52*(4), 595–600.
[49] Simon, P., Kasimir, M.T., Seebacher, G., Weigel, G., Ullrich, R., Salzer-Muhar, U., *et al.* *European Journal of Cardio-Thoracic Surgery* **2003**, *23*(6), 1002–1006.
[50] Vacanti, J.P., Morse, M.A., Saltzman, W.M., Domb, A.J., Perezatayde, A., Langer, R. *Journal of Pediatric Surgery* **1988**, *23*(1), 3–9.
[51] Elmengaard, B., Bechtold, J.E., Soballe, K. *Journal of Biomedical Materials Research. Part A* **2005**, *75A*(2), 249–255.
[52] Li, F.F., Carlsson, D., Lohmann, C., Suuronen, E., Vascotto, S., Kobuch, K., *et al.* *Proceedings of the National Academy of Sciences of the United States of America* **2003**, *100*(26), 15346–15351.
[53] Petrie, T.A., Reyes, C.D., Burns, K.L., Garcia, A.J. *Journal of Cellular and Molecular Medicine* **2009**, *13*(8B), 2602–2612.
[54] Kouvroukoglou, S., Dee, K.C., Bizios, R., McIntire, L.V., Zygourakis, K. *Biomaterials* **2000**, *21*(17), 1725–1733.
[55] Palecek, S.P., Loftus, J.C., Ginsberg, M.H., Lauffenburger, D.A., Horwitz, A.F. *Nature* **1997**, *385*(6616), 537–540.
[56] Rezania, A., Thomas, C.H., Branger, A.B., Waters, C.M., Healy, K.E. *Journal of Biomedical Materials Research* **1997**, *37*(1), 9–19.
[57] Schense, J.C., Hubbell, J.A. *The Journal of Biological Chemistry* **2000**, *275*(10), 6813–6818.
[58] Hubbell, J.A. *Current Opinion in Biotechnology* **1999**, *10*(2), 123–129.
[59] Boehler, R.M., Graham, J.G., Shea, L.D. *Biotechniques* **2011**, *51*(4), 239.
[60] Hume, P.S., He, J., Haskins, K., Anseth, K.S. *Biomaterials* **2012**, *33*(14), 3615–3625.
[61] Lieberman, J., Skolnik, P.R., Parkerson, G.R., Fabry, J.A., Landry, J., Bethel, J. Safety of autologous, ex-vivo expanded HIV-specific cytotoxic T-lymphocyte infusion in HIV-infected patients. *Blood* **1997**, *90*(6), 2196–2206.
[62] Greenberg, P.D. *Advances in Immunology* **1991**, *49*, 281–355.
[63] Tang, Q.Z., Henriksen, K.J., Bi, M.Y., Finger, E.B., Szot, G., Ye, J.Q., *et al.* *The Journal of Experimental Medicine* **2004**, *199*(11), 1455–1465.

[64] Turtle, C.J., Riddell, S.R. *Cancer Journal (Sudbury, Mass.)* **2010**, *16*(4), 374–381.
[65] Zemon, H. *Trends in Biotechnology* **2003**, *21*(10), 418–420.
[66] Rudolf, D., Silberzahn, T., Walter, S., Maurer, D., Engelhard, J., Wernet, D., *et al.* *Cancer Immunology, Immunotherapy* **2008**, *57*(2), 175–183.
[67] Shen, C., He, Y., Cheng, K., Zhang, D., Miao, S., Zhang, A., *et al.* *Immunology Letters* **2011**, *138*(2), 144–155.
[68] van Rensen, A.J., Wauben, M.H., Grosfeld-Stulemeyer, M.C., van Eden, W., Crommelin, D.J. *Pharmaceutical Research* **1999**, *16*(2), 198–204.
[69] van Rensen, A., Taams, L., Grosfeld, M., Bessling, A., Wagenaar, J., Van Eden, W., *et al.* *Pharmaceutical Research* **1997**, *14*(11 Suppl.), 165.
[70] Tham, E.L., Jensen, P.L., Mescher, M.F. *Journal of Immunological Methods* **2001**, *249*(1–2), 111–119.
[71] Lewis, J.S., Zaveri, T.D., Crooks, C.P., Keselowsky, B.G. *Biomaterials* **2012**, *33*(29), 7221–7232.
[72] Bandyopadhyay, A., Fine, R.L., Demento, S., Bockenstedt, L.K., Fahmy, T.M. *Biomaterials* **2011**, *32*(11), 3094–3105.
[73] Tsai, S., Shameli, A., Yamanouchi, J., Clemente-Casares, X., Wang, J., Serra, P., *et al.* *Immunity* **2010**, *32*(4), 568–580.
[74] Rosso, F., Giordano, A., Barbarisi, M., Barbarisi, A. *Journal of Cellular Physiology* **2004**, *199*(2), 174–180.
[75] Sakiyama-Elbert, S.E., Hubbell, J.A. *Journal of Controlled Release* **2000**, *65*(3), 389–402.
[76] Sakiyama-Elbert, S.E., Hubbell, J.A. *Journal of Controlled Release* **2000**, *69*(1), 149–158.
[77] Lee, K.Y., Mooney, D.J. *Chemical Reviews* **2001**, *101*(7), 1869–1879.
[78] Drury, J.L., Mooney, D.J. *Biomaterials* **2003**, *24*(24), 4337–4351.
[79] Richardson, T.P., Peters, M.C., Ennett, A.B., Mooney, D.J. *Nature Biotechnology* **2001**, *19*(11), 1029–1034.
[80] Lucas, P.A., Laurencin, C., Syftestad, G.T., Domb, A., Goldberg, V.M., Caplan, A.I., *et al.* *Journal of Biomedical Materials Research* **1990**, *24*(7), 901–911.
[81] Harris, J.M., Chess, R.B. *Nature Reviews Drug Discovery* **2003**, *2*(3), 214–221.
[82] Veronese, F.M., Pasut, G. *Drug Discovery Today* **2005**, *10*(21), 1451–1458.
[83] Du, J., Du, L.F., Li, F.H., Zheng, X.Z., Li, H.L., Shi, Q.S., *et al.* *Asian Biomedicine* **2011**, *5*(5), 577–587.
[84] Stevenson, D.J., Gunn-Moore, F.J., Campbell, P., Dholakia, K. *Journal of the Royal Society, Interface* **2010**, *7*(47), 863–871.
[85] Yockman, J.W., Kastenmeier, A., Erickson, H.M., Brumbach, J.G., Whitten, M.G., Albanil, A., *et al.* *Journal of Controlled Release* **2008**, *132*(3), 260–266.
[86] Pickering, J.G., Takeshita, S., Feldman, L., Losordo, D.W., Isner, J.M. *Seminars in Interventional Cardiology* **1996**, *1*(1), 84–88.
[87] Xue, J., Peng, J., Yuan, M., Wang, A., Zhang, L., Liu, S., *et al.* *Bone* **2011**, *48*(3), 485–495.
[88] Kesser, B.W., Hashisaki, G.T., Holt, J.R. *The Laryngoscope* **2008**, *118*(5), 821–831.
[89] Phillips, B., Nylander, K., Harnaha, J., Machen, J., Lakomy, R., Styche, A., *et al.* *Diabetes* **2008**, *57*(6), 1544–1555.
[90] Singh, A., Suri, S., Roy, K. *Biomaterials* **2009**, *30*(28), 5187–5200.
[91] Smith, M.F., Ricke, W.A., Bakke, L.J., Dow, M.P.D., Smith, G.W. *Molecular and Cellular Endocrinology* **2002**, *191*(1), 45–56.
[92] Stamenkovic, I. *The Journal of Pathology* **2003**, *200*, 448–464.

[93] Chau, Y., Luo, Y., Cheung, A.C.Y., Nagai, Y., Zhang, S.G., Kobler, J.B., *et al. Biomaterials* **2008**, *29*(11), 1713–1719.

[94] Raeber, G.P., Lutolf, M.P., Hubbell, J.A. *Acta Biomaterialia* **2007**, *3*, 615–629.

[95] Ehrbar, M., Rizzi, S.C., Schoenmakers, R.G., San Miguel, B., Hubbell, J.A., Weber, F.E., *et al. Biomacromolecules* **2007**, *8*, 3000–3007.

[96] Roy, D., Cambre, J.N., Sumerlin, B.S. *Progress in Polymer Science* **2010**, *35*(1–2), 278–301.

[97] Ju, X.J., Xie, R., Yang, L., Chu, L.Y. *Expert Opinion on Therapeutic Patents* **2009**, *19*(5), 683–696.

[98] Zhang, H., Meng, X., Li, P. *Progress in Chemistry* **2008**, *20*(5), 657–672.

[99] Motornov, M., Roiter, Y., Tokarev, I., Minko, S. *Progress in Polymer Science* **2010**, *35*(1–2), 174–211.

[100] Tannock, I.F., Rotin, D. *Cancer Research* **1989**, *49*(16), 4373–4384.

[101] Maeda, H., Bharate, G., Daruwalla, J. *European Journal of Pharmaceutics and Biopharmaceutics* **2009**, *71*(3), 409–419.

[102] He, H., Chen, S., Zhou, J., Dou, Y., Song, L., Che, L., *et al. Biomaterials* **2013**, *34*(21), 5344–5358.

[103] Ganta, S., Devalapally, H., Shahiwala, A., Amiji, M. *Journal of Controlled Release* **2008**, *126*(3), 187–204.

[104] Sawant, R., Hurley, J., Salmaso, S., Kale, A., Tolcheva, E., Levchenko, T., *et al. Bioconjugate Chemistry* **2006**, *17*(4), 943–949.

[105] Foster, S., Duvall, C.L., Crownover, E.F., Hoffman, A.S., Stayton, P.S. *Bioconjugate Chemistry* **2010**, *21*(12), 2205–2212.

[106] Di Zio, K., Tirrell, D.A. *Macromolecules* **2003**, *36*(5), 1553–1558.

[107] Jing, P., Rudra, J.S., Herr, A.B., Collier, J.H. *Biomacromolecules* **2008**, *9*(9), 2438–2446.

[108] Hartgerink, J.D., Beniash, E., Stupp, S.I. *Proceedings of the National Academy of Sciences of the United States of America* **2002**, *99*(8), 5133–5138.

[109] Holmes, T.C., de Lacalle, S., Su, X., Liu, G.S., Rich, A., Zhang, S.G. *Proceedings of the National Academy of Sciences of the United States of America* **2000**, *97*(12), 6728–6733.

[110] Stendahl, J.C., Wang, L.J., Chow, L.W., Kaufman, D.B., Stupp, S.I. *Transplantation* **2008**, *86*(3), 478–481.

[111] Collier, J.H., Rudra, J.S., Gasiorowski, J.Z., Jung, J.P. *Chemical Society Reviews* **2010**, *39*(9), 3413–3424.

[112] Rudra, J.S., Mishra, S., Chong, A.S., Mitchell, R.A., Nardin, E.H., Nussenzweig, V., *et al. Biomaterials* **2012**, *33*(27), 6476–6484.

[113] Rudra, J.S., Sun, T., Bird, K.C., Daniels, M.D., Gasiorowski, J.Z., Chong, A.S., *et al. ACS Nano* **2012**, *6*(2), 1557–1564.

[114] Acharya, A.P., Clare-Salzler, M.J., Keselowsky, B.G. *Biomaterials* **2009**, *30*(25), 4168–4177.

[115] Acharya, A.P., Dolgova, N.V., Moore, N.M., Xia, C.Q., Clare-Salzler, M.J., Becker, M.L., *et al. Biomaterials* **2010**, *31*(29), 7444–7454.

INDEX

Bio-inspired Materials for Biomedical Engineering, First Edition. Edited by Anthony B. Brennan and Chelsea M. Kirschner.